S0-BXD-948

THE ROLE OF SOLAR ULTRAVIOLET RADIATION IN MARINE ECOSYSTEMS

NATO CONFERENCE SERIES

- I Ecology
- II Systems Science
- III Human Factors
- IV Marine Sciences
- V Air—Sea Interactions
- VI Materials Science

IV MARINE SCIENCES

Volume 1 Marine Natural Products Chemistry
edited by D. J. Faulkner and W. H. Fenical

Volume 2 Marine Organisms: Genetics, Ecology, and Evolution
edited by Bruno Battaglia and John A. Beardmore

Volume 3 Spatial Pattern in Plankton Communities
edited by John H. Steele

Volume 4 Fjord Oceanography
edited by Howard J. Freeland, David M. Farmer, and Colin D. Levings

Volume 5 Bottom-interacting Ocean Acoustics
edited by William A. Kuperman and Finn B. Jensen

Volume 6 Marine Slides and Other Mass Movements
edited by Svend Saxov and J. K. Niewenhuis

Volume 7 The Role of Solar Ultraviolet Radiation in Marine Ecosystems
edited by John Calkins

THE ROLE OF SOLAR ULTRAVIOLET RADIATION IN MARINE ECOSYSTEMS

Edited by
John Calkins
Albert B. Chandler Medical Center
University of Kentucky
Lexington, Kentucky

Published in cooperation with NATO Scientific Affairs Division
PLENUM PRESS · NEW YORK AND LONDON

Library of Congress Cataloging in Publication Data

Main entry under title:

The Role of solar ultraviolet radiation in marine ecosystems.

(NATO conference series. IV, Marine sciences; v. 7)
"Proceedings of a NATO conference on the role of solar ultraviolet radiation in marine ecosystems, held July 28–31, 1980, in Copenhagen, Denmark"—Verso t.p.
"Published in cooperation with NATO Scientific Affairs Division."
Bibliography: p.
Includes index.
1. Ultra-violet rays—Environmental aspects—Congresses. 2. Marine ecology—Congresses. I. Calkins, John, 1926– . II. Series.
QH543.6.R64 574.5'2636 82-3797
ISBN 0-306-40909-7 AACR2

Proceedings of a NATO conference on The Role of Solar Ultraviolet Radiation in Marine Ecosystems, held July 28–31, 1980, in Copenhagen, Denmark

A Division of Plenum Publishing Corporation
233 Spring Street, New York, N.Y. 10013

Printed in the United States of America

DEDICATION

This monograph and the preceding NATO Advanced Research Institute, Copenhagen, Denmark, July 28-31, 1980, are dedicated to the two Danish scientists who have contributed so greatly to the understanding of the role of solar ultraviolet radiation in aquatic ecosystems: Professors

N. G. Jerlov, Institute of Physical Oceanography, University of Copenhagen, Denmark

and

E. Steemann Nielsen, Freshwater Biological Laboratory, University of Copenhagen, Denmark

Professors Jerlov and Steemann Nielsen are notable for the attention they have given to the study of the properties and actions of the ultraviolet portion of the solar spectrum on aquatic systems when their contemporary scientists gave this region of the spectrum only passing consideration.

The Organizing Committee, on behalf of the Advanced Research Institute and the authors of this monograph, wish to express our gratitude and respect for their pioneering work in this important area.

PARTICIPANTS

K. S. Baker, U.S.A.
J. A. Barcelo, U.S.A.
D. Berger, U.S.A.
A. Bogenrieder, W. Germany
A. M. Bullock, Scotland
M. M. Caldwell, U.S.A.
J. Calkins, U.S.A.
J. Chavaudra, France
T. P. Coohill, U.S.A.
P. Cutchis, U.S.A.
D. M. Damkaer, U.S.A.
E. C. DeFabo, U.S.A.
B. L. Diffey, UK
A. Eisenstark, U.S.A.
B. Goldberg, U.S.A.
A.E.S. Green, U.S.A.
K. R. Gundersen, Sweden
D. Häder, W. Germany
P. V. Hariharan, U.S.A.
N. K. Højerslev, Denmark
J. R. Hunter, U.S.A.
N. Jerlov, Denmark
R. Klein, W. Germany
G. Kullenberg, Denmark
R. D. Ley, U.S.A.
F. M. Luther, U.S.A.
S. Malmberg, Iceland
D. S. Nachtwey, U.S.A.
D. R. Norris, U.S.A.
M. J. Peak, U.S.A.
M. Polne, U.S.A.
P. Reynisson, Iceland
F. S. Rowland, U.S.A.
A. Ruhland, W. Germany
C. S. Rupert, U.S.A.
R. C. Smith, U.S.A.
R. S. Stolarski, U.S.A.
F. Stordal, Norway
A. H. Teramura, U.S.A.
M. Tevini, W. Germany
R. M. Tyrrell, Brazil
E. Wellman, W. Germany
G. N. Wells, U.S.A.
R. C. Worrest, U.S.A.
C. Yentsch, U.S.A.
R. G. Zepp, U.S.A.
S. Zigman, U.S.A.

PREFACE

The inspiration for this monograph derived from the realization that human technical capacity has become so great that we can, even without malice, substantially modify and damage the gigantic and remote outer limit of our planet, the stratosphere. Above the atmosphere of our ordinary experience, the stratosphere is a tenuous layer of gas, blocked from rapid exchange with the troposphere, some twenty kilometers above the surface of the earth, seldom reached by humans, and yet a fragile shell which shields life on earth from a band of solar radiation of demonstrable injurious potential. It is immediately obvious that if stratospheric ozone were reduced and consequently the intensity of solar ultraviolet radiation reaching the earth's surface were increased, then human skin cancer, known to be related to solar ultraviolet exposure, would also be increased. But how does one even begin to estimate the impact of changed solar ultraviolet radiation on such a diverse, interacting, and complex ecosystem as the oceans?

Studies which I conducted in Iceland focused on this question and were noted to the Marine Sciences Panel of the Scientific Affairs Committee of NATO by Professor Unnsteinn Stefansson, leading to a request to investigate the possibility of organizing a NATO sponsored Advanced Research Institute on this topic. An Organizing Committee was formed including myself and Drs. A.E.S. Green, Per Haldall, Raymond Smith, and Robert C. Worrest. Professor Halldall subsequently resigned from the Organizing Committee due to a scheduling conflict and Drs. Gunnar Kullenberg and Charles Yentsch agreed to serve on the Committee. Because of the leadership of Danish scientists in the area, Copenhagen was always considered the natural site for the meeting. Professor Kullenberg and the Institute of Physical Oceanography generously agreed to host the meeting, an intensive four-day meeting without subdivision of the participants was planned. The grant application was completed, received approval and the arrangements for the conference were made by correspondence or telephone, without an actual meeting of the Organizing Committee.

It should be noted that in addition to developing and conducting the meeting in Copenhagen, the Organizing Committee, Drs:

Alex E. S. Green, The University of Florida
Gunnar Kullenberg, The University of Copenhagen
Raymond Smith, Scripps Institute of Oceanography
Robert Worrest, Oregon State University
Charles Yentsch, Bigelow Laboratory for Ocean Science

also agreed to serve as an Editorial Board for the monograph, reviewing the manuscripts of all participants as well as writing numerous contributions to the monograph themselves.

I wish to express my gratitude to three people who contributed very generously to the success of the conference and the production of the monograph: Ms E. Hallden, Secretary at the Institute for Physical Oceanography was most helpful with the arrangements in Copenhagen; Dr. Jeanne Barcelo served as Executive Secretary for the conference and as Associate Editor in the production of the monograph. I owe a very special debt to my wife Ruth Calkins; she prepared the grant application, typed all correspondence, beginning with the Organizing Committee and later correspondence with the participants of the conference. She has typed the final monograph, often under difficult circumstances; I have no doubt that she more than any other single person is responsible for the issuance of this monograph and the completion of this enterprise.

John Calkins
August, 1981

CONTENTS

ATMOSPHERE AND BASIC BIOLOGICAL DOSIMETRY

THE HYDROSPHERE

THE BIOSPHERE

PREFACE TO SECTION I - ATMOSPHERE AND BASIC BIOLOGICAL DOSIMETRY

The antecedent of this monograph, the NATO Advanced Research Institute on the Role of Solar UV Radiation in Marine Ecosystems, was clearly not an assembly of scientists specializing in this particular corner of ecological science. A conscious attempt was made to bring together scientists whose investigations and disciplines are critical to the evaluation of solar UV action in the oceans. But few if any of the assembled scientists would consider themselves primarily concerned with the evaluation of solar UV effects in aquatic systems. Quantitative evaluation of the ecological role of sunlight must utilize a long chain of data and theories, most of which are, at first sight, not even remotely related to ecological systems.

One must begin by asking: How much solar radiation reaches the earth's atmosphere? Although we have a long sequence of measurements of solar radiation at the earth's surface, ability to predict future solar UV levels can only be attained if we know the flux of solar UV light before it is modified by the atmosphere. One must then ask: What are the photochemical reactions of significance in the atmosphere? While solar radiation can and does produce a multitude of photochemical reactions, it is necessary to focus on the reactions which produce and destroy ozone. It is not just a question of which ozone forming or destroying reactions occur, but the transmission of solar UV to the earth is highly dependent on where these reactions take place; How does one model the dynamics of the atmosphere? thus becomes a relevant question. Even when the solar UV has traversed the atmosphere one encounters the physical problem: How does one measure the solar UV flux?

After coping with the problems of measurement and prediction of solar UV levels at the earth's surface, a new realm of problems arises: living organisms do not respond to radiations in the way that the physicist's instruments perceive radiation. One must determine: How are physical measurements of UV radiation to be translated into measurements which are biologically meaningful? Central to this problem is the action spectrum, a determination of the relative effectiveness of differing wavelengths of radiation in producing particular biological effects. Indeed, a question of great present import is: Is there a typical or representative action spectrum for solar UV? One finds that the unresolved action spectra question lies central to the next problem: What are proper dosage units for measuring biologically significant solar UV radiation dosages? Even if a consensus were developed regarding the proper ways to express dosage: What is a practical system for the scientist working in the ecological area to measure solar UV exposure? One discovers that there are at present several systems of dosimetry which, in principle, are not interconvertible: Is there some system (even if less than perfect) for translating dosages, expressed in one system into another dosimetry system? A final question arises: Is there some simple way to compare long-term exposure at one point on the earth with exposure at other places?

The organizers of the meeting assembled scientists who seemed most capable of addressing these questions. Research on effects of solar UV has received much attention during the last decade and there is no doubt that we have made great progress in our ability to comprehend and address the critical questions compared to the level of understanding available in 1970. For some problems we have very refined answers; other answers still elude us. In this monograph the reader should be able to see the general outlines of the answers to key questions. There is in some cases a difference regarding the merits of a formula, an instrument, the logic of an approach, but through the diversity one finds a broad consensus. More theories, instruments, and much more data are needed, but we do perceive the outlines of the pathway we must follow to comprehend the role of solar UV in aquatic ecosystems and we understand the nature of the problems we must solve to predict the possible effects of ozone depletion on the world's marine resources.

Our first contribution, Green and Schippnick, beginning with the solar flux outside the atmosphere, outlines methods for the calculation of solar ultraviolet reaching the water surface. Numerous parameters (atmospheric ozone, solar elevation, aerosols, cloud cover, etc.) can be incorporated for a more accurate estimate of water surface flux. In the second chapter, Rowland outlines the photochemistry of the atmosphere; Luther and coauthors elaborate on the chemistry of ozone in the stratosphere; Stolarski describes models for calculations of seasonal and latitudinal fluctuations in ozone depletion.

Measurements and model predictions of solar UV reaching the ocean surface are presented by Baker, Smith and Green (Chapter 5), while a less elaborate model is offered by Stordal, Hov and Isaksen in the following chapter.

Chapters 7, by Green, and 8, by Goldberg describe current approaches to measurement of solar UV radiation in physical units at the earth's surface, observations which can then be used to deduce atmospheric ozone concentrations.

We pass from the primarily chemical and physical problems into areas where biological factors tend to predominate. Rupert introduces the basic ideas of photobiological dosimetry in Chapter 9. Some of the key problems which Rupert notes are elaborated in subsequent chapters. Differing aspects of action spectra are presented by Calkins and Barcelo, Caldwell, Eisenstark, and Coohill; viewpoints on dosage units are advanced by Caldwell and by Calkins. It has been possible to construct devices which can measure with varying accuracy biological dose: the most widely used biological solar UV dosimeter, the Robertson-Berger meter, is described by Berger; Diffey, Davis, and Magnus explain a plastic film methodology for such measurements.

One problem Rupert notes is the proliferation of dosimetry systems. Damkaer and Dey render a valuable service by suggesting a system for interconversion of units in current use. The section ends with a description of a methodology, proposed by Cutchis, for converting measurements at one location to estimates of long-term solar UV dose at other points on the earth.

UV-B REACHING THE SURFACE

Alex E. S. Green and P. F. Schippnick

Interdisciplinary Center for Aeronomy and other Atmospheric Sciences, University of Florida
Gainesville, Florida 32611

ABSTRACT

New physical inputs are summarized and placed in analytic forms suitable for calculations of the ultraviolet spectral irradiance reaching the ground. These include (a) recent values of the extraterrestrial solar spectral irradiance based upon the work of Heath et al.; (b) improved analytic characterizations of ozone attenuation coefficients; (c) an improved equation for the Rayleigh optical depth based upon the work of Frohlich and Shaw; and (d) a detailed characterization of aerosol extinction coefficients for various wavelengths and relative humidities based upon the recent work of Shettle and Fenn. In addition, we refine the ratio method of Green, Cross and Smith for characterizing the diffuse spectral irradiance to provide a better interpolation formula between the numerical output of the models of Dave, Braslau and Halpern. Applications of the work are described.

INTRODUCTION

Our knowledge of the ultraviolet irradiance at the earth's surface, particularly the diffuse component (or skylight), has advanced very greatly during the past two decades. The pioneering work of Bener (1963, 1970a,b, 1972) provided the first quantitative body of data for the development of systematics. The theoretical calculations of Dave and Furukawa (1966) for a Rayleigh atmosphere illustrated the impact of the ozone thickness upon direct and diffuse spectral irradiance.

Attempts to allow for aerosol scattering began in 1973 with several parallel efforts. Green, Sawada and Shettle (GSS) (1973 , 1974) developed a phenomonenological analytic formula, which included aerosols, to fit the systematics of Bener's diffuse spectral irradiance data. Simultaneously, Shettle and Green 1974 (SG) used a simple multistream, multilayer multiple scattering calculation to predict UV-spectral irradiances under various representative atmospheric conditions. Shotkin and Thompson (1973) addressed the same problem with two-stream and Monte Carlo calculational methods. About the same time Braslau and Dave (1973, 1975) used a many-layer and many-stream solution to allow for the effects of aerosols on the direct and the diffuse UV spectral irradiance. Thus, by 1974 our knowledge of UV spectral irradiance had been advanced greatly.

The applications of this knowledge to biological problems proceeded rapidly and is probably best summarized in Volume V Part 1 of the CIAP monographs. The work of Green, Mo and Miller (1974a,b), Mo and Green (1974), and Green and Mo (1974) based upon the GSS formula are of particular interest since they used various biological action spectra to calculate the corresponding biological dose rates as functions of the solar zenith angle and various atmospheric parameters. This biological dose rate which was found to be simply dependent upon the sun angle was used to evaluate the daily dose by allowing for the course of the sun across the sky for a particular latitude and solar declination. Such daily doses were used in conjunction with the available systematics of ozone cover and aerosol cover to arrive at clear sky annual doses for various locations for use in epidemiological studies. The influence of clouds, haze and smog was also addressed (Nack and Green, 1974; Borkowski et al., 1977) although the lack of statistical data on a cloud cover and particularly cloud optical depths limited the conclusions which could be drawn.

The same calculational machinery was also used (Johnson, Mo and Green, 1974) to obtain the dose rate in various wavelength blocks as a function of wavelength, latitude, and season for clear sky, and latitudinally averaged ozone amounts. In principle, such data could be combined with any biological sensitivity function as the last step to arrive at annual doses for epidemiological analyses.

The full array of computational machinery described above was used not only to estimate annual UV doses as they are thought to be, but also to estimate what they would be under various ozone reduction scenarios. These calculations led to estimates of so-called atmospheric optical amplification factors for various biological action spectra.

From 1974 to 1978 the original GSS analytic framework for spectral irradiance which was the basis of the derived biological dose rates and amplification factors was subjected to some experimental field tests. It must be noted that field radiometric measurements in the 300 nm region were still relatively primitive and subject to error due to temperature, humidity and pressure variations. However, some suggestions arose concerning the need to improve upon the original GSS diffuse spectral irradiance formula (Caldwell, 1976). In response to these and to the needs of a UV measurement program at Florida, Green, Cross and Smith (1980) (GCS) developed an improved analytic characterization of diffuse spectral irradiance (skylight). Their work achieved a greater accuracy by focusing on certain ratio representations and by adjusting the parameters of their model to the more precise radiative transfer calculations of Braslau and Dave (1973, 1974), and Dave and Halpern (1975). Before describing the GCS model and a new variation of it, we will describe some new inputs which are now available for UV spectral irradiance calculations.

THE EXTRATERRESTRIAL SOLAR SPECTRAL IRRADIANCE

The extraterrestrial solar spectral irradiance $H(\lambda)$ is the spectral irradiance above the atmosphere. It is unfortunate that the uncertainties in $H(\lambda)$ may still be large in the critical region of biological interest. The work of Smith and Baker (1980) (see their Figures 2a-d and Table 2) and Kohl, Parkinson and Zapata (1980) (see their Fig. 5) provide recent illustrations of this problem. Differences of the order of 20% still persist. Most of these data were from rocket flown spectrometers and, of course, only a few spectral scans could be made during the flight. The recent body of measurements (Heath and Park, 1980) with the double monochromator aboard the Nimbus 7 satellite which was launched in October 1978 provides new promise of accuracy in this spectral region. Figure 1 illustrates the data obtained on November 7, 1978 (corrected to a sun earth distance of 1AU). The

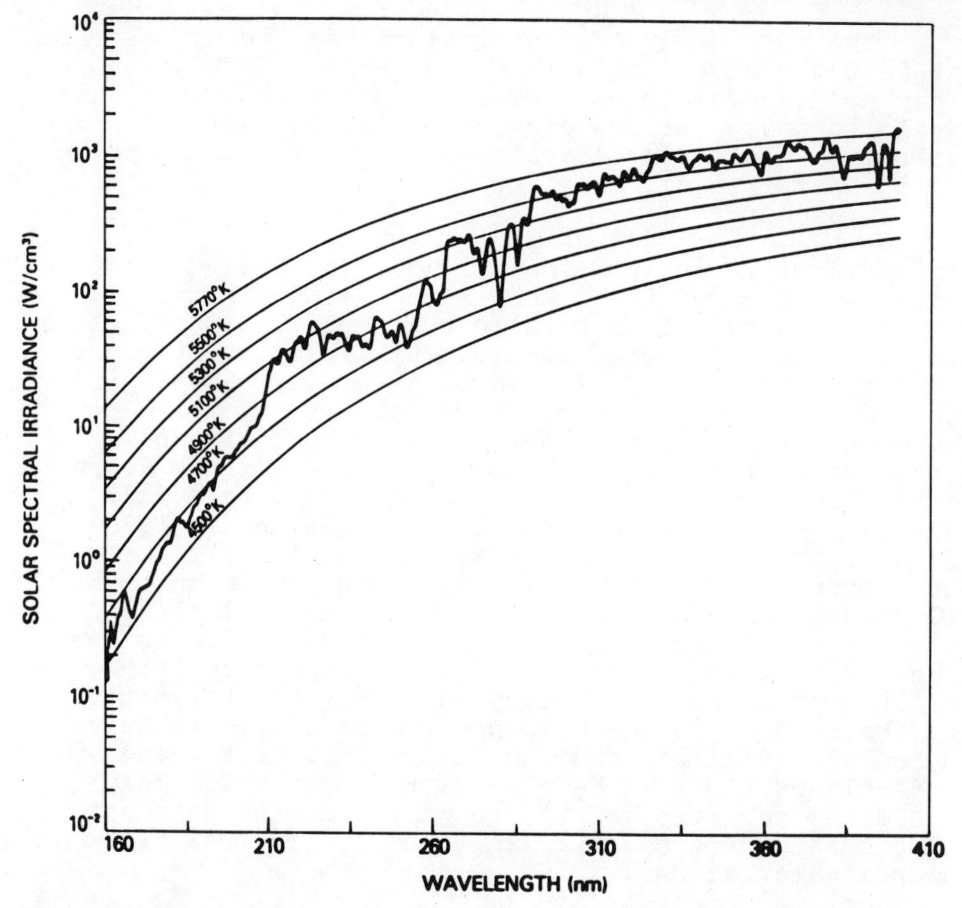

Figure 1. Extraterrestrial solar spectral irradiance obtained by Heath and Park (1980) (private communication) on November 7, 1978, with the Nimbus 7 double monochromator operating in the continuous scan mode.

instrument (Heath et al. 1975) was operating in a continuous scan mode with a slit function bandpass of 1 nm. The data obtained from this satellite will very probably represent the best UV spectral data available until the shuttle experiments begin.

In response to the uncertainties and scatter of the data available in 1973 and the fact that the solar spectral irradiance in the wavelength range of major biological interest (280-340 nm) departed grossly from a black body spectrum, Green, Sawada and Shettle (1974) characterized $H(\lambda)$ with a simple linear formula. To cover the broader range from 270-380 nm Green, Cross and Smith in 1979 resorted to an elaborate transcendental function. The illustrated results in Figure 1 suggest that a black body spectrum may represent the observations reasonably well between 300-380 nm. Unfortunately, shortward of λ_o = 300 nm, where biological activity becomes very important, departures from a black body spectrum are large. For our present work we propose using a black body spectrum in the form

$$H_b(\lambda) = H_o\left(\frac{\lambda_o}{\lambda}\right)^5 \frac{\exp(p) - 1}{\exp\,(p\lambda_o / \lambda) - 1} \qquad (1)$$

where H_o and p are independently adjusted to the data. To allow for major local departures from Eq. (1) in a smooth but realistic fashion we use Gaussian functions to multiply $H_b(\lambda)$ in the important region of biological activity. Thus we multiply $H_b(\lambda)$ by

$$m(\lambda) = 1 + \Sigma_i\ A_i\ \exp - (\lambda-\lambda_i)^2/2\sigma_i{}^2 \qquad (2)$$

The parameters which fit the data of Heath et al. are given in Table 1. Figure 2 illustrates our proposed modified $H(\lambda)$ for UV studies.

We have explored more sophisticated smoothing techniques including the use of truncated Fourier series representations. As more detailed solar spectra become available and as the smoothing most appropriate

Table 1

Parameters for Eqs. (1) and (2) to fit the extraterrestrial solar spectral irradiance measured by the SBUV monochromator on Nov. 7, 1978 (normalized to a mean solar distance of 1 AU) H_o = 0.582 W/m^2 nm = 582 W/cm^3, p = 9.102, λ_o = 300 nm

λ_i(nm)	A_i	σ_i(nm)
279.5	-0.738	2.96
286.1	-0.485	1.57
300.4	-0.243	1.80
333.2	0.192	4.26
358.5	-0.167	2.01
368.0	0.097	2.43

for biological applications becomes clearer it would be well to push such methods to their limits. In the meantime, our representation which now includes the major excursions of $H(\lambda)$ from a monotonic background should add a further degree of realism to spectral irradiance calculations of interest to this conference.

OZONE ATTENUATION COEFFICIENTS

The need to reduce the discrepancies among reported ozone attenuation coefficients is urgent (Klenk, 1980) and new measurements are currently underway (Bass, 1980, private communication). For the moment we might, on the basis of existing data, examine the options available for studies of UV in a marine environment. Figure 2 shows the experimental ozone absorption coefficients of Vigroux (1953). The fact that ozone absorption increases by over four orders of magnitude as one proceeds from 350 nm to 270 nm gives the theoretical and experimental aspects of work in this region a unique quality. Shortward of 310 nm we have, in effect, a continuous function whereas longward of 310 nm the Huggins band structure is very marked. The smooth curve is an analytic representation given by

$$k_3(\lambda) = k_0 \ (\beta + 1)/[\beta + \exp\ (\lambda-\lambda_0)/\delta] \qquad (3)$$

where $\beta = .03679$ and $\delta = 7.309$ and $k_0 = 4.530\ 10^{-19}\ cm^2$ or $12.17\ cm^{-1}$. The constants for this analytic equation, however, depend upon the specific body of data, the range of wavelengths, the temperature range and the weighting scheme chosen. Table 2 lists an assortment of parameters which are still within the realm of plausibility at this time. Hopefully, ambiguities in the ozone absorption coefficient will be greatly reduced by forthcoming measurements. For the present purposes we recommend the starred set which was optimimally tuned to values used in ozone profiling studies with the Nimbus 4 and Nimbus 7 ultraviolet backscatter measurements.

Vigroux (1953) measurements have provided the most extensive body of data to allow explicitly for the temperature dependence of the ozone absorption coefficient. The conclusions which he reached concerning characterization of this temperature dependence can be summarized as follows:

1. In the Hartley region ($245\ nm < \lambda < 311\ nm$)

Table 2

Parameters for Eq. 3 to fit average ozone absorption coefficients used for Ozone Profile Studies (Klenk, 1980) with 255.5 nm value omitted, and other data

	k_o(atm-cm^{-1})	k_o(10^{-19}cm^2)	β	δ
IT*	10.89	4.050	.03554	7.150
Handbook	11.01	4.096	.03710	7.034
SPWB	11.89	4.423	.03730	7.195
Vig. 67	10.63	3.954	.03742	6.841
Vig. 53 (18°)	12.18	4.530	.03679	7.309
GCS	9.517	3.541	.0445	7.294

the temperature dependence is slight, $R_t = K_t/K_{18°}$ typically falls to around 0.90 at t = 92°C over the upper extent of this region, with this dependence diminishing as the absorption peak at 255 nm is reached to $R_t \approx 0.97$. Vigroux noted that the effect can be described by a parabolic function. For persons concerned with details of temperature dependence in the Hartley region we suggest using

$$R_t = 1 + \{(t-18)/642 + [(t-18)/395]^2\} \exp(\lambda-\lambda_o)40 \quad (4)$$

where temperatures are in °C and wavelengths in nm.

2. In the Huggins band region (311 nm < λ < 345 nm) the temperature dependence is of an entirely different character depending on whether the wavelength is in the region of a local maximum or a local minimum. In the region of a maximum, the temperature dependence is relatively slight, and is parabolic in shape, as in the Hartley band and reaches a minimum at -50°C and shows little dependence on wavelength. On the other hand, the temperature dependence of the minima is much greater and its magnitude increases as λ increases. Typically we find $R_t \approx 0.5$ at t = -92°C for λ ≈ 340 nm. It is possible to parametrize the temperature and wavelength dependence $R_t(\lambda)$ for the Huggins region. However, since ozone effects and biological activity are weak in this wavelength region and the formula is cumbersome, we will not give it here.

THE RAYLEIGH OPTICAL DEPTH

The Rayleigh optical depth is another important input to skylight calculations. Very recently Frohlich and Shaw (1980) presented new Rayleigh scattering optical thickness values for the terrestrial atmosphere in the $260 < \lambda < 1500$ nm wavelength range. Their calculations, which are based upon updated optical parameters of atmospheric species, lead to Rayleigh scattering coefficients lower by 4.5% than those listed by Penndorf (1955). Frohlich and Shaw present their results as a formula

$$\tau_R(\lambda) = A \cdot \lambda^{-(B + C\cdot \lambda + D/\lambda)} \quad (5)$$

where the constants B and C are model and elevation independent and have values (when λ is in microns)

$$B = 3.916, \quad C = 0.074 \quad \text{and} \quad D = 0.050$$

Their constant A may be written

$$A = \alpha_o (p/p_o)[1 + a + bH] \quad (6)$$

where p is the pressure in millibars, $p_o = 1013$ mb, $\alpha_o = 8.38 \times 10^{-3}$ and a and b are small model dependent parameters which can be ignored. The formula used in the earlier Florida work for the UV region was of the simple Penndorf type.

$$\tau_R = 1.221 (\lambda_o/\lambda)^{4.27}, \lambda_o = 300 \text{ nm} \quad (7)$$

It fits Frohlich and Shaw's values within 2%. For our applications their proposed formula is unnecessarily cumbersome. However, to within 0.3% in the UV and visible range and about 1% through the IR Eq. 5 may be replaced by

$$\tau_R(\lambda) = \tau_{Ro}(p/p_o) (\lambda_o/\lambda)^4 \exp \alpha[(\lambda_o/\lambda)^2 - 1] \quad (8a)$$

$$= K(p/p_o) (\lambda_o/\lambda)^4 \exp \alpha(\lambda_o/\lambda)^2 \quad (8b)$$

where $\tau_{Ro} = 1.1739$, $\alpha = 0.14620$ and $K = \tau_{Ro} \exp{-\alpha}$
The curve marked τ_R in Fig. 2 represents the fit of Eq. (8) to the Frohlich-Shaw values (illustrated by points) in the 260 to 360 nm spectral range. The crosses represent our previously used Eq. (7). Henceforth Eq. (8) will be used. A recent note by Young

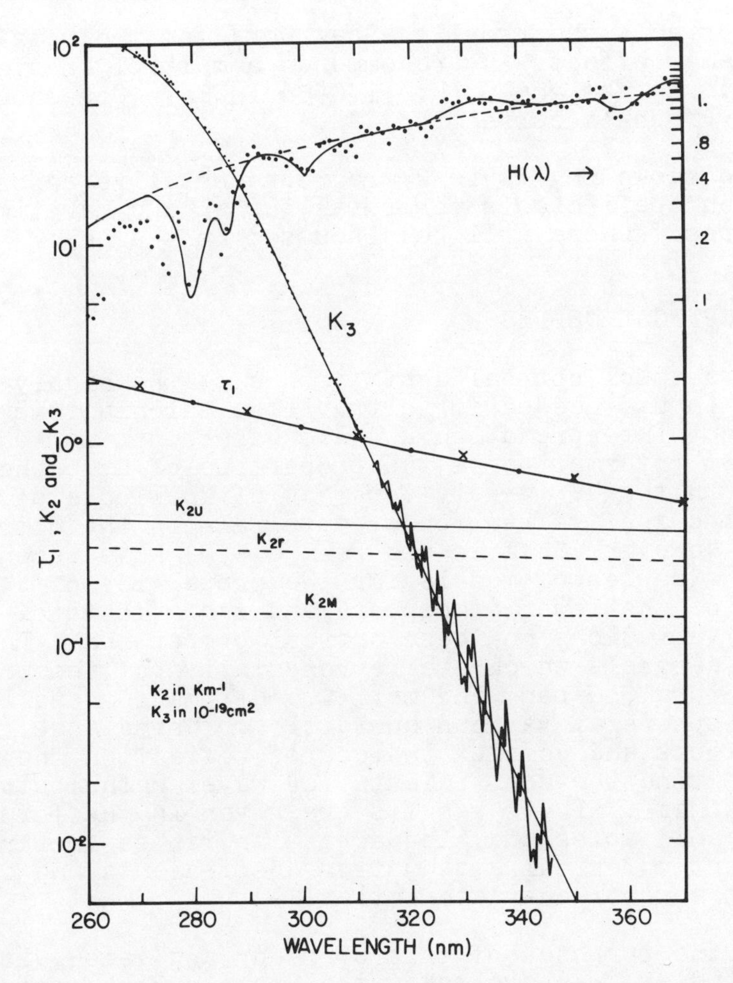

Figure 2. Illustrations of recent physical and analytic inputs available for UV spectral irradiance calculations. The smooth curve H(λ) represents the analytic fit with Eqs. (1) and (2) to the Nimbus 7 SBUV data of Heath and Park (1980) (points). The smooth curve labeled k_3 (λ) represents the fit with Eq. (3) to the ozone absorption coefficient at 18°C of Vigroux (1953) (points). The smooth curve labeled $\tau_R(\lambda)$ represents Eq. (8), the points, the calculated values of Frohlich and Shaw (1980), and the crosses calculated values from our previously used Eq. (7). The lines labeled K_{2u}, K_{2r} and K_{2m} give representative aerosol extinction coefficients (ln km^{-1}urban, rural and maritime aerosols for 80% relative humidity.) They correspond to reasonable optical depths if the equivalent thickness is chosen as 2 km.

(1980) corrects the Frohlich-Shaw work for the rotational Raman effect. He recommends a multiplying factor of 1.031. An assignment K = 1.0456 in Eq. (8b) should allow for Young's correction.

Also shown in Figure 2 are representative aerosol extinction coefficients for rural, urban and maritime atmospheres. These will be discussed in the next section.

AEROSOL OPTICAL DEPTHS

The aerosol optical depth is one of the highly variable inputs needed for ultraviolet calculations and measurement interpretation. There has been a marked absence of information on the properties of atmospheric aerosols in the UV largely for lack of knowledge of the complex index of refraction of materials in the ultraviolet. However, Shettle and Fenn (1979) have recently proposed a series of models for the properties of aerosols of the lower atmosphere and the effects of humidity variations on these optical properties. They consider aerosols which are representative of those found in rural, urban, and maritime air masses and in the troposphere at various humidities. Using model size distributions and complex indices of refraction they compute extinction coefficients for wavelengths between .2 and 40 μm. Their results are given in the form of diagrams and tables and, in particular Tables 12 through 43, present numerical attenuation coefficients and other important aerosol characteristics.

For the purposes of this UV study and later visibility studies, we have taken the Shettle-Fenn extinction coefficients computed for the wavelengths 200, 300, 337, 550, 694, and 1060 nm as input data. This constitutes a total of six wavelengths in 32 tables for a total of 192 numbers. We have fit these numbers by a two-dimensional empirical equation which should be useful for interpolations between the six wavelengths and seven relative humidity values that are represented in the tables. Our proposed formulas arrived at after considerable experimentation are:

$$E(\lambda,r) = E(\lambda_o,r) \exp - (\lambda-\lambda_o)/\lambda_a(r) \tag{9}$$

where λ_o = 300 nm,

$$E(\lambda_o,r) = E(\lambda_o,0)\left[1 + \frac{K\exp-(1/r)^3}{(1-r)^p}\right] \tag{10}$$

and

$$\lambda_a(r) = \lambda_a(0)\left[1 + \frac{Lr}{(1-r)^p}\right] \tag{11}$$

The values of $E(\lambda_o,0)$ and $\lambda_a(0)$ were fixed using the Shettle-Fenn table listings for r = 0 at λ = 300 and 377 nm. The remaining three parameters, K, p, and L were adjusted by non-linear least square fitting to the 40 other data points for each model. Figure 3 illustrates the variation with wavelength and with relative humidity implied by the tables and the types of fits achieved with Eqs. (9), (10), and (11). Table 3 gives the parameters used. Also shown are the particles/cc assumed in the Shettle-Fenn models. To represent other particle number densities one should proportion the parameter $E(\lambda_o,0)$ accordingly.

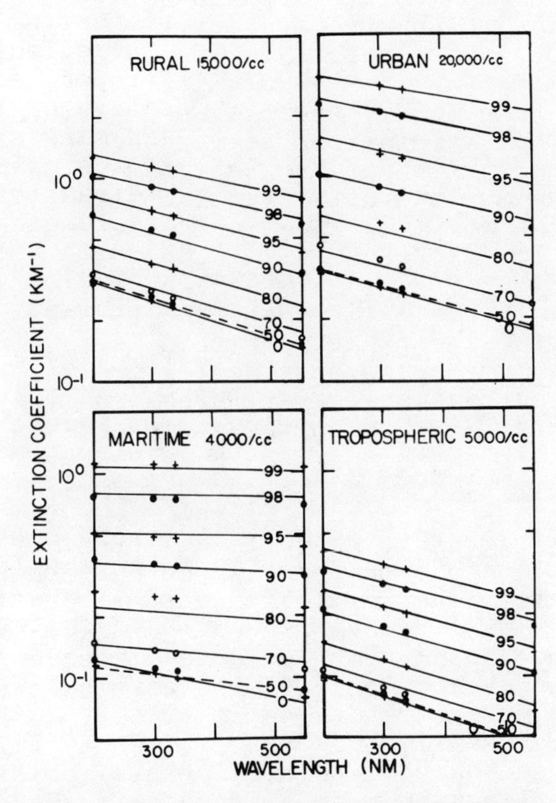

Figure 3. Aerosol extinction coefficients of Shettle and Fenn (1979) and their representations in the 200 to 550 nm region by Eqs. (9), (10), and (11).

Table 3

Parameters for Eqs. 9-11 to Fit Aerosol Models of Shettle and Fenn (1979)

	Rural	Urban	Maritime	Tropospheric
$E(\lambda_o,0)$	0.255	0.288	0.106	0.081
K	1.962	2.758	3.393	2.034
p	0.345	0.471	0.435	0.328
$\lambda_a(0)$	0.439	0.510	0.734	0.412
L	0.122	.0827	1.049	0.102

It is possible to have a more precise representation of the Shettle-Fenn data over a broader range of wavelengths (i. e. from 200 to 2000 nm by using a wavelength dependence of the form given by Eq. (3) rather than the simple exponential used in Eq. (9). However, this elaboration is unwarrented for the narrow range of wavelengths of concern here (i.e. from 270 to 360 nm.)

It might be remarked that it is unfortunate from the standpoint of the topic of this conference that the Shettle-Fenn maritime "data" points have given us the most trouble in our two-dimensional fittings. This may suggest that the maritime model or "data" has an anomaly in the $r \approx .7$ to .8 region.

To calculate the UV at the ocean surface we need the aerosol optical depth rather than the extinction coefficient. This may be obtained by multiplying $E(\lambda,r)$ by the equivalent aerosol layer thickness (in km). Then we may write for the wavelength dependence

$$\tau_a(\lambda) = \tau_a(\lambda_o) \exp - \alpha(\lambda-\lambda_o) \qquad (12)$$

where $\alpha = 1/\lambda_a(r)$. This is quite different from the power law wavelength dependence which has long been used to characterize aerosol extinction. For the narrow UV-B region the difference is academic since over a small wavelength range both forms would be almost constant. However, if one wished to determine aerosol thicknesses from reported turbidities (Flowers et al., 1969; Volz, 1959) which are usually measured at 500 nm, it would be important to use the best physical basis for extrapolation. The Shettle-Fenn work as parametrized here not only suggests that Eq. (12) should be used but also suggests that a knowledge of the character of the air mass would give an approximate value of α. As will be obvious from the next section, the aerosol optical depth is obtainable if absolute radiometry were utilized at longer wavelengths where ozone absorption is small and Rayleigh attenuation can be calculated.

THE RATIO METHOD AND ITS NEW ADAPTATION

The basic philosophy embodied in Green, Cross, Smith (GCS) was to represent the diffuse component in terms of ratios which relate it to the direct component. The purpose in part was to minimize the dynamic ranges of the important dependent variables and to be adaptable to new physical inputs. To obtain the ratios they used as "data" the radiative transfer calculations of Braslau and Dave (1974) and Dave, Braslau and Halpern (1975).

Following GSS, SG AND GCS we write the direct irradiance as

$$D(\lambda,\theta) = \mu H(\lambda) \exp - \Sigma_i (\tau_i/\mu_i) \tag{13}$$

where τ_1, τ_2, and τ_3 denote the air, aerosol and ozone optical depth respectively and μ, μ_1, μ_2, μ_3 denote cos θ and generalized cosine functions which are appropriate to the three species in view of the roundness of the earth. These functions can be conveniently characterized in the form (Green and Martin, 1966)

$$\mu_i = \left[\frac{\mu^2 + t_i}{1 + t_i}\right]^{\frac{1}{2}} \tag{14}$$

where the t_i are small characteristic numbers which depend on the altitude distribution of each species. (See GSS, 1974, for a more complete discussion.) With the improved inputs described in the previous sections we should have a better representation of the direct radiation.

Skylight provides a major component of the total irradiance in the UV in many cases exceeding the direct irradiance itself. To characterize skylight is difficult, however, and other than the two-stream approximation, which is not accurate, no theoretical analytic form is available. The Florida group has therefore developed a number of semi-empirical forms to characterize skylight. Their most recent strategy is to represent the downward diffuse spectral irradiance at the ground in terms of two ratios and the direct spectral irradiance for an overhead sun. Thus, the diffuse sky irradiance is expressed as

$$S(\lambda,\theta) = \mathscr{S}(\lambda,\theta) M(\lambda) H(\lambda) \exp-(\Sigma_i \tau_i) \quad (15)$$

where

$$\mathscr{S}(\lambda,\theta) = S(\lambda,\theta) / S(\lambda,0^o \quad (16)$$

and $M(\lambda) = S(\lambda,0^o) / D(\lambda,0^o)$ (17)

These ratios have relatively limited dynamic range whereas the sky irradiance itself as a function of wavelength and angle can vary over many orders of magnitude. In this work we preserve these basic aspects of GCS. However, the functional forms chosen to represent $\mathscr{S}(\lambda,\theta)$ and $M(\lambda)$ have been generalized somewhat. These generalizations improve the accuracy of the final average formula for use as an interpolative formula between the various Dave-Braslau-Halpern models. In redoing the multidimensional curve fitting, we have proceeded in a more systematic way from the aerosol-free models to the models with aerosols letting the formula successively evolve so that the generalized results reduced to simplified results in appropriate limits. The $\mathscr{S}(\lambda,\theta)$ function of GCS has been simplified slightly to

$$\mathscr{S}(\lambda,\theta) = [F + (1-F)\exp-\gamma_3\tau_3\phi] \exp-(\gamma_1\tau_1 + \gamma_2\tau_2)\phi \quad (18)$$

where we have deleted τ_4 the absorption term of the aerosol extinction coefficient and redefined τ_2 as the scattering term alone, and

where

$$\phi = \frac{(1+t)}{(\mu^2+t)}^{\frac{1}{2}} -1 \quad (19)$$

and

$$F = [1 + A(\tau_3 + \tau_4)^q]^{-1} \quad (20)$$

The M function has been generalized to the form

$$M(\lambda) = [A_{a1}\tau_1^{m_a} F_{a3} + A_{a2}\tau_2^{p_a}(1 + A_{a1}\tau_1^{m_a} F_{a3})] F_{a4} \quad (21)$$

where

$$F_{a3} = [1 + A_{a3}\tau_3^{q_a} w_a^{\nu_a}]^{-1} \tag{22}$$

and

$$F_{a4} = [1 + A_{a4}\tau_4]^{-1} \tag{23}$$

To allow for ground reflectivity we use for the global irradiance

$$G(\lambda,\theta,R) = \frac{G(\lambda,\theta,0)}{1 - r(\lambda)R} \tag{24}$$

where the reflectivity is now expressed by

$$r(\lambda,w_3) = [A_{b1}\tau_1^{m_b}F_{b3} + A_{b2}\tau_2^{p_b}]F_{b4} \tag{25}$$

Figure 4 illustrates the $\mathcal{S}(\lambda,\theta)$ functions. Figure 5 illustrates the $M(\lambda)$ and $r(\lambda,w_3)$ functions. These

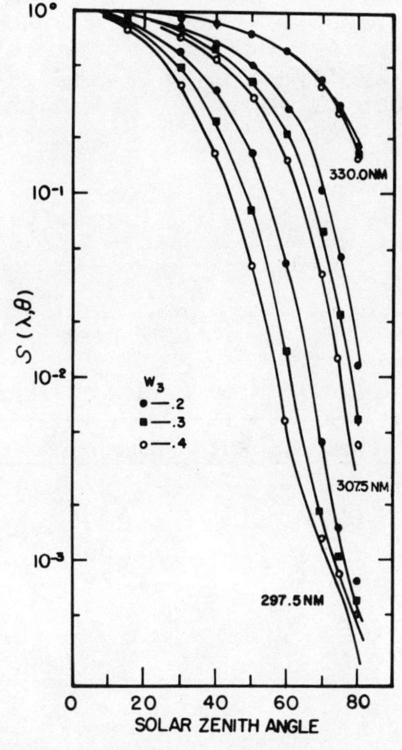

Figure 4. The new skylight $\mathcal{S}(\lambda,\theta)$ function for representative values of λ and w_3 as a function of solar zenith angle.

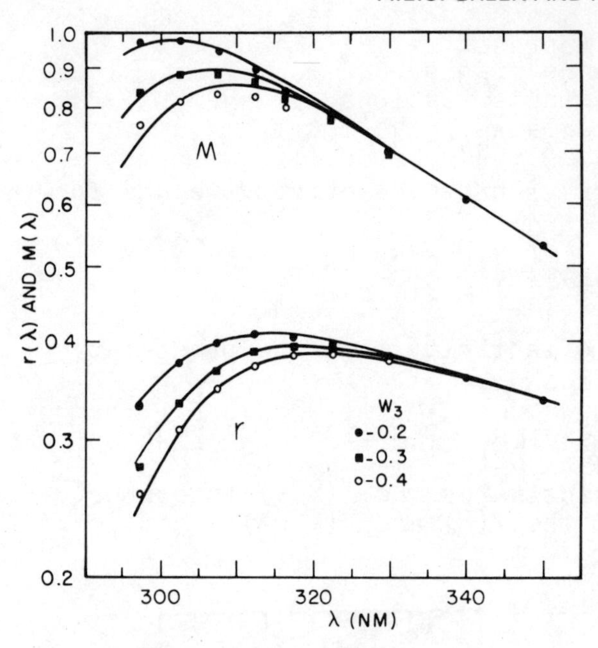

Figure 5. New skylight $M(\lambda)$ and $r(\lambda, w_3)$ functions for representative values of w_3.

Table 4

Parameters for New Skylight Model

	Rayleigh	Part.Sc.	Ozone	Part.Abs.
	i = 1	i = 2	i = 3	i = 4
γ_1	.5346	.6077	1.000	0.0
A_{ai}	.8041	1.437	.2864	2.662
A_{bi}	.4424	.1000	.2797	3.70
	A = 84.37	q = .6776		t = 0.0266

	m_α	p_α	q_α	ν_α
α = a	1.389	1.12	.8244	.4166
α = b	.5626	.878	.8404	.1728

components represent significant improvements with respect to the simpler GCS functions. Indeed, these improved characterizations of skylight represent all of the Dave-Braslau-Halpern models at about the same level of accuracy as the individually adjusted GCS models. Table 4 gives the parameter values for the modified ratio model. It should be clear that the present variation of the ratio method can serve a more useful role to interpolate between and extrapolate from the Dave-Braslau-Halpern atmospheric models.

Finally, we might note that while the Dave-Braslau-Halpern tables were calculated with physical inputs which are now somewhat outmoded, it is not unreasonable to use the skylight functions in conjunction with the new input data given in the previous sections to update the Dave-Braslau-Halpern outputs.

Table 5 summarizes our proposed new physical inputs for a maritime environment. The Gaussian parameters for the extraterrestrial solar irradiance are listed in Table 1. The variable quantities are p_o, τ_2 (λ_o) r and w_3 which must be measured or estimated hourly. The choice $\tau_2(\lambda_o) = 0.1$ may be taken as a nominal clean maritime atmosphere. The absorbing component is generally very small for a maritime atmosphere so we may set $\tau_4 \cong 0$.

Table 3

New physical inputs for maritime Environment ($\tau_4 \approx 0$)

$$H(\lambda)=H_o\left(\frac{\lambda_o}{\lambda}\right)^5 \left[\frac{\exp(p) - 1}{\exp(p\lambda_o/\lambda)-1}\right] \left[1+\Sigma_i A_i \exp-(\lambda-\lambda_i)^2/2\sigma_i^2\right]$$

$\lambda_o = 300$ nm, $p = 9.102$, $K_o = 0.582$ w/m^{2}nm (See Table 1)

$$\tau_1(\lambda) = 1.0142\ (p/p_o)\ (\lambda_o/\lambda)^4 \exp 0.1462(\lambda_o/\lambda)^2, p_o=1013 \text{ mb}$$

$$\tau_2(\lambda) = \tau_2(\lambda_o) \left[1 + \frac{K\exp-r^{-3}}{(1-r)^p}\right] \exp - \left[\frac{(\lambda-\lambda_o)}{\lambda_a(0)}\ \frac{1}{1+Lr(1-r)^{-p}}\right]$$

$K = 3.393$, $p = 0.435$, $\lambda_\theta(0) = 0.734$, $L = 1.049$

$$\tau_3(\lambda) = w_3 k_o\ (\beta+1)/[\beta + \exp(\lambda-\lambda_o)/\delta]$$

$k_o = 10.89$ atm-cm^{-1}, $\beta = 0.0355$, $\delta = 7.150$ nm

TABLE 6

Modified ratio method formulas

$$G(\lambda,\theta,0) = D(\lambda,\theta) + S(\lambda,\theta)$$

$$D(\lambda,\theta) = \mu H(\lambda) \exp - \Sigma_i (\tau_i/\mu_i)$$

$$\mu_i = [(\mu^2 + t_i)/(1 + t_i)]^{\frac{1}{2}}, \ \mu = \cos\theta$$

$$S(\lambda,\theta) = \mathcal{S}(\lambda,\theta) M(\lambda) H(\lambda) \exp - (\Sigma_i \tau_i)$$

$$\mathcal{S}(\lambda,\theta) = S(\lambda,\theta)/S(\lambda,0^o)$$

$$M(\lambda) = S(\lambda,0^o)/D(\lambda,0^o)$$

$$\mathcal{S}(\lambda,\theta) = [F + (1-F) \exp -\gamma_3\tau_3\phi] \exp - (\gamma_1\tau_1 + \gamma_2\tau_2)\phi$$

$$\phi = [(1+t)/(\mu^2 + t)]^{\frac{1}{2}} - 1$$

$$F = [1 + A(\tau_3 + \tau_4)^q]^{-1}$$

$$M(\lambda) = [A_{a1}\tau_1^{m_a} F_{a3} + A_{a2}\tau_2^{p_a} (1 + A_{a1}\tau_1^{m_a} F_{a3})] F_{a4}$$

$$F_{a3} = [1 + A_{a3}\tau_3^{q_a} w_a^{\nu_a}]^{-1}$$

$$F_{a4} = [1 + A_{a4}\tau_4]^{-1} \cong 1$$

$$G(\lambda,\theta,R) = \frac{G(\lambda,\theta,0)}{1-r(\lambda,w_3)R}$$

$$r(\lambda,w_3) = [A_{b1}\tau_1^{m_b} F_{b3} + A_{b2}\tau_2^{p_b}] F_{b4}$$

$$F_{b3} = [1 + A_{b3}\tau_3^{q_b} w_3^{\nu_b}]^{-1}$$

$$F_{b4} = [1 + A_{b4}\tau_4]^{-1} \cong 1$$

For marine atmosphere $\tau_4 \approx 0$.

Table 6 summarizes our ratio method formulas, several of which have been modified from those of GCS. In effect, now Table 5, 6 and 1 contain all the quantitative information needed to calculate approximate spectral irradiances for a maritime atmosphere.

DISCUSSION

Current interest in the UV-B reaching the earth's surface arises primarily because of the biological importance of these radiations. There has been a long-standing interest in the atmospheric science community in UV as an indicator of the total ozone column thickness. Indeed, specialized double monochromators (so-called Dobson's) have been developed to measure direct UV intensities at pairs of wavelengths in the 305 to 330 nm region. Then with the aid of null measurements and difference calculations it is possible to arrive at the ozone optical depth to a 5 - 10% level of accuracy. With a knowledge of the ozone column thickness, Rayleigh column thickness and an estimate of the aerosol thickness it is simple to compute the direct irradiance at the ground. With the aid of our skylight formula it is possible to compute approximately the diffuse irradiance for clear sky conditions and hence, by addition, the global irradiance. While the foregoing overview is satisfying, unfortunately clouds continue to represent a source of difficulty in attempts to assign ground level UV-B irradiance. Cloud corrections have been studied (Nack and Green, 1975; Spinhirne and Green, 1978; Green and Spinhirne, 1978) and the effects of solid homogeneous clouds are reasonably well in hand. Broken clouds which also have been addressed (Borkowski et al., 1977) still remain a problem for quantitative spectral radiometry. In large part this is because meteorological data station reports, while fairly complete in the percentage of sky covered, are very deficient in the matter of cloud optical depths. There has, however, been some progress which is suggested by the fact that clouds are fairly wavelength insensitive in their attenuating and scattering properties. Accordingly, cloud corrections based upon the readings of a broad band UV pyranometer should be adequate to normalize spectral irradiance monitors. The work of Smith and Baker (1979) and Baker, Smith and Green (1980) illustrate this technique.

There are now a large number of Robertson-Berger meters around the world (Berger, 1980). It is generally acknowledged that these dosimeters have a somewhat longer wavelength response than that most suitable for characterizing carcinogenic UV radiation. However, there is no doubt that in conjunction with theoretical modeling the RB meters can serve as a valuable means of correcting for cloud effects in the UV.

Finally, we might note that we now have a massive systematic body of total ozone data assembled from the Nimbus 4 satellite and will soon have additional data from the Nimbus 7 satellite. Both of these sets should be very useful for establishing a global ultraviolet photoclimatology.

ACKNOWLEDGEMENTS

The authors would like to thank J. M. Schwartz and W. E. Bolch for their assistance in the calculations reported in this work and Roxie Mays and Linda Combs for the production of the manuscript. This work was supported in part by contract EPA-R806373010 from the U. S. Environmental Protection Agency and contract Number NAS-5-22908 from the National Aeronautics and Space Administration.

REFERENCES

Baker, K., R. C. Smith and A. E. S. Green. 1980. Middle ultraviolet radiation reaching the ocean surface. Photochem. and Photobiol. 32:367-374.

Bass, A. 1980. Personal communication.

Bener, P. 1963. The diurnal and annual variations of the spectral intensity of ultraviolet sky and global radiation on cloudless days at Davos, 1590 m a.s.l. AFCRL, Contract AF 61 (052)-618. Technical Note No. 2, January.

Bener, P. 1970. Measured and theoretical values of the spectral intensity of ultraviolet zenith radiation and direct solar radiation at 316, 1580 and 2818 m.a.s.l, AFCRL, Contract F 61052-67-C-0029, July.

Bener, P. 1970. Solar intensity and intensity and polarization if sky radiation for 347.0, 488.0 and 533.5 nm at selected points along the sun's vertical and other meridians measured at 2818 m a.s.l., AFCRL, Contract DAJA 37-68-C-1017, August.

Braslau, N. and J. V. Dave. 1973. Effect of aerosols on the transfer of solar energy through realistic model atmospheres, Part III: Ground level fluxes in the biologically active bands, .285-.370 microns, IBM Research Report, TC 4308.

Borkowski, J., A. T. Chai, T. Mo, and A.E.S. Green. 1977. Cloud effects on middle ultraviolet global radiation. Acta Geophys. Pol. 25 (4): 287-301.

Caldwell, M. 1976. Personal communication.

Dave, J. V. and P. M. Furukawa. 1966. Scattered radiation in the ozone absorption bands at selected levels of a terrestrial Rayleigh atmosphere. Meteor. Monographs. 7. No. 29, 353 pp.

Flowers, E. C., R. A. McCormick and K. R. Kurfis. 1969. Atmospheric turbidity over the United States. 1961-1966. J. App. Meteor. 8: 955-962.

Frohlich, C. and G. E. Shaw. 1980. New determination of Rayleigh scattering in the terrestrial atmosphere. App. Opt. 19(11): 1773-1775.

Green, A.E.S., K. R. Cross and L. A. Smith. 1980. Improved analytic characterization of ultraviolet skylight. Photochem. and Photobiol. 31: 59-65.

Green, A.E.S. and T. Mo. 1974. An epidemiological index for skin cancer incidence. Proceedings of the 3rd Conference on CIAP, DOT-TSC-OST-74-15, 518-522.

Green, A.E.S., T. Mo and J. H. Miller. 1974. A study of solar erythema radiation doses. Photochem. and Photobiol. 20:473.

Green, A.E.S., T. Sawada and E. P. Shettle. 1974. The middle ultraviolet reaching the ground. Photochem. and Photobiol. 19: 351.

Green, A.E.S. and J. D. Spinhirne. 1978. Cloud effects on UV photoclimatology in Proceedings of the Twelfth International Symposium on Remote Sensing of Environment, April 20-26, Manila, Philippines.

Green, A.E.S. and J. D. Martin. 1966. A generalized Chapman function, Chapter 7. In: Green, A.E.S. (ed.) The Middle Ultraviolet - Its Science and Technology, J. Wiley.

Heath, D. F., A. I. Krueger, H. A. Roeder and B. D. Henderson. 1975. The solar backscatter ultraviolet and total ozone mapping spectrometer (SBUV/TOMS) for Nimbus G. Opt. Eng. 14: 323.

Heath, D. and H. W. Park. 1980. Amer. Geophys. Union Meeting, Toronto, personal communication.

Inn, E. C. T. and Y. Tanaka. 1953. Absorption coefficient of ozone in the ultraviolet and visible regions. J. Opt. Soc. Am., 43(10):870-873.

Johnson, F. S., T. Mo and A.E.S. Green. 1976. Average latitudinal variation in ultraviolet radiation at the earth's surface. Photochem. and Photobiol. 23:179.

Klenk, K. F. 1980. Absorption coefficients of ozone

for the backscatter experiment. App. Opt. 19:236.

Kohl, J. L., W. H. Parkinson and C. A. Zapata. 1980. Solar spectral radiance and irradiance, 225.2 nm to 319.6 nm, Center for Astrophysics, No. 1289.

Mo, T. and A.E.S. Green. 1974. A climatology of solar erythema dose. Photochem. and Photobiol. 20:438-496.

Nack, L. M. and A.E.S. Green. 1974. Influence of clouds, haze, and smog on the middle ultraviolet reaching the ground. App. Opt. 13: 2405.

Penndorf, R.1957. Tables of the refractive index for standard air and the Rayleigh scattering coefficient for the spectral region between 0.2 and 20.0 μ and their applications to atmospheric optics. J. Opt. Soc. Am. 47: 176-182.

Shettle, E. P. and R. W. Fenn. 1979. Models for the aerosols of the lower atmosphere and the effects of humidity variations on their optical properties, AFGL-TR-79-0214.

Shettle, E. P. and A.E.S. Green. 1974. Multiple scattering calculation of the middle ultraviolet reaching the ground. App. Opt. 13:(7):1567-1581.

Shotkin, L. M. and J. F. Thompson, Jr. 1973. Use of an atmospheric model with aerosols to examine solar UV data. J. Atmos. Sci. 30:1699.

Simons, J. W., R. J. Paur, H. A. Webster, III, and E.J. Bair. 1973. Ozone ultraviolet photolysis. VI. The ultraviolet spectrum. J. Chem. Phys. 59(3): 1203-1208.

Smith, R. C. and K. S. Baker. 1979. Penetration of UV-B and biologically effective dose-rates in natural waters. Photochem. and Photobiol. 29: 311-323.

Smith, R. C. and K. S. Baker. Middle ultraviolet irradiance measurements at the ocean surface (to be published).

Spinhirne, J. D. and A.E.S. Green. 1978. Calculation of the relative influence of cloud layers on received ultraviolet and integrated solar radiation. Atmos. Environ. 12:2449-2445.

Vigroux, E. 1953. Contribution à l'étude expérimentale de l'absorption de l'ozone. Ann. Phys. 12(8): 709-762.

Vigroux, E. 1967. Determination des coefficients moyens d'absorption de l'ozone en vue des observations concernant l'ozone atmospherique à l'aide du spectrometre Dobson. Ann. Phys. 14(2):209-215.

Volz, F. 1959. Photometer mit Selen-photoelement zur spektralen Messung der Sonnenstrahlung und zur Bestimmung der Wellenlängenabhängigkeit der Dunsttrübung. Arch. Meterol. Geophys Bioklimatol. 10(1):100-131.

Young, A. T. 1980. Revised depolarization corrections for atmospheric extinction. Appl. Opt. 19: 3427-3428.

POSSIBLE ANTHROPOGENIC INFLUENCES ON STRATOSPHERIC OZONE

F. Sherwood Rowland

Department of Chemistry
University of California
Irvine, California 92717

THE SOLAR SPECTRUM AND PLANETARY ATMOSPHERES

The intensity of radiation emitted from the sun has its maximum at wavelengths near 500 nm, consistent with its 6000K surface temperature, and is presumed to have done so without large variations for hundreds of millions of years. The natural solar satellites, including the Earth, must therefore have evolved in ways which are responsive to this constant photochemical bombardment. Our usual designations for the various portions of the solar spectrum are oriented toward <u>homo sapiens</u> and the possession by that biological species of an efficient photon detection system which is both extremely sensitive toward solar radiation and capable of discrimination among the various wavelengths of light 760 nm ("red") to 400 nm ("violet"). Not surprisingly, the wavelengths of radiation detected by the eye of man include the broad band covering the most intense solar emission. Our description system further classifies solar radiation <u>not</u> detected by the human eye into those wavelengths which are longer than those in the visible region, i.e. infrared, with $\lambda > 760$ nm, and those wavelengths shorter, i.e. ultraviolet with $\lambda < 400$ nm.

The response of a planetary atmosphere to this perpetual solar photochemical exposure is determined by its interactions with these various wavelengths of light. In general, the energy content of a quantum of infrared light (38 kcal/mole at 760 nm) is insufficient to cause the photodecomposition of molecules which are thermally stable at the temperatures characteristic of the earth's atmosphere. (The troposphere and stratosphere contain all but 0.1% of the atmospheric bulk, at altitudes of 0-50 km and with temperatures between 210-320K.) Molecules with strong infrared interactions are therefore able to persist without destruction despite the constant absorption and re-emission of infrared radiation. Infrared absorption tends to increase rapidly in intensity with increasing molecular complexity; is quite weak for molecules containing two atoms of the same element, such as the major components of the earth's atmosphere, N_2 (78%) and O_2 (21%); and does not occur for monatomic Ar (1%). Consequently, the chief atmospheric absorbers for infrared radiation are the less abundant but more complex triatomic species such as H_2O, CO_2 and O_3. The earth is essentially in balance insofar as radiation is concerned, emitting each day radiation equivalent in total energy to that absorbed from the sun. The maximum intensity of this radiation is determined chiefly by the temperature of the emitting body, and that coming from the earth's surface at approximately 300K has its maximum flux at wavelengths about 20 times longer than that from the sun. Infrared absorption by atmospheric molecules is therefore even more important on the path <u>out</u> from the earth than it is on the way in. The "greenhouse" effect of atmospheric CO_2 is dependent upon its interception of infrared radiation emitted from the earth, followed by re-emission in all directions including some backward toward the ground. The increase in atmospheric CO_2 from greater burning of fossil fuels is expected to enhance this interception of outgoing infrared radiation.

The wavelengths of light toward the blue and violet in the visible spectrum correspond to much higher energies, sufficient with most molecules to cause chemical reactions if absorbed. Consequently, those chemical compounds with absorption spectra in the visible region (i.e. colored gases) have very short atmospheric lifetimes. The most common visible molecular component in the atmosphere is NO_2, a brown gas characteristic of urban photochemical smog. This molecule survives for only about 15 minutes on the average in overhead sunlight before decomposing to NO and O. However, NO is readily oxidized back to NO_2 in the presence of ozone, so that the color can persist for long periods in atmospheres containing both NO_x and O_3.

The major components of the earth's atmosphere are necessarily molecules not easily destroyed by solar visible radiation, and none of the most abundant (N_2, O_2, Ar, H_2O, CO_2) absorb visible radiation at all. The blue color of the atmosphere is caused by the scattering without absorption by O_2 of the violet and blue wavelengths in preference to the scattering of the redder wavelengths. The "reddening" of sunlight near sunset and the yellowness of the rising moon are other results of the preferential loss of blue radiation during passage through the much longer atmospheric pathlengths required when the sun or moon is near the horizon.

As the wavelength of light shortens from the visible into the ultraviolet range, the number of quanta of solar radiation decreases but the energy per quantum increases, and its absorption by a molecule usually decomposes it into atoms or other smaller fragments. The molecular components of the atmosphere are therefore subject to ultraviolet constraints upon their atmospheric lifetimes in addition to those imposed by visible solar radiation. Only a few gaseous molecular species are able to exist unchanged for long if exposed to the full energetic range of solar radiation, even at the sun-earth distance of 93 million miles. Since most molecules absorb extremely energetic ultraviolet radiation with very large cross sections, even a rudimentary atmosphere is sufficient to remove all of this incoming solar flux before it can penetrate to a planetary surface. The energies of the shortest wavelengths cause the detachment of electrons from the absorbing molecule, and are the origin of the earth's ionosphere. Such ionizing events occur for a planetary atmosphere of any chemical composition.

The dominant chemical makeup of the atmosphere and the absorption characteristics of individual molecules become more important for wavelengths longer than about 100 nm. Mars and Venus possess chemically oxidizing atmospheres dominated by CO_2, and solar radiation effects on these planets are best described through examination first of the ultraviolet and infrared absorption spectra of CO_2. On the other hand, the chemically-reducing atmospheres of Jupiter and of the other outer planets, containing He, H_2 and CH_4, can be approached first through consideration of the absorption spectrum of CH_4. In the present atmosphere of the earth, almost all absorption of solar ultraviolet radiation occurs in the various oxygen-containing compounds, including O_2, O_3, and to a lesser extent, H_2O and CO_2.

During the early history of the earth, the atmosphere was probably composed chiefly of reduced chemical forms (i.e., CH_4, NH_3, H_2O, etc.) and the radiation absorption characteristics of such an atmosphere would have been quite different from those existing now. However, with the gradual loss of atomic hydrogen to space, the atmosphere eventually was converted to its present highly oxidizing state containing a substantial concentration of free O_2. The predominant characteristic of solar ultraviolet radiation at the surface of the earth now is the complete absence of wavelengths shorter than about 295 nm. This ultraviolet "cut-off" is dependent upon the existence in the atmosphere of a low concentration of ozone, O_3, whose presence is in turn dependent upon the qualitative existence in the atmosphere of moderate amounts of free O_2.

ULTRAVIOLET ABSORPTION BY O_2; FORMATION OF O_3

The ultraviolet absorption spectrum of O_2 has a very weak band beginning at 242 nm which becomes progressively stronger at shorter wavelengths, and very strong near 190 nm (See Okabe 1978). Absorption of this ultraviolet radiation by O_2 causes photodissociation into two atoms of O, as in reaction (1). Although other chemical reactions remove minor amounts, the overwhelming fate of O atoms released in the earth's atmosphere is combination with

$$O_2 + UV \rightarrow O + O \qquad (1)$$

molecular O_2 to form ozone by reaction (2'). The freshly-formed O_3^* molecule contains sufficient energy to dissociate back to O

$$O + O_2 \rightarrow O_3^* \qquad (2')$$

plus O_2, and will do so within about 10^{-11} second unless the excess energy is removed by collision with another molecule M, as in (2"). Since any molecule can serve in such collisions, M is usually the abundant N_2. The probability of a deexciting

$$O_3^* + M \rightarrow O_3 + M \qquad (2'')$$

collision is proportional to the concentration of M, and therefore to the density of the atmosphere, becoming very unlikely at higher altitudes. Below 50 km the combination of (2') plus (2") has sufficient probability to be important, and this O_3-forming reaction is usually written in the combined form of equation (2). The appearance of free O_2 in the earth's

$$O + O_2 + M \rightarrow O_3 + M \quad (2)$$

atmosphere is thus accompanied by the simultaneous presence in lesser concentration of O_3.

Molecular O_3 is capable of absorption of radiation over the whole solar spectrum of interest (ultraviolet, visible, infrared) and can be dissociated into $O + O_2$ even by some infrared radiation because the O_3 molecule is only weakly-bound. The UV absorption is very strong, while the effects of absorption in the visible region are of minor importance. (Ozone in large concentrations in the laboratory has a faint blue color.) In the ultraviolet range, the absorption coefficients increase from 350 nm toward 260 nm and become so strong that essentially all radiation between 230-290 nm is absorbed by O_3. A typical number of molecules of O_3 in the earth's atmosphere is approximately 8×10^{18} in the total column above 1 cm^2 of the earth's surface. An absorption coefficient of 10^{-19} cm^2 per molecule of O_3 therefore corresponds to removal of 55% of this radiation [fraction penetrating to surface = $\exp -(8 \times 10^{18})(10^{-19}) = \exp(-0.8) = 0.45$] if the sun is directly overhead, and even more (80% for solar zenith angle of 60°) with solar radiation coming in on slanting pathways toward the earth's surface [penetration at 60° =

$\exp -(8 \times 10^{18})(10^{-19})(\sec 60^o) = \exp(-1.6) = 0.20$].

The absorption coefficient for O_3 is 10^{-20} at 325 nm; 10^{-19} at 310 nm; 10^{-18} at 294 nm [$\exp -(8 \times 10^{18})(10^{-18}) = \exp(-8) = 0.0003$]; and 10^{-17} near 260 nm [$\exp -(8 \times 10^{18})(10^{-17}) = \exp(-80) = 10^{-35}$]. Clearly, the upper few percent of the stratospheric O_3 molecules absorb almost all of the 260-290 nm radiation entering the earth's atmosphere.

The absorption of ultraviolet radiation by O_3 causes photodissociation, as in (3) with the release of atomic O.

$$O_3 + UV \rightarrow O + O_2 \quad (3)$$

In most cases, however, the O atom so released recombines with another O_2 molecule by (2) and there is no net change in the O_3 concentration of the atmosphere following the sequence of (3) plus (2). There is, though, a conversion of the incoming solar ultraviolet radiation into heat, as in reaction (2") in which M receives energy, and this heat source produces higher local temperatures than would exist without it. The temperature decreases with increasing altitude from the 270-320K typical of the surface to about 210K near 10-15 km (NOAA) 1976). Above this altitude (designated as the tropopause, the upper limit of the lowest section of the atmosphere, the troposphere), the temperature no longer decreases, and above about 25 km the temperature begins to rise again because of the absorption of UV radiation by O_3. The maximum temperature in this region of the atmosphere is usually near 270-280K, occurring at an altitude of about 50 km. This region of level or rising temperatures with increasing altitude is the stratosphere, and its upper boundary at the temperature maximum is the stratopause. The very existence and location of the stratosphere is therefore closely tied to the distribution with altitude of ozone in the atmosphere. The atmospheric pressure decreases by approximately a factor of 10 for every 15 km altitude, so that the pressure at the top of the troposphere is about 10^{-1} atmosphere, and at the top of the stratosphere is about 10^{-3} atmosphere.

Ozone is thermodynamically quite unstable relative to molecular O_2, and a number of chemical reaction routes exist which can convert O_3 back into O_2. The first of these routes to be considered for the atmosphere was the direct reaction of some of the O atoms with O_3, as in (4), in competition with the usual reaction with O_2. Since most O atoms react in the atmosphere to form O_3 by (2), reaction (4) corresponds to the loss of two ozone molecules, one as O_3 and one as O. The chemistry of the atmosphere is frequently described by grouping together O_1 and O_3 as "odd oxygen", in contrast to the normal, far more abundant O_2. In these terms the net changes in the molecular concentration of "odd oxygen" are +2 for reaction (1), zero for (2), zero for (3) and -2 for (4). The entire cycle of reactions (1) to (4) is often described by the term "pure oxygen reactions", or the "Chapman reactions" after the geophysicist who first rationalized the observed atmospheric concentrations of O_3 about 50 years ago (Chapman 1930).

During the past two decades, and especially during the 1970's, extensive study of the chemical reactions in the stratosphere has identified several additional chemical reaction routes by which various free radical species are able to catalyze the change of O_3 back into O_2. The actual conversion in the atmosphere is a composite of the direct reaction (4) plus contributions from several of these catalytic routes.

$$O + O_3 \rightarrow O_2 + O_2 \qquad (4)$$

CATALYTIC CHAIN REACTIONS IN THE ATMOSPHERE

Most molecular species available both in and out of chemical laboratories possess an even number of electrons, as do all of the most abundant chemical entities in the atmosphere (See McEwan and Phillips 1975). The trace species known as "free radicals", however, have odd numbers of electrons, and are generally quite reactive chemically. The interaction of an odd-electron species with an even-electron species cannot avoid forming at least one odd-electron species and a succession of such reactions constitutes a free radical "chain". When NO (15 electrons) reacts with O_3, as in (5), the

$$NO + O_3 \rightarrow NO_2 + O_2 \qquad (5)$$

free radical NO_2 (23 electrons) is formed. The subsequent reaction of NO_2 with atomic O as in (6) forms NO again, and

$$NO_2 + O \rightarrow NO + O_2 \qquad (6)$$

the net result of (5) plus (6) is the removal of one atom of O and one molecule of O_3, with the formation of two molecules of O_2. The overall result is thus chemically equivalent to reaction (4). Since there is no net change in the concentration of either NO or NO_2 during the sequence (5) plus (6), the process is described as catalytic, and the pair of reactions constitutes the NO_x-catalyzed free radical chain for removal of O_3 from the atmosphere. A similar ClO_x catalytic cycle can be written involving atomic Cl (17 electrons) and ClO (23 electrons) as in (7) and (8), with a net result again equivalent

$$Cl + O_3 \rightarrow ClO + O_2 \qquad (7)$$

$$ClO + O \rightarrow Cl + O_2 \quad (8)$$

to (4). Still other chains in the atmosphere operate through radicals in the HO_x series (H, HO and HO_2, with 1, 9, and 17 electrons, respectively) and with carbon-containing radicals such as CH_3, CH_3O and CH_3O_2 (9, 17 and 25 electrons, respectively).

The NO_x, ClO_x and HO_x free radical cycles in the natural atmosphere furnish additional pathways for the removal of O_3 and result in an overall natural balance between its production by (2) and its removal by (4) to (8), etc. The rates of reactions such as (7) and (8) can be quite rapid in the stratosphere, with the cycle for these two reactions requiring only about one minute near 40 km altitude. Every Cl atom present in the ClO_x cycle in the upper stratosphere can thus catalyze the conversion into O_2 of thousands of molecules of O and O_3 per day. This catalytic property is enormously important in translating the emission of chlorine-containing compounds by man into ozone removal on a scale comparable to the levels found in the atmosphere.

Free radical chains can interact with one another, as in (9) and (10), and be converted from one cycle to another as in (11) and (12). The termination of an odd-electron chain (i.e. formation of even-electron products only) cannot occur except through interaction with another free radical, as in reactions (13) to (16).

$$HO_2 + NO \rightarrow HO + NO_2 \quad (9)$$

$$ClO + NO \rightarrow Cl + NO_2 \quad (10)$$

$$Cl + CH_4 \rightarrow HCl + CH_3 \quad (11)$$

$$HO + HCl \rightarrow Cl + H_2O \quad (12)$$

$$HO + HO_2 \rightarrow H_2O + O_2 \quad (13)$$

$$ClO + HO_2 \rightarrow HOCl + O_2 \quad (14)$$

$$HO + NO_2 + M \rightarrow HONO_2 + M \quad (15)$$

$$ClO + NO_2 + M \rightarrow ClONO_2 + M \quad (16)$$

These chain terminations are usually only temporary because the even-electron products can be photodissociated by solar photons, as in (17), and the free-radical photoproducts start

$$HOCl + UV \rightarrow HO + Cl \qquad (17)$$

the chains again. The ultimate removal of these chemical species from the atmosphere occurs when the water-soluble molecules such as HCl and $HONO_2$ diffuse from the stratosphere down into the troposphere and are taken out in rainfall. This loss of trace material from the atmosphere is balanced by a steady flow into the atmosphere of new molecules originating in biological processes at the surface, including hydrogen as H_2 and CH_4, nitrogen as N_2O, and chlorine as CH_3Cl. All of these stable molecules can be either directly or indirectly decomposed by solar photolytic processes, providing a continuous flux of free radicals. The natural atmosphere thus contains a substantial number of trace species, both free radicals and their chain termination products, which are important for the natural ozone balance. Although ozone is present in the atmosphere at an overall average concentration of about 3×10^{-7}, these trace species at much lower concentrations can strongly influence the ozone balance through the catalytic effect of the free radical chains.

Atmospheric free radical chemistry related to ozone can be conveniently discussed for three different regions of the atmosphere. The troposphere is regularly cleansed by rainfall, and the chemical time scales are typically of the order of hours to days. The concentrations of free radical pollutants, e.g. NO_x from automobiles, are quite variable from one day to the next, dependent upon wind and weather, and most of the observed chemistry occurs in the vicinity of the sources or downwind from them.

The bulk of stratospheric O_3 is found between 15-30 km and the amount above 35 km is insufficient to screen out all ultraviolet radiation below 295 nm, and in particular permits the penetration to 25-30 km of appreciable fractions of the wavelengths between 190-230 nm. When molecules between 30-50 km are exposed to this intense ultraviolet radiation, few can survive for more than tens of minutes, and the number of important chemical species is consequently very limited. The chemistry in this region is therefore comparatively simple, and reasonably well-understood. In the lower part of the stratosphere below 30 km, on the other hand, ultraviolet radiation is effectively limited to wavelengths longer than 295 nm and many polyatomic molecules can survive for appreciable times. The overall chemistry between 15-30 km therefore involves many more chemical species and chemical reactions than are important in the upper stratosphere.

POTENTIAL STRATOSPHERIC POLLUTANTS

The concerns for possible influences of man's activities upon the ozone in the stratosphere center on the introduction into the atmosphere of appreciable quantities of artificial substances capable of altering the concentrations of the various free radicals in the stratosphere (Grobecker et al. 1974). One potential method for such contamination is the direct introduction of the free radicals themselves, as with the NO_x found in engine exhaust gases. The passage of the N_2/O_2 mixture in ordinary air over hot surfaces in an engine converts some of it into NO, which is then released in the exhaust. This process directly introduces NO_x into the atmosphere at ground level from automobiles, and into the upper atmosphere from high-flying aircraft. The current subsonic jet aircraft cruise at altitudes of 10-12 km, which depending upon latitude and season can be either upper tropospheric (tropics; summer temperate) or lower stratospheric (polar; winter temperate). The designed flight altitude of supersonic aircraft is considerably higher (17-20 km), and the exhaust effluents are therefore injected at these higher altitudes. The much larger tonnage of NO_x released at the surface by automobiles is not an important source of <u>stratospheric</u> pollution because these molecules are removed by the physicochemical processes in the troposphere (e.g. rain).

Other potential sources of stratospheric free radicals arise through the release near the surface of stable molecules which are reasonably inert toward tropospheric removal, allowing them to diffuse upward into the stratosphere where they can be decomposed by solar-induced photochemical processes. The naturally-occurring molecules N_2O and CH_4 are destroyed in the region above 25 km by direct photolysis as in (18), or by indirect solar chemical reaction as in (19) and (20). The major

$$N_2O + UV \rightarrow N_2 + O \tag{18}$$

$$O(^1D) + N_2O \rightarrow NO + NO \tag{19}$$

$$HO + CH_4 \rightarrow H_2O + CH_3 \tag{20}$$

fraction of the CH_4 is removed through (20) by chemical attack of HO radicals in the troposphere, but an appreciable amount reaches the stratosphere nonetheless. Any anthropogenic increase in the release rates of N_2O or CH_4 would be followed later by more decomposition in the stratosphere through reactions (18) to (20). A very substantial additional source of free radicals in the stratosphere is the release by man of relatively inert molecules not previously present in the atmosphere, of which the chlorofluorocarbons (CCl_3F, CCl_2F_2, etc.) are the most important (Molina and Rowland 1974, Rowland and Molina 1975). These molecules do not absorb ultraviolet radiation for $\lambda > 240$ nm, and are therefore protected from photodecomposition in the troposphere by the presence of ozone in the stratosphere above. However, when they diffuse to altitudes above 20 km they become exposed to ultraviolet radiation with wavelengths near 200-220 nm, and can be photodecomposed as in (21) and (22).

$$CCl_3F + UV \rightarrow Cl + CCl_2F \quad (21)$$

$$CCl_2F_2 + UV \rightarrow Cl + CClF_2 \quad (22)$$

The inert CCl_2F_2 and CCl_3F molecules are thereby converted into odd-electron species, in each case a Cl atom and a CCl_2F or $CClF_2$ free radical. The radicals then react rapidly with O_2, and subsequent processes release these Cl atoms as well.

Tropospheric measurements during 1980 in remote locations away from urban sources show substantial concentrations in air of several organochlorine compounds, including (in parts per 10^{10} by volume) CH_3Cl (6) CCl_2F_2 (3), CCl_3F (1.7), CCl_4 (1.2) and CH_3CCl_3 (1.4). Among these compounds only CH_3Cl is likely to have existed in the atmosphere in any significant concentration in the year 1900, since the others are almost entirely the product of the technological activities of man during the 20th century.

Worldwide production of CCl_3F, CCl_2F_2 and CH_3CCl_3 increased steadily during the past 25 years, at rates corresponding to doubling of the atmospheric release rate every five to seven years. Although production of CCl_3F and CCl_2F_2 has been roughly constant since the early 1970's, ground level measurements have continued to show rapid increases in their atmospheric concentrations as these long-lived molecules continue to accumulate. The magnitude of this increase is shown

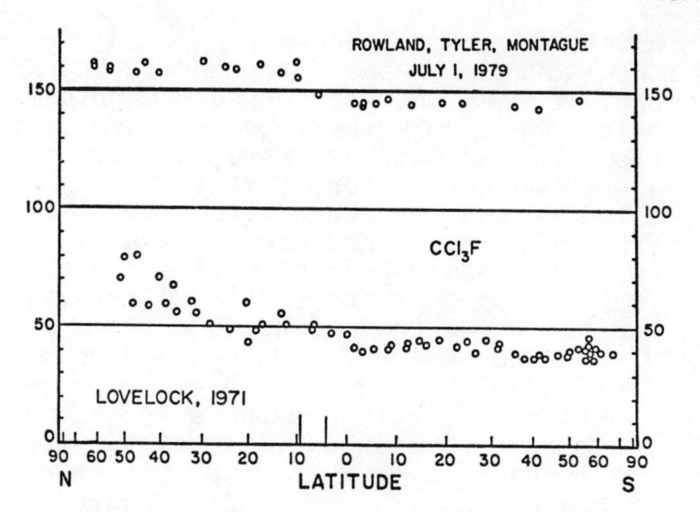

Figure 1. World-wide concentrations of CCl_3F in parts per 10^{12} by volume, as measured by Lovelock in 1971 and Rowland, Tyler and Montague in 1979.

for CCl_3F in Figure 1. Infrared spectrograms made from balloons at an altitude of 30 km have also shown steady increases since 1968 in the CCl_3F/CCl_2F_2 concentrations in the lower stratosphere. The atmospheric lifetimes of CCl_3F and CCl_2F_2 can be directly estimated from these tropospheric concentration increases, and consistently indicate lifetimes of at least several decades (and therefore no important tropospheric removal processes). On the other hand, the average lifetime of the CH_3CCl_3 molecules is only about 7 years because of their reactivity toward HO radicals in the troposphere as in (23). Nevertheless, the atmospheric concentration of CH_3CCl_3

$$HO + CH_3CCl_3 \rightarrow H_2O + C_2H_2Cl_3 \qquad (23)$$

has also continued to rise throughout the 1970's, following the pattern of steady increase in its world production.

Since the concentration of ClO_x radicals in the stratosphere is essentially determined by the total concentration of inert chlorine-containing molecules in the troposphere, the observed increases in CCl_3F, CCl_2F_2 and CH_3CCl_3 must be inevitably followed a few years later by an increase in stratospheric ClO_x, delayed only by the time required for upward mixing into the stratosphere. The intense ultraviolet radiation of the upper stratosphere keeps the chemistry of chlorine simple above 40 km and only ClO and HCl can contain major fractions of the total chlorine. All of the atmospheric model calculations (See NAS 1976, 1979a&b, Department of Environment 1979) are in agreement that continued release to the atmosphere of CCl_3F and CCl_2F_2 at 1981 rates will eventually lead to a large decrease (30-50%) in the average world-wide O_3 concentrations at 40 km. The absorption of solar ultraviolet radiation by O_3 is the primary heat source of the stratosphere, and diminution in its concentration at 40 km will result in less heat released there, tending to reduce temperatures in the upper stratosphere. Much of this ultraviolet radiation would then be absorbed at lower altitudes, releasing heat there, further tending to alter the stratospheric temperature structure.

NUMERICAL MODELS OF THE ATMOSPHERE

The chemical interactions of the stratosphere can be simulated by numerical models of varying degrees of complexity in the treatment of both chemical and meteorological processes (NAS 1976,1979a&b). The computational strain in such models increases rapidly with (a) more chemical species and chemical reactions, with current stratospheric models including more than 50 and 130, respectively; (b) more meteorological grid points; and (c) short time-steps to accommodate the very rapid changes in concentrations of some species near sunrise and sunset (e.g. O atom concentrations are negligible without sunlight for continuous new release). All such present models require appreciable simplification in order to be computationally manageable. The limitations are not simply computer power, however, for the calibrating information from the real atmosphere is often insufficient to distinguish among possible more elaborate models.

One approximation that is frequently used is to consider variations in chemical concentrations and solar irradiation conditions only in one dimension (vertical), using some average atmospheric condition otherwise (often those of 30° N latitude, averaged over all longitudes). The meteorological mixing in these 1-D models is expressed in terms of a single parameter, the "eddy diffusion coefficient", which varies with altitude in a manner calibrated by actual atmospheric measurements of the vertical distribution of molecules such as CH_4 or N_2O. In contrast, the existing 3-D models (latitude, longitude, altitude) can be made self-consistent from a meteorological standpoint only at the expense of chemical simulation at no more than a primitive level. The 2-D models (latitude, altitude) represent intermediate compromises in both chemistry and meteorology. Accurate simulations of the changes from day to night to day ("diurnal models") require very short time-steps in computation, prohibiting their use for long calculations into the future.

The basic premise underlying such computational models is the assumption that the existing concentrations of each chemical species can be rationalized through essentially complete knowledge of the rates of all reactions which either produce or remove it from the system. The individual chemical reaction rates are then measured in the laboratory under conditions of temperature and pressure comparable to those of the stratosphere, and the matrix of reaction rates and concentrations is adjusted until a good fit is obtained to some particular set of current atmospheric measurements. A typical 1-D model will now produce computed vertical concentration profiles of 50 or more chemical species, and in principle offers the possibility of rather precise calibration by comparison with atmospheric measurements. However, the available calibration data normally include substantial information about only a dozen of these species, and these measurements usually have substantial error bars on the accuracy, while frequently being composites of data collected at different times and/or locations. The vertical distribution of chlorine among five chemical species is illustrated in Figure 2 as calculated by a typical 1-D model for the 1980 atmosphere. This distribution is strongly dependent upon the rates of reactions which convert one chlorinated molecular form to another, as in reactions 7, 8, 10, 11, 12, 14, 16 and 17, and these must be known over the temperature range from 210-320K and from 10^{-3} to 1 atmosphere pressure. In general, the various experimental results for the atmospheric concentrations of Cl, ClO and HCl are consistent with such 1-D calculations. The low concentrations calculated for HOCl and $ClONO_2$ are below the currently detectable levels.

Predictions of future effects of added chlorine compounds in the atmosphere can be made by running a model consistent with the present atmosphere into the future, observing the adjustments made in the various concentrations in response to increased tropospheric concentrations of species such as CCl_3F and CCl_2F_2.

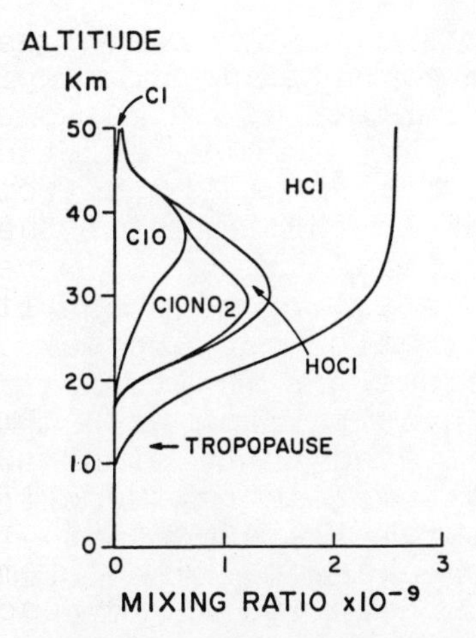

Figure 2. Typical calculated distribution of Cl among various chemical species during daylight in 1980. The calculation is carried out with a 1-D model, and is most appropriate for 30°N latitude, averaged over all longitudes.

The initial predictions made in 1974 suggested an eventual loss of about 10% of the total ozone, with constant release in the future of CCl_3F and CCl_2F_2 at current rates. Subsequent estimates have oscillated about this 10% estimate (NAS 1976, 1979a&b). A typical predictive scenario with constant yearly input of CCl_3F and CCl_2F_2 indicates about a 3% loss of ozone by the year 2000, with 6% by 2035, and depletion approaching 10% near 2100 A.D. Some 2-D model calculations show similar average ozone losses world-wide, while suggesting that percentage losses may be somewhat less for the tropics and during the summer, and greater than average for polar regions and in the winter/spring.

The major causes of fluctuations in these estimates of long-term ozone depletions can be traced to better data on the chemistry and meteorology in the 20-30 km altitude range. The chemical changes have primarily involved improved understanding of the direct interactions among HO_x and NO_x species, e.g. a major re-evaluation of the rate constant for reaction (9), but these shifts also affect the computed concentrations of the ClO_x species because of the many cross-chain reactions. During 1980-81 conflicting laboratory data have been developed for several reaction rate constants (including reactions 13 and 16). The effects of these laboratory inconsistencies are minimal toward the numerical calculation of ozone losses near 40 km, but provide considerable uncertainty in the detailed modeling of the 20-35 km range.

MEASUREMENTS OF OZONE IN THE REAL ATMOSPHERE

The detection of any long-term changes in stratospheric ozone concentration must be carried out against a background of fluctuations in its natural level, and must be considered in comparison with the sum of the various processes believed to be simultaneously affecting such concentrations. Among these possible influences are the natural 11-year solar sunspot cycle, which correlates with changes in the solar flux of ultraviolet radiation near 190 nm; the increasing concentrations of chlorofluoromethane compounds; and other activities of man, including the rising levels of NO_x injection during the past 20 years by subsonic jet aircraft; the observed increase in CO_2; and possible smaller increases in N_2O and CH_4.

Measurements of atmospheric ozone concentrations have been made on a regular basis in a few locations for about 50 years, as with the most complete single record taken in Arosa in the Swiss Alps, illustrated in Figure 3. (Dütsch 1971, NAS 1976, 1979a&b and Department of Environment). These measured O_3 concentrations fluctuate daily and show a regular seasonal variation. Observers have long suspected, too, a relationship with the approximately 11-year solar sunspot cycle. Recently a physical basis for such a correlation with solar activity has been established with the observation of more ultraviolet radiation emitted near 200 nm at sunspot maximum than at the minimum. The maximum sunspot activity during recent solar cycles has been reached during 1980, 1969, and 1958. This additional ultraviolet radiation can enhance the formation of ozone through the usual route of reactions (1) plus (2).

The Arosa data of Figure 3 are expressed both as 1-year and 5-year averages. The presently predicted level of O_3 depletion from the chlorofluorocarbons, calculated in isolation from other variations, is no more than 1%, and is considerably below the level of natural variability shown in the Arosa record. Twenty-year ozone records exist for many additional ground stations, but which are poorly distributed for true global coverage. Global measurements of ozone concentration were begun in 1970 with the Nimbus 4 satellite whose power gave out several years later. They have been resumed again

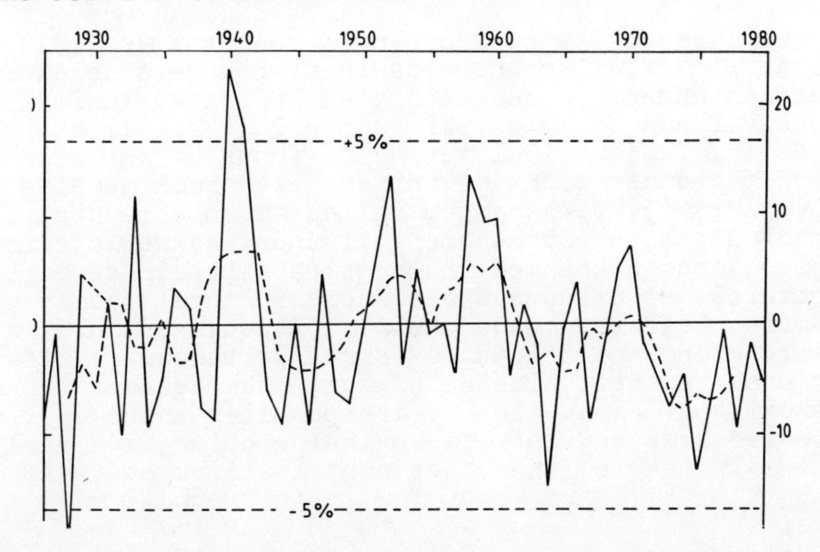

Figure 3. Average ozone concentrations observed at Arosa in the Swiss Alps. The solid line connects points representing the yearly average concentrations. The dotted line is the 5-year running average. The units are Dobson units, and the zero line represents the 50-year average of 337 Dobson units. (Graph from Prof. H. U. Dütsch, ETH, Zürich).

with the Nimbus-7 satellite launched in 1978. Comparison of the vertical ozone profiles from Nimbus-7 and Nimbus-4 indicates an average ozone loss of 5% at 40 km during the 1970's, in agreement with predictions made from the increasing amounts of organochlorine compounds.

ENVIRONMENTAL CONSEQUENCES OF CHANGES IN THE CONCENTRATION OF STRATOSPHERIC OZONE

The possible consequences to the earth and its inhabitants from present and future depletion of stratospheric ozone can be summarized into two physical consequences, and three areas of ecological concern. Ultraviolet radiation in the 290-320 nm range (often called UV-B radiation) is only partially absorbed in the atmosphere by ozone, and a fraction of it reaches the earth's surface. If the total column of O_3 were to decrease by 10%, then there would be an increase of about 20% in the amount of UV-B received at the surface. The second physical consequence, as mentioned earlier, is the anticipated decrease in stratospheric temperatures accompanying the 30-50% ozone depletion near 40 km altitude.

The three major environmental concerns are (NAS 1976, 1979a&b):(1) Will an increase in UV-B have a deleterious effect on human beings through an increased incidence of the various forms of skin cancer? (2) Will an increase in UV-B have deleterious effects of any kind on any of the other plant or animal biological species in the world? (3) Can changes in the amounts or distribution of stratospheric ozone influence the world climate, perhaps through changes in the stratospheric temperature structure? As discussed elsewhere in this volume, estimates have been made that a 10% decrease in stratospheric ozone would lead to larger percentage (≈ 30-50%) increases in the incidence of human skin cancer. The complexity of evaluation of the possible effects of increased UV-B on other biological species is clearly evident in many of the other contributions to this volume.

Satisfactory models do not now exist for prediction of the current characteristics of the world's climate, and the possible climatic perturbations from changes in the concentration of stratospheric ozone therefore cannot be ascertained even on a qualitative basis. About 90% of the atmospheric mass and most of its energy are contained within the troposphere, and changes in the stratospheric temperature structure are unlikely to force direct tropospheric responses. However, the coupling of energy and momentum between the troposphere and the stratosphere has an intricate relationship, and stratospheric changes may well be able either to trigger or to turn off massive tropospheric interactions.

In sum, the depletion of stratospheric ozone has one reasonably well-evaluated direct consequence--an expected increase in the incidence of human skin cancer. Two other effects could also be involved, either potentially far more serious but much more difficult to evaluate: (1) disruption of the ecological cycles of other biological species; (2) changes in the earth's climate. The combination of a projected increased incidence in skin cancer together with these other possible biological and/or climatic effects from predicted future ozone depletion has led to efforts, notably in North America and Scandinavia, to curtail future emissions of CCl_2F_2 and CCl_3F to the atmosphere by regulations restricting their use. As noted earlier, these regulatory activities have not actually led to much reduction in the atmospheric release rates for CCl_3F and CCl_2F_2, but the exponential increases characteristic for two decades into the mid-1970's have disappeared. If the current pattern of approximately constant atmospheric release were to be continued until ozone depletion has been definitely established to have occurred, then the level of depletion would probably need to reach an average ozone loss of several percent because of the large background fluctuations in natural levels. Even if direct release to the atmosphere of the chlorofluorocarbons were to be stopped completely at some point, the level of ozone depletion will always continue to increase for another two decades from later release of that already in use, and from the atmospheric transport time required from ground release to 40 km. The atmospheric repair time after that is then controlled by the lifetime in the atmosphere of the chlorofluorocarbons, i.e. 50-150 years on the average for molecules of CCl_3F and CCl_2F_2. The time required for the atmosphere to return even to the present levels of chlorofluorocarbon concentration would require more than a century. Atmospheric repair of the ozone depletion caused by long-lived chlorinated molecules is a very slow process, and any consequences of such ozone depletion will be with us throughout the 21st century.

ACKNOWLEDGEMENTS

Our research into the atmospheric chemistry of chlorine has been supported originally by D.O.E. Contract No. DE-AT03-76ER-70126 and its predecessor in the A.E.C., and subsequently also by E.P.A. Contract No. 805532010. This manuscript was completed during the tenure of an Alexander von Humboldt Award.

REFERENCES

Chapman, S. 1930. A theory of upper atmospheric ozone Quart. J. Roy. Meteorol. Soc., 3: 103-125.

CIAP Monograph Series. 1974. The Effects of Stratospheric Pollution by Aircraft, A. J. Grobecker, S. C. Coroniti, and R. H. Cannon, Jr., Department of Transportation, DOT-TST-75-50.

Department of Environment. 1979. "Chlorofluorocarbons and their Effect on Stratospheric Ozone", London.

Dütsch, H. U. 1971. Photochemistry of atmospheric ozone. Adv. Geophys. 15: 219-322.

McEwan, M. and L. F. Phillips. 1975. Monographs such as "Chemistry of the Atmosphere". Arnold Ltd. London.

Molina, M. J. and F. S. Rowland. 1974. Stratospheric sink for chlorofluoromethanes -- Chlorine atom catalysed destruction of ozone. Nature 249: 810-812.

National Academy of Sciences, 1976. Halocarbons: Effects on Stratospheric Ozone, Washington, D. C.

National Academy of Sciences, 1979(a). Stratospheric Ozone Depletion by Halocarbons: Chemistry and Transport, Washington, D. C.

National Academy of Sciences, 1979(b). Protection Against Depletion of Stratospheric Ozone by Chlorofluorocarbons, Washington, D. C.

NOAA, NASA, USAF. U. S. Government Printing Office. 1976. The pressure, temperature, density and other characteristics of the atmosphere are given in "U. S. Standard Atmosphere".

Okabe, H. 1978. "Photochemistry of Small Molecules" Wiley-Interscience, New York.

Rowland, F. S. and M. J. Molina. 1975. Chlorofluoromethanes in the environment. Adv. Geophys. Res. 13: 1-35.

OZONE DEPLETION CALCULATIONS

Frederick M. Luther, Julius S. Chang,
Donald J. Wuebbles and Joyce E. Penner

Lawrence Livermore National Laboratory
University of California
Livermore, California 94550

INTRODUCTION

Models of stratospheric chemistry have been primarily directed toward an understanding of the behavior of stratospheric ozone. Initially this interest reflected the diagnostic role of ozone in the understanding of atmospheric transport processes. More recently, interest in stratospheric ozone has arisen from concern that human activities might affect the amount of stratospheric ozone, thereby affecting the ultraviolet radiation reaching the earth's surface and perhaps also affecting the climate with various potentially severe consequences for human welfare. This concern has inspired a substantial effort to develop both diagnostic and prognostic models of stratospheric ozone.

During the past decade, several chemical agents have been determined to have potentially significant impacts on stratospheric ozone if they are released to the atmosphere in large quantities. In the early 1970's it was recognized that oxides of nitrogen (NO_x) and oxides of hydrogen (HO_x) had important roles in regulating stratospheric ozone. The NO_x and H_2O emissions from large fleets of high altitude aircraft were proposed as potential threats to ozone. Atmospheric nuclear explosions produce NO in the fireball which is injected into the atmosphere at the cloud stabilization height. Consequently, a large number of nuclear detonations could lead to stratospheric injections of NO that are several times ambient amounts. Increased use of nitrogen-based fertilizers was postulated to eventually lead to a significant increase of atmospheric N_2O. Rowland and Molina (1975) found that chlorofluorocarbons (CFC's) such as $CFCl_3$ and CF_2Cl_2 could pose a threat to stratospheric ozone by releasing chlorine into the stratosphere via photolysis at high altitudes. More recently, bromine and fluorine have been recognized as having potentially significant effects on ozone if large source rates were to exist. In contrast to the above threats to the ozone layer, increases in the stratospheric concentration of stable IR absorbing gases such as CO_2 could lead to an increase in total ozone by changing stratospheric temperatures which in turn affect chemical reaction rates.

In order to assess the potential impact of these proposed perturbations, mathematical models have been developed to handle the complex coupling between chemical, radiative, and dynamical processes. Basic concepts in stratospheric modeling are reviewed below.

STRATOSPHERIC MODELING

Mathematical models of the chemical processes in the stratosphere are governed by the chemical species conservation equation

$$\partial c_i/\partial t + \Delta \cdot \underline{F}_i(c_i,\underline{x},t) = P_i[c,J(\underline{x},t,c),k(T(\underline{x},t,\rho)] - L_i[c,J(\underline{x},t,c),k(T(\underline{x},t),\rho)]c_i + S_i(\underline{x},t) \tag{1}$$

where $c_i = c_i(\underline{x},t)$ is the concentration of the ith chemical constitutient; c is the general representation of all constituents; P_i and $L_i c_i$ are the production and loss of c_i caused by photochemical interactions; T is the ambient air temperature; ρ is the ambient air density; $\underline{F}_i$ is the transport flux of c_i; S_i represents any other possible sinks or sources of c_i; J represents photodissociation coefficients; k represents chemical reaction rate coefficients; and all of these variables are defined at a given spatial position $\underline{x} = \underline{x}(x,y,z)$ at time t. Equation (1) illustrates the nonlinearity and general complexity of this mathematical system. There is one conservation equation for each chemical species in the stratosphere. Typically, stratospheric models include 30-50 species and 100-150 chemical and photochemical reactions to describe the chemical balances affecting the ozone budget in the stratosphere.

Classification of models

Depending on the nature and the extent of problems to be studied, any particular model may include different levels of detail in its representation of the spatial variation of trace species distributions. The difference in resolution serves as a useful and convenient basis for model classification.

Box models have spatial homogeneity as their fundamental assumption (complete uniform mixing of individual trace species). Consequently, these models are represented by a set of ordinary differential equations describing the time evolution of individual trace species as controlled by chemical interactions only, i.e., Eq. (1) averaged over all space under consideration. Such models have been very useful in the diagnosis of experiments in laboratory kinetics and in the analysis of global budgets of long-lived trace species.

One-dimensional models have been the most widely used diagnostic and prognostic tools in stratospheric research. These models are designed to simulate the vertical distribution of atmospheric trace species. They include a detailed description of chemical interactions and of atmospheric attenuation of solar radiation, but the effect of atmospheric transport is described in a simplified way. In one-dimensional models of the stratosphere, a longitudinal and latitudinal global average of the transport flux is assumed. The resulting net vertical transport flux F_{z_i} of any minor constituent c_i is represented through a diffusion approximation in which F_{zi} is assumed to be proportional to the gradient of the mixing ratio of that trace species:

$$F_i = F_{z_i} = K_z \rho \frac{\partial}{\partial z} (c_i/\rho) \qquad (2)$$

where z is the altitude and K_z is the one-dimensional vertical diffusion coefficient. One of the major assumptions in applying Eq. (2) to a one-dimensional representation of the atmosphere is that the globally-averaged vertical transport can be represented as a diffusive process.

The models are considered to represent either global or midlatitude averages. One-dimensional models can describe the main features of atmospheric chemistry without excessive demands on computer time.

Two-dimensional models with spatial resolution in the vertical and meridional directions and improved representation of transport (mean motion and eddy mixing) are far more realistic than the one-dimensional models. The two-dimensional fluxes in Eq. (1) are now represented by the sum of two terms, $\underline{F}_i = c_i\underline{V} - \rho K\nabla(c_i/\rho)$, where $\underline{V}$ is the vector of mean meridional and vertical velocities and the 2 x 2 matrix K is the eddy diffusion coefficient tensor. These models can simulate both seasonal and meridional variations of trace species distributions. The price for this additional information is a considerable increase in computational cost and required input data. Unfortunately, even for these complex models, the transport representations must still be empirically derived from limited data. A few models have attempted to derive the mean winds numerically while still using empirically determined diffusion coefficients. Feedback between changes in composition and the transport processes can be treated in a limited manner only. This major coupling step can only be accomplished in a realistic sense in a three-dimensional model.

Three-dimensional models give, in principle, the closest simulation of the real atmosphere. The three-dimensional transport fluxes $\underline{F}_i = c_i\underline{V}$ are obtained through the solution, in all three dimensions, of the appropriate equations of continuity for momentum, energy, and mass. These models can, in principle, include most of the important feedback mechanisms in the real world. They are, however, very demanding of computer time and memory, and so far the chemistry has had to be simplified to such a degree that important details may have been lost.

One-dimensional models are generally the most detailed and complete in terms of the treatment of photochemical processes. The one-dimensional models used by the major modeling groups now include O_X, NO_X, HO_X and ClX chemical reactions and cycles. In addition, essentially all one-dimensional models contain reactions for the species resulting from methane oxidation, and many have included bromine and sulfur chemistry in their calculations. Some even treat aerosol formation and loss.

The more extensive reaction sets used in some models include many minor reactions, but a comparison of representative models indicates that the chemical kinetics systems in current use are in essential agreement. However, reaction rate coefficients have been changed and additional reactions have been included over the last few years as a result of new laboratory measurements and as the understanding of the chemistry of the atmosphere continues to evolve.

Transport representation

Globally-averaged vertical transport is represented in one-dimensional models using a vertical diffusion coefficient, K_z. The diffusion coefficients are derived from atmospheric data. Chemical tracers (CH_4 and N_2O) and radionuclides from past atmospheric nuclear tests (^{14}C, ^{90}Sr, ^{95}Zr, and ^{185}W) have been used as source data for deriving and testing the coefficients. In the derivation of the coefficients, it is assumed that the value of K_z is a function of only the transporting motion field.

Vertical diffusion coefficient profiles that have been used by various modeling groups in the past have varied significantly, differing by as much as an order of magnitude in value in the middle stratosphere (National Research Council, 1976). The K_z profiles are shown in Figure 1. In general the profiles are similar in their essential characteristics. They have a rather shown in Figure 1. In general the profiles are similar in their essential characteristics. They have a rather large value in the troposphere, a much lower value in the region near the tropopause, and a large value in the upper stratosphere.

Radiative processes

Photodissociation processes in the atmosphere are often extremely important mechanisms for the production and destruction of chemical species. The photodissociation rate, J_i, for species i to give a product j is defined by

$$J_{i\rightarrow j} = \int_{\lambda} Q_{\lambda,i}\,\sigma_{\lambda,i}\,F_{\lambda}(z)d\lambda \qquad (3)$$

where $Q_{\lambda,i} = Q_{\lambda,i(i\rightarrow j)}$ is the quantum yield for photo-

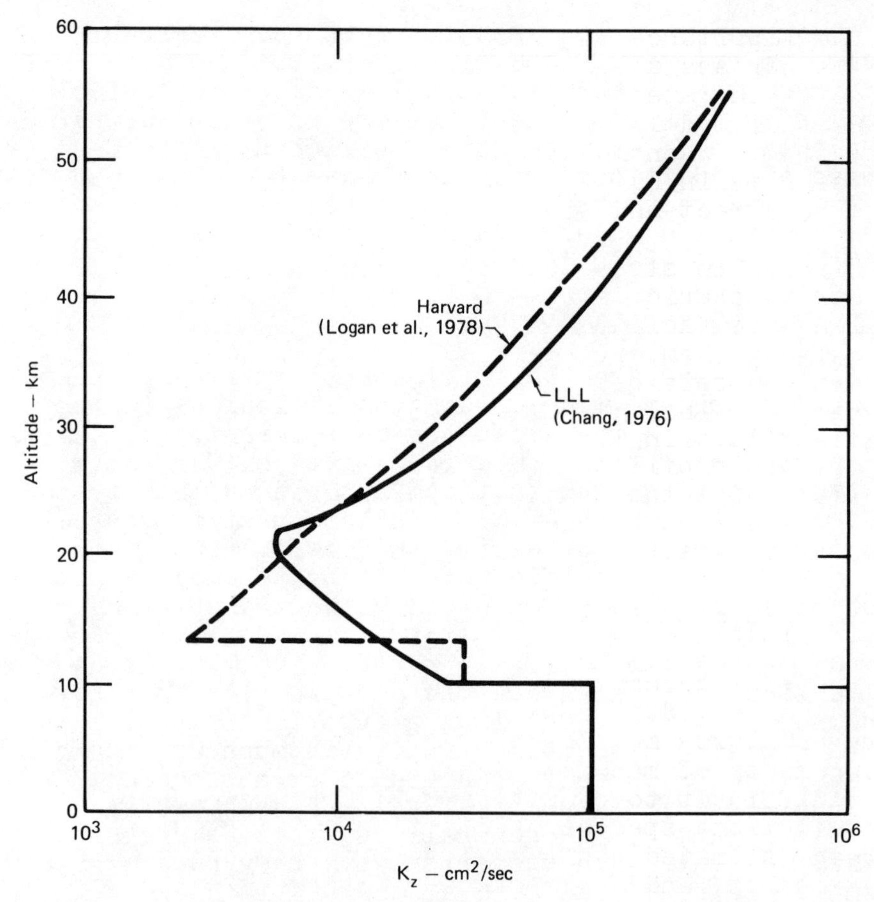

Figure 1. Vertical transport coefficients currently being used by two modeling groups.

dissociation of species i to result in production of species j; $\sigma_{\lambda,i}$ is the photoabsorption cross section; and $F_\lambda(z)$ is the photon flux density. The flux density is affected by changes in the concentration profiles of absorptive species predicted by the model and by the overburden of absorptive and scattering species. Input parameters affecting the calculated photodissociation rate are the wavelength dependence of the quantum yield and absorption cross sections (which are based on laboratory measurements), the solar flux at the top of the atmosphere, the total column of absorptive species above the top of the model (which are based on atmospheric measurements), and the solar zenith angle (which varies by the season and time of day at a given latitude).

The importance of molecular multiple scattering and the surface albedo in determining atmospheric photodissociation rates is now well recognized, and most one-dimensional models now include the effect of multiple scattering. When multiple scattering is treated, the flux density in Eq. (3) includes contributions from both the direct and diffuse flux components.

Changes in stratospheric composition can affect the stratospheric temperature profile via the solar and longwave radiation budgets. Changes in temperature affect chemical reaction rates, which in turn feed back on composition. This temperature feedback mechanism has been included in recent one-dimensional model calculations. In these calculations the stratospheric temperature profile is calculated by using a radiative transfer model that includes solar absorption and longwave interaction by the radiatively important stratospheric species.

Physical domain and boundary conditions

The choice of spatial domain depends on the chemical species of interest and the level of detail desired. For each individual species, either boundary concentrations or fluxes must be prescribed. Because atmospheric, measurements of many chemical species are inadequate, it is difficult to construct reliable boundary conditions for some species. In practice, an iteration between estimated boundary conditions, model simulation results, and appropriate comparisons with available atmospheric data must be carried out. The need for accurate boundary conditions can be reduced by extending the physical domain beyond the minimum required by the problem at hand. Moving the boundary beyond a buffer zone serves to reduce the model sensitivity to uncertainties in boundary conditions. In the vertical direction, many models, especially one-dimensional models, cover the region from the earth's surface to above 50 km.

Averaging processes

In the formulation of models, certain yet unspecified averaging processes must be applied so as to provide a link with physical reality and the means for interpretation of the solutions from such models. Formally $c_i(x,t)$ for Eq. (1) is assumed to be uniform over a unit volume in a unit time interval. But in a representation with reduced dimensionality and/ or coarse spatial resolution, c_i must represent certain spatial and temporal averages. Without detailed data on spatial and temporal variations, it is not possible to devise a totally consistent averaging procedure for the nonlinear photochemical reaction rates. This is likely to be the case for many years to come. Nevertheless, the considerable computational difficulty created by diurnal variations in solar flux can be removed through diurnal averaging procedures. This allows the solution of time-dependent problems over long time durations using a diurnally-averaged model with a considerable savings of computer time.

Sources and sinks

In addition to the sources and sinks caused by the interactive chemistry, the model must include the effect of sources and sinks caused by other processes. Net sources and sinks at the earth's surface for each species must be taken into account in determining necessary flux or concentration boundary conditions. Both wet and dry removal processes for trace atmospheric constituents must be parameterized. The effects of these removal processes are generally approximated using a first-order loss rate that varies with altitude and species. Cosmic ray production of nitric oxide is also included in the models in a parametric way.

Other physical data

Altitude distributions of major constituents such as N_2 and O_2 are generally fixed in the calculations with concentrations based on a reference such as the U. S. Standard Atmosphere (1976). Some trace species distributions may also be fixed based on atmospheric measurements. Because of the difficulty in treating water vapor in the troposphere, many models fix the concentration of water vapor in the troposphere, while calculating its concentration in the stratosphere. Unless temperature feedback effects are included, the temperature profile is specified based on a standard reference such as the U. S. Standard Atmosphere (1976).

Time-dependent and steady-state solutions

Once the mathematical model is fixed (i.e., all parameterizations of the relevant physical variables are determined), the system of differential equations are to be solved. For simplicity and computational economy, steady-state solutions of Eq. (1) are often useful and desirable. Such solutions add additional requirements for time-averaging procedures, since local incident solar fluxes vary both diurnally and seasonally and may vary on even longer time scales. In diagnostic applications, steady-state solutions used in a snapshot manner can yield useful information.

Fully time-dependent models are more useful both diagnostically and prognostically, although their solutions are considerably more complicated and computationally more expensive. For the analysis of atmospheric data on many short-lived species, time-dependent models, in particular diurnal models, must be used.

Model verification

In order to verify a stratospheric model, the various components of the model (chemistry, dynamics, radiative transfer, and numerical methods) need to be tested individually and collectively. Verification tests have included tracer simulation studies (nuclear test debris), comparison of species concentrations and intercomparisons with models of the same or differing dimensionality. Simulations of diurnal variations can be used to test one-dimensional models, and simulations of the seasonal and latitudinal variations can be used to test multidimensional models. Perturbation studies have included simulation of Polar Cap Absorption Events (NO_x produced at high latitudes by solar activity), variations in solar UV flux over the 11-year solar cycle, species variation during a solar eclipse, and the atmospheric nuclear test series during the 1950's and 1960's. Good agreement with observed species profiles is necessary but not sufficient for model verification since models with significantly different sensitivity to perturbations can have similar ambient species profiles.

OZONE PERTURBATION ASSESSMENTS

High altitude aircraft emissions

Assessments of potential changes in ozone due to future large fleets of supersonic transports (SST's) have focused on injection altitudes of 17 and 20 km. With current engine technology, when 1 kg of fuel is burned, approximately 18 g of NO_x and 1250g of H_2O are produced at cruise altitude. It may be possible through future advances in technology to reduce the NO_x emission to 6 g/kg fuel. Although the H_2O emission rate is higher, the NO_x emissions are of greater concern because they cause a larger fractional increase in the stratospheric concentrations of these species (ambient NO_x concentrations being orders of magnitude less than H_2O).

Calculations assuming a constant fuel consumption rate at cruise altitude of 7×10^{10} kg/yr in a hemispheric shell have been used as a standard for assessment and comparison purposes. The corresponding NO_x emission rate is 2000 molecules $cm^{-3} s^{-1}$ over a 1-km-thick shell, which roughly estimates the emissions from a commercially viable fleet of supersonic aircraft (approximately 1000 advanced SST's). Figure 2 shows the historical evolution through 1980 of the Lawrence

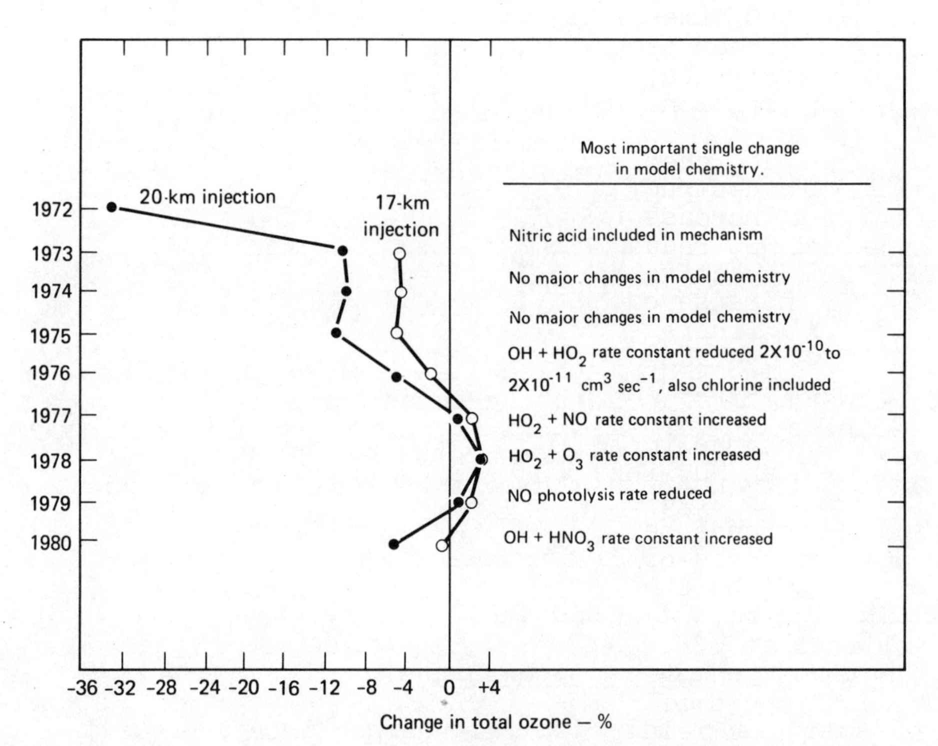

Figure 2. The historical evolution of LLNL model calculations of the change in total ozone due to an NO_x injection of 2,000 molecules $cm^{-3} s^{-1}$ over a 1-km thick layer centered at either 17 or 20 km.

Livermore National Laboratory (LLNL) model calculations of the change in total ozone due to an NO_x injection at either 17 or 20 km altitude. The results in Fig. 2 demonstrate the combined effects of the evolution of our understanding of stratospheric chemistry and the evolution of the treatment of physical phenomena such as multiple scattering of light, the averaging of reaction rates over diurnal cycles, the treatment of boundary conditions, and the transport parameterization. Although many different factors have contributed to the variation shown, the evolution of model chemistry has been the most important single factor.

Prior to 1977 injections of NO_x were estimated to cause a reduction in total ozone due to enhanced NO_x catalytic destruction of ozone in the stratosphere. A major change in the model chemistry occurred in 1977 when the reaction rate for $HO_2 + NO \rightarrow OH + NO_2$ was measured directly for the first time and was found to be about 40 times faster than previously estimated using indirect techniques. As a result of this change, HO_x chemistry became the dominant chemical destruction process for ozone in the lower stratosphere.

When HO_x chemistry is more efficient than NO_x catalytic destruction of ozone, injections of NO_x lead to a net increase in odd oxygen production. Also, the injected NO_x reduces HO_x by concerting OH and HO_2 to the less reactive species H_2O and HNO_3. Both processes act to increase ozone. Consequently, NO_x injections lead to an increase in ozone in the lower stratosphere and to a decrease in the upper stratosphere where NO_x chemistry is dominant.

The change in the local ozone concentration resulting from an NO_x injection of 2000 molecules $cm^{-3}s^{-1}$ is shown in Figure 3. The change in total ozone is the net difference between regions where there are significant increases or decreases in ozone concentration. With the 1980 LLNL model results (shown in Fig. 3, there is again a reduction of total ozone for NO_x injections at 17 or 20 km of 2000 molecules $cm^{-3}s^{-1}$. The recent changes in model chemistry have reduced the effectiveness of the HO_x catalytic cycle in the lower stratosphere relative to the NO_x catalytic cycle.

Current estimates of changes in ozone due to a reduced NO_x emission of 1000 molecules $cm^{-3}s^{-1}$ over a 1-km-thick layer are given in Table 1. The model

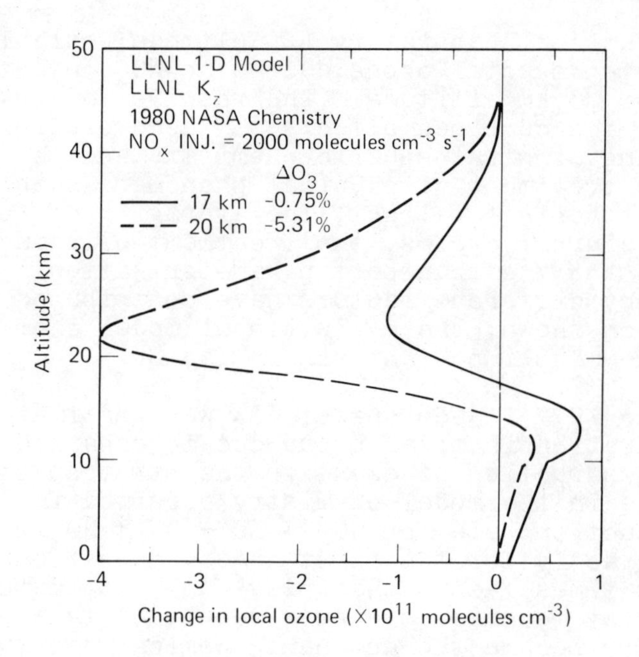

Figure 3. The change in ozone concentration at steady state due to an NO_x injection of 2000 molecules $cm^{-3}s^{-1}$ over a 1-km thick layer, centered at 17 or 20 km.

TABLE 1.

Change in total ozone at steady state computed for an NO_x injection rate of 1000 molecules $cm^{-3}s^{-1}$ over a 1-km-thick layer centered at either 17 or 20 km altitude.

	Change in total ozone (%)	
	17 km injection	20 km injection
Model range (NASA, 1979)	1.3 to 3.4	1.1 to 4.6
LLNL model (1979)	1.4	1.3
LLNL model (1980)	0.01	-2.0

range given for 1979 comes from NASA (1979), where 10 different models (including the LLNL model) were compared. The most recent changes in model chemistry have resulted in an estimated reduction in total ozone instead of an increase for NO_x injections above about 17 km. At lower altitudes ozone may increase or decrease depending upon the magnitude and altitude of injection. The altitude of injection has a significant effect on the computed change in ozone because of the increase in residence time with altitude and because of the variation with altitude of the dominate chemical reactions and cycles.

Estimates have been made of the NO_x emissions (as a function of altitude) expected from fleets of aircraft in 1990 (Oliver et al., 1978). Using these emission estimates, we calculate a net increase in total ozone of 1.3%.

The effect of varying the NO_x injection rate is shown in Figure 4 for injection altitudes of 17 and 20 km. For the 17-km injection altitude, the change in total ozone is nonlinear with NO_x injection rate. There is a slight increase (<0.25%) in total ozone for NO_x injections less than 1000 molecules $cm^{-3}s^{-1}$, switching to a reduction in total ozone for larger NO_x injections. On the other hand, for the 20-km injection altitude the reduction in total ozone is nearly linear with the NO_x injection rate.

Chlorofluorocarbons

It has been firmly established that the chlorofluorocarbons, $CFCl_3$ and CF_2Cl_2, have been increasing in the troposphere and stratosphere, and the observed increase is consistent with model estimates based on the historical production rates. The CFC's are photolyzed in the stratosphere to yield free chlorine which may catalytically destroy ozone.

Quantitative estimates of the depletion of ozone have changed as models, physics, and chemistry have improved. In 1976 the predicted change in total ozone due to steady-state production of CFC's at 1973 rates was estimated to be -7.5% (National Research Council, 1976). In 1977 a major change in the model chemistry occurred when the rate for $HO_2 + NO \rightarrow OH + NO_2$ was measured to be almost 40 times faster than previously thought. In addition, some models at that time adopted

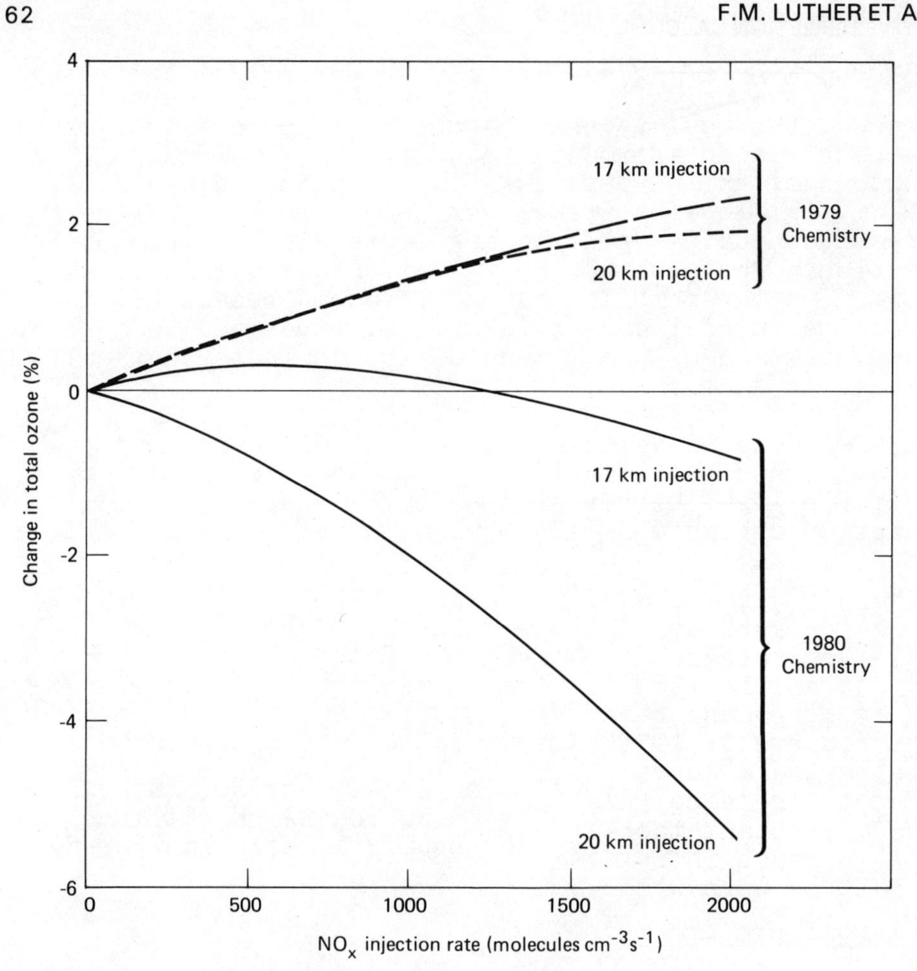

Figure 4. The change in total ozone as a function of NO_x injection rate for injections centered at 17 or 20 km.

diurnally-averaged reaction rates and multiple scattering effects. The depletion estimates in 1977 ranged from -10.8 to -16.5% for the various modeling groups (NASA, 1977); the LLNL estimate was -15.0%.

Figure 5 shows a number of calculated time histories for ozone under various assumptions regarding the future release rates for CFC's using the LLNL model in early 1979. For curve A it was assumed that CFC production continued at the 1976 rate (the 1976 production rate is 3.2% smaller for $CFCl_3$ and 2.9% larger for

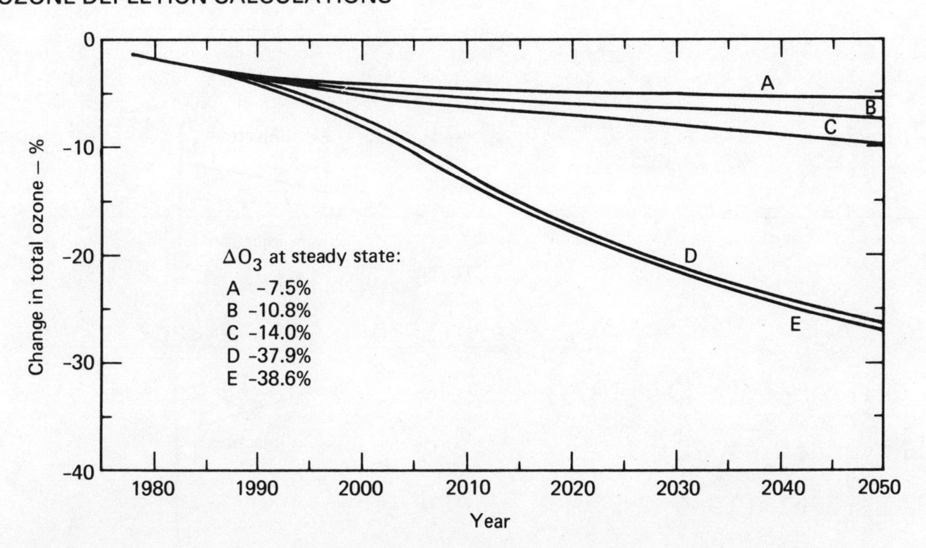

Figure 5. The change in total ozone as a function of time for various CFC release rate scenarios. Curve A: 1976 CFC release rate until 1982, then it is reduced by 25% and in 1987 it is reduced again by 25%. Curve B: 1976 release rate until 1982, then it is reduced by 25%. Curve C: 1976 release rate constant throughout the period. Curve D: 1976 release rate through 1980, then the release rate increased 7%/yr up to the year 2000, constant thereafter. Curve E: same as D except the increase in release rate begins in 1978. All calculations were with 1979 chemical rates.

CF_2Cl_2 than those for 1973) until 1982, then it was lowered by 25%. In 1987, it was lowered again by 25%. The eventual steady-state ozone depletion for this case was -7.5%. For curve B, the 1976 production rate continued until 1982 and was then cut by 25%. The ozone depletion at steady state was -10.8% for this case. For curve C the 1976 production rate was used for the entire time, and the steady-state ozone depletion was -14.0%. For curve D the 1976 production rate was assumed to continue through 1980, then the CFM production was increased by 7%/yr up to the year 2000. The steady-state depletion for this case is -37.9%. Case E is almost the same as case D except that the increased CFM production begins in 1978. The rate of increase is such that the production rate doubles by 1990 and doubles again by 2000. The steady-state ozone

TABLE 2.

Change in total ozone due to release of $CFCl_3$ and CF_2Cl_2. Steady state values are for constant release rates at 1976 levels.

	Change in total ozone (%)	
	January 1980	At steady state
Model range (NASA, 1979)	≈ -2	-15 to -18.3
LLNL model (1979)	-2.0	-15
LLNL model (1980)	-0.9	-9.1

depletion is -38.6%.

Table 2 compares recent calculations of the expected change in total ozone at steady state from constant emissions of $CFCl_3$ and CF_2Cl_2 at 1976 levels. Calculated changes in total ozone expected for January 1980 based on global emissions of these compounds since 1950 are also given. With currently recommended chemistry, the change in total ozone at steady state is -9.1%. The reduced change in total ozone compared to previous assessments is the result of recent chemistry modifications that have reduced HO_x concentrations (and the effectiveness of catalytic cycles involving HO_x) in the lower stratosphere.

Significant uncertainty still remains concerning the magnitude of the ozone reduction expected due to continued release of CFC's at current levels. For example, by varying several key chemical rate coefficients over their range of uncertainty, we obtained ozone reductions ranging from -5.5 to -15.6%.

The computed change in local ozone concentration at steady state is shown in Figure 6. The maximum percent change in ozone concentration occurs near 40 km, whereas the maximum absolute change occurs at a lower altitude (≈35 km with the current model).

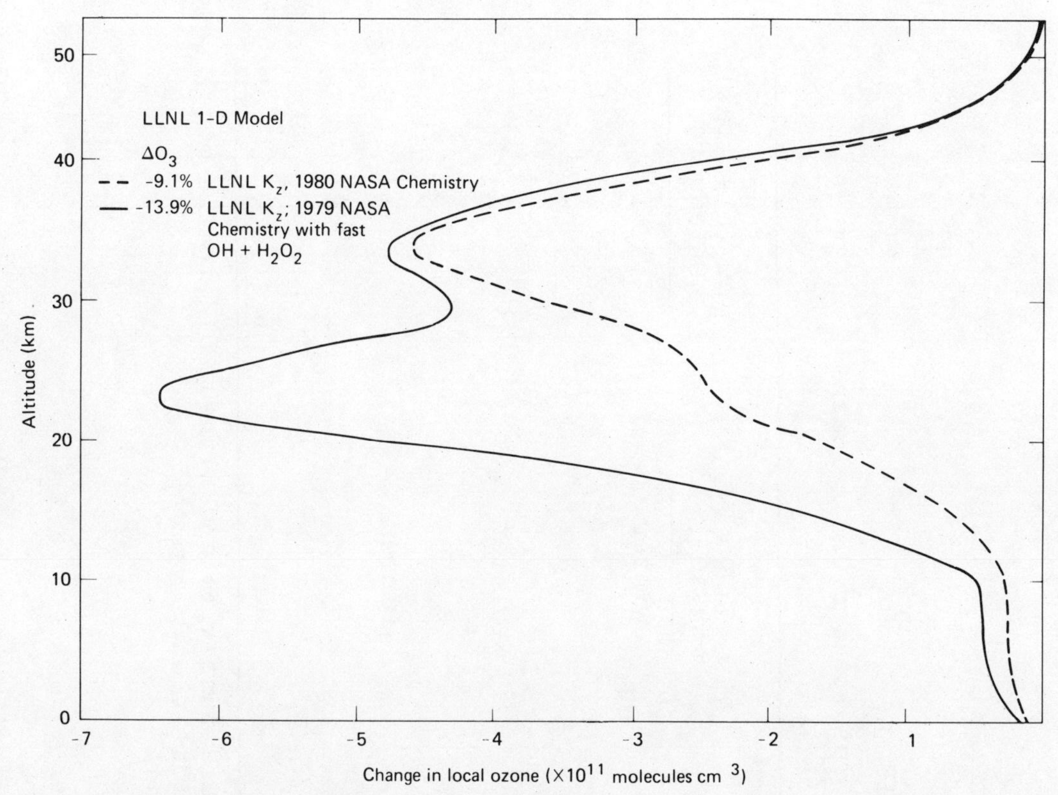

Figure 6. The change in ozone concentration at steady state resulting from CFC's released at the 1976 rate.

Several 2-D models have been used to estimate the impact of CFC releases at current release rates up to several decades into the future. The large computational burden prohibits running the 2-D models to steady state. However, the model runs give an indication of the seasonal and latitudinal variation of the expected ozone reduction. The 2-D model results are summarized in Table 3.

The models are in general agreement that the largest percent reduction in total ozone occurs at high latitudes. There is also reasonable agreement concerning the time of year at which the maximum reduction occurs. The magnitude of the latitude variation of the ozone reduction varies significantly between models. This model intercomparison is only intended to give a qualitative indication of the consistency between models. Differences in the manner in which the CFC perturbation

TABLE 3.

Comparison of 2-D model results for CFC releases

Model	Region of max ΔO_3(%)	Season of max ΔO_3(%)	ΔO_3(high lat) / ΔO_3(equator)
Borucki et al. (1980)	high latitude	winter	1.5
Brasseur and Bertin (1980)	equator	?	2.8
Clough (1979)	high latitude	winter	2.4
Pyle (1980)	high latitude	spring	3
Vupputuri (1979)	high latitude	winter	1.3
Widhopf and Glatt* (1979)	high latitude	spring	2

*These results were inferred from calculations with and without ClX in the model rather than from a CFC perturbation calculation.

was applied and differences in the total years of simulated response make it difficult to compare the model results directly.

Increase of N_2O

Concern that human perturbations to the nitrogen cycle might lead to enhanced concentrations for atmospheric N_2O has stimulated interest in the budget for this gas. Tropospheric N_2O is the major source for stratospheric NO_x so perturbations in N_2O are expected to alter stratospheric chemistry. N_2O is produced by soil and ocean bacteria as well as by combustion. Supplying additional fixed-nitrogen or fertilizer to the biosphere or other perturbations to the nitrogen cycle might lead to increased N_2O concentrations. However, measurements indicate that N_2O is only increasing approximately 0.2%/yr (Weiss, 1981).

In order to test the model sensitivity to N_2O changes, we consider the effect of a doubling of the N_2O concentration at the ground (from 300 to 600 ppbv). Table 4 shows the effect on total ozone of doubling N_2O.

There is a large cancellation in the combined effect of doubling N_2O and constant production of CFC's. The level of atmospheric N_2O affects the atmospheric response to chlorine changes and vice versa, mainly because of the coupling of chemistry by $ClONO_2$. Interferences between the additional NO_x produced by the doubled N_2O and the additional ClO_x produced by the CFC's are particularly important below 30 km where $ClO + NO \rightarrow Cl + NO_2$ and $ClO + NO_2 \rightarrow ClONO_2$ are most important in interfering with the catalytic cycle mechanisms.

TABLE 4.
Change in total zone due to doubling of N_2O

	Change in total ozone (%)	
	$2 \times N_2O$	$2 \times N_2O$ + CFC
Model range (1979)	-1.9 to +3.6	-8.3 to -11
LLNL model (1979)	-2.1	-11
LLNL model (1980)	-11.5	-11.5

Doubling of CO_2

Atmospheric measurements of CO_2 since 1958 have shown a rise in CO_2 concentrations that has been attributed primarily to the use of fossil fuels. Assuming that fossil fuel useage increases by about 2%/yr, it is estimated that atmospheric CO_2 concentrations will double before the end of the next century. An increase in CO_2 is expected to lead to changes in the thermal structure of the atmosphere. The temperature should decrease in the stratosphere (where the infrared opacity is small) and increase in the troposphere. For a doubling of CO_2, the global mean surface temperature is expected to increase by 1.5 to 3 K. Temperatures are estimated to increase 1-2 K near the tropopause and decrease 10-15 K in the upper stratosphere.

Changes in temperature affect stratospheric ozone by altering temperature-sensitive photochemical reactions. Reductions in temperature lead to a reduction in the ozone photochemical destruction rate, and, consequently, an increase in ozone concentration. A doubling of CO_2 is estimated to lead to an increase in total ozone of approximately 6% (Table 5). The effect of CO_2 on total ozone has been computed by many modeling groups. The recent assessments show a larger sensitivity than was computed earlier as a result of new measurements of the temperature dependence of key reaction rates.

TABLE 5.
Change in total ozone computed for a doubling of CO_2

	Change in total ozone (%)
Model range (1975-1978)*	1.2 - 4.7
Model range (1979)*	4.7 - 7
LLNL model (1979)	4.7 - 5.8
LLNL model (1980)	6.2

*From several sources published during these years.

SUMMARY

Understanding of stratospheric chemistry has evolved rapidly over the past decade, and there have been major improvements in the representation of chemical, radiative, and transport processes in stratospheric models. In spite of limitations, these models remain the only available tool for estimating future effects of potential atmospheric perturbations.

Several proposed atmospheric perturbations have been shown to have a potentially significant effect on stratospheric ozone. Chlorofluorocarbons present the most significant threat (the estimated change in total ozone being about -9% at steady state for continued CFC release at the current rate). Two-dimensional model calculations indicate that the largest decreases in ozone would likely occur at high latitudes in winter or spring.

Oxides of nitrogen (NO, NO_2, and N_2O) can have a significant impact on ozone. Estimated 1990 aircraft fleet emissions are computed to lead to an increase of total ozone of 1.3%. The models are sensitive to increases in N_2O, but this species is not projected to increase significantly over the next few decades.

A doubling of CO_2 might lead to an increase in total ozone of about 6%, which would tend to offset the effect of a CFC perturbation. The coupled effects on stratospheric ozone of CO_2 and CFC perturbations have been shown to be nonlinear (Penner, 1980). The ozone decrease at steady state is larger than the decrease calculated by linearly combining the results of separate CO_2 and CFC perturbations. The decrease in ozone estimated to occur over the next several decades would be strongly dominated by the CFC perturbation because of the difference in the projected rates of increase of CFC's and CO_2 in the stratosphere.

Although stratospheric models have been successful in diagnostic applications, significant uncertainties remain concerning their accuracy for prognostic applications. Uncertainties in chemical rate coefficients contribute most significantly to the uncertainty in model results. In spite of these uncertainties, there is good agreement between independent model predictions. Because important decisions will be made based on model results, improving the models will remain a matter of high priority.

ACKNOWLEDGEMENT

This work was performed under the auspices of the U. S. Department of Energy by the Lawrence Livermore National Laboratory under contract No. W-7405-Eng-48, and supported in part by the High Altitude Pollution Program of the Department of Transportation, Federal Aviation Administration.

REFERENCES

Borucki, W. J., R. C. Whitten, H. T. Woodward, L. A. Capone, C. A. Reigel, and S. Gaines. 1980. Stratospheric ozone decrease due to chlorofluromethane photolysis: Predictions of latitude dependence. J. Atmos. Sci. 37:686-697.

Brasseur, G. and M. Bertin. 1978. The action of chlorine on the ozone layer as given by a zonally averaged two-dimensional model. Pure Appl. Geophys. 117: 436-447

Chang, J. S. 1976. Eddy diffusion profile described in "First Annual Report of Lawrence Livermore National Laboratory to the High Altitude Pollution Program", F. M. Luther [ed.], Lawrence Livermore National Laboratory Report UCRL-50042-76.

Clough, S. A. 1979. Chlorofluoromethanes and their effect on stratospheric ozone. Pollution Paper No. 15, H.M.S.O., London, U.K.

Logan, J. A., M. J. Prather, S. C. Wofsy, and M. B. McElroy. 1978. Atmospheric chemistry: Response to human influence, Phil. Trans. Roy. Soc. London. 290: 187-234.

NASA Reference Publication 1010. 1977. Chlorofluoromethanes and the Stratosphere. R. D. Hudson [ed.], National Aeronautics and Space Administration.

NASA Reference Publication 1049. 1979. The stratosphere: Present and future. R. D. Hudson and E. I. Reed [eds.] National Aeronautics and Space Administration.

National Research Council. 1976. Halocarbons: Effects on stratospheric ozone. National Academy of Sciences,

Oliver, R. C., E. Bauer, and W. Wasylkiwskyj. 1978. Recent developments in the estimation of potential effects of high altitude aircraft emissions on ozone and climate. Report No. FAA-AEE-78-24. U. S. Department of Transportation, Federal Aviation Administration, Washington, D.C.

Penner, J. E. 1980. Increases in CO_2 and chlorofluoromethanes: Coupled effects on stratospheric ozone. Lawrence Livermore National Laboratory Report UCRL-84058, in Proceedings of the Quadrennial International Ozone Symposium, Boulder, Colorado, August 4-9.

Pyle, J. A. 1980. A calculation of the possible depletion of ozone by chlorofluorocarbons using a two-dimensional model. Pure Appl. Geophys. 118: 355-377.

Rowland, F. S. and M. J. Molina. 1975. Chlorofluoromethanes in the environment. Rev. Geophys. Space Phys. 13: 1-35.

U. S. Standard Atmosphere. 1976. NOAA/S/T 76-1562, U.S. Government Printing Office, Washington, D. C.

Vupputuri, R. K. R. 1979. The structure of the natural stratosphere and the impact of chlorofluoromethanes on the ozone layer investigated in a 2-D time dependent model. Pure Appl. Geophys. 117: 448-485.

Weiss, R. W. 1981. The temporal and spatial distribution of tropospheric nitrous oxide. preprint.

Widhopf, G. F. and L. Glatt. 1979. Two-dimensional description of the natural atmosphere including active water vapor modeling and potential perturbations due to NO_x and HO_x aircraft emissions. Report No. FAA-EE-79-07, Federal Aviation Administration, Washington, D. C.

ON THE LATITUDINAL AND SEASONAL DEPENDENCE OF OZONE PERTURBATIONS

Richard S. Stolarski

NASA/Goddard Space Flight Center
Laboratory for Planetary Atmospheres
Greenbelt, MD 20771

ABSTRACT

Seasonal and latitudinal variations in model predicted chlorofluorocarbon (CFC) ozone reductions are not well understood. Two-dimensional model results indicate a maximum reduction in the high-latitude winter (i.e. a reduction in the seasonal fluctuation), but there are some indications that this result may be chemistry dependent. This opens the possibility that as our knowledge of chemistry changes the seasonal-latitudinal fluctuation of the predicted ozone depletion may change as well as the global average magnitude.

INTRODUCTION

When considering possible changes in the ultraviolet solar radiation received at the surface of the earth due to projected changes in the total column of overhead ozone it is important to know more than the global average change as is usually given from one-dimensional model calculations. The ultimate effect of ozone induced UV changes certainly depends on both the seasonal and latitudinal dependence of the change. In the extreme case, if all of the change is concentrated in the polar night or regions of high solar zenith angle, little significant effect can occur on the surface UV because there was essentially none there in the first place. If, on the other hand, the change is concentrated at low to mid-latitudes during the summer, when the ozone is normally minimum and UV a maximum, the change takes on added significance. Recent 2-dimensional stratospheric photochemical model calculations (Borucki, et al. 1980; Vupputuri, 1978; Pyle and Derwent, 1980; Brasseur and Bertin, 1978; Crutzen et al. 1979) of the potential ozone perturbation due to chlorofluorocarbon (CFC) release have indicated a tendency toward larger depletions at high latitudes in the winter, thus tending to reduce the effect of the change in UV as compared with the numbers deduced from a global average model.

The purpose of the following discussion is to outline some of the factors involved in determining the atmosphere's seasonal and latitudinal response to a chemical perturbation in the ozone layer. Much of this consideration was prompted by a question asked by some of the biologists as to whether the computed seasonal and latitudinal response was inevitable in any model or was sensitive to the assumed chemistry and was likely to change when significant changes occurred in our chemical knowledge of the atmosphere. Unfortunately, the answer is that we don't know the sensitivity of the seasonal and latitudinal effect to chemistry so that the best I can do is outline some of the more important considerations.

TWO-DIMENSIONAL MODEL RESULTS

Most of the published model projections of the impact of chlorofluorocarbon release on the ozone layer have been the result of one-dimensional model calculations. The effects of the transport by global circulation are lumped into a single parameterized function, the vertical diffusion coefficient. The transport of long-lived species via this diffusion approximation therefore represents, in some sense, the global average of the actual transport effects while the shorter-lived species which are near to photochemical equilibrium are generally treated with the solar flux varying with time of day in a manner appropriate to 30° latitude at equinox. By their nature these models cannot be used to determine the seasonal and latitudinal dependence of the thickness of the ozone column or the perturbation of that ozone column. The complexity that must be introduced is the latitude coordinate which combined with the existing vertical coordinate makes the model two-dimensional. The major advantages of the two-dimensional model are that latitude dependence is explicitely calculated and that the driving force for seasonal variations, the latitude change of the subsolar point, can be explicitely represented. Another important improvement in the 2D representation is the possibility of including mean velocity fields in addition to the diffusion coefficient which can now be expanded to be a matrix field. This particular improvement is somewhat mitigated by the fact that the data base and theoretical understanding are not yet developed to the point of being able to uniquely specify all of these variables. This greatly complicates the task of comparing results from different models. As an example, a typical 2D model usually has at least 20 altitude levels (0 to 60 km in 3 km steps), 18 latitudes (pole to pole in 10° steps), and 4 seasons. At each of these points a vertical and meridional (north-south) velocity is assigned in addition to 3 components of a diffusion matrix (vertical, horizontal, and a cross-term). Altogether this makes a minimum of 20 x 18 x 4 x 5 = 7200 transport values to be either specified or somehow calculated.

Despite these difficulties much progress is now being made on the development of such models and their application to the problem of assessing possible changes in the ozone layer. The results for injection of chlorofluorocarbons generally agree in the global average with the magnitude of one-dimensional models but exhibit a variety of seasonal and latitudinal responses. The models of Borucki, et al. (1980) and Crutzen (1979) both assume fixed transport parameters (i.e. not changed by the perturbation) and obtain a maximum ozone depletion during the winter at high latitudes. The models of Vupputuri (1978) and Pyle and Derwent (1980) include a dynamical feedback when ozone is changed by recomputing the mean velocity of the circulation in response to the change in the heating rate by ozone absorption. The eddy-coefficients are not self-consistently changed in this formulation. Both of these models exhibit a winter, high-latitude maximum in the CFC induced ozone depletion with Pyle's being more pronounced than the fixed transport models and Vupputuri's being less pronounced. In contrast to these results, Brasseur and Bertin (1978), have published a fixed transport model calculation in which they multiplied the total chlorine (ClX) in the stratosphere by a factor of 5 (rather than increasing the CFC injection rate causing a ClX change) which resulted in a low latitude, summer maximum in ozone depletion. Additionally, Borucki et al. (1980) state that they obtained a similar result when using an "older" chemistry set (presumably similar to Brasseur and Bertin). This older chemistry set is one which resulted in a lower OH concentration in the lower stratosphere and hence lower average reductions as calculated in the models. Recent chemical kinetics results, obtained since the conference, are also resulting in smaller calculated OH concentrations in the models' lower stratosphere. Although these are due to different mechanisms they might also yield lower summer-winter and low latitude-high latitude contrasts in ozone depletion.

CONCLUSIONS

The evaluation of the mean amount of change in the atmospheric ozone column due to CFC or other releases is a difficult task requiring an atmospheric model with relatively complex chemistry and dynamics. Answers have been obtained for the effect of continued release at today's rates which range from depletions of about 5% to 20%. The changes have resulted from increased knowledge of reaction kinetics obtained from laboratory experiments and have increased and decreased with no obvious convergence other than that they have thus far remained depletions with no obvious mechanisms available to lead to a change of sign.

The additional problem of the latitudinal and seasonal dependence of the calculated ozone depletion is also crucial to the determination of any possible biological effects and is seemingly even more complex (or perhaps just less well understood). Since the seasonal high-latitude maximum is caused by poleward downward motion and the depletion is chemically driven then the seasonal-latitudinal variation of ozone depletion must result from a strong interplay between the dynamics and the chemistry. Although much has been learned about atmospheric dynamics and chemistry, separately their interaction is less well understood. The conclusion, at this time, must be that we do not understand the two-dimensional structure of ozone perturbations well enough to determine their sensitivity to chemistry changes and that future chemistry changes may lead to a change in the seasonal-latitudinal gradient but then again they may not.

REFERENCES

Borucki, W. J., R. C. Whitten, H. T. Woodward, L. A. Capone, C. A. Riegel, and S. Gaines, 1980, "Stratospheric Ozone Decrease Due to Chlorofluoromethane Photolysis: Predictions of Latitude Dependence", J. Atmos. Sci., 37, 686-697.

Brasseur, G., and M. Bertin. 1978. "The action of chlorine on the ozone layer as given by a zonally averaged two-dimensional model" Pure Appl. Geophys. 117: 436-447.

Crutzen, P.J., L. T. Gidel, and J. Fishman. 1979. "Numerical investigations of the photochemical and transport processes which affect ozone and other trace constituents in the atmosphere". Colorado State University Report, Fort Collins, Colorado.

Pyle, J.A., and R.G. Derwent. 1980. "Possible ozone reductions and UV changes at the earth's surface". Nature. 286: 373-375.

Vupputuri, R.K.R. "The structure of the natural stratosphere and the impact of chlorofluoromethanes on the ozone layer investigated in a 2-D time dependent model". Pure Appl. Geophys. 117: 448-485.

MIDDLE ULTRAVIOLET IRRADIANCE AT THE OCEAN SURFACE: MEASUREMENTS AND MODELS

Karen S. Baker (1), Raymond C. Smith (1) and A.E.S. Green (2)

Scripps Institution of Oceanography
University of California, San Diego
La Jolla, Ca., 92093 (1) and
University of Florida, Gainesville, Fl., 32601 (2)

ABSTRACT

Measurements of the downward spectral middle ultraviolet (MUV, 280-380 nm) irradiance $E(0^+,\lambda\theta)$ have been made just above the ocean surface in the central equatorial Pacific. These data were obtained for sun zenith angles in the range 12° to 70°. Analytic models have been fit to these data. The data and the models, their use and their limitations, will be summarized here. Most of this work including the data (Smith and Baker, 1980; hereafter referred to as SB) and the model(Baker et al. 1980; hereafter referred to as BSG) has been explored in detail in previously published works.

In order to work with analytic calculations dealing with irradiance underwater, it is necessary to be able to calculate the spectral irradiance incident on the surface in various ocean locations. For this purpose, $E(0^+, \lambda,\theta)$ has been modeled as a function of wavelength, solar zenith angle, ozone thickness, aerosol thickness, and surface albedo.

EXTRATERRESTRIAL SOLAR IRRADIANCE

The extraterrestrial solar irradiance, $E_\odot(\lambda)$, is the spectral irradiance reaching the top of the earth's atmosphere at mean solar distance. Most models, including those discussed below, assume that $E_\odot$ is a constant. This is a simplifying assumption. A predictable variation in the solar irradiance is due to the eccentricity of the earth in its orbit which causes a periodic change of ± 3% during the year (Anon. 1971). Currently unpredictable variations due to solar events such as sun spots (White, 1977) may cause 1 to 2% variations in $E_\odot$. These events, as well as the difficulty in measuring the extraterrestrial solar irradiance, explain the lack of agreement among researchers as to the values of $E_\odot(\lambda)$.

A summary of measurements of $E_\odot$ in the MUV region is presented in Figures 1. The measurements of Johnson (1954), Arvenson (1969), and Thekaekara (1971) are plotted in Figure 1a. Also shown on these graphs are analytic fits made to $E_\odot$ data in the 280-340 nm region (Green et al., 1974; hereafter referred to as GSS; Green et al., 1980; hereafter referred to as GCS). More recent measurements by Broadfoot (1972), Donnelly and Pope (1973), and Labs and Neckel (1970) are shown in Figure 1b again with the GSS and GCS analytic fits. Comparison of all the data shows a variation of approximately 8% but with disagreements of as much as 15% in the middle ultraviolet region of the spectrum. The models discussed below assume $E_\odot$ is a constant equal to the GSS and GCS values, respectively. Since $E_\odot$ is the input to these models from which the irradiance at the ocean's surface is derived, any user of the models must be cautioned to recognize the resultant uncertainty built into the first step of the model.

Green and Schippnick in this monograph have presented more up-to-date information on $E_\odot$ that fits a unique set of satellite data (Heath and Park, 1980) now available. The comparison of this new fit with that used in the GSS and GCS models is shown in Figure 1c.

DOWNWELLING SPECTRAL IRRADIANCE DATA

Downward spectral irradiance, $E(0^+,\lambda,\theta)$, measurements were obtained using an underwater spectroradiometer (Smith et al., 1979) designed specifically to obtain accurate data in the MUV portion of the spectrum

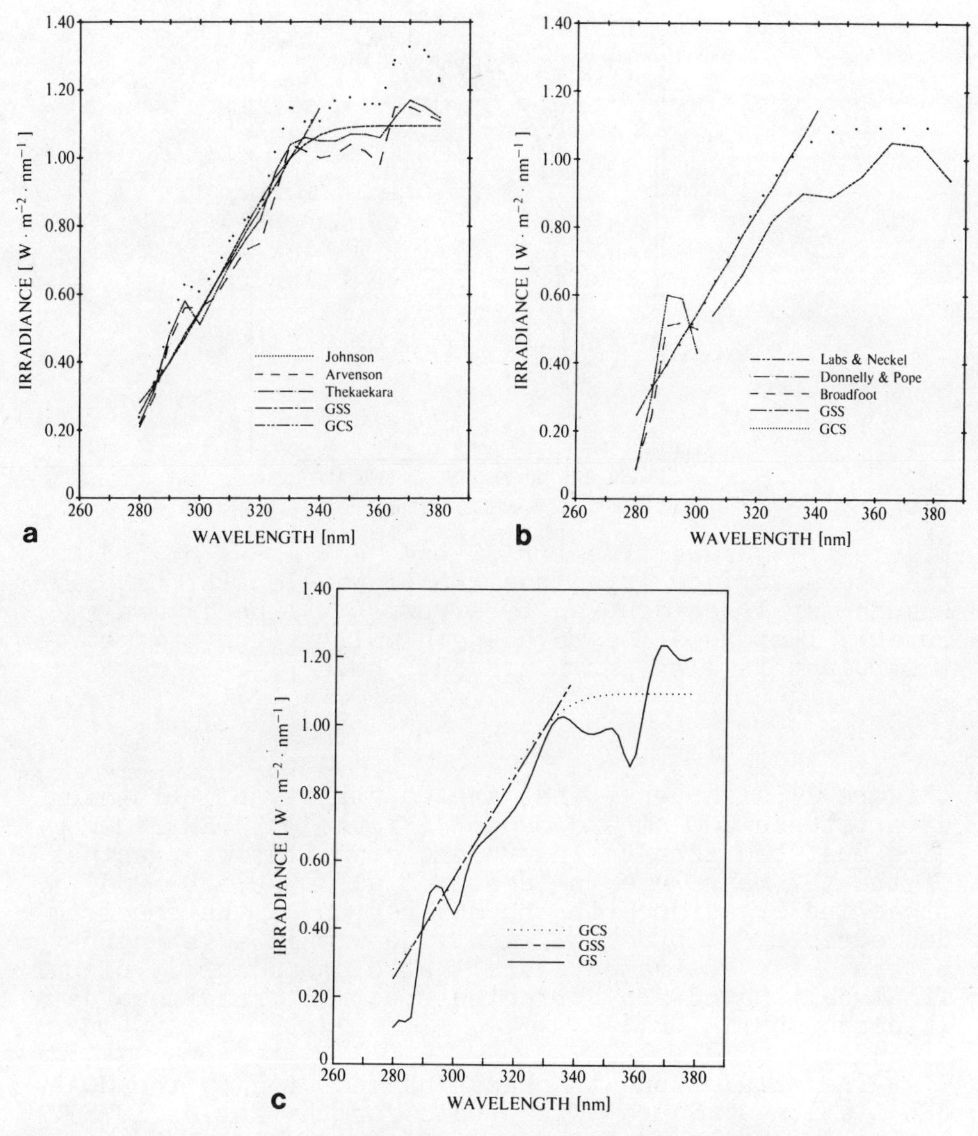

Figure 1, a,b,c. Mean extraterrestrial solar irradiance [W $\cdot$ m^{-2} $\cdot$ nm^{-1}] vs. wavelength [nm] as measured by several investigators and for analytic fits to these data by Green and co-workers.

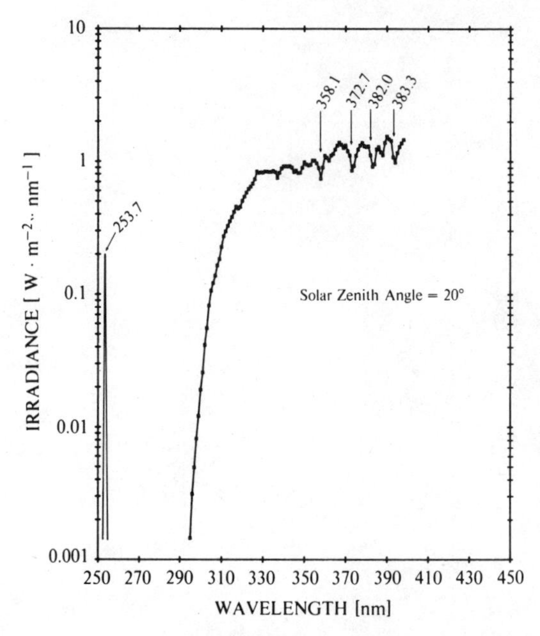

Figure 2. Downwelling irradiance [W · m^{-2} · nm^{-1}] at the ocean surface as a function of wavelength [nm] with Fraunhofer lines denoted by arrows. A line from a mercury lamp used for wavelength calibration of the instrument is also shown at 253.7 nm.

(Figure 2) although it has the capability of obtaining data between 200 nm and 800 nm (Figure 3). Extreme care must be taken to insure the quantitative exactness of the MUV measurements (Smith et al. 1979; SB; BSG) since the irradiance in the MUV region of the spectrum decreases by an order of magnitude within a few nanometers. The estimated overall absolute accuracy of our final calibrated and corrected spectral irradiance data is discussed in SB.

The total ozone thickness as measured by the NOAA-ARL station at Samoa (Ozone Data for the World, 1978) during our cruise was 0.269±0.006 (atm-cm). This is the mean (and standard deviation) for the month during our cruise and assumes a constant stratospheric ozone thickness over the area traveled (see SB Fig. 1).

Also, for the cruise time and location, sun zenith angles at noon ranged from 12° to 20°. Reasonably accurate extrapolations to a sun zenith angle of 0°

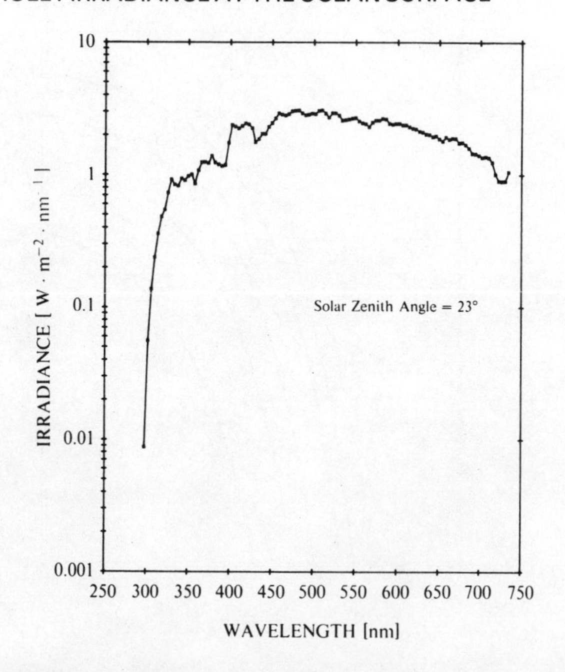

Figure 3. Downwelling irradiance [W · m^{-2}·nm^{-1}] at the ocean surface as a function of wavelength [nm] throughout both the ultraviolet and visible region of the spectrum.

can be made from these sun zenith angles.

The experimental data is shown in Figure 4. The dots represent measured values of downward spectral irradiance for the solar zenith angles indicated. These raw data have been tabulated and are available in reference SB.

ANALYTIC REPRESENTATION

Green et al. have developed semi-analytic models that calculate the global solar middle ultraviolet radiation reaching the earth's surface in the 280-380 nm region. The formulae accommodate variations in environmental variables including solar zenith angle (θ), ozone thickness (w_{OZ}), and surface albedo (A). Parametric fits to the oceanic data have been made first using the original GSS model and subsequently using the GCS model as is described in detail in BSG. Since the notation has been varied, Table 1 is included as a summary of terms used. The notation used here (SB in Table 1) is

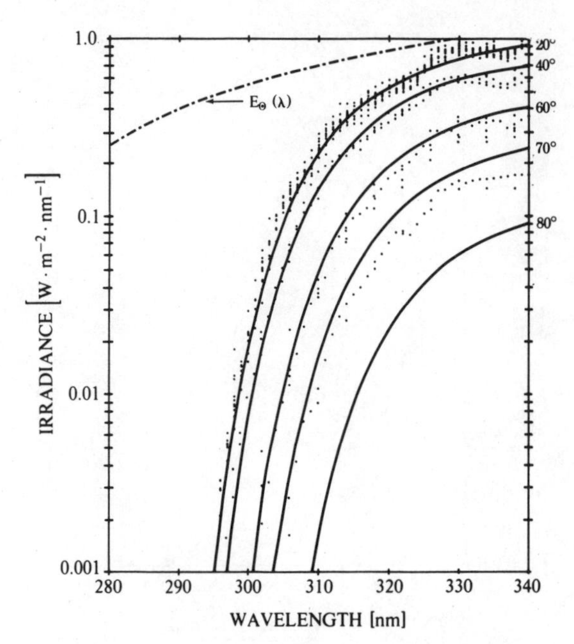

Figure 4. Least-squares fit of Green, Cross, Smith (GCS) model to Smith, Baker (SB) data. Downwelling spectral irradiance at the surface of the ocean at equatorial latitudes. The dots represent experimental data for the solar zenith angle indicated. The solid lines are the fit given by the GCS model. The dot-dashed line is an analytic (GCS) fit to the mean extra-terrestrial solar irradiance. The calculated irradiance for a solar zenith angle of 80° is also shown for comparison; measurements were not made at this zenith angle.

that currently accepted by the oceanographic community (Smith and Tyler, 1976; Smith and Baker, 1979; Morel and Smith, 1981).

The two models deal with the irradiance in two components: a direct component and a diffuse component,

$$E(0^+, \theta, \lambda) = E_{direct}(0^+, \theta, \lambda) + E_{diffuse}(0^+, \theta, \lambda). \quad (1)$$

The models give similar results for small solar zenith angles where the direct component of irradiance dominates, but they are significantly different for large solar zenith angles where the diffuse components of irradiance

TABLE 1

	SB	GSS	GCS
TOTAL DOWNWELLING IRRADIANCE	$E(0+,\theta,\lambda)$	$G(\theta,\lambda)$	$G(\lambda,\theta)$
DIRECT COMPONENT	$E_{direct}(0+,\theta,\lambda)$	$B_s(\theta,\lambda)$	$D(\lambda,\theta)$
DIFFUSE COMPONENT	$E_{diffuse}(0+,\theta,\lambda)$	$B_d(\theta,\lambda)$	$S(\lambda,\theta)$
EXTRATERRESTRIAL	$E_0(\lambda)$	$H(\lambda)$	$H(\lambda)$
OPTICAL DEPTHS			
AIR-RAYLEIGH		τ_a	τ_1
PARTICLE		τ_p	--
PARTICULATE SCATTERING		--	τ_2
PARTICULATE ABSORPTION		--	τ_4
OZONE		τ_{oz}	τ_3

are appreciable. The GCS model uses an improved analytic characterization of the diffuse spectral irradiance and consequently gives a more accurate representation in this region. Hence, although both models are discussed in BSG, only the GCS model will be further discussed here.

The GCS model determines the direct component of surface irradiance beginning with the use of an analytic fit to the extraterrestrial solar spectral irradiance data of Thekaekara (1971). The diffuse component is handled by working with the ratio of diffuse to direct irradiance. The GCS model thus achieves a numerical fit to the diffuse irradiance at a 5% level of accuracy.

A summary of the GCS analytic formulae are given in Table 2a. The formulae have been written for an altitude at sea level (Y=0) which applied to our situation. For this model there are fixed parameters (Table 2b) and fit parameters (Table 2c). The fixed parameters are the original GCS constants except for the three

TABLE 2a

GCS Model Equations

$G(\lambda,\theta) = D(\lambda,\theta) + S(\lambda,\theta)$

$D(\lambda,\theta) = \mu\ H(\lambda) \exp[-\Sigma_j(\tau_j/\mu_j)]$

$S(\lambda,\theta) = \mathcal{S}(\lambda,\theta)\ M(\lambda)\ H(\lambda) \exp[-\Sigma_j(\tau_j)]$

$H(\lambda) = K\ (1 - \exp[-k \exp((\lambda-\lambda_o)/\delta)])$

$\tau_1 = \tau_{10}\ (\lambda_o/\lambda)^{\nu_1}$

$\tau_3 = \omega_3 k_3$

$k_3 = k_o\ (\beta + 1)/[\beta + \exp((\lambda-\lambda_o)/\delta_3)]$

$\tau_4 = 0.15\tau_2$

$\mu_1 = [(\mu^2 + t_i)/(1 + t_i)]^{\frac{1}{2}}$

$\mu = \cos\theta$

$\mathcal{S}(\lambda,\theta) = S(\lambda,\theta)/S(\lambda,0^o) = \{F + (1 - F) \exp[-\gamma_3 (\tau_3 + \tau_4)\ \phi]\} \exp[-(\gamma_1\tau_1 + \gamma_2\tau_2)\phi]$

$M(\lambda) = S(\lambda,0^o)/D(\lambda,0^o) = (A_{a1}\ \tau_1 + A_{a2}\tau_2{}^{P_a})/[1 + A_{a3}\omega_3 (\tau_3 + \tau_4)^{q_a}]$

$F = [1 + A_{03}(\tau_3 + \tau_4 + g_o k_3)^{q_o}]^{-1}$

$**\phi = [(1- t_o)/(\mu^2 + t_o)]^{\frac{1}{2}} - 1$

$*G(A,\lambda,\theta) = G(0,\lambda,\theta)/[1 - r(\lambda,\omega_3)\ A]$

$r(\lambda,\omega_3) = (A_{b1}\tau_1 + A_{b2}\tau_2{}^{P_b})/[1 + A_{b3}(\tau_3 + \tau_4)^{q_b}]$

*NOTE: This equation had typographical errors as originally printed in Green et al. (1980)

**NOTE: This equation had a typographical error as printed in Baker et al. (1980)

TABLE 2b
GCS Fixed Parameters

Extraterrestrial Solar Irradiance	Rayleigh (air) i=1	Particulate i = 2	Ozone i = 3	
$K = 1.095\ w/m^2\ nm$	$t_1 = 1.8x10^{-3}$	$t_2 = 3.0x10^{-4}$	$t_3 = 7.4x10^{-3}$	$\lambda_o = 300nm$
$k = 0.6902$	$\tau_{10} = 1.22$	$t_4 = t_2$	$K_o = 9.517(atm\ cm)^{-1}$	
$\delta = 23.74\ nm$	$*\ \nu_1 = 4.27$		$\beta = 0.0445$	
			$\delta_3 = 7.294\ nm$	
Equatorial Ocean		$\tau_2 = 0.1$	$\omega_3 = 0.27$	$A = 0.08$

*NOTE: This equation had a typographical error as printed in Baker et al. (1980)

TABLE 2c
GCS Fit Parameters

Species Parameters	Rayleigh $i = 1$	Particulate $i = 2$	Ozone $i = 3$
γ_i	0.5777	0.427	0.977
A_{ai}	0.7879	12.8*	0.1978
A_{bi}	0.5399	0.0230	0.6710
A_{oi}			3.285
Alpha Parameters	$\alpha = 0$	$\alpha = a$	$\alpha = b$
q_α	1.104	1.079	0.587
P_α	--	1.523	0.785
g_α	1.433	--	
t_α	0.020		

* In "average" model this number is 2.96.

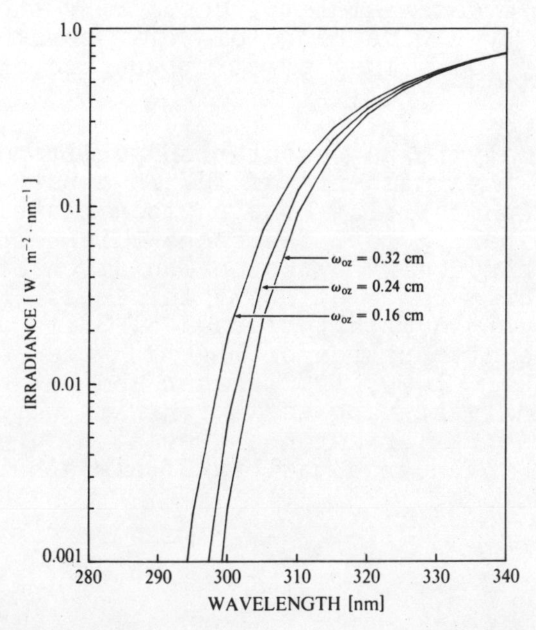

Figure 5. Spectral irradiance at the ocean surface for a range of ozone thicknesses as calculated from the analytic formulas of Tables 1 and 2.

which were chosen based upon our best estimate of an equatorial marine atmosphere. τ_2 was based on knowledge of the marine atmosphere (Green, unpublished). ω_{oz} is based on the NOAA-ARL data as mentioned previously. A is based on an estimate from clear natural waters (Smith et al., 1973) and unpublished work by Green. The GCS model fit parameters to our marine atmosphere data are listed in Table 2c. These parameters were originally obtained as the best fit parameters for a series of atmospheric models (Braslau and Dave, 1973; Dave and Halpern, 1975). As might be expected, the best fit to the oceanic data was obtained by varying a particulate variable, A_{a_2}, from what was found for terrestrial atmospheres. The resulting model fit to the data is shown in Figure 4.

CONCLUSIONS

Measurements of downward spectral irradiance reaching the ocean surface in the wavelength region from 280 nm to 380 nm have been made. While the data are limited, they represent the only spectral irradiance data we are aware of for an open ocean marine atmosphere. These data taken at equatorial latitudes during the summer are distinguished by a wide range of solar zenith angles and relatively low values of stratospheric ozone thickness.

The GCS analytic representation permits quantitative assessment of the influence of MUV in aquatic ecosystems to be made by photobiologists, photochemists, and oceanographers. In particular, the ozone thickness is a variable in this model so that calculation of $E(0^+,\lambda,\theta)$ can be made for a range of ozone thickness (Figure 5). This model combined with the model of Smith and Baker that calculates the attenuation of light in the water column (Smith and Baker, 1979; 1980; Baker and Smith, in press) provides an analytic representation with which to calculate the spectral irradiance present at any point in the water column under a variety of atmospheric and oceanic conditions.

ACKNOWLEDGEMENTS

This work was supported by the United States Environmental Protection Agency, Stratospheric Impact and Assessment Program, Grant No. R 806489010 (RCS and KSB) and Grant No. R 806373010 (AFSG).

REFERENCES

Anon. 1971. Solar Electromagnetic Radiation, NASA Space Vehicles Design Criteria. NASA, Washington, D. C. SP8005.

Arvenson, J.C., R.N. Griffin, Jr., and B.D. Pearson, Jr. 1969. Determination of extraterrestrial solar spectral irradiance from a research aircraft. App. Opt. 8:2215-2232.

Baker, K.S. and R.C. Smith. Bio-Optical classification and model of natural waters II. Limnol. Oceanogr. (In press)

Baker, K.S., R.C. Smith, and A.E.S. Green. 1980. Middle ultraviolet radiation reaching the ocean surface. Photochem. Photobiol. 32:367-374.

Braslau, N. and J.V. Dave. 1973. Effect of aerosols on the transfer of solar energy through realistic model atmosphere III ground level fluxes in the biologically active bands 0.2850-0.3700 microns. IBM Research Report, RC 4308.

Broadfoot, A.L. 1972. The solar spectrum 2100-3200 angstrom. Astrophys. Journ. 173:681-689.

Dave, J.V. and P. Halpern. 1976. Effect of changes in ozone amount on the ultraviolet radiation received at sealevel of a model atmosphere. Atmos. Env. 10: 547-555.

Donnelly, R.F., and J.H. Pope. 1973. NOAA Tech. Rep. ERL276-SEL, U.S. Gov. Printing Office, Washington, D. C.

Green, A.E.S., K. R. Cross, and L. A. Smith. 1980. Improved analytic characterization of ultraviolet skylight. Photochem. Photobiol. 31:59-65.

Green, A.E.S., T. Sawada, and E. P. Shettle. 1974. The middle ultraviolet reaching the ground. Photochem. Photobiol. 19:251-259.

Heath, D. and H.W. Park. 1980. Amer. Geophys. Union Meeting, Toronto, personal communication.

Johnson, F.S. 1954. The colar constant. J. Meteor. 2: 431.

Labs, D., and H.L. Neckel. 1970. Transformation of the absolute total solar radiation data. Solar Physics 15: 79-87.

Morel, A. and R. C. Smith, 1981. Terminology and units in optical oceanography. Mar. Geodesy 5 (4)

Smith, R. C. and K. S. Baker. 1979. Penetration of UV-B and biologically dose rates effective in natural waters. Photochem. Photobiol. 29: 311-323.

Smith, R. C. and K.S. Baker. 1980. Middle ultraviolet irradiance measurements at the ocean surface. SIO Ref. Report No. 80-12.

Smith, R. C., R.L. Ensminger, R.W. Austin, J.D. Bailey, and G.D. Edwards. 1979. Ultraviolet submersible spectroradiometer. S.P.I.E. Ocean Optics VI: 127-140.

Smith, R.C., and J.E. Tyler. 1976. Transmission of solar radiation into natural waters. Photochem. Photobiol. Rev. 1, Plenum Press Pub. Co., N.Y. 117-155.

Smith, R.C., J.E. Tyler and C.R. Goldman. 1973. Limnol. Oceanog. 18 (2): 189-199.

Thekekara, M.P. 1971. Solar electromagnetic radiation NASA SP-8005.

White, O.R., [ed.] 1977. The solar output and its variations. Assoc. Univ. Press, Boulder, Colorado..

THE EFFECT OF PERTURBATION OF THE TOTAL OZONE COLUMN DUE TO CFC ON THE SPECTRAL DISTRIBUTION OF UV FLUXES AND THE DAMAGING UV DOSES AT THE OCEAN SURFACE: A MODEL STUDY

Frode Stordal (1), Øystein Hov (2) and
Ivar S.A. Isaksen (2)
Norwegian Meteorological Institute (1) and
Institute of Geophysics
University of Oslo (2) Norway

INTRODUCTION

The amount of ozone in the atmosphere is a key factor in controlling the UV radiation reaching the earth's surface. A reduction of the ozone layer due to anthropogenic release of chlorofluorocarbons (CFC) will therefore give rise to increased UV radiation. Based on our 1-D model estimates of future ozone depletion by CFC's, and on different action spectra for solar UV radiation, we will estimate the magnitude of the effect of possible future ozone depletions. In these estimates we utilize the detailed solar flux calculation which is included in the chemical model where ozone fluctuations are calculated (Isaksen and Stordal, 1981). The UV radiation model takes into account absorption by ozone and nitrogen dioxide. Rayleigh scattering and multiple scattering is modelled by assuming that half of the scattered light proceeds in the direction of the incoming radiation, the other half in the opposite direction.

Increased UV light at the earth's surface is a matter of concern because of the sensitivity of biological systems towards changes in the UV dose. The effect of UV radiation on biological systems is strongly wavelength dependent, and the sensitivity of various biological processes may exhibit different wavelength dependencies.

The solar UV spectrum at the earth's surface as a function of ozone column density is discussed, together with the total biological effect (the damaging ultra-violet dose-DUV dose). The DUV dose and the effect of ozone column perturbations is examined for different latitudes and seasons. For one particular latitude, 45° N, and for the summer season we have calculated the fluxes and the DUV doses with ozone profiles derived from the chemistry-diffusion model.

RADIATION TRANSFER AND DUV DOSES

The radiation transfer model

A simple radiation transfer model is used (Isaksen et al. 1977). The solar fluxes are, for the purpose of studying biological effects, calculated in the UV spectral region 280-400 nm. The fluxes are computed from the ground and up to 50 km at 1 km intervals.

The solar radiation is attenuated by molecular scattering and by absorption by O_3 and to some extent NO_2. The attenuation for a given wavelength is according to the equation

$$F_z = F_{z+\Delta z} \cdot \exp[\{-\sigma_{O_3} \cdot [O_3] - \sigma_{NO_2} \cdot [NO_2] - \sigma_R [M]\} f(\theta) \Delta z] \qquad (1)$$

where the flux F_z at one height level z is derived from the flux $F_{z+\Delta z}$ at a height level Δz above (Δz = 1 km). σ_{O_3} and σ_{NO_2} are the absorption cross sections for O_3 and NO_2 (NASA, 1979), σ_R is the Rayleigh scattering coefficient (Penndorf, 1957). $[O_3]$, $[NO_2]$ and $[M]$ are mean densities of O_3, NO_2 and M (U. S. Standard Atmosphere) in the actual layer. $f(\theta)$ is equal to $1/\cos\theta$ for solar zenith angles less than 75°, and equal to the Chapman function for a θ greater than 75°, where the curvature of the atmosphere is accounted for. Fluxes of incoming solar radiation are taken from Ackerman (1971).

The molecular scattering is treated in a simplified way. It is assumed that one half of the radiation proceeds in the direction of the direct beam, the other half in the opposite direction. The model calculations have previously been validated against the results obtained with more elaborate methods (Fig. 1), and have

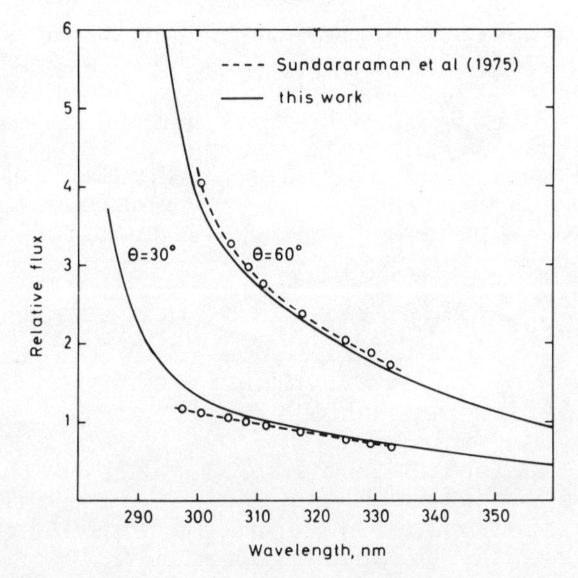

Figure 1. The ratio of the downward multiple scattered flux to the direct incoming solar flux versus wavelength at ground level for the two solar zenith angles of 60° and 30°. The curves denoted Sundararaman et al. (1975) are taken from their tabulated values of the same ratio for an ozone column density of 350 Dobson Units. For θ = 30° the numbers are derived by interpolation between the solar zenith angles 25° and 40° in their Table. The fully drawn lines show the ratios when the fluxes are calculated in the simplified way adopted in this work (see Isaksen et al., 1977).

shown good agreement (less than a few per cent deviation, see Isaksen et al. 1977). Multiple scattering is accounted for. It is found that scattered fluxes above the third order are negligible compared to the total fluxes. In the calculations described in the following, fluxes are calculated for every 15 minutes, and 24 - hour mean values are computed.

Action spectra

The biological effect of solar radiation depends strongly on the wavelength. We have adopted three different types of wavelength dependence, i.e., we have used three action spectra, taken from the NAS report "Protection against Depletion of Stratospheric Ozone by Chlorofluorocarbons" (1979b). These spectra account for different aspects of biological activity.

Spectrum A is a generalized plant action spectrum developed by Caldwell (1971) from a number of action spectrum studies with plant material.

Spectrum B is a DNA damage spectrum derived by Setlow (1974).

Spectrum C gives an expression of photosynthesis inhibition. It gives the influence of UV and visible radiation on the photosynthetic reaction and is derived by Jones and Kok (1966).

The total biological effect is described by the DUV dose which is computed as the product of mean fluxes and relative efficiency (values from the action spectra), integrated over the UV spectral region:

$$\text{DUV dose} = \int_{UV} A(\lambda) \cdot F(\lambda)\, d\lambda \qquad (2)$$

The action spectra A and B (Fig. 2) give the greatest weights for the fluxes in the shortwave region. In this region the absorption by ozone is strong and therefore sensitive to changes in total ozone. The DUV doses corresponding to these two spectra are therefore much more sensitive to ozone depletion than the dose derived from spectrum C (Fig. 2) which gives large contributions from the longwave region where the fluxes are rapidly increasing and where the ozone absorption is small. As much as 50% of the photosynthesis inhibitory effect results from radiation of wavelengths greater than 390 nm.

MODEL CALCULATIONS

Seasonal and geographical variation of solar fluxes and DUV doses

The 24-hour mean fluxes are calculated for different latitudes, with a resolution of 10 degrees, and for four seasons. Solar declinations for 15th January, 15th April, 15th July and 15th October are used. The total ozone column densities are as reported by Dütsch (1978), representing zonal means.

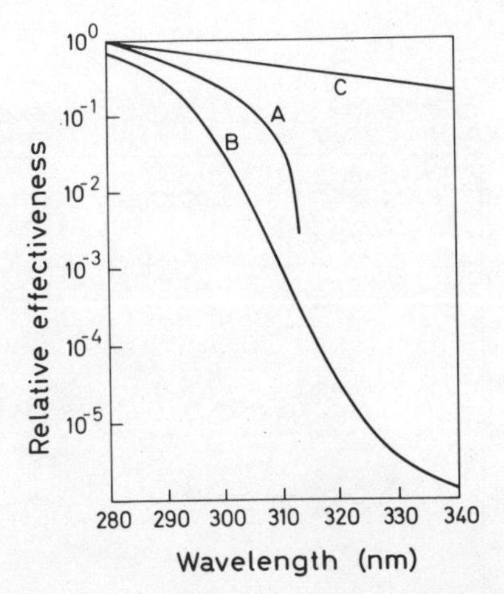

Figure 2. Action spectra, i.e., spectral weighting functions for biological effects.

Spectrum A is a generalized plant action spectrum developed by Caldwell (1971) from a number of action spectrum studies with plant material.

Spectrum B is a DNA damage spectrum derived by Setlow (1974).

Spectrum C gives an expression of photosynthesis inhibition. It gives the influence of UV and visible radiation on the photosynthetic reaction and is derived by Jones and Kok. (1966)

The distribution of DUV doses, corresponding to the calculated fluxes and the action spectra A, B and C, are shown in Table 1. The two former have maximum values at low latitudes in summer, with decreasing values during the winter and towards the poles. The lower values towards the poles are due to the lower solar altitude and increasing ozone column, affecting especially the UV-B fluxes where O_3 absorbs strongly and where the contributions to the DUV doses attain their maxima (for spectrum A and B). The UV and the visible spectral region contribute to the doses corresponding to spectrum C. The distribution of DUV doses has maximum values at high latitudes in the summer season where most of (or all of) the day is sunlit.

The effect of ozone depletion is studied by reducing the ozone column densities by 1%, 10% and 40% in order to identify any nonlinear effects as the ozone column is perturbed. The corresponding increases in DUV doses

TABLE 1

Seasonal and geographical distribution of DUV doses corresponding to the action spectra A, B and C. The values are given relative to the values for 30°N, summer season

DUV doses

Latitude, degrees		A				B				C				
		JAN	APR	JUL	OCT	JAN	APR	JUL	OCT	JAN	APR	JUL	OCT	
Northern hemisphere	80		.01	.13			.03	.19			.40	1.11		80
	70		.05	.24			.08	.26	.01		.57	1.00	.07	70
	60		.17	.35	.01		.19	.35	.03	.03	.70	.99	.26	60
	50		.33	.53	.10	.02	.32	.50	.11	.20	.79	1.00	.45	50
	40	.04	.50	.76	.31	.06	.47	.72	.29	.37	.85	1.01	.60	40
	30	.14	.64	1.00	.63	.15	.61	1.00	.62	.53	.89	1.00	.72	30
	20	.32	.76	1.05	.80	.30	.74	1.07	.81	.66	.91	.97	.80	20
	10	.56	.83	.99	.86	.53	.81	1.00	.86	.77	.91	.93	.86	10
	0	.81	.83	.81	.84	.80	.82	.80	.83	.86	.89	.86	.89	0
Southern hemisphere	10	.92	.73	.58	.93	.91	.71	.56	.93	.92	.84	.77	.91	10
	20	.92	.57	.35	.96	.91	.55	.33	.97	.96	.78	.66	.92	20
	30	.86	.40	.17	.90	.84	.38	.17	.90	.99	.70	.53	.91	30
	40	.89	.25	.05	.65	.87	.24	.07	.63	1.01	.60	.38	.87	40
	50	.87	.11	.01	.44	.84	.12	.02	.42	1.03	.46	.20	.82	50
	60	.75	.02		.27	.72	.04		.27	1.04	.28	.03	.74	60
	70	.34			.15	.34	.01		.16	1.02	.08		.63	70
	80	.12			.04	.18			.06	1.10			.45	80

are shown in Table 2 a - c. The figures are given as per cent change in DUV dose per per cent ozone depletion (Radiation Amplification Factor - RAF). The relative changes in fluxes are greater at higher than at lower latitudes (Fig. 3), again because of the difference in solar altitudes and ozone column densities. This gives rise to an increase in RAF values from equator towards the poles for the spectrum A doses. For spectrum B the situation is somewhat different. Even if the changes in fluxes for all wavelengths are greater at higher than lower latitudes, the RAF values are smaller. This can be explained by a weak shifting in effectiveness towards longer wavelengths with less relative change in flux as we move towards higher latitudes (Fig. 4). This shows that the geographical distribution of the RAF values depends critically on the shape of the action spectrum. The spectrum C doses are only weakly affected by the ozone depletion, because of the great importance of the large and unperturbed fluxes in the UV-A and the visible region.

In all cases the relative change in DUV doses increases more than the relative change in ozone. Tables 2 a - c show that the greater the ozone depletion, the greater the RAF values. This is due to the exponential form of the radiation transfer Eq. (1).

The estimates presented so far are made under the assumption of no surface albedo and without clouds. Comparative calculations of fluxes and DUV doses are made with a 10% surface albedo, which is representative for the UV spectral region at the ocean surface and a 30% cloud albedo, according to a global average when a cloud fraction of about 50% is taken into account. The cloud cover is modelled as an infinite, homogeneous, horizontal and partially reflecting layer. These calculations show that the magnitude of the fluxes is somewhat altered. The RAF values are, however, unchanged.

Prognostic ozone profiles and DUV doses

The effect from release of chlorofluorocarbons on the atmospheric content of ozone is estimated with a one-dimensional time dependent chemistry model with vertical diffusion. The model is the same as that used by Isaksen and Stordal (1981). Our calculations are for 45° N latitude. The constituents Freon-11 and Freon-12 (F-11 and F-12) are emitted at rates given by NAS (1979a) during the years 1955 to 1977. After 1977

TABLE 2a

Seasonal and geographical distribution of the Radiation Amplification Factors (RAF values). The values give the per cent change in DUV dose per per cent ozone reduction, for 1%, 10% and 40% depletion. The figures are for spectrum A doses.

Latitude, degrees		RAF(1%) values				RAF(10%) values				RAF(40%) values				
		JAN	APR	JUL	OCT	JAN	APR	JUL	OCT	JAN	APR	JUL	OCT	
Northern hemisphere	80		4.07	2.71			4.99	3.13			11.44	5.60		80
	70		2.95	2.38	4.82		3.44	2.71	6.11		6.39	4.58	16.15	70
	60	6.67	2.37	2.20	3.13	9.20	2.70	2.48	3.68	35.83	4.55	4.09	7.04	60
	50	3.87	2.09	1.98	2.36	4.70	2.36	2.22	2.68	10.34	3.82	3.55	4.49	50
	40	2.86	1.93	1.81	1.93	3.32	2.16	2.01	2.16	6.06	3.43	3.15	3.42	40
	30	2.32	1.83	1.67	1.69	2.63	2.04	1.85	1.87	4.40	3.19	2.85	2.87	30
	20	2.01	1.76	1.64	1.65	2.25	1.96	1.82	1.83	3.61	3.04	2.78	2.80	20
	10	1.81	1.72	1.64	1.67	2.02	1.91	1.82	1.85	3.15	2.96	2.79	2.84	10
	0	1.70	1.71	1.69	1.70	1.88	1.90	1.88	1.89	2.90	2.93	2.89	2.92	0
Southern hemisphere	10	1.68	1.74	1.79	1.67	1.86	1.94	2.00	1.85	2.86	3.00	3.11	2.85	10
	20	1.70	1.81	1.96	1.65	1.88	2.02	2.19	1.83	2.91	3.16	3.49	2.81	20
	30	1.74	1.92	2.21	1.67	1.93	2.15	2.50	1.85	3.00	3.40	4.11	2.85	30
	40	1.73	2.07	2.63	1.79	1.92	2.33	3.02	2.00	2.97	3.75	5.30	3.12	40
	50	1.73	2.31	3.37	1.95	1.93	2.62	4.01	2.18	2.99	4.36	8.01	3.47	50
	60	1.79	2.79	5.39	2.13	2.00	3.23	7.02	2.40	3.11	5.80	20.81	3.90	60
	70	2.17	4.29		2.29	2.45	5.30		2.59	4.02	12.60		4.30	70
	80	2.77			2.71	3.20			3.12	5.77			5.51	80

TABLE 2b. As in Table 2a, for spectrum B.

Latitude, degrees		RAF(1%) values				RAF (10%) values				RAF (40%) values				
		JAN	APR	JUL	OCT	JAN	APR	JUL	OCT	JAN	APR	JUL	OCT	
Northern hemisphere	80		.68	1.44			.78	1.66			1.45	3.14		80
	70		1.33	1.69	.28		1.53	1.90	.31		2.89	3.75	.51	70
	60	.14	1.69	1.82	1.08	.15	1.96	2.10	1.24	.22	3.75	4.07	2.30	60
	50	.75	1.85	1.92	1.63	.86	2.15	2.23	1.88	1.57	4.16	4.35	3.57	50
	40	1.37	1.94	2.00	1.88	1.59	2.25	2.32	2.18	2.99	4.39	4.54	4.20	40
	30	1.72	1.99	2.05	2.01	1.99	2.31	2.39	2.33	3.80	4.52	4.63	4.51	30
	20	1.89	2.02	2.06	2.04	2.19	2.35	2.40	2.37	4.24	4.59	4.65	4.59	20
	10	1.98	2.04	2.06	2.05	2.30	2.37	2.40	2.38	4.49	4.62	4.64	4.62	10
	0	2.04	2.04	2.04	2.04	2.37	2.37	2.37	2.38	4.60	4.62	4.61	4.63	0
Southern hemisphere	10	2.05	2.02	1.99	2.05	2.38	2.35	2.31	2.39	4.63	4.59	4.50	4.64	10
	20	2.04	1.99	1.90	2.06	2.38	2.31	2.21	2.39	4.63	4.50	4.29	4.64	20
	30	2.03	1.92	1.76	2.04	2.36	2.23	2.04	2.38	4.60	4.34	3.91	4.61	30
	40	2.03	1.83	1.48	1.99	2.35	2.12	1.71	2.31	4.59	4.09	3.23	4.50	40
	50	2.01	1.66	.92	1.91	2.34	1.92	1.05	2.22	4.55	3.64	1.94	4.31	50
	60	1.97	1.25	.17	1.80	2.29	1.44	.18	2.09	4.45	2.68	.28	4.02	60
	70	1.78	.35		1.65	2.06	.40		1.91	3.97	.68		3.62	70
	80	1.41			1.23	1.63			1.41	3.07			2.60	80

TABLE 2c. As in Table 2a, for spectrum C.

Latitude, degrees		RAF (1%) values				RAF (10%) values				RAF (40%) values				
		JAN	APR	JUL	OCT	JAN	APR	JUL	OCT	JAN	APR	JUL	OCT	
Northern hemisphere	80		.05	.06			.05	.06			.06	.07		80
	70		.06	.06	.03		.06	.06	.03		.07	.08	.04	70
	60	.03	.06	.06	.05	.03	.06	.07	.05	.03	.07	.08	.06	60
	50	.05	.06	.06	.06	.05	.06	.07	.06	.06	.08	.08	.07	50
	40	.06	.06	.06	.06	.06	.06	.06	.06	.07	.08	.08	.07	40
	30	.06	.06	.06	.06	.06	.06	.06	.06	.07	.08	.07	.07	30
	20	.06	.06	.06	.06	.06	.06	.06	.06	.07	.08	.07	.07	20
	10	.06	.06	.06	.06	.06	.06	.06	.06	.07	.08	.07	.07	10
	0	.06	.06	.06	.06	.06	.06	.06	.06	.07	.08	.07	.08	0
Southern hemisphere	10	.06	.06	.06	.06	.06	.06	.06	.06	.07	.08	.07	.08	10
	20	.06	.06	.06	.06	.06	.06	.06	.06	.08	.07	.07	.07	20
	30	.06	.06	.06	.06	.06	.06	.06	.06	.08	.07	.07	.07	30
	40	.06	.06	.06	.06	.06	.06	.06	.06	.07	.07	.07	.07	40
	50	.06	.06	.05	.06	.06	.06	.05	.06	.07	.07	.06	.07	50
	60	.06	.05	.03	.06	.06	.05	.03	.06	.07	.06	.03	.07	60
	70	.06	.03		.05	.06	.03		.06	.07	.04		.07	70
	80	.06			.05	.06			.05	.07			.06	80

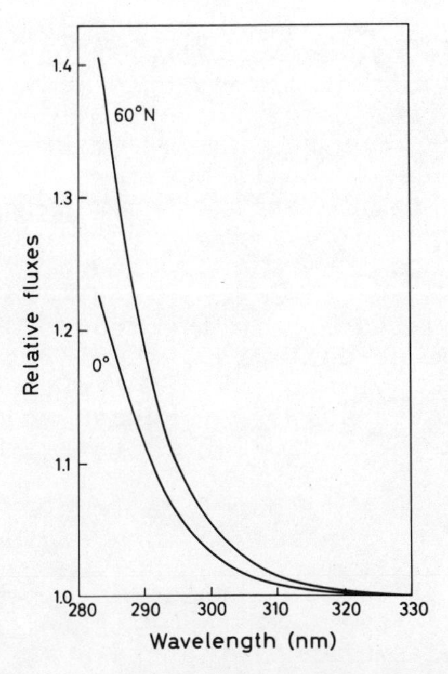

Figure 3. Spectral distribution of relative increase in fluxes due to a 1% reduction of total ozone. Conditions as for summer 60° N latitude and equator.

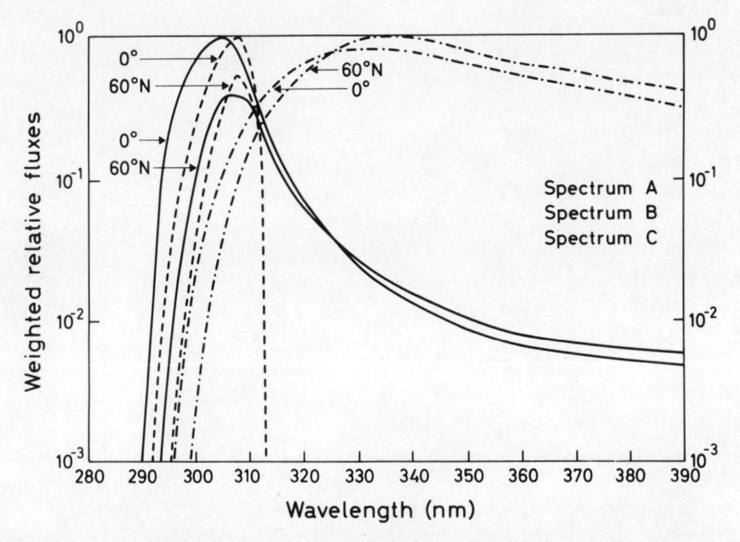

Figure 4. The biologically effective fluxes, represented by the product of the flux and the values of the weighting functions (action spectra A, B and C). Conditions as for summer 60° N latitude and equator.

constant release rates, equal to the 1977 production are adopted. The freons, being rather resistant in the troposphere, are slowly transported to the stratosphere where they are dissociated, mainly by photodissociation, to free chlorine compounds which attack the ozone by catalytic reactions. The temperature in the stratosphere is controlled by ozone through the warming that accompanies the absorption of solar radiation. Reduction of stratospheric ozone lowers the temperature which influences the chemical reaction rates. This temperature feedback is accounted for in the model. Figure 5 shows the calculated ozone depletion assuming the faster of the two sets of K_z-values (vertical diffusion) adopted in our above referenced work (Isaksen and Stordal, 1981). The depletion in the year 2040 is estimated to be 8.0%.

The stratospheric temperature is also controlled by the CO_2 content. A general cooling of the stratosphere is predicted to take place when CO_2 increases. Anthropogenic release of carbon dioxide may therefore influence the ozone layer. We have shown (Isaksen et al. 1980) that the estimates of the temperature effect on ozone depend markedly on the level of chlorine in the stratosphere, as the catalytic ozone reduction by nitrogen and chlorine reactions respond differently to changes in temperature. The ozone-reducing catalytic nitrogen cycle is less effective when the temperature is lowered. This leads to a 5.1% increase in total ozone when a doubling of the CO_2 content is assumed. The depletion of ozone due to the chlorine cycle will, in contrast, be more effective. With the chlorine content computed for the year 2040, the increase in total ozone due to a doubling of CO_2 is calculated to be only 2.8%.

Some ozone profiles computed with the one-dimensional chemistry model are used to study the relative spectral changes in the UV doses. We have adopted the profiles as calculated for the years 1977, 2000, 2020 and 2040 when O_3 is affected by release of F-11 and F-12.

In the reference year 1955, the ozone content is assumed not to be influenced by releases of CFC. The calculations are for the summer season. Figure 6 shows the spectral distribution of relative changes in fluxes, and Figure 5 illustrates the magnitude of the increase in DUV doses at different years associated with the ozone depletion. In the year 2040 the 8.0% ozone depletion is accompanied by a 16.2% and 16.8% increase in DUV doses related to the spectra A and B. The spectrum C dose shows only a modest increase.

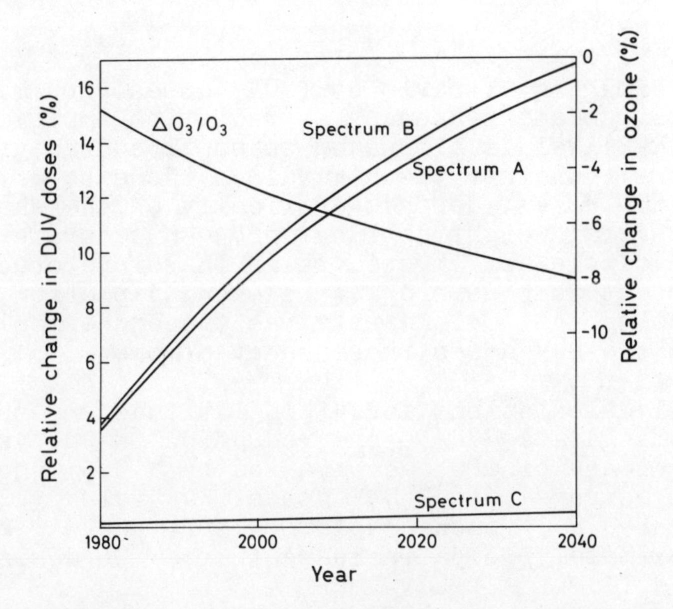

Figure 5. Calculated ozone depletion due to release of CFC. Release rates up to 1977 are taken from NAS (1979a). Constant release rates on 1977 level are adopted for the years thereafter. The figure also shows the relative change in DUV doses related to the action spectra A, B and C due to the ozone depletion.

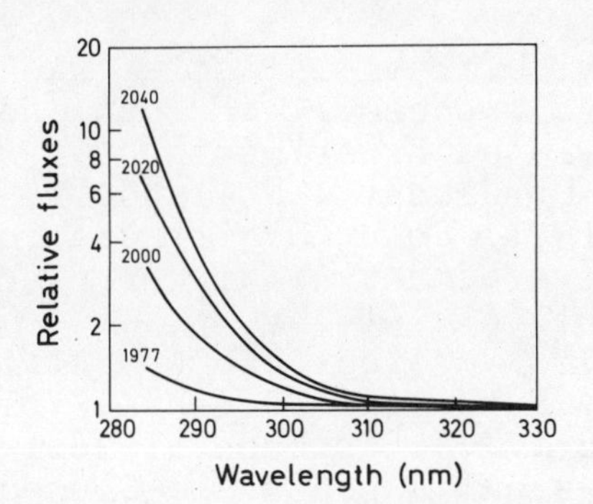

Figure 6. Spectral distribution of fluxes under the assumption of ozone reduction due to releases of Freon-11 and Freon-12. Computed ozone profiles for the years 1977, 2000, 2020 and 2040 are adopted. The flux for a given wavelength is given relative to the value in the reference year 1955. The values represent 45° N latitude and summer conditions.

CONCLUSIONS

The damaging ultraviolet (DUV) doses vary markedly with latitude and season. The radiation amplification factor (RAF) values also show seasonal and geographical variations. Whether the RAF values increase or decrease towards the poles, depends critically on the shape of the biological weighting function. An accurate description of the action spectra is therefore crucial. With respect to possible future ozone depletion, it is worth noting the nonlinearity, as the ozone depletion worsens, the DUV doses increase even more.

Model calculations for 45° N latitude and summer conditions give a 4.8% ozone reduction in the year 2000 due to release of CFC, accompanied by a 9.5% increase in the spectrum A and B DUV doses (RAF value ≈ 2.0). The computed 8.0% ozone depletion for the year 2040 leads to approximately 16.5% increase in the two doses (RAF-value ≈ 2.1).

ACKNOWLEDGEMENTS:

This work is partly sponsored by the Norwegian Research Council for Science and the Humanities (NAVF).

REFERENCES

Ackerman, M. 1971. Ultraviolet solar radiation related to mesospheric processes. Fiocco [ed.], Mesospheric Models and Related Experiments. D. Reidel Publishing Company, Dordrecht - Holland. 149-159.

Caldwell, M. M. 1971. Solar UV irradiation and the growth and development of higher plants. Photophysiology Vol. 6: 131-177.

Dütsch, H. U. 1978. Vertical ozone distribution on a global scale. Pure Appl. Geophys. Vol. 116: 511-529.

Hesstvedt, E. and I.S.A. Isaksen. 1978. WMO Symposium on the Geophysical Aspects and Consequences of Changes in the Composition of the Stratosphere, Toronto. June 1978 (WMO No. 511).

Isaksen, I.S.A., K. H. Midtbø, J. Sunde, and P. J. Crutzen. 1977. A simplified method to include molecular scattering and reflection in calculations of photon fluxes and photodissociation rates. Geophys Norv. Vol. 31: No. 5.

Isaksen, I.S.A., E. Hesstvedt and F. Stordal. 1980. Influence of stratospheric cooling from CO_2 on the ozone layer. Nature 283 No. 5743: 189-191.

Isaksen, I.S.A. and F. Stordal. 1981. The influence of man on the ozone layer; Readjusting the estimates. Ambio. Vol. 10, No. 1: 9-17.

Jones, L. W. and B. Kok. 1966. Photoinhibition of chloroplast reactions. I. Kinetics and action spectrum, Plant Physiol. 41: 1037-1043.

National Academy of Sciences. 1979a. Stratospheric Ozone Depletion by Halocarbons: Chemistry and Transport.

National Academy of Sciences. 1979b. Protection against Depletion of Stratospheric Ozone by Chlorofluorocarbons.

NASA 1979. The Stratosphere: Present and Future. NASA Reference Publication 1049.

Penndorf, R. 1957. Tables of the refractive index for standard air and the Rayleigh scattering coefficient for the spectral region between 0.2 and 20.0 μ and their application to atmospheric optics. J. Opt. Soc. Am. 47: 176-182.

Setlow, R. B. 1974. The wavelength in sunlight effective in producing skin cancer: a theoretical analysis. proc. Nat. Acad. Sci. U.S. 71: 3363-3366.

Sundararaman, N., D. E. St. John and S. U. Venkateswaran. 1975. DOT-TST-75-101. U. S. Dept. of Transportation.

U. S. Standard Atmosphere Supplement. 1966. Environmental Science Services Administration, National Aeronautics and Space Administration, United States Air Force.

MULTI-WAVELENGTH DETERMINATION OF TOTAL OZONE AND ULTRAVIOLET IRRADIANCE

Alex E. S. Green

University of Florida
Gainesville, Florida 32611

ABSTRACT

An improved analysis of our multi-wavelength least square method of total ozone determination is given for various monochromator band passes. The refinements include use of new input data as described in another conference paper, improved instrument slit function characterization and absolute calibration. The requirements for a system which will simultaneously provide accurate ozone monitoring and ultraviolet spectral irradiance monitoring are defined.

INTRODUCTION

The traditional method of determining the total ozone column utilizes the Dobson double monochromator to measure spectral irradiances at a certain array of wavelengths in the 305-335 nm region. Particular wavelength pairs are chosen for the purpose of cancelling the effects of aerosol scattering by the different techniques used in the analysis. Multi-wavelength methods of ozone column thickness determination have been suggested by Kutznetsov in Moscow (1975) and Green et al. (1973, 1979)in Florida. The latter multi-wavelength method is largely based upon recent advances in spectroscopic technology which permits the measurement of wavelengths shortward of those used by the Dobson instrument.

The work at the University of Florida evolved from some exploratory (experimental) work carried out under the CIAP program (1973) in which our primary assignment was to model the spectral irradiance at the ground for biological investigators. It became clear from a number of analytic and experimental studies (e.g. Green et al. 1974 (a); Shettle and Green, 1974; Green et al. 1974 (b); Green et al. 1974 (c); and Sutherland et al. 1975) that the ozone thickness provides a sharp determinant of both the direct and diffuse relative spectral irradiance at the ground. Our first experimental verification of this was made using a single monochromator with a solar blind photomultiplier and a UV-blazed ruled grating (Green et al. 1974 (c), with a simple heliostat arrangement to bring direct sunlight through an open window into a laboratory spectrographic arrangement. A more detailed verification was reported in 1975 (Sutherland et al. 1975) with the measurements of direct sunlight using an altazimuth mount on the rooftop of our Space Sciences Research Building.

The development of double holographic grating monochromators greatly advanced the possibility of total ozone monitoring by multi-wavelength measurements in the UV-B region. The idea of using a finite number of wavelengths in a least square mode to fix the ozone thickness was examined (Green, 1977) and mathematical simulations (Green, Riewe, 1978) suggested that a sharp determination of the total ozone column was possible. The first practical implementation of this least square method was reported by Garrison et al. (1979).

One problem with the multi-wavelength method of ozone determination has been the lack of reproducibility of wavelength drives of belt coupled double monochromators. To utilize short wavelengths imposes high wavelength accuracy and reproducibility and a belt-type wavelength drive such as used in many instruments does not lend itself to such requirements. In the ultraviolet double monochromator aboard the Nimbus 7 satellite (the solar backscatter experiment, Heath et al. 1975) wavelength reproducibility and grating coupling problems are overcome by mounting both gratings on a single axis. This device is also used in a new field instrument developed by Fastie et al. (1978) which is under study by the National Bureau of Standards for UV work (Kostkowski et al. 1980). Parallelogram linkages might also be used for such purposes when a twin axis double monochromator is used. However, without doubt the ultimate solution to this problem is to use fixed gratings and a multichannel spectrographic arrangement where wavelength calibration once fixed does not change.

The purpose of the present analysis is to address the question of the optimum least square method that can be used when wavelength reproducibility has been established. This could be by using a very precise scanning system with high wavelength reproducibility, e.g., the new NBS instrument, or by using fixed gratings and a multichannel detection system.

The analysis assumes that we are looking directly at the sun, presumably with a telescope and does not allow for the aureole effect. This could be introduced as a correction in an improved analysis.

Multi-wavelength analysis with sharp slit function

Given a spectrometer of infinite wavelength resolution, the direct irradiance at the ground as measured by an instrument pointed at the sun is given by

$$D(\lambda) = H(\lambda)R(\lambda) \exp - [k_i(\lambda)w_i/\mu_i],$$
$$\mu_i = [\mu^2 + t_i)/(1 + t_i)]^{\frac{1}{2}} \qquad (1)$$

where $d(\lambda)$ is in the instrumental units, $\mu = \cos\theta$, $H(\lambda)$ is the extraterrestrial solar spectral irradiance, $R(\lambda)$ is a response function of the measuring instrument, $k_i(\lambda)$ is the extinction coefficient, w_i the column thickness and t_i characteristic parameters for Rayleigh scattering, aerosol scattering and ozone absorption. Our multispectral method of total ozone determination is based upon the concept that relative radiometric measurements can be made with a good degree of precision at this time and that the relative input values at this point of time are fairly precisely known. The method is described in Garrison, Doda and Green (1979) (GDG). Essentially in GDG the slit function is assumed to be sharp. The extraterrestrial solar flux is represented as

$$H(\lambda) = H(\lambda_o)F_h(\lambda) \qquad (2)$$

where F_h is the relative solar spectral function appropriate to the instrumental band pass and the response function is in the form

$$R(\lambda) = R(\lambda_o)F_r(\lambda) \qquad (3)$$

where $F_r(\lambda)$ is the relative response function, R_o defines the absolute response and λ_o is a convenient reference wavelength. Then for each of a series of wavelengths λ_k we may write

$$D(\lambda_k) = H(\lambda_o)R(\lambda_o)F_h(\lambda_k)F_r(\lambda_k) [\exp - \Sigma\, w_i k_i(\lambda_k)/\mu_i] \tag{4}$$

If now we take the logarithm of both sides we obtain as shown in GDG an expression in the form suitable for a linear least square determination of w_3, the ozone column thickness, without the requirement of knowing the absolute values of $H(\lambda_o)$, $R(\lambda_o)$ or the aerosol column thickness w_2. We further assume $\tau_1 = w_1 k_1(\lambda)$ is fully known as a function of wavelength.

$$\tau_2 = w_2 k_2(\lambda) = \tau_{20} \exp -\alpha\ (\lambda-\lambda_o) \cong \tau_{20}[1-\alpha(\lambda-\lambda_o)] \tag{5}$$

$$\tau_3 = w_3 k_3(\lambda) = k_{30} f_3(\lambda) w_3 \tag{6}$$

where $k_{30}f_3(\lambda)$ is precisely known. It follows

$$\ln D(\lambda) - \ln F_h(\lambda) - \ln F_r(\lambda) + \tau_1(\lambda)\ /\mu_1 = K_o + \tau_{20}\alpha(\lambda-\lambda_o)/\mu_2 - k_{30} f_3(\lambda) w_3/\mu_3 \tag{7}$$

where

$$K_o = \ln R_o + \ln H_o - \tau_{20}/\mu_2 \tag{8}$$

Since the left hand side of Eq. (7) is known for each λ_k a least square fit may be performed to determine K_o, τ_{20} and w_3. If R_o and H_o are known τ_{20} and hence α may be separately determined. However, even if we do not have R_o and H_o we may determine w_3.

Multiwavelength analysis for finite slit function, $S(\lambda-\lambda_k)$

Let us now consider the effect of having a finite slit function which varies over a range in which the wavelength dependent extinction functions,

$k_i(\lambda)$ undergoes only small variation. In this

case the direct reading at any specific wavelength center λ_k as determined by the monochromator is given by

$$D(\lambda_k) = H_o R_o \int F_h(\lambda) F_r(\lambda) S(\lambda_k, \lambda)[\exp -\sum_i k_i w_i / \mu_i] d\lambda \tag{9}$$

Let us now assume that each of the attenuation functions may be expanded about the wavelength point, i.e.

$$k(\lambda) = k(\lambda_k) + s(\lambda_k)(\lambda - \lambda_k) + q(\lambda_k)(\lambda - \lambda_k)^2 \tag{10}$$

it follows that we may write

$$D_o(\lambda_k) = H_o R_o F_r(\lambda_k) [\exp - \Sigma w_i k_i (\lambda_k)/\mu_i] I \tag{11}$$

where

$$I = \int d\lambda S(\lambda_k, \lambda) F_h(\lambda) \exp - \{ \Sigma_i [w_i s_i (\lambda_k)/\mu_i] (\lambda - \lambda_k) + [w_i q_i (\lambda_k)/\mu_i](\lambda - \lambda_k)^2 \} \tag{12}$$

Let us now drop all quadratic terms and assume that terms in the exponential are small, so that we may expand the exponential

$$\exp - [\Sigma w_i s_i/\mu_i](\lambda - \lambda_k) = 1 - [\Sigma w_i s_i/\mu_i](\lambda - \lambda_k) \tag{13}$$

It follows that

$$I = I_o - I_1(\Sigma w_i s_i/\mu_i) \tag{14}$$

where

$$I_o = \int S(\lambda, \lambda_k) F_h(\lambda) d\lambda \tag{15}$$

and

$$I_1 = \int S(\lambda, \lambda_k) F_h(\lambda) (\lambda - \lambda_k) d\lambda \tag{16}$$

We may now reconstitute Eqs. 13 and 14 to the exponential form by writing

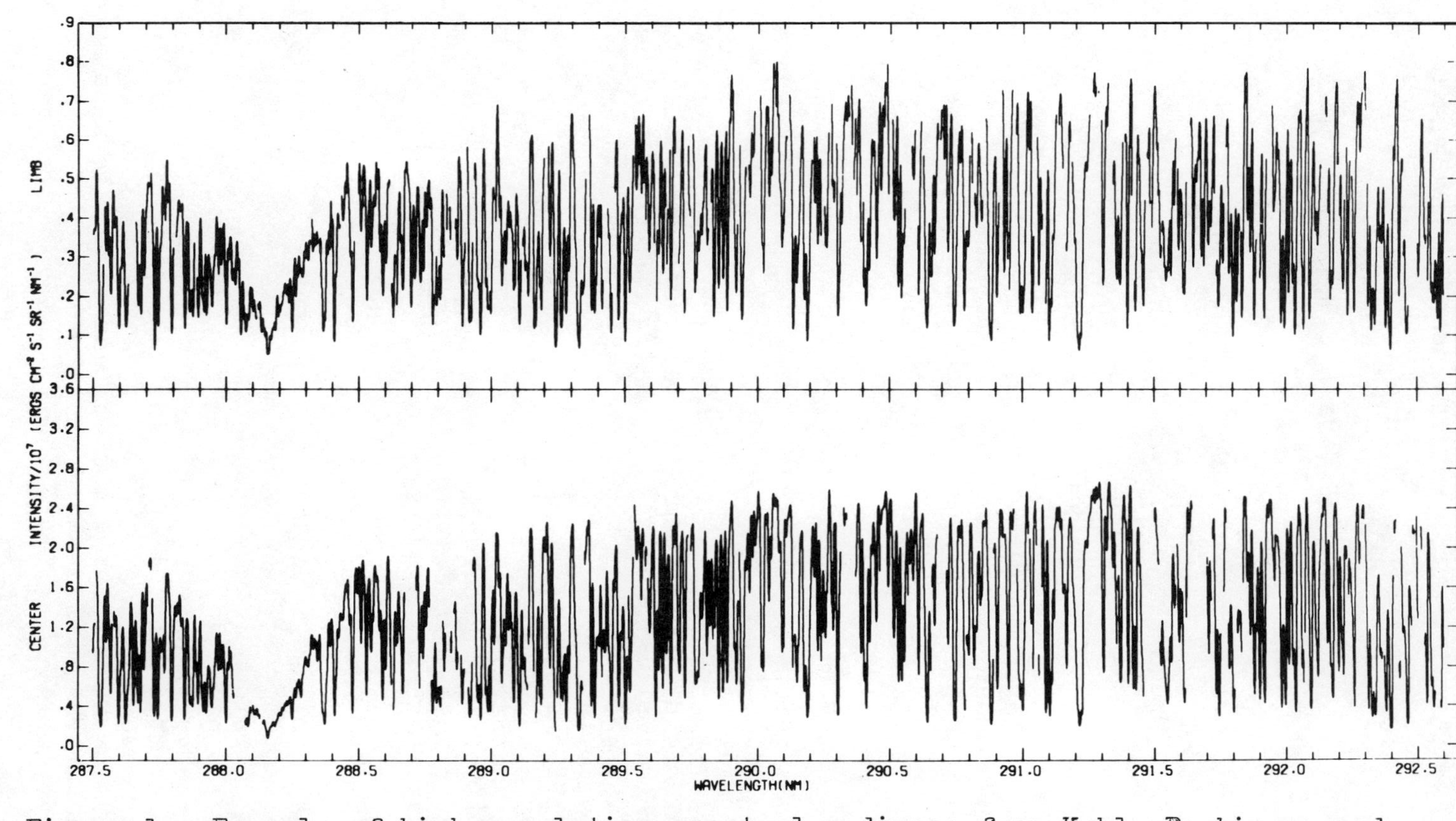

Figure 1. Example of high resolution spectral radiance from Kohl, Parkinson and Kurucz, Harvard Smithsonian Center for Astrophysics.

$$I = I_o \left[1 - \Sigma_i \frac{w_i s_i}{\mu_i} \frac{I_1}{I_o}\right]$$

$$= I_o \exp - \Sigma w_i s_i r/\mu_i \qquad (17)$$

where $r = I_1/I_o$. Using this in Eq. 11 we may now combine the two exponentials to arrive at a result identical in form with Eq. 4 except that we replace each attenuation coefficient by

$$\tilde{k}_i(\lambda_k) = k_i(\lambda_k) + r_k s_i(\lambda_k) \qquad (18)$$

Accordingly we may use the procedure described in GDG for a sharp slit function if after choosing a set of wavelengths we assign the corresponding set of pseudo-attenuation coefficients $\kappa_i(\lambda_k)$ using the slopes $s_i(\lambda_k)$ of the attenuation coefficients and the ratio $r(\lambda_k)$ obtained from the extraterrestrial solar irradiance for the chosen set of wavelengths.

The nature of $H(\lambda)$ is of important consequence to our method. Figure 1, a selection of central and limb solar radiances measured by Kohl, Parkinson and Kurucz (1978), illustrates the extraterrestrial solar spectrum at high resolution. Clearly the Fraunhofer structure is very complicated in the UV and one should be careful in one's choice of wavelengths lest one sit on the side of a rapidly varying envelope.

Fig. 2 illustrates recent irradiances derived by Heath and Park 1980 from continuous scan data using the SBUV experiment aboard the Nimbus 7 satellite. Even at their resolution (1 nm band pass) there are rapid fluctuations.

Returning to the question of slit function corrections, based upon our studies to date we estimate that if the sampling is done with less than 0.1 nm bandpass no corrections will be needed for multi-wavelength sampling below 310 nm. If the wavelength sampling below 310 nm. If the sampling is done with 1 nm band pass, corrections on k_3 are needed. If the sampling is done say at 5 nm bandpass, linear corrections are probably needed for k_1 and second order corrections may be needed for k_3. With the recent new physical inputs as characterized by Green and Schippnick (1980) and with the availability of new high and moderate resolution

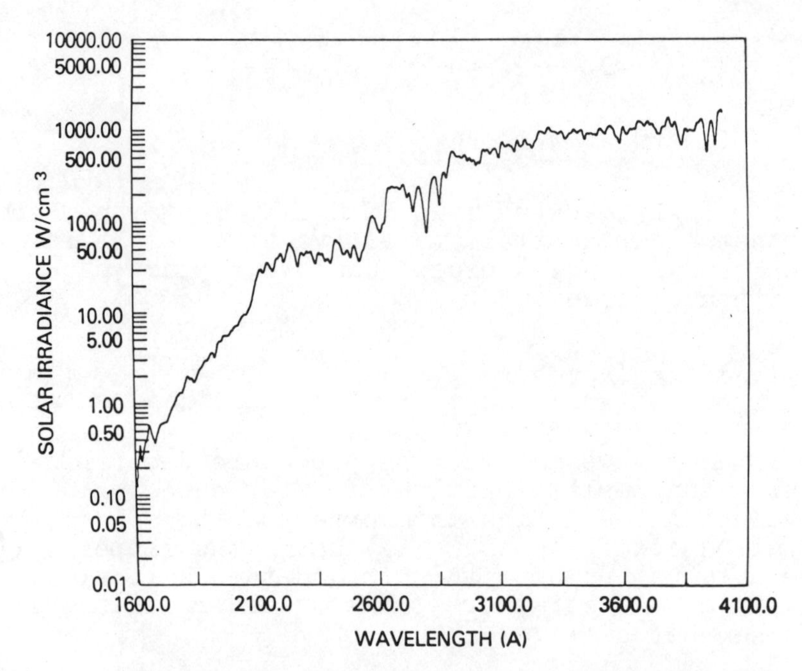

Figure 2. Extraterrestrial Solar Spectral Irradiance from continuous scan SBUV double monochromator on Nimbus 7 satellite from Heath and Park (private communication).

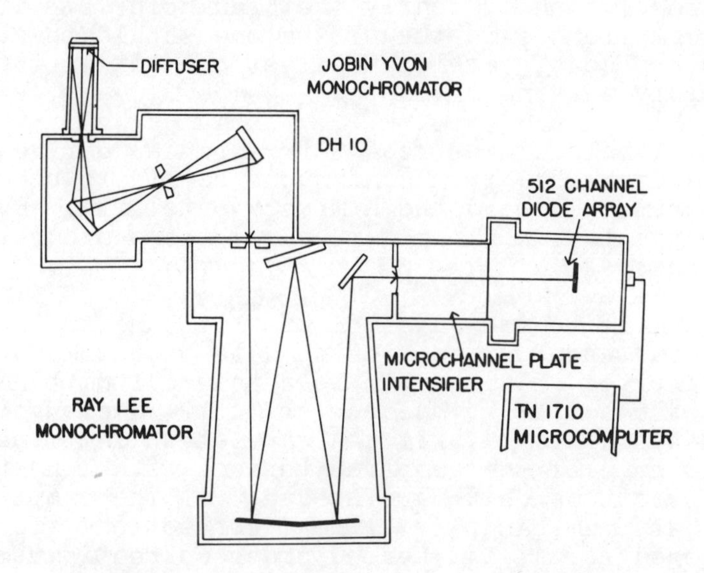

Figure 3. Schematic diagram of an experimental system for use in the determination of total ozone and ultraviolet irradiance.

solar spectra (Kohl et al. 1980, Heath et al. 1980) it should be possible to estimate the errors due to the neglected terms with precision.

DISCUSSION

It should be clear from the present analysis that the slit function must be known with considerable accuracy for multi-wavelength sounding below 310 nm. The solar spectrum must also be known at a higher resolution so that the values of I_0 and I_1 and the ratios $r(\lambda)$ can be computed.

In the past (GDG) we have used 1 nm band passes in our ozone soundings. We are currently investigating the magnitude of the correction terms appropriate to this band pass. In work underway in collaboration with the National Bureau of Standards Radiometry Group (Kostkowski et al. 1980) we have acquired ground based data with 1.8Å resolution with state of the art radiometry. To estimate corrections for this band pass we will utilize the Kohl, Parkinson and Kurucz (1978) spectral radiances. These are available to 319 nm for the center of the solar disc but only to 305 nm for a point on the limb. Nevertheless, since our correction method requires only ratios it is not unreasonable to use $r(\lambda)$ values calculated from central radiances.

We have recently initiated work using a variety of triple holographic grating monochromator arrangements and an intensified multichannel array detector which permits us to work with band passes below 1 Å. Fig. 3 illustrates a schematic sketch of such an arrangement. We plan to use the KPK central radiances to estimate our slit function corrections. One important problem in this work is to reduce the dynamic range of intensities in the 280-320 nm region which reach our 512 channel detector array. Thus far our most promising solution has been to use the long wavelength side of the slit function of our double monochromator predisperser to level the incident intensities. Various techniques are being explored which will permit both measurement of the global spectral irradiance for UV monitoring and direct spectral irradiance measurement for total ozone column measurement. Our previous work (Garrison et al. 1978) measuring diffuse-direct ratios provides an excellent point of departure.

In connection with UV monitoring it is interesting to note that a spectrometer slit function has many features of a biological action spectra. Indeed the convolution of a biological action spectrum with the solar spectral irradiance follows the same mathematical formula used in slit function analysis (see Eq. 9). While triangular slit functions are customarily used, differential Gaussian slit functions of the form

$$S(\lambda-\lambda_p) = \exp - [(\lambda-\lambda_p)/\delta]^2 \tag{19}$$

can also be used. A Gaussian response curve with $\lambda_p = 560$ nm and $\delta = 59.4$ nm provides a fairly good representation of the relative spectral response used to define photometric units (Green and Hedinger, 1978). Biological action spectra which define UV-B dose units may also be represented analytically (Green et al. 1974). Equation 19 could be used to define by simple specification of parameters λ_p and δ a dose unit in the UV-B region. It is only important that the fit to the data be good longward of the peak since the ozone attenuated radiation shortward of the peak contributes weakly. The two parameter differential Verhulst function

$$S(\lambda-\lambda_p) = \frac{4 \exp(\lambda - \lambda_p)/\delta}{[1 + \exp(\lambda - \lambda_p)/\delta]^2} \tag{20}$$

which falls off more slowly than the Gaussian could also be used for UV dose studies. For example the erythema action spectra is well represented if $\lambda_p = 297$ nm and $\delta = 3.2$ nm. If the biological action spectra is asymmetric about λ_p we could add $\gamma dS/d\lambda$ to Eq. 19 or 20 where γ is a third adjustable parameter. This device would not change the unit value at λ_p. To allow for solar-atmospheric effects one could follow the same treatment as given in the present study in going from Eq. 9 to Eq. 18.

A mathematical characterization of a biological action spectrum prescribes the dose unit and thus removes any ambiguity in the specification of biological dose. By avoiding a Tower of Babel the science of photobiology could be advanced significantly if agreement could be reached on the specification of a few UV-B dose units in this fashion.

ACKNOWLEDGEMENTS

This work was supported in part by a State of Florida Equipment Grant and in part by Contract EPA-R806373010 from the U. S. Environmental Protection Agency. The author would like to thank Dr. Henry J. Kostkowski for helpful discussions on the multi-wavelength method. The work of Ralph E. Foster in assembling our trial experimental arrangement shown in Fig. 3 is gratefully acknowledged.

REFERENCES

CIAP Monograph Series, Editor in Chief A. J. Grobecker, Department of Transportation, Climatic Impact Assessment Program, Washington, D.C., 20590. Sept. 1975, DOT-TST-75-51, Monographs 1-6.

Garrison, L.M., D.D. Doda and A.E.S. Green. 1979. Total ozone determination by spectroradiometry in the middle ultraviolet. App. Opt. 18(6):850-855.

Garrison, L.M., L.E. Murray, D.D. Doda and A.E.S.Green. 1978. Diffuse-direct ultraviolet ratios with a compact double monochromator. App. Opt. 17:827-835.

Green, A.E.S. 1977. A least square technique for ozone determination. Internal Report.

Green, A.E.S. and F.E. Riewe. 1978. An analysis of the least square techniques for ozone determination. Internal Report.

Green, A.E.S., K.R. Cross and L.A. Smith. 1980. Improved analytic characterization of ultraviolet skylight. Photochem. and Photobiol. 31:59-65.

Green,A.E.S. and T. Mo. 1974. Proceedings of the 3rd Conference on CIAP, DOT-TSC-OST-74-15, 518-522.

Green, A.E.S., T. Mo and J. H. Miller 1974 (b). A study of solar erythema radiation doses. Photochem. and Photobiol. 20:473.

Green, A.E.S. and R. A. Hedinger. 1978. Models relating ultraviolet light and non-melanoma skin cancer incidence. Photochem. and Photobiol. 28: 283

Green, A.E.S., T. Sawada and E. P. Shettle. 1974 (a). The middle ultraviolet reaching the ground. Photochem. and Photobiol. 19: 351.

Green, A.E.S., R. A. Sutherland and G. Ganguli. 1974 (c). A sun glint heliostat for atmospheric spectroscopy. Rev. Scientific Instruments, 45: 60-63.

Heath, D. F., A.J. Drueger, H. A. Roeder and B. D. Henderson. 1975. The solar backscatter ultraviolet and total ozone mapping spectrometer (SBUV/TOMS) for Nimbus G. Opt. Eng. 14: 323.

Heath, D.F. and H.W. Park, Amer. Geophys. Union Meeting, Toronto. 1980. Personal communication.

Kohl, J.L., W.H. Parkinson and R.L. Kuruca. 1978. Center and limb solar spectrum in high spectral resolution 225.2 to 319.6 nm. Harvard Smithsonian Center for Astrophysics, Cambridge.

Kostkowski, H.J., R.H. Saunders, C.M. Popenoe, J.F. Ward and A.E.S. Green, New state of the art in solar terrestrial spectroradiometry below 300 nm, Abstract Optical Soc. of America, Chicago 1980.

Kuznetsov, G.A. 1975. A multi-wavelength method and apparatus for the study of atmospheric ozone and aerosols. Atmospheric and Ocean Physics, 11:647-651.

Mo, T. and A.E.S. Green. 1974. A climatology of solar erythema dose. Photochem. and Photobiol. 20:438-496.

Shettle, E.P. and A.E.S. Green. 1974. Multiple scattering calculations of the middle ultraviolet reaching the ground. App. Opt. 13:1567-1581.

Sutherland, R.A., R.D. McPeters, G.B. Findley and A.E.S. Green. 1975. Sun photometry and spectral radiometry at wavelengths less than 360 nm. J. Atmos. Sci. 32:427-436.

RADIOMETRIC MEASUREMENTS IN THE UV-B REGION OF DAYLIGHT

Bernard Goldberg

Radiation Biology Laboratory
Smithsonian Institution
12441 Parklawn Drive
Rockville, Maryland 20852

ABSTRACT

The Smithsonian Radiation Biology Laboratory has developed a low level light-monitoring device which can be operated in the natural environment with little maintenance. Such a device has been functioning for over four years at Rockville and for about a year and a half at NASA-Langley in Virginia. The device at Rockville is an original analog version of the integrating digital unit used at Langley. These units are monitoring solar ultraviolet radiation from 280-325 nm in 5 nm bands, using interference filters in pairs. These filters are centered at 285, 290, 300, 305, 310, 315, and 320 nm.

THE DETECTION AND RECORDING SYSTEM

A general description of the low level monitoring device was published in 1974 (Goldberg and Klein, 1974). Numerous changes have been made since then, and a more precise and accurate instrument has evolved. The most recent version is a digital unit which can integrate over various periods of time and is complete with its own recording system. There are three basic sections: (1) the sensing unit, (2) the integrating unit, (3) the recording and control unit. A block diagram of the system is shown in Figure 1.

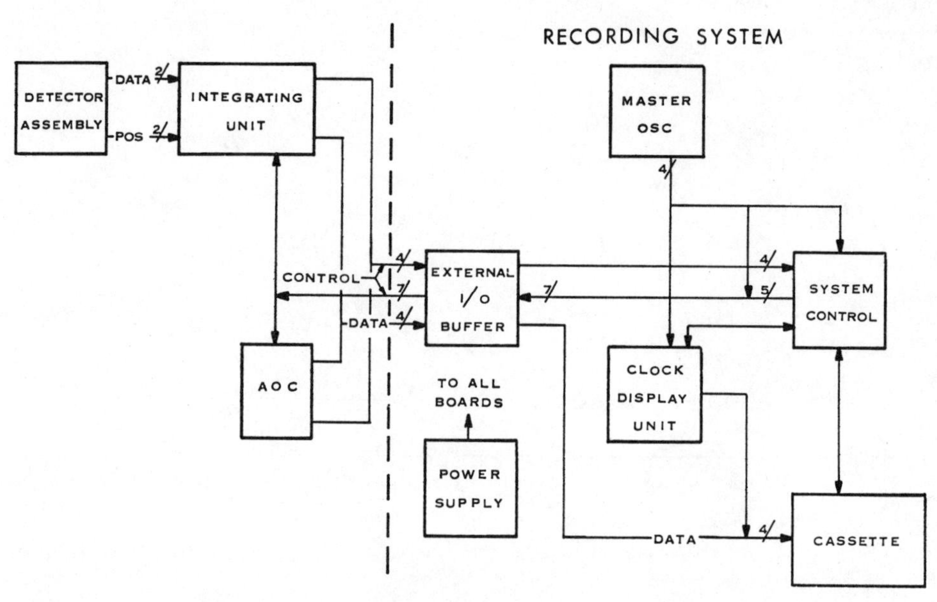

Figure 1. A block diagram of a low level digital scanning radiometer. This instrument can integrate energy over various intervals of time. Usually, the selected times are for one minute.

The detector assembly (sensing unit) consists of an optoelectronic detecting system and a digital positioning system. A diffuser is used at the top of the optical chain to obtain global irradiance values. The diffused light is collimated and passed through a pair of interference filters to obtain the desired spectral band. The interference filters are used in pairs to give 10^8 blocking outside the pass band. At the present time, the UV-B is being monitored in bands 5 nm wide. Figure 2 is a plot of the solar irradiance data obtained on a clear summer day. The detector to be used in each system is determined by the spectral bands being monitored. For the UV region, it is generally a 1P28 PMT that has been solar-blinded, using a nickel sulfate crystal. In the visible region, various detectors can be used. The units use only eight interference filter positions, which are sampled seventy-two times a minute. The sampling rate is usually controlled by a synchronous 72-RPM motor. The position of the filter wheel is determined by a magnetic sensor and a digital counting system. If the filter wheel is operated with a d.c. motor, some form of motor rotation control is needed to

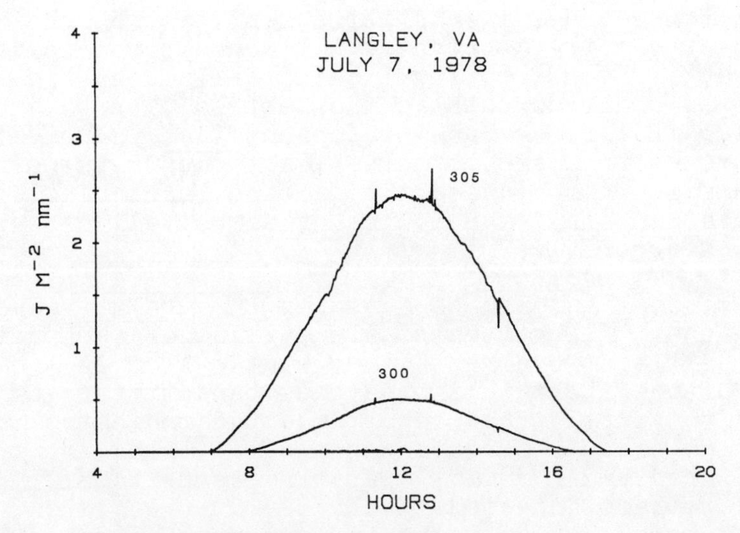

Figure 2. A graphic demonstration of the relationship between the various 5 nm bands on a clear day in Hampton VA (Langley-NASA).

maintain the 72 RPM rotation rate and the integrity of the clocking.

The sensor data and filter position information are fed from the detector assembly to the integrating unit in the form of 5-volt pulses. The data are stored in the integrating unit until a command is given to record the accumulated sensor data into the cassette recorder and restart the integration cycle. The record cycle timing is switch-selectable on the front panel of the instrument. The cassette recorded data utilizes the following format:

1. The last digit of the year.
2. The day of the year (1 to 366).
3. The time in hours, minutes, and seconds.
4. The filter position.
5. The integrated counts from the previous record cycle which can be directly related to the energy input to the sensor.

Instrument errors

There are inherent design defects in actual instruments as well as deficiencies arising from the present state of the art. The low level scanning device introduces errors because of limitations in the component parts of the device; the establishment of the magnitude of these errors is important. All monochromators admit some stray light, stray light being defined as the amount of light transmitted outside the desired pass-band of a particular filter. An unwanted count that is not attributable to solar energy, "noise", also contributes to the instrumental error. These two factors, noise and stray light, are, in fact, the source of almost all of our instrument errors.

The stray light can be measured in a relatively simple manner. Light that falls outside of the pass-band of a filter is directed into the system in the absence of the light which the pass-band filter is designed to admit. The signal detected indicates the stray light error. No detectable stray light error has been measured under normal operating conditions; however, we have detected levels of 10^{-10} watts cm^2 which is, normally, at least three orders below our lowest monitoring levels.

Noise in the signal is detected in a different manner. Because the unit is a digital system, we measure noise as counts, and these counts are sometimes difficult to relate to the true energy input. Noise levels are measured in various ways: (1) When there is no signal applied to the unit, noise is simply the net count arising from the various electronic parts; (2) When there is a precision signal simulating a known input. We perform this test using a highly stable current source with drift rates small enough to permit measurements better than ± 0.1% over long periods of time. Any variation in the counts is the measure of the noise error; (3) A count is made using the detector as the input source, but without a light signal.

Total system noise is generally on the order of ± 10 counts. This represents a very small energy level, about 10^{-11} joules m^{-2}. The best estimate, at this time, for instrument performance is an instrument with a precision of 1 part in 10^4, and accuracy which is limited only by the available standards.

CALIBRATION

The radiometers are calibrated against standard lamps traceable to NBS. These are the usual spectral irradiance standards, 1000 watt quartz halogen lamps. Because these lamps are not very stable in the UV-B region, we have found it necessary to check them frequently, using standard detectors. It has been found that a 1% current change will vary the irradiance of the lamps from 10% to 15%, depending on the wavelength used. Our power supply is stable to 1 part in 800 or 0.125%. The irradiance, then, should not vary by more than 2%. The greatest problem, we find, is the rapid loss of calibration of a lamp after about 25 hours of use. To combat this, we keep two or more standards available.

FIELD OBSERVATIONS

The UV-B monitoring units have collected data in Barrow, Alaska (71°N), Rockville Maryland (39°N), and in Panama (9°N). Examples of the information that has been collected were presented in New Delhi, India (Klein and Goldberg, 1978). It is felt that this instrument can not only monitor the UV-B, but can also be used to monitor changes in ozone. The use of radiometric measurements for ozone computation has been simplified so that computation time is greatly reduced. The simplified form is:

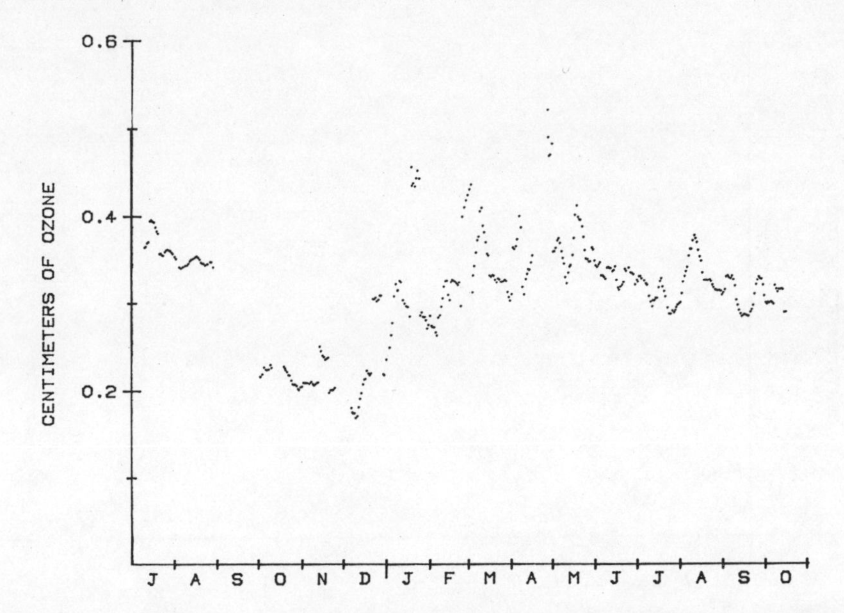

Figure 3. The ozone thickness as computed from the 305 nm and 310 nm bands at Hampton VA. Daily ozone from ratio of 305 nm to 310 nm 7 day running average Jul. 1978 - Oct. 1979.

$$\ln \frac{(G_1)}{(G_2)} \frac{(I_2)}{(I_1)} - (R_{s2} - R_{s1}) = x\,(a_2 - a) \tag{1}$$

where I = intensity outside the atmosphere
G = global irradiance in a 5 nm band as measured
M = air mass
R_s = Raleigh scattering
a = ozone absorption coefficient

Figure 3 shows the calculated values at Hampton, VA, and Figure 4 shows the Dobson measurements from Wallops Island, VA. (The calculated values have not been reduced to STP.) This method of computation is still tentative. The consistency of the calibration and detection are evident in Figure 5, representing UV-B as monitored at two sites about 180 miles apart, and having almost the same weather conditions and O_3 levels.

The radiometric data can be used to determine erythemal energy incident at the earth's surface.

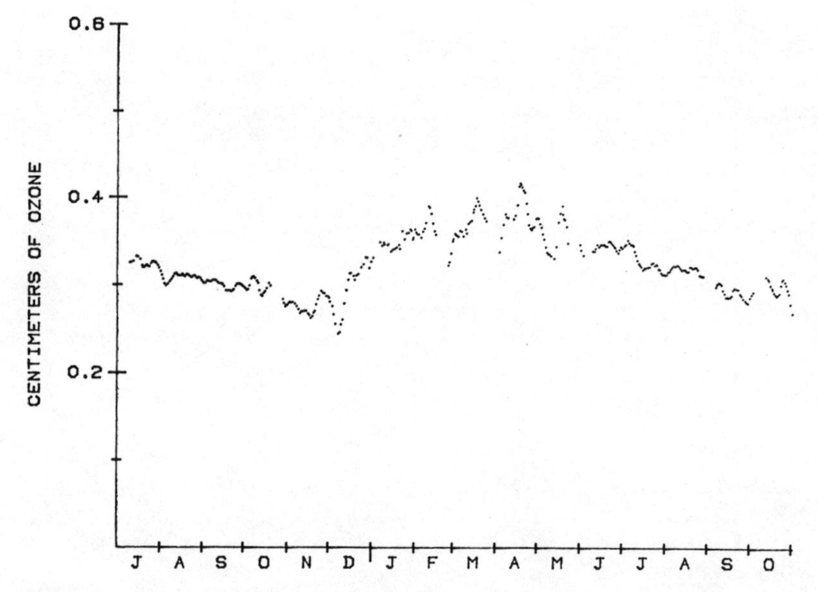

Figure 4. The ozone thickness as measured by a Dobson at Wallops Island VA. This site is located only a few miles from Hampton. Daily ozone from Dobson measurements 7 day running average Jul. 1978 - Oct. 1979

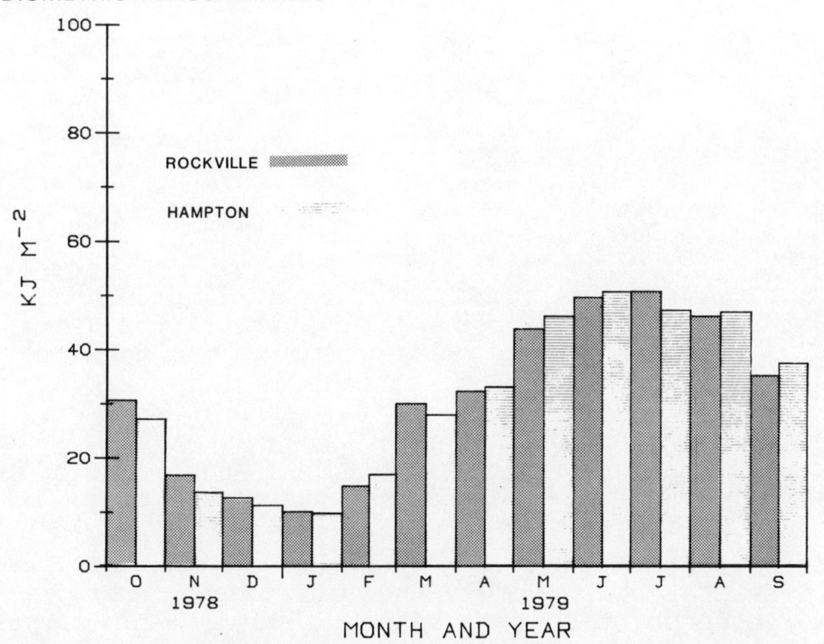

Figure 5. A comparison of the data collected at Rockville MD and Hampton VA. The two sites are located about 180 air miles apart.

Figure 6 shows graphically the average results of the comparison to the Robertson-Berger dosimeter made at Rockville. Calculations based on this comparison yielded an average value of 68 joules m^{-2}(weighted by the erythema action spectra) for 440 counts, the Robertson-Berger equivalent of 1 MED. Weighting the radiometric UV-B data to the spectral sensitivity of the Robertson-Berger dosimeter, the comparison is quite different; the variability in the counts/joule m^{-2} is greatly decreased (Figure 7. Using the data from the second set of comparisons, 440 counts are now equal to about 800 joules m^{-2}.

Because there are significant differences in proposed action spectra, our instrument is being used by our laboratory to radiometrically monitor the spectral quality of daylight wherever we are performing biological experiments in the natural environment. The use of this instrument allows us to collect data which can subsequently be weighted to any chosen action spectrum and also be used to determine other environmental factors, such as ozone variations, which help to characterize the environment.

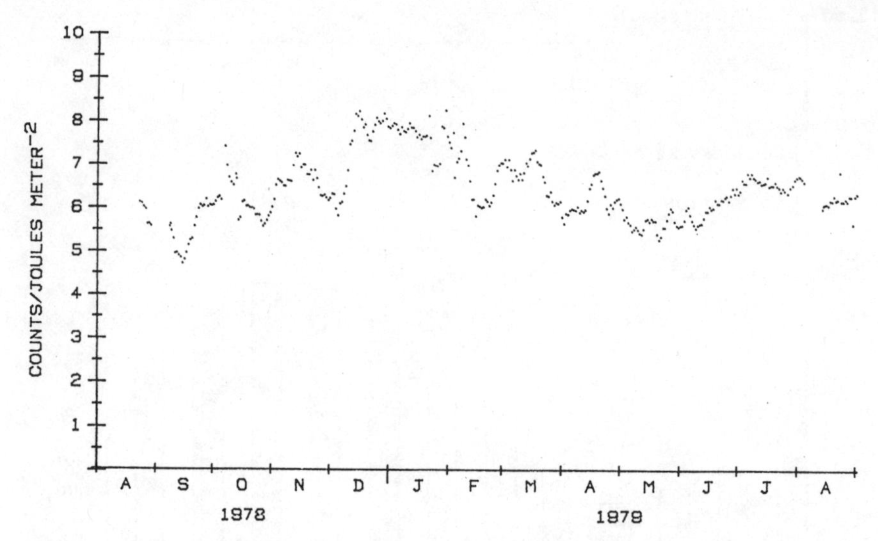

Figure 6. A comparison of radiometric data and dosimeter data at Rockville MD. The radiometric data have been weighted to effective erythemal energy, using the standard curve found in the CIAP Report 1975. Comparison UV dosimeter vs available erythemal energy 7 day running average for Aug. 1978-Aug. 1979.

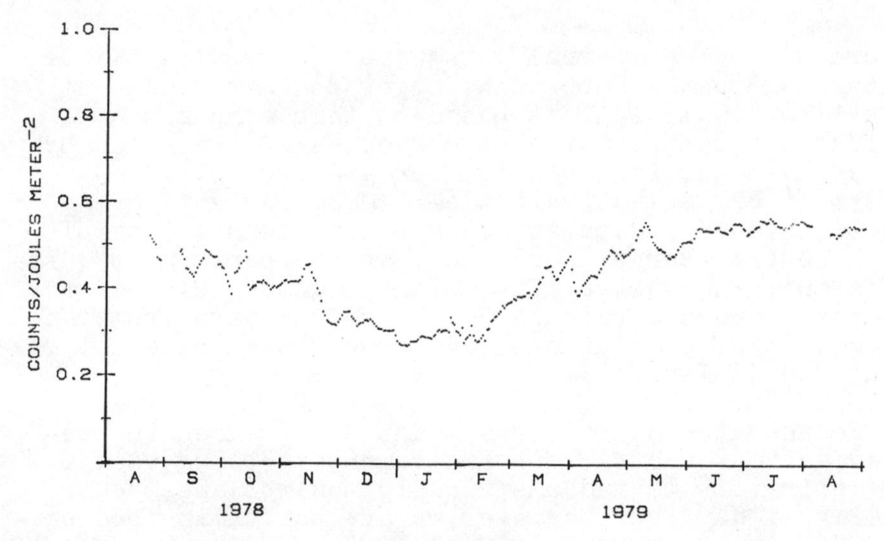

Figure 7. Using the sensitivity curve of the Robertson-Berger dosimeter, a comparison with the same radiometric data as Figure 6 is shown. The data used in this comparison are weighted to the dosimeter's spectral sensitivity.Comparison UV-B energy with dosimeter spectral response curve 7 day running average Aug. 1978-Aug.1979.

REFERENCES

Berger, D. S. 1976. The sunburning ultraviolet meter: design and performance. Photochem. Photobiol. 24: 587-593.

Goldberg, B. and W. H. Klein. 1974. Radiometer to monitor low levels of ultraviolet irradiance. Appl. Opt. 13: 493-496.

Green, A.E.S. and J. H. Miller. 1975. Impacts of Climatic Change on the Biosphere. CIAP Monogr. 5, Chapter 2: 2-61 and 2-70.

Klein, W. H. and B. Goldberg. 1978. Monitoring UV-B Spectral Irradiances at Three Latitudes. Proceedings of the International Solar Energy Society Congress, New Delhi, India, January 1978. Vol. 1. Pergamon Press, NY: 400-413.

PHOTOBIOLOGICAL DOSIMETRY OF ENVIRONMENTAL ULTRAVIOLET RADIATION

Claud S. Rupert

Programs in Biology
University of Texas at Dallas
Dallas, Texas 75080

The physics and photochemistry underlying light absorption and UV-induced alteration of macromolecules are understood well enough to establish clearly the basic dosimetric quantities which govern the biological effects of an irradiation. However, practical considerations, stemming both from the characteristics of the biological systems themselves (including their often inconvenient optical properties), and the lack of control over many parameters in natural situations, may prevent an investigator from dealing with environmental ultraviolet radiation in simple, direct terms. Questions about what quantities one ought to be trying to measure, and their relation to the process under investigation, can be answered only by experience, coordinated with deliberate inquiry. These answers will determine the measuring devices and dosage units appropriate for describing the biological effects of solar UV-B.

CREATION OF PHOTOCHEMICAL DAMAGE

The photochemical effect of a unidirectional beam of monochromatic radiation which falls on a nearly transparent sample containing photoreactive molecules depends on the number of quanta (photons) transmitted per unit area across a surface normal to the beam direction. If this number is P, the fraction of the photoreactive molecules raised to an excited state will be equal to $P.\sigma$, where σ -- the molecular cross section for photon absorption -- is proportional to the molar absorption coefficient as measured spectrophotometrically. If the quantum yield for the reaction is ϕ, the fraction of the molecules photochemically altered (when this fraction is small) is then

$$f = P \cdot [\sigma\phi] \tag{1}$$

The quantity inside the brackets depends only on the properties of the irradiated molecules, while P describes the light treatment they have received. Although photochemical and photobiological events are more easily described in photon -- rather than energy -- terms, this latter quantity is frequently expressed in terms of the energy F transmitted per unit area normal to the beam. In that case, the right hand side of Eq. 1 must be divided by the quantum energy to give the slightly more complicated Eq. 2:

$$f = F \cdot (\lambda/hc) \cdot [\sigma\phi] \tag{2}$$

(λ being the wavelength, h Planck's constant and c the velocity of light).

Where the radiation comes simultaneously from a number of different directions, instead of from only one, P (or F) is equal to the sum of its values for all the elementary constitutent beams. For a widely divergent cross-fire irradiation this is not equal to the number of photons (or to the energy) per unit area crossing any particular surface. Instead it can be described as the number of photons (or the energy) entering a small sphere surrounding the region of interests, divided by the sphere's cross sectional area. This quantity is completely analogous to the fluence defined for ionizing radiation (ICRU, 1971). Although objections have been raised to use of the term for light and ultraviolet radiation, we employ it here simply for lack of any other agreed-upon name, distinguishing between the photon fluence P and the energy fluence F where necessary. We may note that the energy fluence rate dF/dt and photon fluence rate dP/dt respectively are identical with the scalar irradiance and photon scalar irradiance respectively, as defined by the International Association of Physical Oceanography (see Smith and Tyler, 1976), and with the energy flosan and photon flosan suggested by the Joint Working Group on Radiation Quantities of four international organizations concerned with radiations (see Rupert and Latarjet, 1978).

BIOLOGICAL EFFECTS OF PHOTOCHEMICAL CHANGE

In biological systems we, of course, do not usually observe the actual fraction f of a molecular species which is photochemically altered, or the associated number of photoproduct molecule, but rather some genetic or physiological consequence of such change. Therefore, for many purposes, an exact knowledge of $[\sigma\phi]$ may be dispensed with, the factor being simply lumped in with others determining the overall responsiveness of the system to radiation of the wavelength in question.

The way in which photochemically altered molecules in an organism affect its functions depends on where they fit into its life processes. For a non-metabolizing system, like an extracellular virus exposed to UV-B, the reduction in its subsequent infectivity depends simply on the number of finally altered molecules (in that case, altered nucleotides in the viral nucleic acid), regardless of the rate at which these alterations have been produced. However, for effects on metabolizing cells, the involvement of both light-dependent and light-independent processes frequently makes the irradiation rate significant. Indeed, in an extreme case, like the visual system, the response is primarily determined by the rate of photoproduct generation. The rate dependence may also fall in between these two examples. Thus for some microorganisms the effects of UV damage inflicted rapidly (i.e., over a few minutes' time) is essentially independent of the exact rate, but when the same amount of damage is spread out over a number of hours it is much less effective (Harm, 1968) -- probably because of the success of DNA repair processes in dealing with injury inflicted slowly. In general, the exact way in which a biological effect depends on the total amount and rate of photochemical damage can be determined only by experiment. Until this behavior is known at least approximately, one does not know whether to describe an irradiation treatment in terms of the cumulative fluence, the fluence rate or some combination of the two.

SELF SHIELDING

Small structures, like viruses and microorganisms,

as well as the exterior cells of larger organisms, do not shield their sensitive structures significantly from the UV-B wavelengths present in daylight, but the interior cells of large structures can be wholly or partly protected, and the amount of any partial protection usually varies with wavelength. This changes somewhat the parameters on which the amount of photochemical damage may depend. In the skin of humans, for example, the living cells all lie below a dead, optically absorbing and scattering layer -- the corneum -- and the cells themselves lie at various depths. Near the base of the epidermis the value of P (or F) for daylight UV-B wavelengths is therefore much reduced from the value at the skin surface (see Blum, 1975), while the flow of the radiation is extensively redirected by multiple scattering -- somewhat like the sunlight reaching ground level on an overcast day. The absolute value of the photon (or energy) fluence in, say, the region of the basal cells then becomes very difficult to determine, but is approximately proportional to the number of photons (or energy) per unit area entering the outer skin surface. This latter quantity has long been defined as the dose D, of ultraviolet radiation (ICP, 1954.

It should be noted that the word "dose" has been used with quite different meanings in different contexts. The one given here is internationally sanctioned for ultraviolet radiation, but differs greatly from official of the same word for ionizing radiation (ICRU, 1971). For ultraviolet radiation, the quantity dD/dt is identical with the irradiance as defined for physical radiometry (CIE, 1970). (Note that this latter quantity is different from the scalar irradiance of the physical oceanographers referred to in Smith and Tyler, 1976). Some coordination of the terminologies of various groups would be helpful. Eq. 2 would therefore become

$$f = K(\lambda) \cdot (\lambda/hc) \cdot [\sigma\phi] = C(\lambda)D \qquad (3)$$

where the wavelength-dependent factors K and C usually have unknown values.

Lack of knowledge of K and C usually causes no particular problem because they simply become incorporated into the overall biological responsiveness to the applied dose. However, any variations of K from subject to subject, (with skin type, tanning, etc.) will change the relation between the radiation measured and the amount of photochemical damage inflicted, complicating comparisons between individual cases. The dose D delivered by a given beam of radiation changes with the cosine of its angle of incidence (angle with the normal to the skin surface) -- unlike the fluence, for which no such orientation factor occurs. Thus, as a practical matter, the damage to living cells lying deep enough to be partly shielded is described by a different dosimetric quantity (the UV dose) than the one describing damage to isolated or surface cells (the fluence). For effects on a leaf, or a larva of a marine animal, only experiment can determine which of these quantities is more relevant.

There is a tendancy to use UV dose and fluence interchangeably in some laboratory situations. Not only do dose and fluence have the same physical dimensions, but when a unidirectional beam falls normally on a surface it delivers a dose to it which is numerically equal to the fluence at the same location. However, these quantities are not the same with a widely divergent irradiation -- such as may be encountered in the environment -- nor with oblique irradiation of the surface. Basically, fluence concerns the radiation <u>field</u> in a locality, the fluence rate being a measure of the field intensity, and of the associated ability to produce photochemical change in molecules located there. The UV dose, on the other hand, is a practical description of the exterior treatment of an object, giving the amount of radiation per unit area entering some portion of its surface. This radiation, after partial absorption and scattering, can give rise to effects which depend on the fluence at sites of interior photochemical change. For a fixed interior geometry of the object, the dose can provide a relative measure of fluence at these sites.

POLYCHROMATIC DOSIMETRY

The quantities σ and ϕ change with wavelength, a fact which must be taken into account in dealing with the effects of broadband sources, like daylight. A moderate change of wavelength changes only the efficiency with which any single photochemical reaction is produced, without altering the nature of reaction itself, or its final product. Consequently, where only one single photoreaction is involved, one can simply add up the effect of each constituent wavelength to determine the total fraction of photosensitive molecules changed, obtaining in place of Eq. 1

$$f = \int \sigma(\lambda)\phi(\lambda)\ dP(\lambda) = \int \sigma(\lambda)\phi(\lambda)\ (dP/d\lambda)d\lambda. \quad (4)$$

The spectral distribution $(dP/d\lambda)$ is simply the photon fluence contributed per unit wavelength interval by radiation near the wavelength λ.

Eq. 4 can be put in the form of Eq. 1 by choosing a convenient reference wavelength λ_r at which $\sigma = \sigma_r$ and $\phi = \phi_r$, and expressing $[\sigma(\lambda)\phi(\lambda)]_r = \sigma_r\phi_r A(\lambda)$. In these terms,

$$f = \langle P \rangle [\sigma_r \phi_r] \quad (5)$$

where

$$\langle P \rangle = \int A(\lambda)\ (dP/d\lambda) d\lambda \quad (6)$$

is the photon fluence weighted by the relative photochemical effectiveness of each incident wavelength, as expressed by the action spectrum $A(\lambda)$. An analogous definition exists for $\langle F \rangle$.

There is no generally agreed-upon convention for picking λ_r -- the wavelength at which $A(\lambda)$ is taken = 1 -- and different people have been free to do this differently, even when working on the same system. This results in different values of $\langle P \rangle$ (or $\langle F \rangle$), matched with correspondingly different values of $[\sigma_r\phi_r]$; i.e. with a different magnitude of the responsiveness of the system to a given action weighted fluence. This points up a fundamental peculiarity of polychromatic dosimetry: neither the radiation treatment applied to the system, nor the system's own responsiveness can be fixed until $A(\lambda)$ is known, and the wavelength at which its value is set equal to unity has been decided. Determination of this weighting function in all particulars must therefore precede any polychromatic dosimetry. Another consequence of this same peculiarity is that, since the quantity expressing the radiation treatment depends on properties of the system being irradiated, as well as properties of the light source, its value will be different if exactly the same radiation treatment is applied to two different systems which have different action spectra.

Otherwise, everything said about single wavelength illumination has its counterpart in the polychromatic case. For effects which arise from damage to cells below the surface of a multicellular organism an action-weighted polychromatic dose $\langle D \rangle$, analogous to $\langle F \rangle$, can be defined. In such a case the wavelength dependence of the scattering and absorption factor K appearing in

Eq. 3 is hard to separate from the wavelength dependence of $A(\lambda)$. However, since $A \cdot K$ (or $A \cdot C$) is usually determined experimentally, this affects only interpretation, not utilization of the quantities.

One further complication of changing wavelength when irradiating a multi-cellular organism is the change in amount of shallow vs. deep damage as the wavelength changes. The longer wavelengths of UV-B are both less effective and more penetrating (i.e., correspond to smaller values of $\sigma\phi$ and larger values of K in Eq. 3) than the shorter wavelengths. Consequently longer wavelengths can produce more nearly equal damage to outer and inner cells than the shorter ones. (The latter, conversely, can damage outer cells more severely relative to those inside). This means that a change in wavelength may cause a qualitatively different pattern of injury to the organism as a whole, even if essentially the same photochemical process is taking place in all the irradiated cells at all the wavelengths. Although such a phenomenon, arising from the disposition of cells in the organism, may affect nature of the response, it is hard to incorporate into the definition of a dosimetric quantity.

We have assumed in developing Eqs. 4 and 5 that we are dealing with a single photochemical process, but there is no automatic assurance that this is so. Not only might more than one kind of photochemical damage produce damages giving similar observable effects, but the amount or effectiveness of product produced by one photochemical reaction could be altered by another. (Picture, for example, DNA damage created by solar UV being simultaneously repaired by photoreactivation, or the interaction between photosystems I and II in photosynthesis). The validity of the simple summation assumed in Eqs. 4 and 5 can (in principle) be checked by testing different wavelengths, or bands of wavelengths, singly and in combination, to see if they add directly in producing the observed effect, but this has rarely been done due to the labor involved. Nevertheless it is an important check. The concept of an action spectrum is clear only in the simple case where light of each constituent wavelength in an illumination band is doing more or less the same thing, but with different efficiencies. Where more than one photoprocess is taking place, they must be resolved and treated separately in analysis.

DAYLIGHT IRRADIATION

The spectral distribution function of daylight (i.e., $dP/d\lambda$, $dF/d\lambda$ or $dD/d\lambda$ and the action spectrum $A(\lambda)$ for UV-B damage to cells interact in a very inconvenient way for making estimates of radiation effectiveness. Both functions change rapidly with wavelength over the only region in which their product differs appreciably from zero, and each one has its highest value at the wavelengths where the value of the other is lowest. Consequently, to evaluate Eq. 6 satisfactorily more complete information about both functions is necessary than might at first appear. Furthermore, the UV-B spectral distribution in sunlight continually changes with the concentration of stratospheric ozone and the path length of light through it (higher ozone concentration and longer path length reducing the proportion of shorter, more effective wavelengths). Since ozone concentration normally changes with the latitude and seasons, as well as with the passage of weather systems, while the path length changes with solar altitude above the horizon, the important UV-B features of daylight change from hour to hour and from day to day. This considerably complicates the UV-B dosimetry of daylight illumination.

Spectroradiometric measurements to determine ($dP/d\lambda$) as a function of λ, combined with a knowledge of $A(\lambda)$, will, of course, permit $\langle P \rangle$ (or the analogous dosimetric quantities $\langle F \rangle$ or $\langle D \rangle$) to be determined as often as necessary. This is the ideal procedure in situations where the instrumental cost, size, speed and flexibility permit. On the other hand, there will be circumstances in which this is inconvenient-to-impossible, and one must rely on other procedures. One such method is the use of analog systems whose relative response to different wavelengths resembles the $A(\lambda)$ for particular biological effects. Two examples are described elsewhere in the present volume: (1) an analog photochemical system (Diffey this volume), and (2) a fluorescent system whose excitation spectrum has the desired form (see Berger, 1981). Measurements on radiation made with these analog devices, suffer from their only approximately correct action spectra, but they are enormously better than measurements involving no spectral weighting for effectiveness at all. Because only limited kinds of spectral distributions and distribution changes occur with daylight illumination it is possible that the dosimetric errors caused by the faulty action spectra of these devices could be partly corrected for. (Berger, this volume)

REMAINING QUESTIONS

1. Some very basic information is still needed about most daylight UV-B effects on organisms before the potentially well-controlled experiments of a laboratory can be applied efficiently to interpretation of natural situations. First, the effects of continuous and intermittent exposure at different rates encompassing the range of natural daylight intensities must be known in order to determine whether the fluence (or dose) alone is important, or whether some weighting of peak rates is also involved. Unless the action spectrum for the effect in question is known, such studies really require artificial light sources that reasonably simulate the spectral distribution and intensities of normal daylight (e.g., a xenon lamp with suitable filters). This can become expensive, except where it is possible to maintain and irradiate the organisms in a small, controlled space, and for many studies might require special purpose facilities.

2. Much more needs to be known about action spectra for the biological effects of UV-B. As has been indicated, these spectra are indispensible for dealing with solar UV damage, yet we are probably not even making best use of the information already available. Many different action spectra, determined with varying degrees of certainty, have been expressed in terms of different reference wavelengths (making them hard to compare), and are published in a wide variety of places. (See NAS, 1979, or Nachtwey and Rundle, 1980, for one small collection). Workers find little help in settling on a current "best" spectrum for practical dosimetric purposes. Some effort should therefore be made to organize and evaluate this important literature.

To this end a suitable study committee might be set up under the auspices of one or more appropriate scientific organizations to collect and evaluate the action spectra for daylight UV-B effects, express them in terms of some common reference wavelength and publish them in the form of a review in a readily available place. Such a review would probably require periodic updating. The purpose would be not to eliminate disagreements by edict, but rather to make clear the present state of knowledge, and bring into focus the reasons lying behind differences. Workers could choose whether or not to be guided by such "standards" but their existence would help determine when different conclusions have arisen solely from the use of different action spectra. As part of the critique of the action spectrum information, the various complications mentioned above (possible interactions of different wavelengths, or changes in the pattern of deep vs. shallow injury, etc.) could be dealt with, leading perhaps to improved understanding of, and better determinations of action spectra.

3. Finally, the appropriate dosimetric quantity for characterizing the daylight UV-B environment needs clarification. At present this is not as important as the other unresolved questions, because of the larger uncertainties they introduce, but until the matter is settled literature will accumulate, instruments will be built and thought patterns will solidify which finally may or may not be considered desirable.

The question is whether the general UV-B environment is best described in terms of the dose rate (or irradiance) on a horizontal surface, or in terms of the fluence rate (the scalar irradiance, or flosan) at the point of interest. The latter involves no reference surface, and thus no cosine factor. Most terrestrial work has described the UV environment in terms of the irradiance on a horizontal plane, but as noted by Smith and Tyler (1976) there is much to favor use of the fluence rate, which has apparently been adopted in some aquatic work. Our question does not concern which quantity is most reliably measured or how to measure it. It concerns which one most nearly characterizes the average UV-B effects on an organism inhabiting the locale.

Organisms small enough not to shield their internal structures will of course experience photochemical damage proportional to the fluence. In large organisms, which can provide self-shielding, absorption and scattering in the outer structures tend to make the fluence at inside points proportional to the dose at the outer surface. There are, however, many orientations of outer surfaces, all aimed in different directions relative to the sun and skylight, and these can change repeatedly as an animal moves about. The rationale for focussing only on a <u>horizontal</u> surface, and weighting everything with the cosine of the angle of incidence on it, is not entirely obvious. On land, such a procedure gives extra weighting to the radiation arriving vertically, relative to the oblique radiation, whose spectral distribution is more strongly affected by any ozone concentration changes.

The difference between the dose rate on a horizontal surface and the fluence rate for diffuse skylight on land is a factor of exactly two. The difference is less under smooth, clear water because the entire hemisphere of UV irradiation has then been compressed by refraction into a cone with a total vertex angle (2θ) of only about 95°. On the other hand, a rough water surface, and diffuse scattering within the water, or back scatter from a clean, sandy bottom all tend to restore the difference again.

We cannot dispose of the matter here, but it should not be ignored. The two dosimetric quantities are different, and can lead to some differences in predictions. One of them must be better than the other, but little attention has been given to deciding which.

REFERENCES

Blum, H. F., 1955. In Radiation Biology [Ed. by A. Hollaender] Vol. II, McGraw-Hill, N. Y.
CIE, 1970. International Lighting Vocabulary, 3rd Edn., Publication CIE No. 17 (E-1.1). International Commission on Illumination, and International Electrotechnical Commission, Bureau Central de la CIE, 4 Av. du Recteur Poincare, 75-Paris 16 France.
Harm, W., 1968. Photochem. and Photobiol. 7:73-86.
ICP, 1954. Proceedings of the First International Photobiology Congress (Amsterdam). H. Veenman en Zonen, Wageningen.
ICRU, 1971. Report 19: Radiation Quantities and Units. International Commission on Radiological Units and Measurements. Washington, D. C.
Nachtwey, D.S. and R.D. Rundel, 1980. In Man and Stratospheric Ozone [Ed. by F. Bower and R. Ward] CRC Press, West Palm Beach, FL. Chapter 10.
National Academy of Sciences, 1979. Protection against Depletion of Stratospheric Ozone by Chlorofluorocarbons. (Report of Committee on Impacts of Stratospheric Change and of Committee on Alternatives for the Reduction of Chlorofluorocarbon Emissions). National Academy of Sciences, Washington, D.C.
Smith, R.C. and J.E. Tyler, 1976. Photochemical and Photobiological Reviews 1:117-155.

ACTION SPECTRA

John Calkins (1) & (2) and Jeanne A. Barcelo (2)

Department of Radiation Medicine (1) and
School of Biological Sciences (2)
University of Kentucky, Lexington, Kentucky

An "action spectrum" is a determination of the relative response upon exposure to a series of monochromatic radiations sampling a given spectral range (For review see Loofbourow, 1948, Giese, 1968). Action spectra may be determined for many different purposes. By determining action spectra for the killing of microorganisms exposed to ultraviolet radiation, it was observed that the most efficient wavelengths were those corresponding to the maximum absorption of DNA. Through the close correspondence of the action spectrum and DNA absorption, the critical importance of DNA in living organisms was deduced by action spectroscopy before the biological role of DNA was established by non-optical means. It is reasonable to expect that the primary chromophore of many biological processes,induced by solar ultraviolet radiation,may be revealed by action spectra.

Aside from their use in basic biological science, action spectra are needed for establishing practical systems for dosimetry of solar ultraviolet (UV) radiations for ecological purposes and this function may be clarified by analogy with ionizing radiation. "Action spectra" of various ionizing radiations (termed "Relative Biological Effectiveness" (RBE) are basic to the proper expression of tolerance limits of environmental ionizing radiation exposures when the ionizing radiation is not "monochromatic". There has been extensive research on RBE with problems and complications arising in solar UV dosimetry. Two radiations producing exactly the same biological effect with differing efficiency will produce dose-response curves which can be superimposed if one selects the proper linear transformation of the dose scale (Eq. 1); the transformation coefficient (E) would express the relative efficiency of the two radiations.

$$D_1 = ED_2 \qquad (1)$$

where D_1 is the measure of the first radiation and D_2 the measure of the second radiation and E is a constant.

It is well known that response to many pairs of radiations cannot be made to superimpose by Eq. 1. For instance, the dose-response relations of mammalian cells to alpha particles and gamma rays are fundamentally different, alpha particles producing a simple exponential while gamma rays produce a shouldered response. Differing dose-response relations for two radiations implies a basic difference in biological response to the two radiations and the designation of the difference in biological effectiveness by a single numerical parameter will be impossible. The RBE for alpha killing of mammalian cells is high (often 5-15) relative to gamma rays if measured at 90% survival; the RBE will be much more modest (2 or 3), evaluated at low survival levels. Although the choice of a single number to express the RBE of various ionizing radiations contains arbitrary elements, it has been found necessary to select a single number RBE value to establish useful and practical criteria for human radiation safety in situations of mixed ionizing radiation exposure.

Solar UV radiation exposure shares the difficulties noted for ionizing radiations. Solar UV arrives at the water surface with variable amounts of biologically injurious radiations. The monochromatic components of solar UV demonstrate a much larger range of differences in biological effectiveness than ionizing radiations as well as showing different dose-response relations when wavelengths shift a few tens of nm. The spectral composition of solar ultraviolet radiation in nature is highly variable. It is essential to even the most elementary quantitation of solar effects that some parameter of exposure be generated which compensates for the different capacities of solar UV components to produce biological effects. In addition, one can envision complex interactions of different spectral components such that the biological effect of mixed radiations would be dependent upon subtle interactions of various components as expressed by Eq. 2.

$$dA_{B/dt} = \Sigma_{ij=\lambda_S}^{ij=\lambda_L} W_i I_i + F_{ij} I_i I_j + F'_{ij} I_i I_j \quad (2)$$

where $A_{B/dt}$ is the instantaneous biological action (dose rate), $I_i I_j$ are the irradiance of the spectral components ranging from the shortwave limit λ_S to the longwave λ_L. A primary weighting factor appropriate to the spectral component is designated by W_i, while the additional terms, F_{ij}, $I_i I_j$ ... are needed to represent interactions of components which might contribute to the total effect through synergistic effects not evident through simple addition of primary weighted components. The total biological action parameter A_B (dose) would be the time integral of Eq. 2 over the time of exposure.

There is reason to believe that the interactive components are significant if organisms are exposed to several monochromatic radiations of rather different wavelengths (Jagger, 1958, Mackay et al. 1976). The interactive terms lead to exposure parameters of great complexity since F_{ij}, F'_{ij} may be complex functions of I_i and I_j. The formal nature of F_{ij}, F'_{ij} has not yet been reported for even the most extensively studied organism and it is at present uncertain as to the necessity of including interactive terms for purposes of general ecological dosimetry of sunlight. In essence, action spectra determined using single monochromatic radiations establish only the primary (linear) weighting factor W_i.

Figure 1 shows a recent compilation of action spectra for the lethal effect of ultraviolet radiations from 254 to 365 nm (Calkins and Barcelo, 1979). The data points are essentially limited to wavelengths in the emission spectrum of mercury vapor, the source of radiation for most of the effectiveness determinations. Clearly, Figure 1 does not provide the resolution needed to suggest the chromophore responsible for UV-A lethality or similar purposes. If there were a detailed action spectrum for the actual species of organism under investigation, it would be much more useful for purposes of evaluation of the ecological impact of solar UV on this organism than the data plotted in Figure 1.

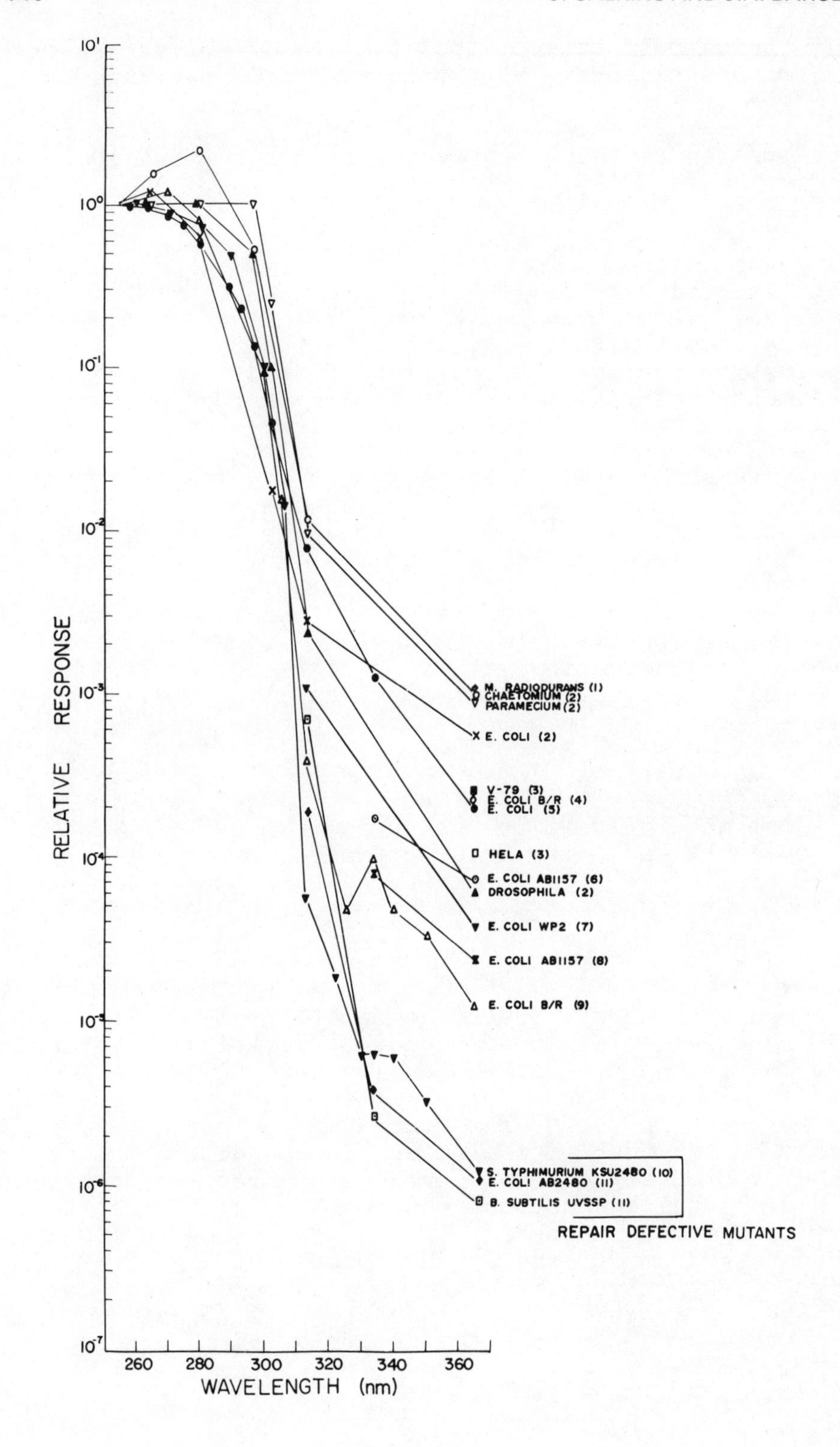

RELATIVE RESPONSE
WAVELENGTH (nm)
260
280
300
320
340
360
M. RADIODURANS (1)
CHAETOMIUM (2)
PARAMECIUM (2)
E. COLI (2)
V-79 (3)
E. COLI B/R (4)
E. COLI (5)
HELA (3)
E. COLI AB1157 (6)
DROSOPHILA (2)
E. COLI WP2 (7)
E. COLI AB1157 (8)
E. COLI B/R (9)
S. TYPHIMURIUM KSU2480 (10)
E. COLI AB2480 (11)
B. SUBTILIS UVSSP (11)
REPAIR DEFECTIVE MUTANTS

Nevertheless, if the repair defective organisms are excluded, the remainder of the species show responses with much the same trend. Considering the trend and variety of organisms plotted in Figure 1, it is reasonable to estimate that (pending more specific information) photons of wavelength 300 nm will be about 1000 times as biologically injurious as 365 nm photons.

It is important to note that A_B may be significant only within a limited wavelength range. In particular, it should be noted that solar irradiance falls rapidly below 300 nm, while biological effectiveness is very low at wavelengths above 360 nm (Fig. 1). Under present conditions the precise value and wavelength dependence of the weighting factor (W_i) for wavelengths less than

←

Figure 1. A compendium of action spectra for UV-induced lethality assembled from various published action spectra of diverse microorganisms and the trend (dotted line) of the response of "wild type" organisms. All spectra include observations of response to the two mercury lines 254 nm (far UV) and 365 nm (near UV). Results have been normalized to the response at 254 nm: where survival curves were available the 90%-lethal effects were utilized as the criterion of biological response; if survival data was not reported, action spectra were included as published. The indicated observations have been replotted from the following sources as noted by the numbers: (1) Webb et al., 1978; (2) McAulay and Taylor, 1939; (3) Danpure and Tyrrell, 1976; (4) Peak, 1970; (5) Luckeish, 1946; (6) Tyrrell, 1978; (7) Webb and Lorenz, 1970; (8) Tyrrell, 1976; (9) Webb and Brown, 1976; (10) Mackay et al. 1976; (11) Tyrrell, 1978. Action spectra of very sensitive repair-defective organisms are noted and are similar to each other while clearly different in relative response compared to the remaining organisms which are presumed to be wildtype in radiation response. Modified from Calkins and Barcelo, 1980).

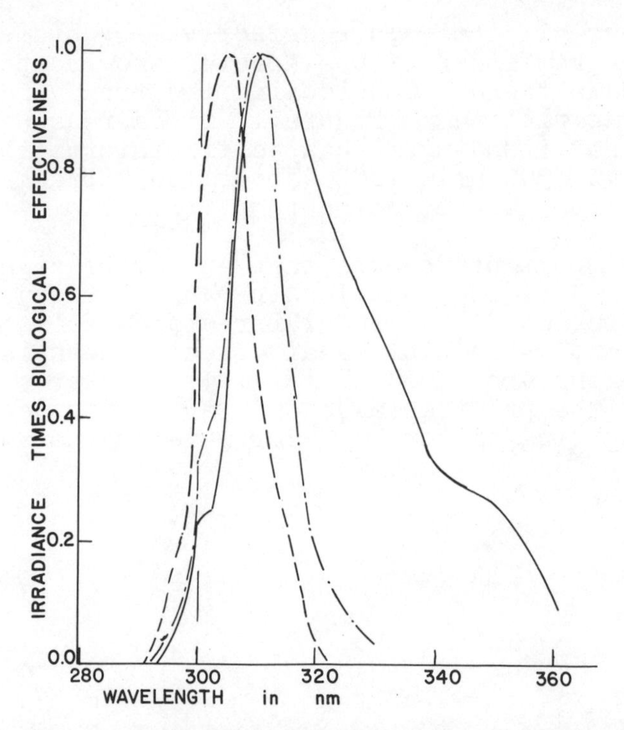

Figure 2. The wavelength dependence of biological action produced by temperate zone midsummer noontime sunlight. The global UV irradiance for O_3 = .32 cm and sun position 15° from vertical was deduced from Shettle and Green (1974) and Diffey (1977). Solar irradiances were multiplied by weighting factors (W_i) arising from three action spectra, i.e., (1) the Setlow (1974) "DNA" action spectra (---); (2) the fish larvae lethality action spectrum (Hunter et al., this volume (- . -); and (3) the Luckiesh (1946) action spectrum for killing Escherichia coli (solid line). All computed spectra were normalized to the wavelength of peak effectiveness and all indicate that the components of sunlight around 305-310 nm constitute the most injurious part of the solar spectrum. The total biological action is proportional to the time integral of the areas under the curves. Biological action curves such as these are frequently plotted with a logarithmic ordinate which may lead to an erroneous impression as to the contribution of various wavelength components to the net biological action of sunlight. Under the assumptions noted here (which maximize the short wavelength component of temperate zone sunlight), wavelengths less than 300 nm contribute less than $\frac{1}{4}$ of the biological action for responses of the "DNA" type and much less for the other two type action spectra.

300 nm may only have a slight effect on A_B. Figure 2 (computed using only W_iI_i, i.e. assuming $F_{ij} = 0$, a common assumption) illustrates the limited range of wavelengths which may contribute to A_B.

It is evident that if interactive parameters (F_{ij}) are essential to expressing solar UV dosage that exceedingly complex data and expensive processing equipment will be required in computation of biological dosimetry. The lack of a suitable and widely accepted dosimetry system is, even at present, a great hindrance to ecological research on solar UV. The elimination or discouragement of the use of a simple linear weighting factor derived from monochromatic action spectra as the basis of a practical dosimetry system (such as the Robertson sensor) should be rejected only upon the most compelling evidence.

ACKNOWLEDGEMENTS

This work was supported in part by the Office of Water Research and Technology, U. S. Department of Interior, under the provisions of Public Law 88-379.

REFERENCES

Calkins, J. and J. A. Barcelo. 1979. Some further considerations on the use of repair-defective organisms as biological dosimeters for broad band ultraviolet radiation sources. Photochem. and Photobiol. 30: 733-737.

Danpure, H. J. and R. M. Tyrrell. 1976. Oxygen-dependence of near-UV (365 nm) lethality and the interaction of near-UV and x-rays in two mammalian cell lines. Photochem. Photobiol. 23: 171-177.

Diffey, B. L. 1977. The calculation of the spectral distribution of natural ultraviolet radiation under clear day conditions. Phys. Med. Biol. 22: 309-316.

Giese, A. C. 1968. Ultraviolet action spectra in perspective: with special reference to mutation. Photochem. and Photobiol. 8: 527-546.

Jagger, J. 1958. Photoreactivation. Bacteriological Reviews 22: 99-142.

Loofbourow, J. R. 1948. The effect of ultraviolet radiation on cells. Growth Supplement to Vol. 12: 77-149.

Luckeish, M. 1946. Application of Germicidal Erythemal and Infrared Energy. Van Nostrand, New York.

McAulay, A. L. and M. C. Taylor. 1939. Lethal and quasi-lethal effects produced by monochromatic ultraviolet irradiation. J. Exp. Biol. 16: 474-482

Peak, M. J. 1970. Some observations on the lethal effects of near-ultraviolet light on Escherichia coli, compared with the lethal effects of far-ultraviolet light. Photochem. Photobiol. 12: 1-8.

Shettle, E. P. and A.E.S. Green. 1974. Multiple scattering calculation of the middle ultraviolet reaching the ground. App. Opt. 13: 1567-1581.

Tyrrell, R. M. 1976. Synergistic lethal action of ultraviolet-violet radiations and mild heat in Escherichia coli. Photochem. Photobiol. 24: 345-351.

Tyrrell, R. M. 1978. Solar dosimetry with repair deficient bacterial spores: action spectra, photoproduct measurements and comparison with other biological systems. Photochem. Photobiol. 27: 571-579.

Webb, R. B. and J. R. Lorenz. 1970. Oxygen dependence and repair of lethal effects of near-ultraviolet and visible light. Photochem. Photobiol. 12: 283-289.

Webb, R. B. and M. S. Brown. 1976. Sensitivity of strains of Escherichia coli differing in repair capability to far-UV, near-UV and visible radiations. Photochem. Photobiol. 24: 425-432.

Webb, R. B., M. S. Brown and R. M. Tyrrell. 1978. Synergism between 365- and 294- nm radiations for inactivation of Escherichia coli. Radiat. Res. 74: 298-311.

SOME THOUGHTS ON UV ACTION SPECTRA

Martyn M. Caldwell

Department of Range Science and the Ecology Center, Utah State University
Logan, Utah 84322

A very convincing case for the pivotal role of action spectra in assessing potential consequences of ozone reduction has been made (e.g., National Academy of Sciences 1979, Nachtwey and Rundel 1981, Caldwell 1981) and Smith and coworkers (Smith et al. 1980, Smith and Baker 1980) have applied this very appropriately in the case of photoinhibition of marine phytoplankton. Their data of short term photoinhibition of plankton under various combinations of polychromatic radiation are basically consistent with an action spectrum developed by Jones and Kok (1966) for the photoinhibition of isolated spinach chloroplasts. Since they have recently presented this information in the literature, as well as further consideration in this volume, it would be pointless for me to describe their experiments. Instead, I would like to offer a few comments on the importance of action spectra in assessment of the consequences of ozone reduction, the use of polychromatic radiation in determining action spectra, the nature of photoinhibition, and finally, a few comments on the limitations in the employment of many action spectra for ecological purposes.

Traditionally, biological action spectra have been determined by exposing organisms to monochromatic radiation at the same irradiance and scoring the biological response or by supplying the amount of monochromatic flux necessary to elicit a certain threshold response. In either case, the emphasis has been to determine the fine structure of the action spectra with as much spectral resolution as possible in order that potential chromophores might be identified. Seldom have the tails of these action spectra been elucidated, as usually they contribute little to identification of chromophores. Since solar spectral irradiance changes by orders of magnitude within the UV spectrum, tails of action spectra can be quite important even though they may represent greatly diminished biological effectiveness compared to the maxima.

ECOLOGICAL UTILIZATION

For assessing consequences of ozone reduction, action spectra serve as weighting functions in the calculation of radiation amplification factors, (sometimes referred to as optical amplification factors). Briefly, this is a unit change in the integrated UV radiation weighted with a particular weighting function (i.e., an action spectrum) which will result from a unit change in the atmospheric ozone layer at a particular location. (This has been fully explained most recently in the report of the National Academy of Sciences (1979) and in Smith and Baker (1980).) Most biological UV action spectra that involve UV-B radiation exhibit similar characteristics in that radiation at shorter wavelengths elicits a greater response than that at longer wavelengths, especially in the UV-A. Nevertheless the tails of action spectra in the UV-A are of considerable significance because solar UV spectral irradiance at the earth's surface increases by orders of magnitude from the UV-B to the UV-A. For the case of assessing responses in marine environments, further spectral changes of effective irradiance at different depths in water can occur because of the filtering action of the water itself. The diffuse attenuation coefficients for natural waters compiled by Smith and Baker (1979) from their own measurements and those of other workers suggest that the greatest spectral shifts in water occur in the UV-B portion of the spectrum--this is particularly the case for very productive waters. (The diffuse attenuation coefficient is an optical parameter that relates spectral irradiance just beneath the water surface to the downward spectral irradiance at some depth in the water.) As diffuse attenuation coefficients for various waters are refined and appropriate action spectra are identified, calculation of radiation amplification factors for depth profiles in the euphotic zone of marine systems will be feasible. This should constitute a key element in assessing consequences of ozone reduction.

Weighting functions derived from action spectra also play a key role in the comparison of effective UV-B radiation delivered by lamp systems with solar UV-B radiation. Since exact solar simulation is usually not feasible with lamp systems, it is only possible to compare the biologically effective UV-B from lamps with that from the sun under specified conditions. Depending on the action spectrum used for such calculations, appreciably different conclusions may result. This has been discussed in detail in recent papers (National Academy of Sciences 1979, Nachtwey and Rundel 1981).

USE OF POLYCHROMATIC RADIATION

As mentioned above, action spectra have traditionally been determined using monochromatic irradiance. It is also, however, possible to estimate action spectra using different combinations of polychromatic irradiation (e.g., Harm 1979). This approach allows less spectral resolution of action spectra yet it provides much more ecologically meaningful action spectra for use as weighting functions. Smith et al.(1980) have also used this approach in evaluating the applicability of the Jones and Kok (1966) action spectrum for photoinhibition. If only monochromatic irradiation is used in action spectra determinations, the assumption must be made that radiation at different wavelengths does not interact when presented simultaneously.

There is an apparent interaction of radiation at different wavelengths for some DNA-mediated UV-B damage as has been described recently (National Academy of Sciences 1979, Nachtwey and Rundel 1981). A similar interaction of radiation at different wavelengths may likely be involved with photosynthetic inhibition. For higher plants, UV-B radiation appears to have much greater depressive effects on photosynthesis in the presence of low visible irradiance compared to UV-B irradiation with intense visible flux (Sisson and Caldwell 1976, Teramura et al. 1980). The same may well apply to phytoplankton. Although it is unlikely that photoreactivation *per se* is effective in repairing photosynthetic damage, some mechanism is operating to mitigate the inhibition. Thus, such interactions that might alter the expression of UV-B damage in plants diminish the usefulness of biological action spectra determined with monochromatic radiation unless such interactions are fully understood.

PHOTOINHIBITION AND POTENTIAL PRODUCTIVITY

Inhibition of photosynthesis in phytoplankton or higher plants by solar UV-B radiation appears to be one of the most important avenues of potential damage that might ensue from atmospheric ozone reduction. Thus the action spectrum, and appropriate weighting function, for photoinhibition is of particular importance. Another complication apart from those already raised involves the degree to which photoinhibition is representative of lasting depressions in the primary production potential in both aquatic and terrestrial ecosystems. The manifestation of photoinhibition depends considerably on the type of assay employed and the time following the actinic irradiation. For example, in the original work of Jones and Kok (1966) they noted that if the photosynthetic assay (in this case dichlorophenolindophenol reduction of isolated chloroplasts) was conducted with weak as opposed to saturating visible irradiance very different characteristics emerged. If weak light was used,the evidence of photoinhibition was immediate and was independent of the intensity of the actinic irradiation. However, if the assay was conducted with saturating radiation, there was a characteristic lag in the photoinhibition and the degree of photoinhibition was lessened as the assay irradiation increased. The same observations were also made with algal cells (Kok 1956). We have recently noted similar phenomena with intact leaves of higher plants. Following the actinic radiation treatment, exposure of leaves to strong visible irradiance resulted in a lessening of the photoinhibition. It is thus imperative that short term phenomena be distinguished from permanent damage and that ecologically realistic conditions be used in laboratory assays, since it is the long term productive potential of a higher plant leaf or a population of phytoplankton that is of ultimate concern. With phytoplankton this would be rather complicated by population characteristics of phytoplankton. (Smith and Baker (1980) were careful to point out that immediate photoinhibition may not be related to long term declines in productivity.)

LIMITATIONS OF ACTION SPECTRUM

Action spectra determinations, whether with monochromatic or polychromatic radiation, are time-consuming undertakings, especially if the tails of the action spectra are pursued. It is then not surprising that relatively few action spectra for plant processes exist.

The generalized action spectrum for DNA-mediated damage developed by Setlow (1974) is an exception in that the tail of this action spectrum has been pursued far into the UV-A portion of the spectrum even though the effectiveness declines by orders of magnitude. (Calkins and Barcelo (1979) have also recently compiled action spectra for lethality which span several orders of magnitude.) The tails of most action spectra for nonlethal phenomena have not been pursued.

Most UV-B action spectra for plant processes do have a common tendency in that the biological responsiveness increases rather sharply with decreasing wavelength. The similar shape of these spectra made it tempting to devise a generalized plant UV action spectrum (Caldwell 1971). This generalized spectrum was based on a variety of plant responses for both lower and higher plants including induction of mutations and chromosomal aberrations, cessation of cytoplasmic streaming, etc. Most of the action spectra determinations did not include assessment of the small responses, i.e., the tails of these spectra that occur at wavelengths greater than 313.3 nm (which is a convenient emission wavelength of a mercury vapor lamp). An analytical equation by Green et al. (1974) designed to describe this generalized action spectrum truncated the spectrum at 313.3 nm which, of course, is biologically unreasonable. Yet, since data are not available to describe the tails of these spectra, it is not worthwhile at this point to pursue another analytic equation. Unless the effects of solar radiation wavelengths greater than 313 on these processes are truly negligible, this generalized spectrum is not appropriate for calculation of radiation amplification factors. The generalized spectrum may be still useful for some comparative purposes.

Certainly the need for appropriate action spectra determinations by photobiologists with an eye to the ecological utility of spectral weighting functions is becoming apparent.

REFERENCES

Caldwell, M. M. 1971. Solar UV irradiation and the growth and development of higher plants. In A. C. Giese, [ed.] Photophysiology, 131-177, Vol. 6 Academic Press. New York.

Caldwell, M. M. 1981. Plant response to solar ultraviolet radiation. In O.L. Lange, P. S. Nobel, C. B. Osmond and H. Ziegler [eds.] Encyclopedia of plant physiolog, Vol. 12A Interaction of plants with the physical environment, Springer-Verlag, Berlin, (In press).

Calkins, J. and J. A. Barcelo. 1979. Some further considerations on the use of repair-defective organisms as biological dosimeters for broadband ultraviolet radiation sources. Photochem. Photobiol. 30:733-738.

Green, A.E.S., T. Sawada and E.P. Shettle. 1974. The middle ultraviolet reaching the ground. Photochem. Photobiol. 19:251-259.

Harm, W. 1979. Relative effectiveness of the 300-320 nm spectral region of sunlight for the production of primary lethal damage in E. coli cells. Mutat. Res. 60:263-270.

Jones, L. W., and B. Kok. 1966. Photoinhibition of chloroplast reactions. I. Kinetics and action spectra. Plant Physiol. 41: 1037-1043.

Kok, B. 1956. On the inhibition of photosynthesis by intense light. Biochim. Biophys. Acta 21:234.

Nachtwey, D.S. and R.D. Rundel. 1981. Ozone change: biological effects. In F. Bower and R. Ward [Eds.] Man and stratospheric ozone. CRC Press, Inc., West Palm Beach, FL. (In Press).

National Academy of Sciences. 1979. Protection against depletion of stratospheric ozone by chlorofluorocarbons. Washington, D. C.

Setlow, R.B. 1974. The wavelengths in sunlight effective in producing skin cancer; a theoretical analysis. Proc. Nat. Acad. Sci. USA. 71(9):3363-3366.

Sisson, W.B. and M.M. Caldwell. 1976. Photosynthesis, dark respiration, and growth of Rumex patientia L. exposed to ultraviolet irradiance (288 to 315 nanometers) simulating a reduced atmospheric ozone column. Plant Physiol. 58:563-568.

Smith, R. C. and K. S. Baker. 1979. Penetration of UV-B and biologically effective dose-rates in natural waters. Photochem. Photobiol. 29:311-324.

Smith, R.C. and K. S. Baker. 1980. Stratospheric ozone, middle ultraviolet radiation, and carbon-14 measurements of marine productivity. Science 208:592-593.

Smith, R. C., K.S. Baker, O. Holm-Hansen and R. Olson. 1980. Photoinhibition of photosynthesis and middle ultraviolet radiation in natural waters. Photochem. Photobiol. 31: 585-592.

Teramura, A. H., R. H. Biggs and S.V. Kossuth. 1980. Effects of ultraviolet-B irradiances on soybean. II. Interaction between ultraviolet-B and photosynthetically active radiation on net photosynthesis, dark respiration, and transpiration. Plant Physiol. 65:483-488.

ACTION SPECTRA AND THEIR ROLE IN SOLAR UV-B STUDIES

A. Eisenstark

Division of Biological Sciences
University of Missouri
Columbia, Missouri 65211

Action spectrum studies have provided historic milestones in the identification of key chromophores for biological activity. Perhaps the best early example was the identification of photoreceptors responsible for photosynthetic reactions. Other milestones of special note were Gates'(1930) observation that the absorption maximum peak for DNA was the same as the maximum wavelength for killing bacteria and Stadler's (1942) discovery that the action spectrum for mutation in maize was the same as the absorption spectrum for DNA.

We have noted that action spectra can be important in understanding the mechanisms of DNA damage and repair when bacteria and their viruses are exposed to UV-B radiation.

The importance of the UV-B region was demonstrated by a careful study of action spectra for lethality in recombinationless strains of <u>Salmonella typhimurium</u> and <u>Escherichia coli</u> (Mackay et al. 1976). As may be noted from Figure 1, maximum sensitivities are observed at 260 nm, as expected, since this corresponds to the maximum absorption of DNA. However, a shoulder occurs in the 280-320 nm range that departs significantly from the absorption spectra of DNA. This departure informs us that 280-320 nm irradiation has a profound biological effect, which we are currently exploring. One of our findings is that the special sensitivity of recombination deficient (<u>recA</u>) mutants of bacteria to UV-B

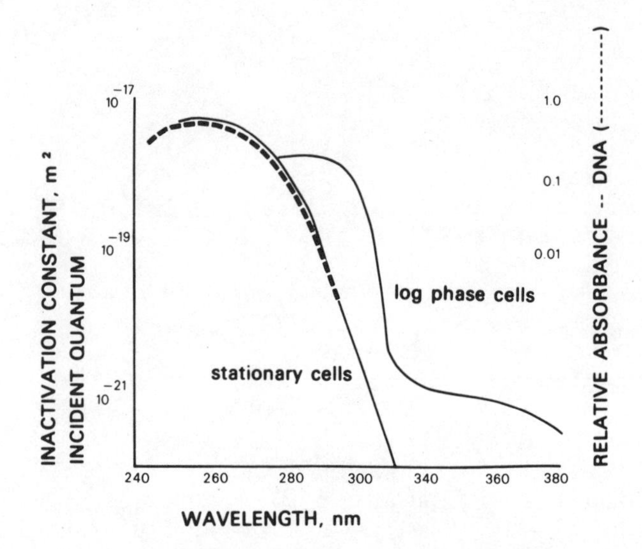

Figure 1. Action spectra for exponential and stationary phase cells of Salmonella typhimurium. When interference filters are used to obtain greater monochromicity, exponential curve becomes similar to that for stationary phase cells. However, in all cases, there is a shoulder between 320 and 380 nm. Redrawn from ref. 3. Note accompanying report on Bacterial Life Cycles (Eisenstark, this volume). Stationary cells are in a single (resting stage, and far less sensitive to UV-B; exponentially growing cells are mostly in "DNA replication" stage and far more sensitive to UV-B.

radiation may be due to synergistic effects of different wavelengths. In these studies, interference filters were used for monochromatic irradiation. The results were quite different than when the monochromater was used without these filters; without interference filters, a small amount of polychromicity contaminates the chosen irradiation wavelength. These observations of monochromatic inactivation of recombinationless mutants suggests that (a) there is a synergistic effect of multiple wavelength irradiation, and/or (b) a photosensitizer (now known to be hydrogen peroxide) might be involved in the high sensitivity of these strains to broad spectrum near-UV radiation. The role of peroxide as a sensitizing agent is discussed in Ananthaswamy et al., 1979 and Eisenstark et al. 1980.

Action spectra can be particularly useful in determining whether the sensitizing agent alters the chromophore, thus creating a new absorption peak. One particular example can be cited. Hydrogen peroxide in non-lethal doses enhances the killing of phage T7 by UV-B radiation (Ananthaswamy et al 1976). There is no synergistic action when irradiation takes place below 290 nm. Upon addition of hydrogen peroxide, there is a marked increase in the number of single-strand DNA breaks and in DNA-protein cross-linkages. The important point to be emphasized is that the sensitizing agent can indeed create a new chromophore, and action spectra can provide information as to the nature of this new chromophore (Eisenstark et al. 1980).

In summary, careful action spectra determinations can lead to identification of (a) photoreceptors, (b) altered chromophores, (c) synergistic activity of additional wavelength irradiation, and/or (d) other photochemical and photobiological properties of bacteria and their viruses.

REFERENCES

Ananthaswamy, H. M., P. S. Hartman, and A. Eisenstark. 1979. Synergistic lethality of phage T7 by near-UV radiation and hydrogen peroxide: an action spectrum. Photochem. Photobiol. 29: 53-56.

Eisenstark, A., M. Schrodt, S. Klita, and J. P. McCormick. DNA-protein cross-links produced by near-ultraviolet radiation (NUV). Proceedings of the VIIIth International Congress on Photobiology, July 20-25, 1980, Strasbourg, France.

Gates, F. L. 1930. A study of the bacteriocidal action of ultraviolet light. III. The absorption of ultraviolet light by bacteria. J. Gen. Physiol. 14: 31-42.

MacKay, D., A. Eisenstark, R. B. Webb, and M. S. Brown. 1976. Action spectra for lethality in recombinationless strains of Salmonella typhimurium and Escherichia coli. Photochem. Photobiol. 24: 337-343.

Stadler, L. D. and F. M. Uber. 1942. Genetic effects of ultraviolet radiation in maize. IV. Comparison of monochromatic radiations. Genetics 27: 84-118.

ACTION SPECTRA: EMPHASIS MAMMALIAN CELLS

Thomas P. Coohill

Biophysics Program
Western Kentucky University
Bowling Green, Kentucky 42101

The utility of action spectroscopy to study responses of biological cells and tissues is firmly established (Jagger, 1967). At a minimum, a carefully completed action spectrum can assess the amount of biological effect as a function of wavelength. While action spectra are of intrinsic value, providing experimental assessment of the consequences of exposure to electromagnetic radiation, in addition, regulations that set standards for radiation exposure must be based on such biological data. Beyond the need for action spectra in health and regulation lies a more fundamental question frequently addressed by workers in this field, i.e., what is the "target" molecule for each observed effect? The "target" is not always obvious from the details of the action spectrum observed. Although it is possible that the measured effect has the same wavelength dependence as the absorption spectrum for an identifiable, important cellular molecule (Gates, 1930) action spectra, especially in complex systems, often cannot be simply superimposed on the absorption spectrum of a known molecule (Todd et al., 1968).

I will limit my discussion to those action spectra obtained by irradiating mammalian cells in culture. The action spectra of simpler organisms (e.g. bacteria (Jagger 1967) are well known; action spectra of intact animals and human cells are more complicated and can be difficult to interpret (Freeman et al., 1970). Before presenting some of the limited data available for wavelengths above 300 nm, I will briefly mention those results obtained between 230 and 300 nm.

In 1965, E.H.Y. Chu reported an action spectrum for chromosome breakage and other chromosomal effects on Chinese hamster cells. In 1968, Todd et al. reported the first action spectrum for reproductive death in mammalian cells (also Chinese hamster cells). Both of these spectra exhibited a broad peak in the wavelength region 260 to 275 nm. These data caused an immediate dilemma in the choice of a target molecule for these effects. Nucleic acids (in particular DNA) absorb maximally at 260 nm. Proteins absorb maximally at 280 nm. It was thought that the broad peaks obtained for the first action spectra suggested that a combination of nucleic acid and proteins were most likely the target chromophores for cell death and chromosome aberrations (Todd et al., 1968).

Recently, however, Rothman and Setlow (1979) have re-determined the action spectrum for mammalian cell lethality and correlated it with the wavelength dependence of thymine dimer formation in the irradiated cells. Their spectrum for lethality closely resembled that of Todd et al. (1968) but also followed the production spectrum for dimer formation. This experiment seems to point to DNA as the major, or most important, chromophore for cell killing by ultraviolet radiation.

Several points should be kept in mind when attempting to correlate cell killing with DNA absorption in mammalian cells. First, the accuracy of the collected data may limit the confidence in the result. As an example, the dimer production spectrum of Rothman and Setlow (1979) reported standard deviations ranging from 6% (at 265 nm) to 45% (at 313 nm). Second, several action spectra involving mammalian cells show a more rapid decrease in value at wavelengths below the peak. (i.e. shorter than 260 nm) than does the absorption spectrum for DNA (Coohill et al., 1977); Jacobson, 1980; Todd et al., 1967). Whether this indicates the involvement of another molecule (or combination of molecules) at these shorter wavelengths is open to question. Third, the same group of action spectra reported above continue to decline in value at wavelengths below 250 nm. At 250 nm and below proteins begin to absorb heavily. Whether this argues against protein involvement is also undetermined at the present time. In summary, mammalian cells exhibit a broader peak of response upon far ultraviolet radiation (for several biological end points) than is found in bacteria. The data seem to be indicative of DNA, especially thymine dimer formation, as the major target molecule. A conclusive identification of the absorbing chromophore is complicated by several factors and may await further experimentation.

In the near UV (300-380 nm) less data has been collected. Monochromatic studies in the near UV are at present limited by the exposure rates available with current monochromatic systems (Coohill and James, 1979; Kantor et al., 1980). Some data with repair defective cell lines have been possible, for example: Kantor et al., (1980) have extended the spectrum for cell lethality out to 313 nm (for XP cells) and the correlation of lethality with pyrimidine dimer formation still holds.

Other data are available if photosensitizing chemicals are used. An example of the latter is the use of the photosensitizing drug 8-methoxypsoralen and other associated derivatives. At exposure levels of UV alone that did not effect cellular function, irradiating in the presence of 8-MOP caused a substantial effect and allowed an action spectrum to be completed for such cellular parameters as capacity for viral growth (Coohill and James, 1979), radiation enhanced reactivation (James and Coohill, 1979) and SV40 activation from transformed cells (Moore and Coohill, 1980). In general, the effect followed the absorption spectrum for the drug in the near UV region.

One way to conduct experiments in the near UV is to forego monochromatic sources and work with polychromatic lamps (e.g. "Sunlamps", "Blacklights", etc.). Data obtained with such sources is both interesting and limited, interesting in the fact that natural exposures (e.g. sunlight) are polychromatic and may give different effects than those predicted from monochromatic experiments (e.g. synergism) (Tyrrell, 1980; Elkind, et al., 1978), limited in that the target molecule is difficult to identify since a true action spectrum is not available by these methods. Both monochromatic and broadband types of radiation sources are needed for an accurate assessment of risk from environmental radiation.

For the immediate future it would appear that near UV studies may be limited to (1) monochromatic studies in the near UV with photosensitive cell lines or studying cellular functions that are especially sensitive to near UV (2) synergistic studies, various UV wavelengths, UV + heat, UV + photosensitizers (3) the availability of a powerful near UV source that is also monochromatic, e.g., a tunable dye laser that is continuous band and capable of producing adequate irradiance at wavelengths from 300 nm to 400 nm.

As public awareness of the possible effects of environmental radiation exposures on human, animal, and plant populations increases, further experimentation in the near UV region will become essential.

REFERENCES

Coohill, T. P., S. P. Moore and S. Drake. 1977. The wavelength dependence of ultraviolet inactivation of host capacity in a mammalian cell-virus system. Photochem. Photobiol. 26: 387-391.

Coohill, T. P. and L. James. 1979. The wavelength dependence of 8-methoxypsoralen photosensitization of host capacity inactivation in a mammalian cell-virus system. Photochem. Photobiol. 30: 243-248.

Chu, E. H. Y. 1965. Effects of ultraviolet radiation on mammalian cells. I. Induction of chromosome aberrations. Mutat. Res. 2: 75-94.

Elkind, M. M., A. Han, and C. Liu. 1978. Sunlight-induced mammalian cell killing: A comparative study of ultraviolet and near ultraviolet inactivation. Photochem. Photobiol. 27: 709-715.

Freeman, R. G., H. T. Hudson, and K. Carnes. 1970. Ultraviolet wavelength factors in solar radiation and skin cancer. Int. J. Dermatol. 9: 232-235.

Gates, F. L. 1930. A study of the bacteriocidal action of ultraviolet light. I. The reaction of monochromatic radiations. II. The effect of various environmental factors and conditions. J. Gen. Physiol. 14: 31-42.

Jacobson, E. 1980. Personal communication.

Jagger, J. 1967. Introduction to Research in Ultraviolet Photobiology, Prentice-Hall, Englewood Cliffs, N. J.

James, L., and T. P. Coohill. 1979. The wavelength dependence of 8-methoxypsoralen photosensitization of radiation enhanced reactivation in a mammalian cell-virus system. Mutat. Res. 62: 407-415.

Kantor, G., J. Sutherland and R. Setlow. 1980. Action spectra for killing non-dividing normal human and xeroderma pigmentosum cells. Photochem. Photobiol. 31: 459-464.

Moore, S. P. and T. P. Coohill. 1980. UV radiation activation of latent tumor viruses from mammalian cells: A wavelength dependence and enhanced activation with 8-methoxypsoralen. 2nd symposium of Light Effects in Biological Systems. HEW-FDA-BRH-DBE. Washington, D. C.

Rothman, R. and R. B. Setlow. 1979. An action spectrum for cell killing and pyrimidine dimer formation in Chinese hamster V-79 cells. Photochem. Photobiol. 29: 57-61.

Todd, P., T. P. Coohill and J. A. Mahoney. 1968. Responses of cultured chinese hamster cells to ultraviolet light of different wavelengths. Radiat. Res. 35: 390-400.

Tyrrell, R. 1980. 8th Int. Cong. of Photobiol., Strausbourg, France.

DOSAGE UNITS FOR BIOLOGICALLY EFFECTIVE UV-B: A RECOMMENDATION

Martyn M. Caldwell

Department of Range Science and the Ecology Center, Utah State University
Logan, Utah 84322

In the earlier panel discussion on action spectra, a case has been made for the pivotal role played by action spectra when used as weighting factors in the calculation of biologically effective UV-B radiation. This is of import not only for the determination of radiation amplification factors but also for relating the dosage received from lamp systems to that received from the sun. Because both biological photochemical reactions and the spectral irradiance from the sun change so rapidly as a function of wavelength, all would agree that total irradiance, or fluence rate, is of little value in determining UV-B dosage. Instead, it is preferable to speak of "weighted" or "effective" irradiance, either at a single wavelength, or integrated over a particular waveband. A familiar analogue is the use of a "standard eye" action spectrum as a weighting function to calculate luminous energy. When integrated with respect to wavelength, this constitutes illuminance, normally expressed in units of lux. Although special units have been contrived in the case of illuminance, confusion in the literature could be greatly minimized for other types of biologically effective radiation if one speaks of "weighted" or "effective" irradiance, and employs SI units. Thus, the integrated weighted irradiance can be expressed as effective Wm^{-2} or photons m^{-2} s^{-1}. Different unitless weighting functions can be substituted as the occasion warrants as long as these are clearly specified.

Special ultraviolet units based on standardized weighting functions such as E-viton or G-viton coined several decades ago have long since fallen into disuse (Luckiesh 1946). New attempts to define "standard" UV units (such as standard "sun" or "solar" units) will likely meet the same fate and in the meantime only serve to confuse readers unfamiliar with UV photobiology.

Once a particular weighting function has been chosen, it is, of course, helpful to relate the effective radiation provided by lamps to solar UV radiation for a particular location and time. For this purpose I suggest the use of the Green et al. (1980) model which appears to provide a reasonable representation of global UV spectral irradiance and can be relatively easily computed. Although the model has not been widely validated it appears to be a very reasonable approximation (e.g., Baker et al. 1980, Caldwell et al. 1980). Use of the model would also allow better intercomparison between laboratories than actual spectral irradiance measurements. The model also would facilitate calculation of daily doses which in most cases are more meaningful than irradiance at any one point in time since global irradiance is so dependent upon solar angle.

REFERENCES

Baker, K.S., R.C. Smith, and A.E.S. Green. 1980. Middle ultraviolet radiation reaching the ocean surface. Photochem. Photobiol. 32: 367-374.

Caldwell, M.M., R. Robberecht, and W. D. Billings. 1980. A steep latitudinal gradient of solar ultraviolet-B radiation in the arctic-alpine life zone. Ecology 61: 600-611.

Green, A.E.S., K.R. Cross, and L.A. Smith. 1980. Improved characterization of ultraviolet skylight. Photochem. Photobiol. 31:59-66.

Luckiesh, M. 1946. Applications of germicidal, erythemal, and infrared energy. D. Van Nostrand, New York.

MEASURING DEVICES AND DOSAGE UNITS

John Calkins

Department of Radiation Medicine (1) and
School of Biological Sciences, (2)
University of Kentucky, Lexington, Kentucky

RATIONALE FOR A BIOLOGICAL DOSE UNIT

The choice of a system of units is ultimately an arbitrary agreement for the facilitation of communication among different investigators. It is evident that units appropriate for the measurement of biologically injurious exposure to solar UV should be available and comprehensible to workers in the field. Any decision regarding units requires very careful consideration; systems of units must be practical and useful as well as scientifically correct to receive general acceptance and avoid frequent changes of unit. Units for specialties may arise from the larger framework of units, i.e., the SI system. In particular, it would be most desirable if photobiological exposures could be expressed in the SI system as joules/m^2; however, this is impossible and a unit incorporating biological factors is required.

High-sun sunlight is quite injurious to a wide variety of common organisms (Calkins, 1975). If wavelengths shorter than 360 nm are removed from solar radiation injurious actions on common organisms essentially disappear. The relative importance of the sunlight components of wavelengths less than 360 nm has been established by inference from responses using optical filters to remove components of incident radiation or by exposing organisms to monochromatic ultraviolet radiations; both methods agree that the relative effectiveness of a photon of the short-wave components of solar UV (wavelengths less than 300 nm) is at least

500 times greater than the effectiveness of a 360 nm photon. Because solar irradiance falls rapidly below 300 nm due to ozone absorption and photons of wavelength greater than 360 nm are of low potency, the injurious action of solar UV primarily arises from wavelengths from 300-360 nm. (Note Calkins and Barcelo, this volume.)

Comprehension of solar UV effects and the establishment of tolerable limits to manmade ozone depletion are presently hindered for lack of an accepted unit suitable for expressing the exposure (dosage) of solar UV incident upon ecosystems or their component organisms. The expression of exposure in physical units for laboratory experiments presents technically difficult but conceptually tractable problems. Laboratory radiation sources producing relevant monochromatic or bands of UV may be measured in absolute physical units of irradiance (joules/m^2, etc.). With a precisely defined spectral output from the source and appropriate information about the measiring device, the irradiance of all wavelengths components incident on the subject can be deduced.

Deducing exposure (expressed in physical units) of aquatic organisms or ecosystems to real sunlight is far more difficult. Spectral composition varies with solar elevation (functions of time of day, season and latitude) in a regular way. In addition, there are large irregular variations introduced by atmospheric ozone variability and especially by local weather. Selective attenuation of various wavelength components upon penetration into natural waters further compounds the problems of aquatic dosimetry.

Aside from difficulties of measurement, the problem with reliance on purely physical dosimetry is that it leads to unacceptable levels of ambiguity in the exposure parameter, i.e., 10^{-5} joules/m^2 of early morning (low-sun) sunlight can be entirely without injurious biological effect on a given organism, while the same number of joules/m^2 of high-sun exposure may produce extensive lethality.

Thus, even assuming data on incident UV radiation of adequate spectral and temporal resolution, the most difficult problem remains, how are physical dose parameters to be converted into biologically meaningful dose parameters? i.e., how to properly "weight" the spectral components of dose. Obviously, one cannot simply add the components of physical dose; exposures to low activity wavelengths cannot be directly added to exposures at wavelengths where each photon is two or three orders of magnitude more effective. It must be clearly understood that biological weighting is not a scheme to avoid or simplify physical dosimetry; wavelength weighted response is an observable response of living organisms, a fact of nature, which a meaningful dosimetry system must somehow simulate or otherwise accommodate in a suitable way.

The essence of a biological dosimetry system is a transformation function for physical units such that equal exposures expressed in the chosen dosimetric units produce equal biological responses regardless of the spectral composition of the incident radiation. Two major limitations of biologically adjusted dosimetry become immediately evident; 1) different solar UV wavelength bands produce qualitatively different effects; 2) species vary in their relative sensitivity to the various components of solar UV. Recognizing that a practical biological dosimetry system must of necessity be less than perfect, then there should be a search for points of general consensus as to which attributes should be sacrificed and which retained. For instance, if it is insisted that the transformation function must fit each species response very precisely, then special dose units for almost every species may be required. If a more general transformation function is selected, it will fit the responses of some species better than it fits others and it must also be expected that there will be variability in response to the same nominal dose when the wavelength distribution changes radically.

THE PRIMARY FUNCTION OF A BIOLOGICAL DOSAGE UNIT

While many basic aspects of the biological effects of solar UV may be established from laboratory studies with dosages expressed in physical units, the larger significance of solar UV for life on earth can only be appreciated by quantitative studies which can be freely translated from laboratory to natural environments, including a variety of locations and times. A dosimetry system is essential to such quantitation. The more adequate the dosimetry system, the more accurately can interpretation of solar UV effects be made; however, to fail to reach a consensus on suitable techniques and units of dosimetry, even though they are less than perfect, will greatly inhibit research on this most ubiquitous environmental agent.

CRITERIA FOR EVALUATING THE MERIT OF POTENTIAL DOSIMETRY SYSTEMS

A number of systems for dosimetry have been proposed. As a first step toward reaching a consensus as to the merit of various systems, some factors which can be evaluated are proposed below.

Functional ability of the system

Using radiation of a variety of spectral composition (including high-sun and low-sun spectral distributions) test organisms should be exposed to measured doses. The more nearly the same nominal dose produces the same biological response independent of the spectrum of the radiation source, the better the dosimetry system.

Biological generality of the system

Systems should be tested for functional ability with diverse organisms. A system could have high functional ability for one or a few species and thus be an excellent dosimeter for that species but respond poorly when tested with other organisms.

Availability of the system

In principle, radiation exposures (irradiances) should be recorded with high resolution of wavelengths and temporal fluctuations. The recorded physical data could then be subjected to computer analysis with transformation factors especially designed for each biological component. Not only would this system be the most adequate of any now envisioned but would be capable of improving the transformation factors by comparison of predicted and observed responses. However, to suggest that such a system (of obvious merit) be the criterion of adequate dosimetry would do a great disservice to environmental research. The basic outlines of the ecological actions of solar UV will doubtless be elucidated long before such an ideal system for dosimetry becomes generally available. There are, or will be developed, methods of dosimetry which, although less than ideal are still adequate for biological and ecological purposes without requiring inordinate technical and financial resources.

Number of dosimetry systems

The proliferation of units for many narrow specialties is generally undesirable, the more disciplines a dosimetry system can adequately serve the more meritorious it would be, but environmental scientists should also consider the possibility that the most reasonable way to accommodate the needs of aquatic biology, terresterial plant science, and skin cancer research might be through three distinctly different, non-interconvertible dosimetry units.

Number of parameters required to express dose

One possible way to improve the ability of measured "dose" to delineate biological response is to express dose using multiple parameters, for instance, energy above 300 nm and energy below 300 nm, etc. The simplicity and clarity of expressing dose as a single parameter needs to be balanced against the possibility that biological effects might be more directly related to events requiring expression of two or more separate dose factors.

THE TRANSFORMATION FUNCTION

The essence of biological dosimetry is the operator, TF, which converts physical dose into biological dose; it is presumed that TF is a function of wavelength. Two approaches have been used to establish TF (λ) 1) deducing TF (λ) from general biological principals, 2) fitting TF (λ) to the available data relating biological response to exposure at different wavelengths. It is unfortunate that consideration of proposed dosimetry systems have focused on the methods used to generate TF (λ) rather than how well the proposed systems satisfy various criteria of quality as measuring devices. The methods used to generate TF (λ) are irrelevant so long as the ultimate product, the measuring system, functions as desired.

The primary current approach to biological dosimetry is the concept of simple additivity of the various wavelength components of exposure but incorporating a multiplicative constant appropriate for each wavelength (weighting factor), an approach proposed by Luckiesh and Holladay (1931), Caldwell (1971), Setlow (1974), Billen and Green (1975), Damkaer (1980), and many others. The concept is basic to the Robertson-Berger Meter Robertson, 1969), (Berger, 1976). The selection of

TF (λ) for the various proposals has largely derived from observed action spectra.

There have been suggestions that simple additivity is an inadequate basis for TF (λ) and that different wavelength components of solar UV interact, thus requiring a more complex operator to properly express biological dosage. Interactions of long and short wavelength components of solar UV are well known. Photoreactivation requires radiation of wavelengths from 330-450 nm to reverse damage induced by short wavelength solar UV (Harm, 1969). Also, interactions through destruction of repair systems by long wavelength solar UV needed to cope with shortwave induced injury have been reported (Webb, 1977). The precise mathematical formulation of TF (λ) required to express such interactions is not clear.

CONSIDERATION OF ACTUAL OR PROPOSED DOSIMETRY SYSTEMS

The Robertson-Berger (R-B) Meter

At present, the most highly developed biological dosimetry system is based on the Robertson-Berger (R-B) Meter. There is an extensive network of these meters (Berger, this volume) and data from the network has been published (Scotto et al.,1976) or is available from NOAA. The Robertson-Berger system was designed to simulate the human erythema action spectrum and its primary purpose is for analysis of production of human skin cancer. Robertson-Berger meters can be purchased for a reasonable price. There is a growing body of data based on this metering system, including biological response, attenuation of UV-B radiation in natural waters and incident radiation at many locations.

The major point of criticism of the R-B system for general biological dosimetry has focused on the inappropriate wavelength response of the sensor; it should be especially noted that this criticism is largely based on comparison with DNA absorption spectra and not systematic study of the meter for purposes of evaluation of its performance with actual sunlight. A subtle point regarding this meter and all others which present biologically weighted dose as a single parameter, is that there is no possibility of re-evaluating TF (λ) at any subsequent time. It is inherently impossible to convert R-B data on exposure into any other system, although estimates of conversion factors for certain standard conditions can be made (Damkaer and Dey, this volume.

The test for the functional adequacy of a biological dosimetry system for polychromatic radiations is the constancy of response to a measured dose when organisms are exposed to radiation sources of variable spectral composition. The Robertson sensor is a dose meter which utilizes a particular linear weighting function (W_i) and makes no allowance for interactive terms (Berger 1976), Machta et al., 1975). The Robertson sensor has been used to measure dosage for lethality experiments using 10 small organisms (Calkins, et al. 1978). Two sources of UV radiation were used, natural sunlight and a "simulator" of solar UV-B which is quite rich in the shortwave component of UV (equivalent to > 50% ozone depletion) compared to real sunlight. Responses of the same organism to the two sources when dose was measured by the Robertson sensor agreed within a factor of about 2 (Table 1). It should be noted that the deviations of response for equal measured dose were not random but tended to show greater sensitivity for the same measured dose of real sunlight. It thus appears that if the Robertson sensor could be made to weight the longwave components somewhat more heavily, then an even closer agreement of given dose for equal response to the two sources might be obtained. The trend of the action spectra plotted in Calkins and Barcelo (1980) also suggests the Robertson sensor underweights the longwave components (the "tails") of solar UV.

In summary, the R-B system seems to function well in providing a quantitative measure of solar UV exposure of organisms, it is reasonably consistent for differing wavelength distributions, it is readily available, it expresses dose as a single parameter which is desirable for general use, but makes it impossible to interconvert data based on R-B readings to other systems. It would seem reasonable that dosage based on the R-B system be accepted as one valid way to express solar UV exposure.

Calculated Weighted Dose

Caldwell (1971), Billen and Green (1975) and Green and Hedinger (1978) have noted that a dose parameter can be generated by multiplying spectroradiometric (or calculated incident solar UV spectral distribution) by a weighting factor determined from standard action spectra. This approach to specifying exposure has been

TABLE 1

Relative Sensitivity to Equal Exposures As Measured By The RB Meter*

ORGANISM	LONGWAVE SOURCE (REAL SUNLIGHT) SHORTWAVE SOURCE (Simulated Solar UV)
Staphylococcus aureus	1.2
Neisieria gonorrhoeae	1.6
Flavobacterium	1.4
Pseudomonas	1.5
Tetrahymena (Kentucky Strain)	2.2
Tetrahymena pyriformis	2.2
Nematode	2.6
Cyclops	1.2
Cypris	1.23
Aedes aegypti	1.0

*NOTE: Although the RB meter weights the longwave tail more heavily than most other weighting systems used for lethality, this table suggests that the RB meter still *underweights* the tail (longer wavelengths) present in natural sunlight.

used to some extent. It is clearly a very versatile solution to the problems of solar UV dosimetry and should doubtless be an approved way of specifying biological dose.

The problems with calculated dosimetry arise on the practical side rather than from the basic principles involved. The spectroradiometer recording system and its continual operation represent a large investment of resources; accurate calculation of incident radiation is a formidable procedure (Green et al. 1980). Ideally, the weighting factor should be determined for the end point and organism under study, which represents another major project in itself. If standard action spectra are used to generate weighting functions, then the dosimetry will be limited by the suitability of action spectra adopted as standard. Dosimetry by calculation can accommodate more complex TF (λ) if such functions become known. A serious problem arises from the nomenclature involved proposed for this system. It was proposed that units be termed joules/m^{-2} weighted or effective (Caldwell, this volume). It is clear that these units have been interpreted as joules/m^2, failing to recognize there is a biological factor of large uncertainty incorporated into the physical unit. Since there have been so many variants of the weighting schemes utilized, it is very important that full details of the dosimetry be stated, including spectral distribution of weighting coefficients, source of unweighted irradiance data, reference wavelength for weighting, etc. Users should be aware that change of weighting coefficients from previously used values introduces a completely new system of units.

Dosimetry systems based on DNA absorption

It is widely recognized that the biological action of ultraviolet radiations arises to a major extent from lesions in DNA. Some dosimetry proposals have regarded the absorption characteristics of DNA as a scheme to generate TF (λ) (Setlow, 1974) (Tyrrell, 1978). As noted previously, the source of TF (λ) is immaterial and values of TF (λ) deduced from "the DNA absorption spectrum" may produce an entirely satisfactory dosimetry system. However, there are certain flaws in the logic leading to a DNA-absorption based dosimetry system which potential users should recognize and thus realize that "DNA-absorption-based dosimetry" could suggest a very poor set of values of TF (λ).

If TF (λ) is deduced from the absorption spectrum of DNA, it is implicitly assumed that all relevant lesions are direct photochemical lesions in DNA. There is abundant evidence that sunlight injury may arise indirectly through photosensitizing chromophores. If such indirect actions play a part in the UV injury of an organism, then the DNA absorption spectrum may be quite inappropriate for definition of TF (λ). Secondly, it must be considered that DNA is not a chemically defined substance. One can only speak of *the* absorption spectrum of even a defined substance because there are conventions as to concentration (molar, etc.). With a defined substance, the relative absorption of two different wavelengths varies with absorber concentration (See Loofbourrow, 1948). There is no *a priori* reason to believe that the relative absorption of different UV wavelengths in the DNA of a virus or a bacterium will be identical to the DNA absorption spectrum in other organisms.

Biological dosimeters

Harm, 1969, Billen and Green, 1975, and Tyrrell, 1978 have proposed the use of radiation sensitive bacteria as biological dosimeters for solar UV. This form of biological dosimetry would seem to offer advantages for investigators with limited financial resources to invest in the dosimetry system. Biological dosimetry appears to be manpower intensive and to involve considerable delay between administration of an exposure and the determination of the magnitude of the exposure. Biological dosimetry systems may be quite useful and should be an accepted form, but their full characteristics should be documented and the validity of such dosimetry should not be accepted entirely on the basis of biological reasoning.

ACKNOWLEDGEMENTS

This work was supported in part by the Office of Water Research and Technology, U. S. Department of Interior, under the provisions of Public Law 88-379.

REFERENCES

Berger, D. S. 1976. The sunburning ultraviolet meter: design and performance. Photochem. Photobiol. 24: 587-593.

Billen, D. and A.E.S. Green. 1975. Comparison of germicidal activity of sunlight with the response of a sunburning meter. Photochem. Photobiol. 21: 449-451.

Caldwell, M. M. 1971. Solar UV irradiation and the growth and development of higher plants. In Photophysiology, Vol. 7, 131-177. A. C. Giese [ed]. Academic Press, N. Y.

Calkins, J. 1975. Effects of real and simulated solar UV-B in a variety of microorganisms - possible implications of elevated UV irradiance, CIAP Monogr. 5, [Eds. D. S. Nachtway, M. M. Caldwell, and R. H. Biggs], DOT TST 75-55, U. S. Dept. of Transportation, Washington, D. C., 5-33.

Calkins, J., J. A. Barcelo, P. Grigsby and S. Martin. 1978. Studies of the role of solar ultraviolet radiation in "natural" water purification by aquatic ecosystems. Research Report 108,

University of Kentucky Water Resources Institute, Lexington, Ky.

Calkins, J. and Jeanne A. Barcelo. 1980. Some further considerations on the use of repair-defective organisms as biological dosimeters for broad-band ultraviolet radiation sources. Photochem. Photobiol. 30: 733-737.

Damkaer, D., D. B. Dey, G. A. Heron and E. F. Prentice. 1980. Effects of UV-B radiation on near-surface zooplankton of Puget Sound. Oecologia (Berl.) 44 : 149-148.

Green, A. E. S. and R. A. Hedinger. 1978. Models relating ultraviolet light and non-melanoma skin cancer incidence. Photochem. Photobiol. 28: 283-291.

Green, A. E. S., K. R. Cross and L. A. Smith. 1980. Improved analytic characterization of ultraviolet skylight. Photochem. Photobiol. 31: 59-65.

Harm, W. 1969. Biological determination of the germicidal activity of sunlight. Radiat. Res. 40: 63-69.

Luckiesh, M. and L. L. Holladay. 1931. Nomenclature and standards for biologically effective radiation. J. of the Opt. Soc. of Amer. 21: 420-427.

Machta, L., G. Cotton, W. Hass, and W. Komhyr. 1975. Measurement of solar ultraviolet radiation. 3rd Ann. Meeting, American Society for Photobiology.

Robertson, D. F. 1969. Long term field measurements of erythemally effective natural ultraviolet radiation. In Biological Effects of Ultraviolet Radiation [F. Urbach, ed.], Pergamon Press, Oxford. 432-436.

Scotto, S., T. R. Fears, and G. B. Gori. Measurement of ultraviolet radiation in the United States and comparison with skin cancer data. U. S. Department of Health Education and Welfare (NIH) 761029.

Setlow, R. B. 1974. The wavelengths in sunlight effective in producing skin cancer: a theoretical analysis. Proc. Natl. Acad. Sci. U.S.A. 71: 3363-3366.

Tyrrell, R. M. 1978. Solar dosimetry with repair deficient bacterial spores: action spectra, photoproduct measurements and a comparison with other biological systems. Photochem. Photobiol. 27: 571-579.

Webb, R. B. 1977. Lethal and mutagenic effects of near-ultraviolet radiation. In Photochemical and Photobiological Reviews, Vol. 2 [K. L. Smith, ed], Plenum Press, New York. 169-261.

THE SUNBURN UV NETWORK AND ITS APPLICABILITY FOR BIOLOGICAL PREDICTIONS

Daniel Berger

Temple University School of Medicine
Center for Photobiology
3322 N. Broad Street
Philadelphia, Pennsylvania 19140

HISTORY AND DISTRIBUTION

Since 1973 a global network of Meters has been measuring the half-hourly dose of sunburn ultraviolet. From the original 10 stations there had been an increase by 1980 to almost 30 providing a Sunburn UV (SUV) climatology at a wide variety of latitudes from 38° S to 71° N, at altitudes from sea level to 3.4 KM., and in climates ranging from cold wet to hot arid. (Fig. 1)

The results of this network were published for its first full year of operation of 1974 (HEW 1976). A summary of the first 5 years should be available in this calendar year from the National Oceanic and Atmospheric Administration.

The data is being presented in a variety of forms to facilitate use: daily fluctuations, weekly daily averages (less erratic than daily fluctuations), seasonal patterns, monthly patterns, exposure by time of day, average half-hourly dose, etc.

OPERATING PRINCIPAL

The meter spectral response was selected because of its similarity to the skin's erythema action spectrum. The detector is a phosphor whose excitation spectrum is the meter's spectral response.

Global energy strikes the meter at the black glass filter surface. This filter transmits the UV and absorbs the visible. The transmitted UV causes fluorescence of the magnesium tungstate phosphor. The visible component of this fluorescence is detected by a photodiode resulting in a current. This current is integrated to a preset charge and then counted as a unit of dose.

The fluorescent sunburn meter was selected as a global monitor because it monitors the entire UV-B with a sharply rising characteristic similar to the erythema action spectrum, and is also insensitive to UV-A; it has long-term stability; it requires little maintenance; its initial costs are low; its lifetime is long; its temperature variability is not considered excessive; the experience from its use in Australia was favorable as an instrument and for correlation with skin cancer incidence (Berger 1976).

OPERATION OF METER SYSTEM

Monthly data received from each station shows that for the whole network 95 to 98% of all possible data is captured per year. The validity of field data must be maintained; all meters are checked on a regular schedule. Each year each field station is compared against a meter which is sent out of the central station in Philadelphia. The results of these comparisons is that on an average each station has varied by 4% per year. Two meters in the Philadelphia Central station are regularly compared to an NBS traceable standard. Three circulating meters are run against these central station meters. The variation of each field station meter from the comparison meter allows the results from the field station to be mathematically adjusted so that all station results are intercomparable.

In any meter network it is important to calibrate both sensitivity and spectral response. Sensitivity is determined by comparing daily total counts. We have been utilizing a novel technique for evaluating spectral correspondence between pairs of detectors. This method may be of interest to others who obtain a series of pairs of dosages in the course of a day from any pair of meters, be they spectroradiometers, fluorescent meters, skin reddening, the killing of bacterium, or

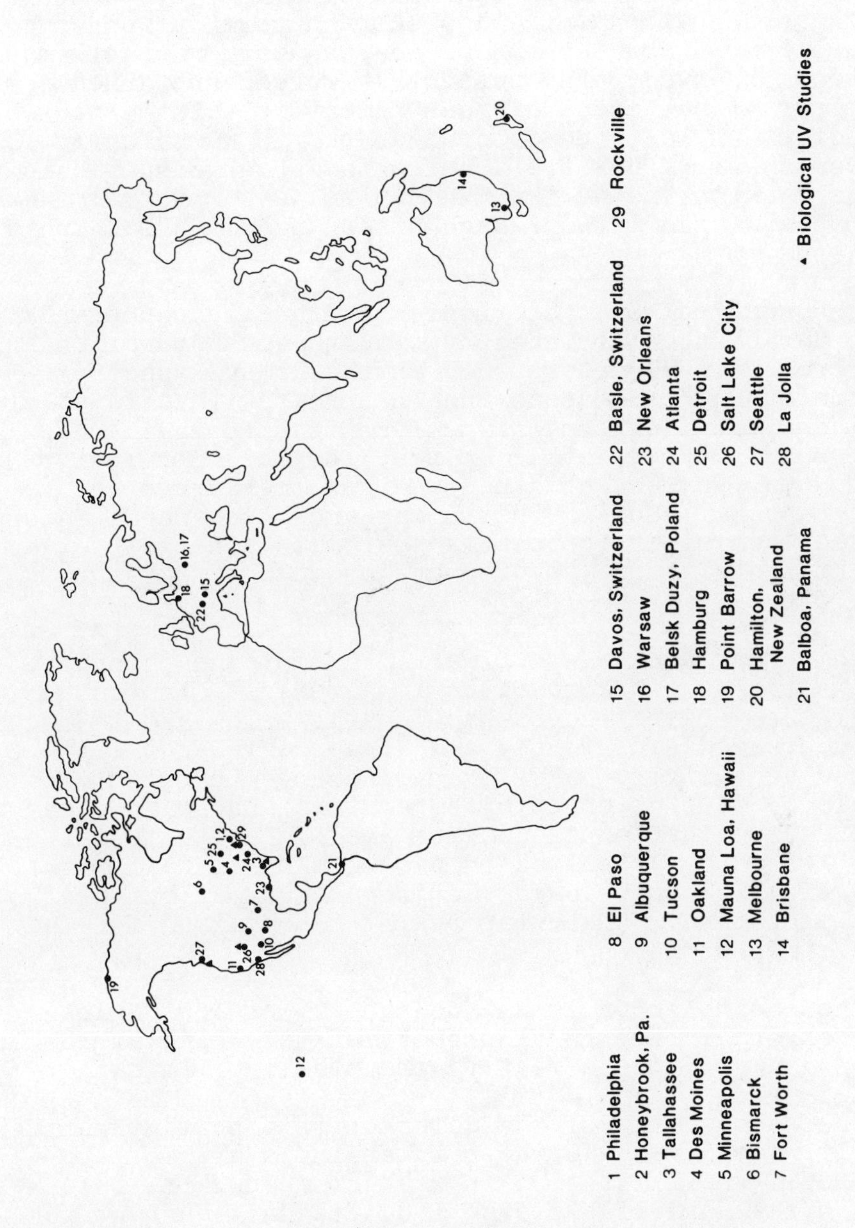

Figure 1. Location of Robertson-Berger meter stations.

the damage of DNA by sunlight. For all detectors a numerical evaluation of each response must be obtained. For the fluorescent SUV meter, Figure 2 indicates a set of field data from which spectral similarity is computed. For these operational field spectra determinations the variability is about ± 1% and no long term variability has been observed. The percent difference obtained from this operational spectral measurement is related to the correction angle to be described, but is numerically greater. All of our field units have been compared to standard units and we find the units have spectral responses within ± 2% of each other and of Philadelphia standards.

The spectral characteristics of the fluorescent UV meter has been established by simultaneous operation with precision UV meters such as the Komhyr spectroradiometer in Tallahassee and in Bismarck (Machta et al. 1975) and the Smithsonian instrument in Rockville (described by Goldberg, this volume. Correlation factors of .96 and .98 were obtained against the Komhyr instrument. The Smithsonian instrument and fluorescent meter tracked each other for over a year with a slight summer to winter difference apparently due to a temperature coefficient.

Temperature changes primarily affect the phosphor. The excitation characteristic is altered by increased temperature so that short wavelengths below about 310 nm have decreased sensitivity and longer wavelengths increased sensitivity. The overall change in output, therefore, depends on both temperature and solar spectrum. Running a weather exposed meter against a thermostated meter for a year resulted in a variation of daily total count of + 1.2% per degree Centigrade of maximum daily air temperature. A similar result is obtained for the Smithsonian comparison.

The cosine error is as much as 20% at low angles. Since sky radiation is disproportionately high at low sun angles, much of the effect is practically eliminated.

The sensitivity of the detector is important particularly when weak underwater intensities are to be measured. The limitation on weakest measurable intensity is the unavoidable signal due to dark current in the detector. Typically, this dark current is about 1 picoampere. For an overhead sun the maximum signal current is about 40,000 picoamps. Thus a 40,000 to 1

FIELD METHOD FOR COMPARING SUV METER SPECTRA

1. RECORD HALF HOURLY PRINTOUTS OF TWO METERS BEING COMPARED
2. NORMALIZE ONE IN RESPECT TO THE OTHER TAKEN AS REFERENCE
3. RECORD THE ARITHMETICAL DIFFERENCE OF EACH HALF HOURLY COUNT
4. SUM THE ARITHMETICAL DIFFERENCE AND DIVIDE BY DAILY (NORMALIZED) COUNT X100 FOR PERCENT
5. IF MIDDAY DIFFERENCES ARE + IN RESPECT TO REFERENCE THEN + SIGN USED FOR %, - SIGN OTHERWISE

Example: June 22nd, 1977, Philadelphia, DETECTOR C1 COMPARED TO C9

TIME	C9 Counts	C1 Counts	C1 Normalized	C1 Norm. -C9
5:00		1	1	+1
5:30	5	4	4	-1
6:00	14	15	15	+1
6:30	33	35	34	+1
7:00	63	66	65	+2
7:30	104	108	106	+2
8:00	156	161	158	+2
8:30	217	224	220	+3
9:00	**289**	**295**	**290**	**+1**
9:30	**363**	**371**	**365**	**+2**
10:00	**432**	**440**	**433**	**+1**
10:30	**505**	**513**	**505**	**0**
11:00	**490**	**496**	**488**	**-2**
11:30	**566**	**571**	**562**	**-4**
12:00	563	567	558	-5
12:30	585	587	578	-7
13:00	564	566	557	-7
13:30	497	500	492	-5
14:00	358	364	358	0
14:30	328	333	328	0
15:00	203	208	205	+2
1530	214	219	216	+2
16:00	143	148	146	+3
16:30	135	139	137	+2
17:00	97	101	99	+2
17:30	66	69	68	+2
18:00	37	38	37	0
18:30	18	19	19	+1
19:00	7	7	7	0
19:30	1	1	1	0
SUM	7053	7166	7052	|61|

61 / 7053 x 100=-.86%

Figure 2. Spectral comparison of 2 meters.

range of sensitivity is possible. Allowing for the fact that lower than maximum intensity is the more common field situation, a sensitivity range between 1000 to 10,000 should be obtainable.

If the meter spectral response were identical to the erythema action spectrum the meter output could be calibrated in units of minimal erythema dose per hour. Since the two spectra are not identical the meter is calibrated in terms of its own spectrum in a unit called Sunburn Units per hour. For the solar spectrum of an overhead sun, taken as a reference, the numerical value of the meter's output and the average skin's response to sunburn energy are made to be identical. The reference spectrum is for a clear sky and an ozone thickness of about 2.6 mm. An untanned average Caucasian will develop a minimum erythema from this reference spectrum in about 12 minutes. One MED in 12 minutes is a sunburn intensity of 5 MED per hour. For this reference solar spectrum the meter output is, therefore, made equal to 5 Sunburn Units per hour. For the network meter 440 counts is equal to one Sunburn Unit.

Since 1 MED is 18 millijoules per cm^2 of 297 nm energy, one Sunburn Unit for the reference solar condition is likewise equivalent to 18 millijoules per cm^2 of radiation at 297 nm. These energies are calculated by the convolution of the response spectra with a solar spectrum. For an overhead sun the convoluted integrals are made equal. For all other solar spectra the convolution of the meter spectrum results in an effective energy greater than that for erythema, which remains constant at 18 millijoules per cm^2.

The previous discussion indicates that the use of the meter output in sunlight as an indicator of a biologic action would be in error if the spectral response of the meter were not identical to the biological spectral response. If both response spectra were known then by convoluting each against a series of solar spectra the error between the two responses could be generated so that the meter output could either be shown to adequately represent the biological response within an acceptable error or a correction factor found which could be used to correct the error. While this sounds like a formidable task, in fact, it is less difficult than it appears, especially for the case of erythema where the action spectrum is well established.

Table I and Figure 3 illustrate the generation and use of the correction method. In Table I the erythema action spectrum is convoluted with global radiation. For particular ozone thicknesses a series of convolutions is generated from zenith angles 0° to 70°. Substituting the meter spectral response for the erythema action spectrum a convolution is produced at every zenith angle previously used. Each set of convolutions is normalized.

Plotting normalized erythema convolutions against normalized meter convolutions results in a straight line

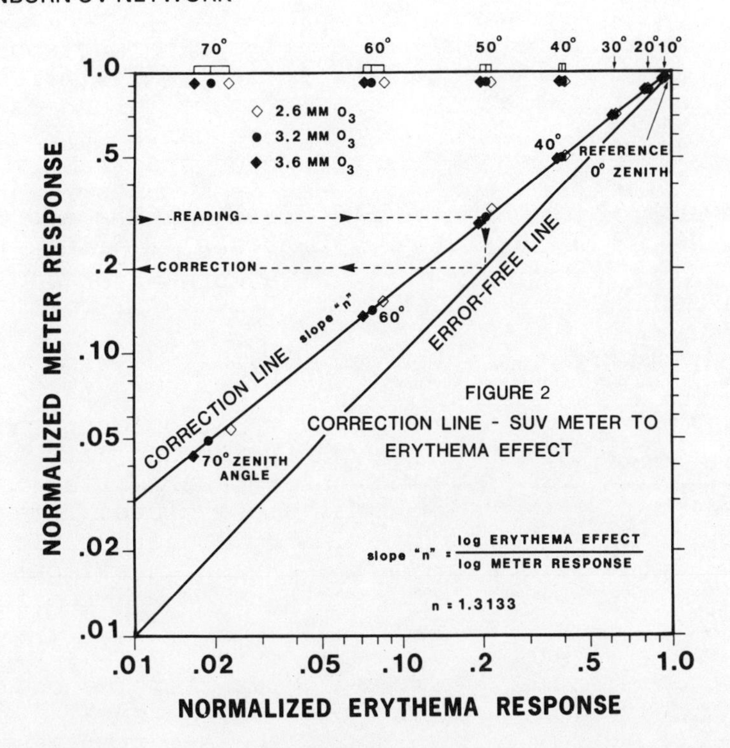

Figure 3. Correcting meter for erythema.

on log-log coordinates going through point (1,1). The straight line relationship can be analytically defined as follows:

$$(\text{normalized meter response})^n = \text{normalized erythema effectiveness}$$

$$n = \frac{\text{log normalized erythema effect}}{\text{log normalized meter response}}$$

n is the slope of the line. Slope variation is over the narrow range of 1.3605 to 1.3088. These small slope variations are more affected by sun angle than by ozone.

Variation of ozone doesn't affect slope but does cause the zenith angle point (ZAP) on the correction line to move towards the reference point for decreasing ozone. The ratio of coordinates at the zenith angle point is the correction which allows the meter reading to more accurately indicate erythema effectiveness.

The variation of ZAP due to ozone thickness changes and the variation of slope with sun angle each introduce

TABLE 1

CONVOLUTIONS OF ERYTHEMA VERSUS SOLAR SPECTRA AND CONVOLUTIONS OF METER RESPONSE VERSUS SOLAR SPECTRA

	2.60 mm O_3			2.88 mm O_3			3.04 mm O_3		
Zenith Angle	Convol. Erythema X Solar (Norm.)	Convol. Meter X Solar (Norm.)	(n) log-log slope	Convol. Erythema X Solar (Norm.)	Convol. Meter X Solar (Norm.)	(n) log-log slope	Convol. Erythema X Solar (Norm.)	Convol. Meter X Solar (Norm.)	(n) log-log slope
0°	1.000	1.000	--	1.000	1.000	--	1.000	1.000	--
10	.952	.964	1.3417	.951	.963	1.3326	.951	.963	1.3326
20	.815	.859	1.3460	.813	.857	1.3415	.812	.857	1.3495
30	.619	.698	1.3341	.615	.696	1.3414	.613	.694	1.3398
40	.403	.505	1.3302	.398	.501	1.3330	.395	.499	1.3362
50	.213	.311	1.3241	.208	.306	1.3260	.205	.304	1.3309
60	.0841	.152	1.3142	.0806	.148	1.3181	.079	.145	1.3145
70	.0227	.055	1.3051	.0211	.0519	1.3042	.0201	.0502	1.3059
			(1.3279)*			(1.3281)*			(1.3299)*
MED/hr. @0° = 4.936				= 4.303			= 3.982		

	2.30 mm O_3			3.60 mm O_3		
Zenith Angle	Convol. Erythema X Solar (norm.)	Convol. Meter X Solar (Norm.)	(n) log-log slope	Convol. Erythema X Solar (Norm.)	Convol. Meter X Solar (Norm.)	(n) log-log slope
0°	1.000	1.000	--	1.000	1.000	--
10	.950	.963	1.3605	.949	.962	1.3512
16.5°	.868	.900	1.3436	.807	.854	1.3587
20	.810	.856	1.3553	.604	.688	1.3482
30	.610	.692	1.3425	.383	.490	1.3454
40	.392	.496	1.3356	.194	.293	1.3359
50	.202	.301	1.3322	.0707	.136	1.3279
60	.076	.143	1.3250	.0166	.0439	1.3111
70	.019	.484	1.3088			
MED/hr. @ 0° = 3.688			(1.3379)*	= 3.049		(1.3398)*

(meter reading)^n = erythema effectiveness
n = slope of line on log-log paper = 1.3338 ave.
* = average slope

errors which are much smaller than the correction afforded at all ozone thicknesses when an average slope and average ozone thickness are used.

As an example of the use of correction line: with the zenith angle at 50° the normalized meter coordinate is .301 and the normalized erythema coordinate is .202. Therefore, the meter reading must be multiplied by .202/ .301 = .671 to obtain the erythema effectiveness. If the meter under a clear sky were reading 1.1 Sunburn Units per hour, multiply by 0.671 to obtain a .74 MED/hr. which means 1 hour and 21 minutes would be required for a minimal erythema dose. If the sky were cloudy and 0.5 Sunburn Units per hour was the meter indication the same factor of 0.67 would be used since the sun angle is the same. The erythema effectiveness would calculate to be 0.33 MED/hr.

This same correction technique can be used for any biological effect whose action spectrum is known such as DNA, bacterial killing, etc.

Figure 2 can be used for all latitudes and all zenith angles up to 70°. Taking the highest absolute value of the reference point as equal to 5 MED/hr or 5 S.U./hr the ozone thickness is implied as being 2.55 mm exactly, 2.6 mm approximately. The position for this ozone thickness at 10° increments of zenith angle is indicated on the correction line. For sites outside the tropics the minimum zenith angle is always greater than zero and the ozone thickness is generally more than 2.6 mm. Both of these changes reduce the erythemogenicity of sunlight. Table 2 included on Figure 2 allows the absolute reference value at 0° zenith angle to be determined if the ozone thickness is known.

The ozone thickness changes with latitude, season, and there is a longitudinal as well as a daily variability. Average ozone as a function of latitude and season has been published (London and Kelly 1974, CIAP 1975) and allows an approximate guide for selecting probable ozone thickness. However, even gross changes in ozone thickness cause only a relatively small change in the correction. For example, at a 70° zenith angle, where ozone variability causes larger effects than for all lesser zenith angles, the correction at 2.6 mm multiplies the meter reading by $\frac{2.27}{5.5} = .41$.

Whereas 3.6 mm of ozone results in a correction factor

of 0.38. The correction factors in this extreme example cause about a 10% difference in correction, while the correction itself is about 150%. Consequently, even ignoring ozone differences does not significantly alter the accuracy of the correction.

Plotting correction factors obtainable from the log-log coordinate correction line in relation to zenith angles presents the data (Figure 4) in convenient form. The corrections at all zenith angles are determined and the effect of ozone can also be accommodated.

Based on the results obtained from preparing a correction line for erythema in respect to the meter the following basic attributes would be anticipated for correction lines for other UV-B affected biological phenomena:

1) There is a straight line slope on log-log coordinates between normalized meter and normalized photobiological effect.

2) The slope is independent of ozone.

3) The correction point is relatively independent of ozone.

4) The absolute reference value, for 0° zenith angle and clear skies, is dependent on ozone only.

Note that errors are small for high sun altitudes when UV intensity is also high, and large only where UV output is low. Thus uncorrected meter results can often be used as an indicator of certain biological effects - erythema, vitamin D effective radiation, bacterial killing, etc.

GENERAL CONSIDERATIONS REGARDING UV-B METERS

The correction method is applicable to all UV-B meters, regardless of type. Since any particular biologic action spectrum will, in general, not be identical with the response of another meter or biologic response, higher correlation is obtainable when this correction is used.

When the meter response is the ordinate, as in Figure 3, a shallower slope for an action spectrum

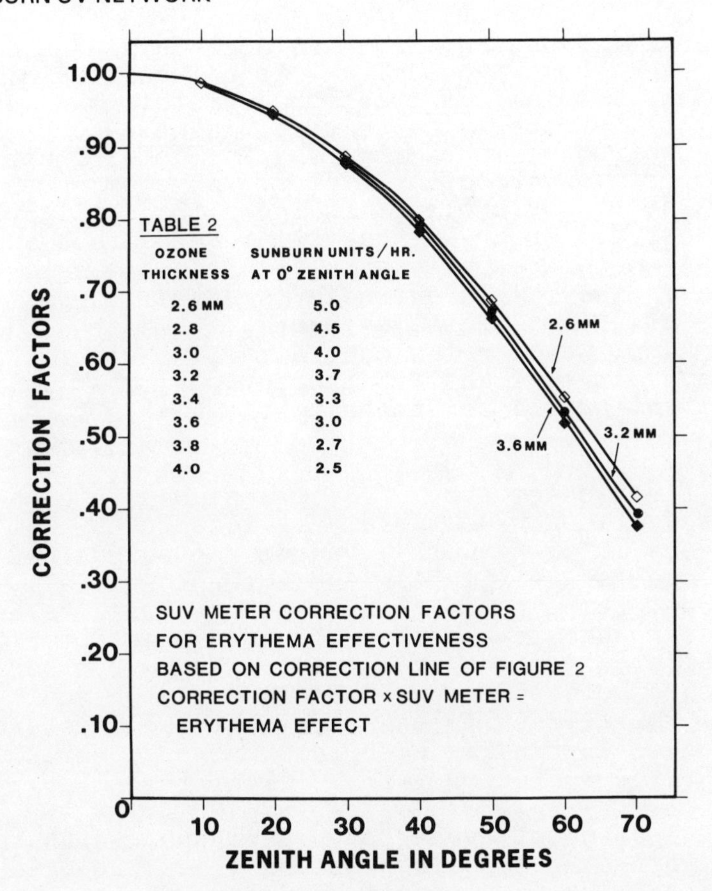

Figure 4. Correction factors for erythema from previous figure.

indicates a stronger short UV response. Quantitative ordering or comparison of different biologic spectra can be made by comparing exponentials or slopes.

Artificial UV sources generally have considerably different wavelength composition than does sunlight with the same biologic effect. Combinations of artificial UV and sunlight also present the same problem. If the spectral distribution output of an artificial light source is stable, then its effect on a biological system should be proportional to the meter reading. However, the relationship of meter reading to biologic response needs to be determined for the particular lamp and meter combination. Meters responding to any part of the stable radiation source's spectrum can be used as monitors.

A variable spectrum from the artificial source alone or in conjunction with sunlight is a much more difficult measurement problem. A spectroradiometer measurement of the total UV energy convoluted against the biological action spectrum would be required for determination of the appropriate correction factor. A meter whose output in terms of biologic effect is known for sunlight could be seriously in error when combination sources are used, unless the meter response simulated the biologic response over the lamp's entire UV spectral range. In the case of an artificial UV source, like the fluorescent sunlamp, wavelengths considerably shorter than those found in sunlight are present. A meter calibrated for sunlight should be applied with caution to a mixed source.

The fluorescent type of UV transducer is capable of being used with a wide variety of detectors. In the case of the meter network a vacuum photodiode has appropriate sensitivity for the intensity usually encountered. For a personal dosimeter we will be deploying, electro-chemical cells will be the detector allowing integrated dosages over the long time periods to be evaluated. Detectors for UV-B can be quite small and rugged. Portable units operating with self-generating photodiodes can provide UV-B information as simply as a light meter indicates visible light intensity.

A UV-B Climatology has been developed which should provide insight for ecologists and other biological scientists. The development of meters which closely correlate their response with any desired biologic UV-B action spectrum should be possible.

REFERENCES

Berger, D. 1976. The sunburning ultraviolet meter: Design and performance. Photochem. Photobiol. 24: 587-593.

Climatic Impact Assessment Program. 1975. Unperturbed Ozone Column, Monogr. 5, Part 1, Chapter 2: 2-4.

London, J. and J. Kelley. 1974. Global trends in total atmospheric ozone. Science 184: 987-989.

Machta, L., G. Cotton, W. Hass, and W. Komhyr. 1975. Measurement of solar ultraviolet radiation, 3rd Ann. Meeting, American Society for Photobiology.

US Dept. HEW. 1976. Measurements of Ultraviolet Radiation in the U.S. and comparisons with Skin Cancer Data. DHEW No. (NIH) 76-1029.

DESCRIPTION AND APPLICATION OF A PERSONAL DOSIMETER FOR MEASURING EXPOSURE TO NATURAL ULTRAVIOLET RADIATION

Brian Diffey (1), Tony Davis (2) and
Ian Magnus (3)
Medical Physics Department, Dryburn Hospital
Durham, England (1), Propellants, Explosives
and Rocket Motor Establishment
Waltham Abbey, England (2) and Photobiology
Department, Institute of Dermatology,
London, England (3)

ABSTRACT

Personal doses of natural UVR have been measured using the polymer film, polysulphone. Applications of personal UVR dosimetry have included studies of UVR exposure to people in different environments, the relationship of UVR and drug photosensitivity of the skin, and the measurement of the anatomical distribution of solar UVR, particularly on the face.

INTRODUCTION

It is well established that exposure to ultraviolet radiation (UVR) can have both beneficial and detrimental effects on humans.

Although only relatively low doses of UVR are required to form measurable levels of Vitamin D in the skin, people deprived of UVR are more liable to develop osteomalacia, particularly when there is impaired utilization or dietary insufficiency of Vitamin D. Groups which have been found to show low levels of Vitamin D associated with chronic under-exposure to UVR are the elderly, submariners and Asiatic immigrants.

On the other hand, excessive repeated exposure to UVR is well known to induce both skin cancer and ageing effects. Also the increasing use of drugs, which have photosensitivity side effects (e.g. phenothiazines, nalidixic acid and certain tetracyclines), have made a greater percentage of the population more liable to show abnormal sensitivity to UVR.

At present, environmental conservationists are becoming increasingly concerned that changes in the atmospheric ozone mantle may be induced by both high-flying supersonic aircraft and by the build-up of freons and related compounds. Should significant changes occur in the depth of the ozone layer, then it can be expected that the amount of UVR reaching the Earth's surface will be altered.

DESCRIPTION OF A PERSONAL UVR DOSIMETER

Natural UVR is generally measured with solid state detectors often used in conjunction with optical filters. In particular, the Robertson-Berger meter, which measures those wavelengths in the global spectrum less than 320 nm, has been used to monitor continuously natural UVR at several sites throughout the world. A different, yet complimentary approach, is the use of various photosensitive films as UVR dosimeters. The principle is to relate the degree of deterioration of the films, usually in terms of changes in their optical properties, to the incident UVR dose. The principle advantages of the film dosimeter are that it provides a simple means of continuously integrating UVR exposure and also that it allows numerous sites, inaccessible to bulky and expensive instrumentation to be compared simultaneously. A film which has received wide application in the medical context because it has a spectral response similar to the erythema action spectrum for human skin, is the polymer, polysulphone (Davis et al. 1976).

The polysulphone film, approximately 40 μm thick,is mounted in cardboard holders, 30 mm square, with a central aperture of 16 x 12 mm (Kodak type 110 mounts) which constitutes the film badge (see Fig. 1). The badges are worn by subjects, much as photographic film badges are worn as ionizing radiation monitors.

Figure 1. A polysulphone film badge.

OPTICAL PROPERTIES OF POLYSULPHONE FILM

When exposed to UVR polysulphone film readily degrades. As a measure of the degree of degradation of the polymer, the change in absorbance measured at 330 nm (ΔA_{330}) is noted. This change is proportional to UV exposure and is shown in Figure 2 as induced by monochromatic irradiation at 297 nm (bandwidth at half maximum intensity of 10 nm). The UV dose in this Figure can be calculated from the empirical relationship,

$$\text{UV Dose} = 4.3\ (\Delta A_{330})^{1.38}\quad \text{Jcm}^{-2} \tag{1}$$

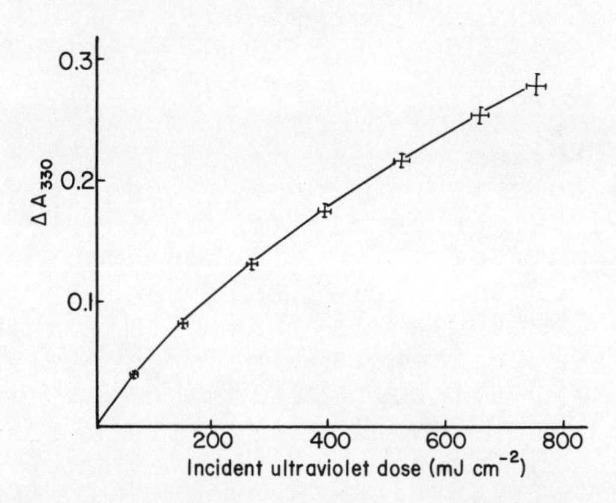

Figure 2. The variation of ΔA_{330} with incident UV dose at 297 nm.

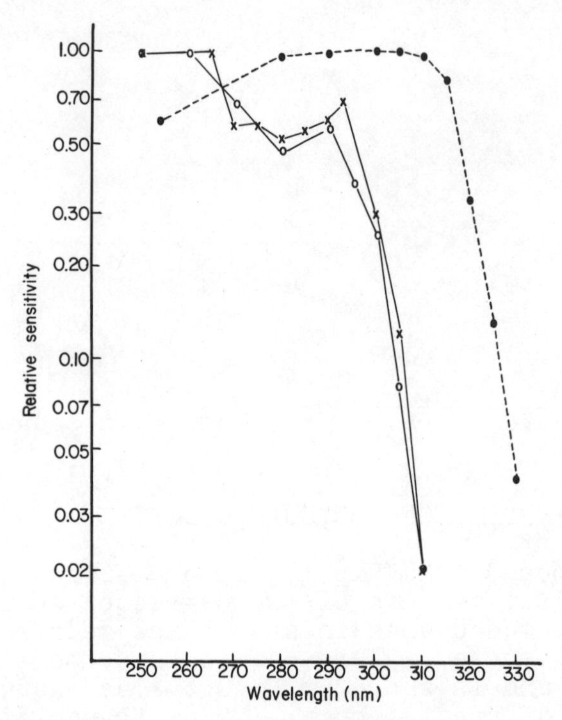

Figure 3. Spectral response of polysulphone (----) and erythema action spectrum of human skin (——).

The dose-response curve of the film to UVR is non-linear, probably because of the filtering effect of new absorption centers in the polymer from photolytic breakdown of the sulphone linkages.

The spectral sensitivity of the film, shown in Fig. 3 is the reciprocal of the dose required to produce a ΔA_{330} of 0.1 in the wavelength range 254 to 330 nm, and normalized to unity at the most sensitive wavelength. In medical or physiological applications the wavelength response of the dosimeter should match the action spectrum of the biological response to be studied. The wavelength response of polysulphone film is similar to the erythema action spectrum for human skin (see Fig. 3), although it responds to wavelengths up to about 330 nm. Nevertheless, it is possible to relate polysulphone response to "an effective erythemal dose" by employing published data of the relative spectral distribution of the UVR source, e.g. the global UVR spectrum. The effective erythemal dose D_E, at 297 nm is related to the equivalent polysulphone dose D_P (obtained from Fig. 2 or Eq. 1), by the equation

$$D_E = D_p \cdot Q \text{ Jcm}^{-2} \quad (2)$$

where Q is a correction factor which allows for the difference between the erythema action spectrum and the wavelength response of polysulphone, and may be expressed as

$$Q = \int I(\lambda)\ E(\lambda)\ d\lambda \ / \int I(\lambda)\ P(\lambda)\ d\lambda \quad (3)$$

where $I(\lambda)$ is the relative spectral distribution of the incident radiation at wavelength λ, $E(\lambda)$ is the erythemal effectiveness at wavelength λ relative to that at 297 nm, assuming the mechanism of response does not change with wavelength,and $P(\lambda)$ is the spectral sensitivity of polysulphone in terms of absorbance change, ΔA_{330}, at wavelength λ relative to that at 297 nm, as the biological effects produced by this wavelength have been studied extensively.

The numerical value of Q depends, of course, on the source of radiation, and even for global UVR will vary with factors such as solar altitude and ozone layer thickness. For the summer months in the U.K. the value of Q is approximately 0.05. Some of the uses of polysulphone film as a personal UVR dosimeter will be outlined in subsequent sections.

STUDIES OF UVR EXPOSURE TO HUMANS IN DIFFERENT ENVIRONMENTS

Three groups of people were examined; two hospital gardeners, nine laboratory workers and six long-stay geriatric patients. The geriatric patients were further divided into two groups; those who were able to sit on a balcony in the sun and those who were restricted to the ward. The gardeners wore two badges per day each over a period of 4 - 5 hours, the laboratory workers and the mobile patients wore one badge per day, while the ward-fast patients wore one badge per 'working week' and one badge at the weekend. Observations were made over a 2-week period (23 June - 6 July, 1975) which proved to be particularly sunny. With the exception of the gardeners, all of the subjects were in the same urban area. Each day a badge was placed on the balcony, where the geriatric patients were able to sit in the sun, to record the total daily UVR. However, as the badge was shaded during part of the day, a direct comparison of the polysulphone UVR dosimeter was made with an electronic UVR detecting system (Robertson-Berger meter). This equipment is placed in such a position that total UVR incident on a horizontal plane during the day is recorded. A linear relationship for UVR exposure was demonstrated between the two systems and results were expressed in $mJcm^{-2}$ of effective erythemal dose equivalent to 297 nm (297 eq.)

As would be expected, the groups were fairly clearly separable in terms of their UV exposure. The gardeners were comparable to the sunshine exposed patients, and received by far the largest doses [$\approx$ 20 $mJcm^{-2}$ (297 eq)per day]. The laboratory workers received less [$\approx$ 4 $mJcm^{-2}$ (297 eq)per day] and the ward-fast patients least [$\approx$ 0.2 $mJcm^{-2}$ (297 eq)per day]. The outdoor groups received a dose approximately 10% of the mean daily total recorded on a horizontal plane.

The very low doses received by the ward-fast patients is not surprising since normal window glass has negligible transmission below 310 nm. This may help explain the high incidence of Vitamin D deficiency in such patients; it is known that exposure to UVR in the range 290 - 310 nm is required for the photosynthesis of cholecalciferol.

A longer term investigation with the polysulphone film on a larger group of normal subjects over a two year period involved about 30 office personnel monitored for 8 fortnightly periods throughout the winter and summer solstices, autumn and spring equinoxes and periods half-way between these.

It was found that the office personnel received about 40 $mJcm^{-2}$ (297 eq) per week in the summer and only a few $mJcm^{-2}$ (297 eq) per week in the winter. The office personnel received 3 – 4 % of the total incident UVR and there was no noticeable variation in this fraction with season. As expected this value is similar to the fraction received by laboratory workers in the previous study.

All groups in these two studies showed relatively large standard deviations, which may have been due to the position and plane in which the badge was worn and the normal working postures, i.e. stooping and turning away from the sun.

UVR AND DRUG PHOTOSENSITIVITY OF THE SKIN

The polysulphone film badge has been used to investigate personal UVR doses in abnormal skin photosensitivity due to therapeutic drugs. This was in groups of five to six institutionalized psychiatric patients on phenothiazine therapy. Chlorpromazine was the most commonly prescribed drug and commonly causes skin photosensitivity as an adverse side effect. It was assumed that the photosensitive action spectrum of this drug is 320 - 350 nm, which is at least partly within the action spectrum of polysulphone. The project was done in five separate hospitals, four in England, one in Eire. The medical and nursing staff recorded daily the hours that each of the patients under study spent out of doors in bright sunlight over July and August, 1976. They also recorded symptoms of photosensitivity on a special form provided; from this a simple numerical scoring system for the severity of photosensitive symptoms was derived. All patients wore the polysulphone badge, which was returned for spectrophotometric assay at the end of each week.

On analysis of the results, it was found that badge dose and the recorded hours as spent in direct sunlight had a strong positive correlation ($p < 0.001$). What was especially interesting was that the badge dose also correlated positively with the symptom score of the patients ($p < 0.001$), but that symptom score and hours outdoors did not correlate; this is plausible, for the time outdoors in our climate will include many periods of overcast weather. These results suggest that, given suitable situations, this method is a valid approach for collecting objective data where previously results were subjective and conclusions had to be tentative.

THE ANATOMICAL DISTRIBUTION OF SUNLIGHT

The simplicity of film dosimetry allows numerous sites to be compared simultaneously and this lends itself readily to measuring the anatomical distribution of sunlight. One purpose of this work was to look at the supposed correlation between sunlight exposure and the distribution of skin cancer on the body surface. The relative distribution of shortwave, natural UVR at ten sites on the surface of an unclothed manikin was measured using polysulphone film as the dosimeter. Although the situation chosen is somewhat artificial, the results indicated well the variation of UVR over the anatomy and provide useful guidelines until more elaborate measurements become possible.

The manikin was positioned in a normal upright posture with arms at the side on an unshaded lawn and rotated on a turntable at 0.5 rev min^{-1} for 2 hours, one hour either side of solar noon (12.00 to 14.00 BST) on 19 different days during August and September 1976, a period which proved to have large variations in cloud cover. The measurements were made at Canterbury, England (latitude 51°N, altitude 40 m above sea level). Since the change of solar altitude is negligible during one revolution of the manikin, its rotation at a constant velocity may be regarded as equivalent to the "random" motion of a human subject outdoors.

If the dose on the vertex is taken at 100%, it was found that on approximately horizontal surfaces such as the top of the shoulders (epaulet region) and the dorsa of the feet, where there will be some shading from the rest of the body, the value is about 80% of the vertex. Approximately vertical planes on the body received about 60%; the major component of UVR is scattered skylight not direct (unscattered) radiation from the sun. It was also observed that the relative dose at each site was approximately independent of cloud cover.

The results obtained may be used to give a qualitative indication of erythemal or carcinogenic UVR at different anatomical sites since the anatomical distribution of the waveband 295-310 nm, which includes most of the normal sunburning wavelengths and presumably those causing skin cancer, will be very similar to that of 295-330 nm which encompasses the action spectrum of the polysulphone film.

These measurements indicate that the hands receive roughly the same UVR dose as the face. This is in harmony with the clinical observation that the skin changes in the commoner idiopathic photodermatoses are mostly located on these regions. This is also in agreement with the distribution of squamous cell epitheliomas. Per contra the incidence of basal-cell epitheliomas on the face is about an order of magnitude higher than on the hands. This apparent discrepancy strongly suggests that sunlight is not the only factor in the aetiology of carcinoma of the skin, particularly basal-cell carcinomas; a conclusion which has been voiced by several workers.

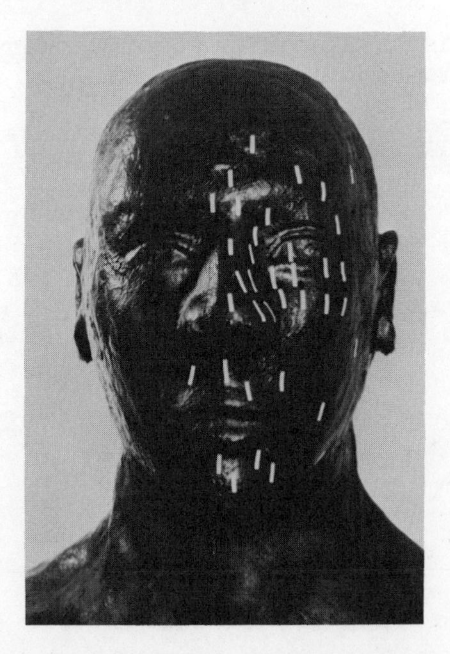

Figure 4. The fiber glass head used to measure the distribution of solar UR radiation.

Since more than 90% of basal cell epitheliomas (BCC) occur on the head and neck, a further study was carried out during the summer of 1978 to measure in detail the distribution of global UVR on the face and to relate the findings to published data on the facial distribution of BCC.

In order to allow a sufficiently high degree of spatial resolution of measurement across the face, a large fiberglass model of a head was constructed with all linear dimensions increased by a factor of approximately three. It was felt that this approach is valid since it was required to measure only relative UV doses which are principally functions of head geometry and solar altitude. Polysulphone film badges were located on 41 numbered sites on the head by means of elastic bands. This method of attachment of badges was simple, quick and reproducible, and allowed the badges to be as close to the 'skin' as possible (see Fig. 4). Measurements on the reflectance of UVR from the fiberglass head at a wavelength of 300 nm gave negligible results, as from skin.

The whole assembly was placed outdoors with low buildings on two sides and open grass on the other two and the head rotated at 10 rev min^{-1} between the hours of 12.00 and 15.00 BST. The measurements were made at Canterbury, England on nine different days during August 1978.

The results indicated at 100-fold range of UV dose over the face, the vertex always being the site of maximum dose. It was evident that regions such as the orbits, philtrum, sublabial area, and submental area receive very little UVR. The results were correlated with the distribution of BCC in various cutaneous sites on the head given by Brodkin et al. (1969). A least squares fit of the logarithm of tumor density against the logarithm of UV dose as the dependent variable gave

$$\text{tumor density proportional to } (\text{UV dose})^{p} \qquad (4)$$

where $p = 1.71 \pm 0.33$ with a correlation coefficient of 0.69.

The poor correlation clearly questions the validity of the present comparison. The tumors were recorded at a different geographical location (New York: latitude 41°N) from the measurements (Canterbury: latitude 51°N) and their occurrence is presumably related to all-year-round sunlight exposure for many years. The solar altitude averaged over the year at a location of 41°N between 07.00 and 17.00 hours for each day of the year has been calculated to be 35° which is substantially lower than the range of solar altitude (46-55°) encountered in the present experiment. Even though the spectral distribution of global UVR is, to a large extent, dependent on solar altitude, there are also other factors which influence the quality and quantity of UVR which reaches the face, e.g., posture, ground reflection, atmospheric conditions, and so on. Furthermore, it is apparent that skin tumors originate in living cells in the viable layer of the epidermis, and so the tumor frequency would be expected to depend upon either the UV flux which reaches the critical cells or on products of UV damage to more superficial skin which reach the critical cells by diffusion, or both. Whatever the mechanism the thickness of the stratum corneum could well be critically important and there is no reason to suppose that the thickness of this layer is constant for all the sites chosen in the present comparison. Nevertheless, the results are compatible with the hypotheses that (a) human skin cancer incidence increases with environmental UVR exposure, and (b) sunlight is not the only factor in the aetiology of basal cell carcinomas of the face.

It is concluded that a polymer film badge is useful as a personal dosimeter in field trials, epidemiological and similar studies, where information on UVR exposure to the skin is required. Such studies, apart from the more obvious dermatological and industrial applications, would include observations in geographical regions, industrial or sociological situations, where skin cancer epidemiology and surveys on Vitamin D status are of importance. Work is currently in hand to produce a polymer film with a closer match to the erythema action spectrum, together with the production of films for use in other regions of the electromagnetic spectrum.

REFERENCES

Brodkin, R. H., A. W. Kopf and R. Andrade. 1969. Basal cell epithelioma and elastosis: a comparison of distribution . In 'The Biologic Effects of Ultraviolet Radiation with Emphasis on the Skin' [Ed. F. Urbach (Oxford: Pergamon Press.

Challoner, A.V.J., D. Corless, A. Davis, G.H.W. Deane, B. L. Diffey, S. P. Gupta and I. A. Magnus. 1976. Personnel Monitoring of exposure to ultraviolet radiation . Clinical Exp. Derm. 1: 175-179.

Corbett, M. F., A. Davis and I. A. Magnus. 1978. Personnel radiation dosimetry in drug photosensitivity . Brit. J. Derm. 98: 39-46.

Davis, A., G.H.W. Deane and B. L. Diffey. 1976. Possible dosimeter for ultraviolet radiation . Nature. 261: 169-170.

Diffey, B. L., K. Kerwin and A. Davis. 1977. The anatomical distribution of sunlight . Brit. J. Derm. 97: 407-410.

Diffey, B. L., T. J. Tate and A. Davis. 1979. Solar Dosimetry of the Face . Phys. Med. Biol. 24: 931-939.

Leach, J. F., V. E. McLeod, A. R. Pingstone, A. Davis, G.H.W. Deane. 1978. Measurement of the ultraviolet doses received by office workers . Clin. Exp. Derm. 3: 77-79.

NOMOGRAMS FOR BIOLOGICALLY EFFECTIVE UV

David M. Damkaer (1),(2) and Douglas B. Dey (2)

University of Washington WB-10,
Seattle, WA 98195 (1) and
National Marine Fisheries Service/NOAA
Manchester, WA 98353 (2)

ABSTRACT

Regressions were determined establishing relationships between absolute UV-B irradiance (285-315 nm) and commonly used weighting functions for biological effectiveness. Nomograms were prepared from these regressions for both artificial and solar UV-B irradiance. Under well-defined conditions the nomograms may be used to rapidly compare data-sets which are derived from different weighting functions.

INTRODUCTION

Inadequate knowledge of proper weighting functions for evaluating the relative biological effectiveness of different UV-B wavelengths remains a major obstacle in understanding the effects of incident solar ultraviolet irradiation. All quantitative discussions and predictions regarding ambient or enhanced levels of UV-B irradiation are greatly affected by the choice of a weighting function (National Academy of Sciences, 1979) Different investigators have employed different weighting functions for calculating effective doses of UV-B. Sometimes different normalization wavelengths, the wavelength where biological effectiveness is 100%, have been used for the same action spectrum. While each weighting function, based on an observed action spectrum, might be justified under particular conditions, the most appropriate weighting function for most groups of aquatic organisms has not been determined.

Because of the variety of action spectra for biological effectiveness, and the frequent need to compare different authors' data-sets, we found it useful to establish a practical relationship between absolute UV-B irradiance (285-315 nm) and some common weighting functions.

METHODS

Absolute UV-B irradiance measurements were made with an Optronic Laboratories® Model 741 spectroradiometer coupled with a HP® 9815A computer. The spectroradiometer was periodically recalibrated using a National Bureau of Standards lamp of standard spectral irradiance. Measurements were also made using the Robertson-Berger (R-B) Sunburn Ultraviolet Meter (Berger et al., 1975; Billen and Green, 1975). Ambient and enhanced solar irradiance at 285-315 nm was simulated with double-lamp fixtures, each holding one Westinghouse® FS-40 fluorescent "sunlamp" and one "cool white" (CW) fluorescent lamp, transmitting through cellulose triacetate plastic sheets (CTA). UV-B intensity was adjusted by varying the CTA thickness (combinations of 5-, 10-, and 20-mil sheets) and the distance between the UV-B source and the spectroradiometric sensor. Other general techniques for the measurement of artificial and solar UV-B were as described by Damkaer et al. (1980).

The analytical representations of Green and Miller (1975) were used in calculations with two of the biological weighting functions: (1) the DNA action spectrum (Setlow, 1974),

$$\epsilon_{DNA}(\lambda) = \exp \{ k [\frac{1}{1 + \exp[(\lambda-\lambda_o)/\lambda_f]} -1] \} \quad (1)$$

where $k = 13.82$, $\lambda_o = 310$, and $\lambda_f = 9$,

and (2) the generalized action spectrum for plants (Caldwell, 1968),

$$\epsilon_{PLANT}(\lambda) = A[1-(\lambda/\lambda_c)^n] \exp - [(\lambda-\lambda_o)/\lambda_f] \quad (2)$$

where $A = 2.618$, $n = 2$, $\lambda_c = 313.3$, $\lambda_o = 300$, and $\lambda_f = 31.08$.

The analytical representation of Green et al. (1974) was used for the erythema (sunburning) action spectrum,

$$\epsilon_{ERYTHEMA}(\lambda) = \frac{\alpha}{1 + \exp[(\lambda - \lambda_o / \Delta)]} + \frac{4\alpha' \exp[(\lambda - \lambda_o') / \Delta']}{(1 + \exp[(\lambda - \lambda_o') / \Delta'])^2} \quad (3)$$

where $\alpha = 0.04485$, $\Delta = 3.130$, $\lambda_o = 311.4$, $\alpha' = 0.9949$, $\Delta' = 2.692$, and $\lambda_o' = 296.5$.

The R-B meter Sunburn Units (SU) were read directly from the instrument. The spectral response of this meter is shown with several other commonly used action spectra in NAS (1979).

Polynomial regressions were derived using the HP computer with standard statistical software for regression analysis. The range of absolute UV-B irradiance (285-315 nm) values in the nomograms includes normal laboratory levels as well as ambient levels measured over four years at our experimental site at Manchester, Washington. Absolute irradiance was limited to wavelengths no longer than 315 nm since all of the weighting functions sharply decline above this point. However, absolute irradiance between 285-320 nm may be estimated by adding 40% and 100%, respectively, to values reported here for the lamps and the sun.

RESULTS

Because of the spectral differences in UV-B irradiance between the lamps and the sun (Damkaer et al., 1981), a separate set of regressions is required for each. Given B, the absolute irradiance (285-315 nm) in Wm^{-2} from one FS-40 lamp combined with one cool-white lamp and both transmitting through CTA, the respective weighted irradiance (285-315 nm) using each of the functions can be estimated:

$$\text{DNA(Setlow)} = (1.4\text{x}10^{-2})B^2 - (1\text{x}10^{-2})B + (4\text{x}10^{-3})$$

$$\text{Plant(Caldwell)} = (3.1\text{x}10^{-2})B^2 + (2.7\text{x}10^{-2})B + (9\text{x}10^{-3})$$

$$\text{Erythema} = (-3.3\text{x}10^{-3})B^2 + (1.2\text{x}10^{-1})B + (4\text{x}10^{-3})$$

$$\text{R-B Meter(SU)} = (1.2)B^2 + (5.3\text{x}10^{-1})B + (3.9\text{x}10^{-1})$$

Given S, the absolute incident solar irradiance (285-315 nm) in Wm^{-2}, the weighted solar UV-B irradiance can be estimated:

$$\text{DNA(Setlow)} = (-2\text{x}10^{-3})S^2 + (9\text{x}10^{-3})S - (1\text{x}10^{-3})$$

$$\text{Plant(Caldwell)} = (-7.9\text{x}10^{-3})S^2 + (6\text{x}10^{-2})S - (6.2\text{x}10^{-3})$$

$$\text{Erythema} = (2\text{x}10^{-4})S^2 + (6.4\text{x}10^{-2})S - (1\text{x}10^{-2})$$

$$\text{R-B Meter(SU)} = (1.8\text{x}10^{-2})S^2 + (2.8)S - (2.8\text{x}10^{-1})$$

Using the relationships established by these regressions, nomograms were constructed which provide rapid conversions among weighting functions most commonly applied to biological effectiveness of UV-B in aquatic ecosystems (Figs. 1 and 2).

DISCUSSION

It must be emphasized that nomograms can provide only an estimate of equivalent irradiance values among the weighting functions. The applicability of nomograms is limited by difficulties in exactly duplicating experimental conditions and techniques. Among the factors to consider when using the nomograms are number, type, and age of artificial light sources, type, thickness, and condition of light-filters, and wavelength band. Nevertheless, the nomograms are useful not only in comparing the levels of UV-B irradiance used by other investigators but, by using the two nomograms together, one can also compare artificial laboratory levels of irradiance with natural levels.

While the FS-40 lamps with CTA filters provide a reasonable simulation of solar UV-B irradiance, there are well-documented differences in spectral irradiance between this artificial source and the sun. These spectral differences account, in general, for the differences in scale between the two nomograms. They also account, in part, for the differences in shape of

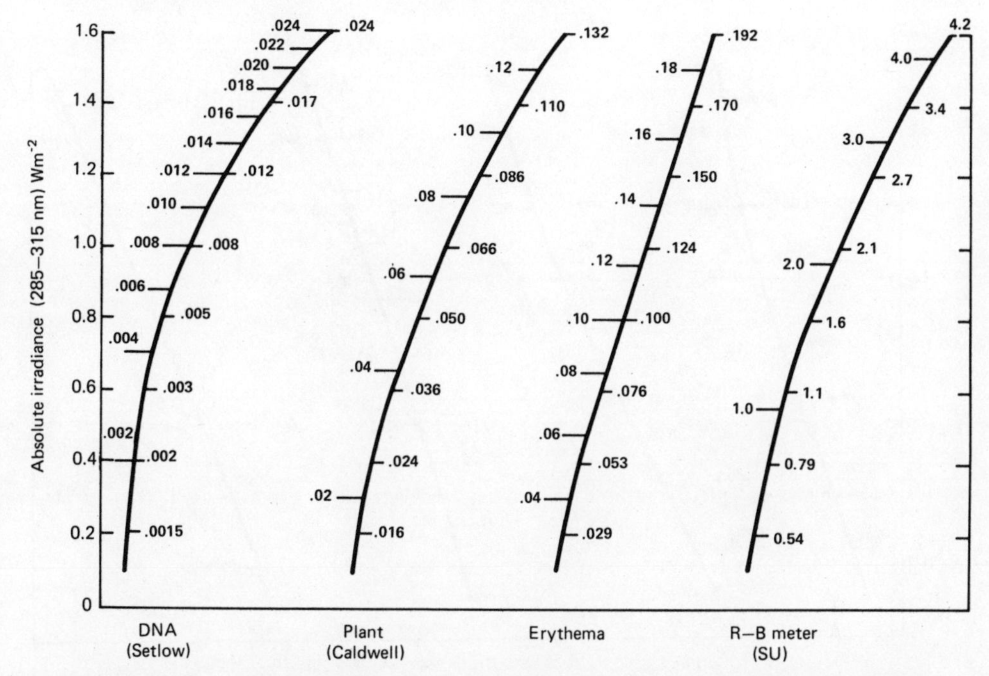

Figure 1. Nomogram for artificial (FS-40 + CW + CTA) absolute irradiance (285-315 nm), to estimate biologically effective UV-B irradiance from four weighting functions. A horizontal line connecting the left and right scale of absolute irradiance will intersect the other scales at the estimated equivalent values.

the two DNA curves and the two Plant curves (Figs. 1 and 2). With the lamps, absolute irradiance is primarily regulated by the thickness of CTA filters. High absolute irradiance values are achieved with a very thin filter which also allows more of the shorter UV-B wavelengths to penetrate. Because the DNA and Plant weighting functions increase sharply at the lower end of the UV-B range (NAS, 1979), high levels of absolute irradiance from the lamps lead to accelerated increases in DNA-weighted irradiance and, to a lesser extent, in Plant-weighted irradiance (Fig. 1). On the other hand, with increases in total solar absolute irradiance the intensity of shorter wavelength UV-B does not increase in the same proportion as with the lamps. At the same time the large increases in longer wavelength solar UV-B are given extremely low weightings in the DNA and Plant action spectra (NAS, 1979). Therefore, with increases

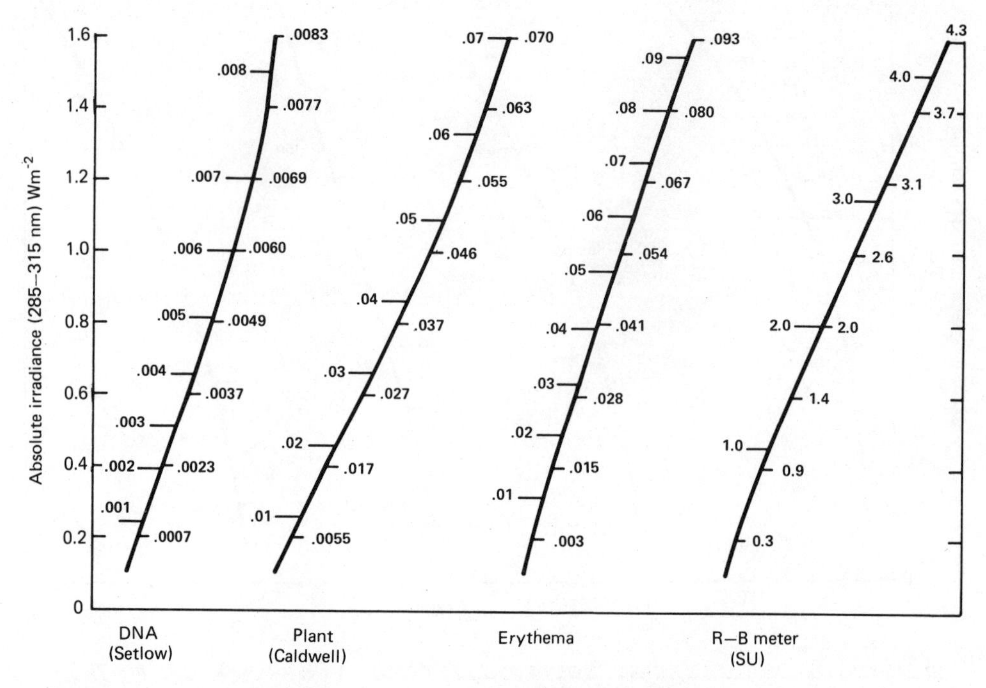

Figure 2. Nomogram for solar (Sun + Sky) absolute irradiance (285-315 nm), to estimate biologically effective UV-B irradiance from four weighting functions. A horizontal line connecting the left and right scale of absolute irradiance will intersect the other scales at the estimated equivalent values.

at high levels of absolute irradiance in nature, DNA- and Plant-weighted irradiances increase at a much lower rate than with the artificial sources of UV-B (Fig. 2).

The determination of appropriate UV-B weighting functions of biological effectiveness for a variety of aquatic organisms must be one of the priorities of future investigations. Without this work there will be difficulties in relating and comparing data and conclusions.

REFERENCES

Berger, D., D. F. Robertson, and R. E. Davies. 1975. Field measurements of biologically effective UV radiation. Impacts of Climatic Change on the Biosphere, CIAP Monogr. 5, 1(2): 235-264.

Billen, D., and A. Green. 1975. Comparison of germicidal activity of sunlight with the response of a sunburning meter. _Photochem_. _Photobiol_. _21_: 449-451.
Caldwell, M. M. 1968. Solar UV radiation as an ecological factor for alpine plants. _Ecol_. _Monogr_. _38_: 243-268.
Damkaer, D. M., D. B. Dey, and G. A. Heron. 1981. Dose/dose-rate responses of shrimp larvae to UV-B radiation. _Oecologia_ (_Berl._) _48_: 178-182.
Damkaer, D. M., D. B. Dey, G. A. Heron, and E. F. Prentice. 1980. Effects of UV-B radiation on near-surface zooplankton of Puget Sound. _Oecologia_ (_Berl._) _44_: 149-158.
Green, A., and J. H. Miller. 1975. Measures of biologically effective radiation in the 280-340 nm region. Impacts of Climatic Change on the Biosphere, CIAP Monogr. _5_ 1(2): 60-70.
Green, A., T. Sawada, and E. Shettle. 1974. The middle ultraviolet reaching the ground. _Photochem_. _Photobiol_. _19_: 251-259.
National Academy of Sciences. 1979. Protection against depletion of stratospheric ozone by chlorofluorocarbons. Washington, D. C.
Setlow, R. B. 1974. The wavelengths in sunlight effective in producing skin cancer: a theoretical analysis. _Proc_. _Nat_. _Acad_. _Sci_. _71_: 3363-3366.

A FORMULA FOR COMPARING ANNUAL DAMAGING ULTRAVIOLET (DUV) RADIATION DOSES AT TROPICAL AND MID-LATITUDE SITES

Pythagoras Cutchis

Institute For Defense Analyses
Science and Technology Division
400 Army-Navy Drive
Arlington, Virginia 22202

ABSTRACT

A formula is presented in this paper for the relative annual damaging ultraviolet radiation (DUV) dose at different tropical and mid-latitude sites. The "DUV dose" signifies the solar energy incident on a horizontal surface, weighted by the erythemal response spectrum (Robertson 1975, Scotto et al. 1976, Cutchis 1974). The DUV dose is assumed equivalent to the "dose" which would be measured by a Robertson-Berger meter unit (Robertson 1975, Scotto et al. 1976). The DUV formula consists of six multiplicative factors which include the effects of amount of ozone, latitude, altitude, cloudiness, ground albedo, and amount of aerosols. A seasonal ozone variation factor is introduced to modify the tropical relative DUV formula for application to mid-latitude sites. The formula has been developed by mathematically fitting sometimes sparse data, and remains to be validated in the general sense.

The formula should be useful, in the absence of more exact data, in studies of the effects of solar ultraviolet radiation and its possible increase from a reduction in stratospheric ozone on land and marine ecological systems and skin cancer incidence in Caucasian populations.

INTRODUCTION

A formula has been developed which can be used to compare the annual damaging ultraviolet radiation (DUV) doses at different geographical sites (Cutchis 1980). Relative DUV dose plays an important role in studies of the effects of solar ultraviolet radiation, and its possible enhancement by stratospheric ozone depletion on marine ecosystems, agricultural crops, skin cancer incidence in human populations, etc. Such a formula also can be used to suggest sites for future solar UV-B measurements.

The formula consists of six multiplicative factors which include the effects of amount of ozone, latitude, altitude, cloudiness, ground albedo, and amount of aerosols. The basis of the formula is the assumption that D, the relative annual damaging ultraviolet radiation dose (DUV) at a given site can be expressed as the product of six separable multiplicative factors, i.e.,

$$D = D_\tau \, D_L \, D_h \, D_c \, D_A \, D_\beta , \qquad (1)$$

where the subscripts τ, L, h, C, A, and β refer, respectively, to the average amount of ozone, latitude, altitude, average cloud amount, ground albedo, and amount of aerosols. The reference value of D is unity, corresponding to an equatorial sea-level site with an average annual amount of ozone of 240 m atm-cm, no clouds, zero ground albedo, and standard atmosphere. The two most significant factors, ozone and latitude have been shown to be independent of each other. The remaining factors are of lesser importance and could involve some interdependency effects. While the magnitudes of these effects are probably of second order the validity of the formula remains to be demonstrated by either calculation or measurements.

METHOD

The method of establishing the empirical formula will be illustrated using the factors ozone and latitude for tropical sites. Methods for developing the remainder of the formula are fully explained in Cutchis (1980).

TROPICAL SITES

Ozone

The thickness of the ozone column, τ, varies with latitude. Fig. 1 (from London et al. 1976) shows the global distribution of total ozone averaged over the 10-year period starting in July, 1957. The worldwide distribution of the stations used to obtain data is also indicated in the Figure. The contour map indicates that the northern part of South America and the central part of Africa had the minimum average total ozone column of approximately 240 x 10^{-3} cm, or 240 m atm-cm, for the period 1957-1967. If only the ozone factor is considered, the highest ultraviolet radiation levels could be expected to be found within the 240 m atm-cm contour of Fig. 1.

The use of an average value of total ozone in the formula derived below for the tropics would be inappropriate for regions outside the tropical zone because of the large ozone fluctuations with season. The effect of season on amount of ozone is minimal near the equator, as indicated in Fig. 2 (from London et al. 1976). Also, differences in total average ozone of only approximately 10 m atm-cm are to be found at the equator, while a difference of 70 m atm-cm is found between longitudes 130°E and 40°E at a latitude of 50°N (Fig. 1).

The ozone function D_τ at the equator is shown as a function of τ in Fig. 3. The five circled points shown are based on the sum of the 12 monthly DUV tabulated values, including both scattered and direct solar radiation incident on a flat horizontal surface, calculated by Mo and Green (1975) at sea level for a standard amount of aerosols in a clear atmosphere with zero ground albedo. A parabolic fit to the three calculated points at 256, 288, and 320 m atm-cm was made, leading to

$$D_\tau = 9.424 \times 10^{-4}\ \tau^2 - 0.830\ \tau + 206.2 \qquad (2)$$

However, a simpler linear approximation can be used here to cover the narrow band of expected values of τ near the equator. The following linear equation is adequate for the purpose of this paper to cover the tropical range of 240 - 280 m atm-cm (10-year average).

$$D_\tau = 62 - 0.30\ (\tau - 240) \qquad (3)$$

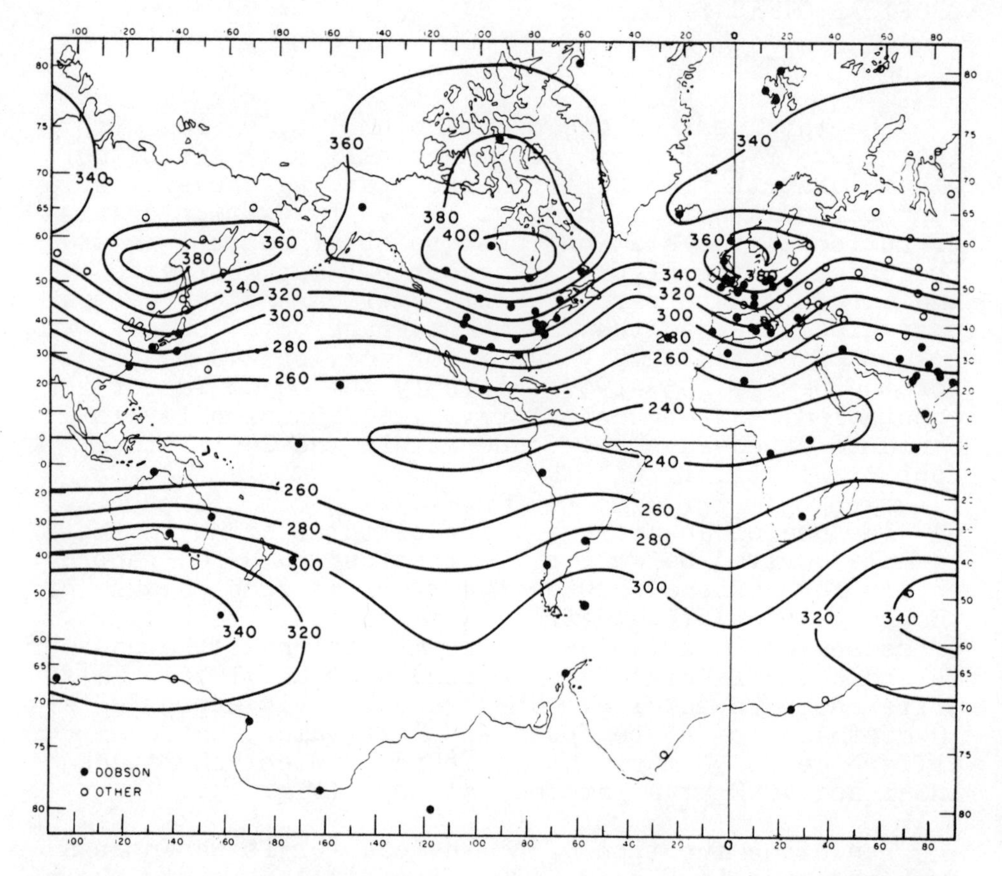

Figure 1. Global distribution of total ozone averaged over the period 1957-1967. (Source: London et al 1976).

Normalizing, we have the relative ozone value function

$$D_{\tau} = 1 - 0.00484\ (\tau - 240) \qquad (4)$$

Eq. 4 can be used to compare the annual DUV at two sites of different longitude along the equator ($L = 0^{o}$) if they differ in their τ value and all of the other four factors are assumed to be equal.

The second term in Eq. 4 gives the fractional decrease in moving from a site having 240 m atm-cm of ozone to one with a higher value of τ. In agreement with Fig. 4.4.81 in The Report of the Committee on Meteorological Effects of Stratospheric Aircraft

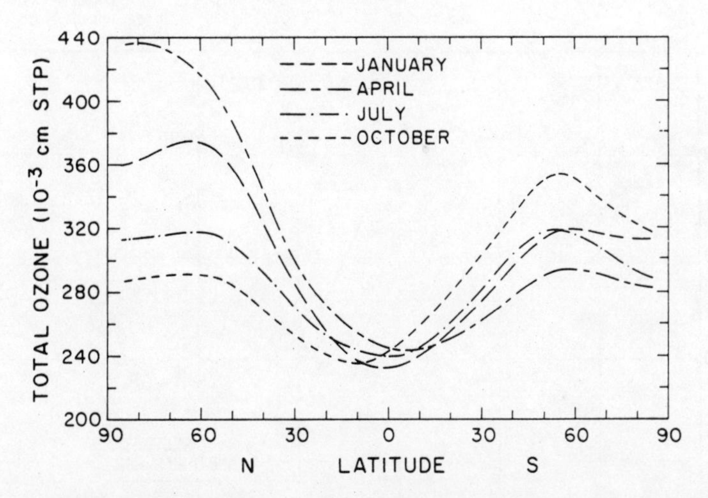

Figure 2. Latitudinal variation of average total ozone averaged over the period 1957-1967. (Source: London et al. 1976)

(Comesa) 1972-1975, Part 2, 1975, for every 1 percent increase in ozone along the equator, there is a 1.16 percent decrease in DUV (as compared to approximately 2 percent decrease for mid-latitude sites).

LATITUDE

In Fig. 4 the variation of the log of relative DUV is plotted against the square of the latitude (L^2); for values of τ between 256 and 320 m atm-cm. The circles points are again based on the calculations of Mo and Green (1975). There are two significant empirical observations to be noted in Fig. 4: (1) for a given value of τ the relative DUV values fall almost exactly on a straight line for latitudes less than 25°, and (2) the slopes of the lines are almost exactly equal. With these two fortuitous characteristics, it is possible to represent the relative annual DUV, considering only ozone and latitude, as the product D_τ D_l where

$$D_L = \exp(-\ 3.74 \times 10^{-4} L^2) \qquad (5)$$

and L is in degrees.

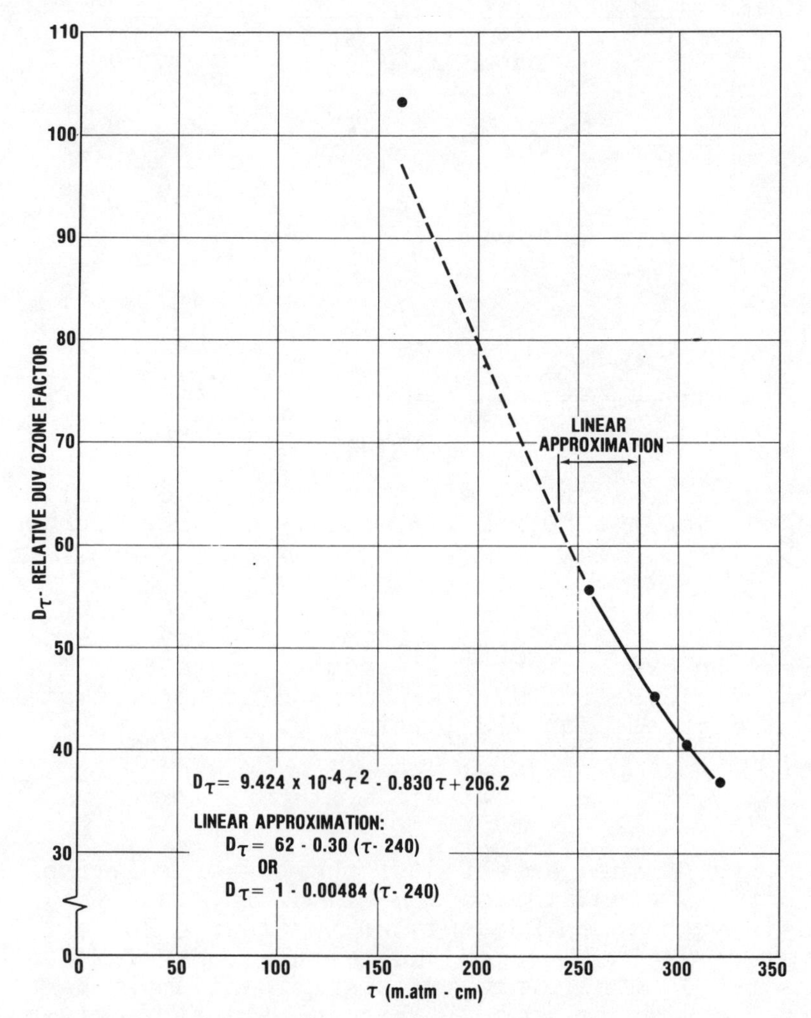

Figure 3. Relative DUV vs. ozone thickness at the equator

With Eqs. 4 and 5 it is possible to compare the relative DUV for any two tropical sites, assuming the other four factors are equal. Thus, for example, the coastal town of Townsville in Queensland, Australia at a latitude of 19° S with a τ of 260 m atm-cm (Fig. 1) had, over the period 1957-1967, an average DUV, relative to an equatorial sea-level site in South America or Africa, of

$$(1 - 0.0968)\ \exp\ (-0.135) = 0.78 \qquad (6)$$

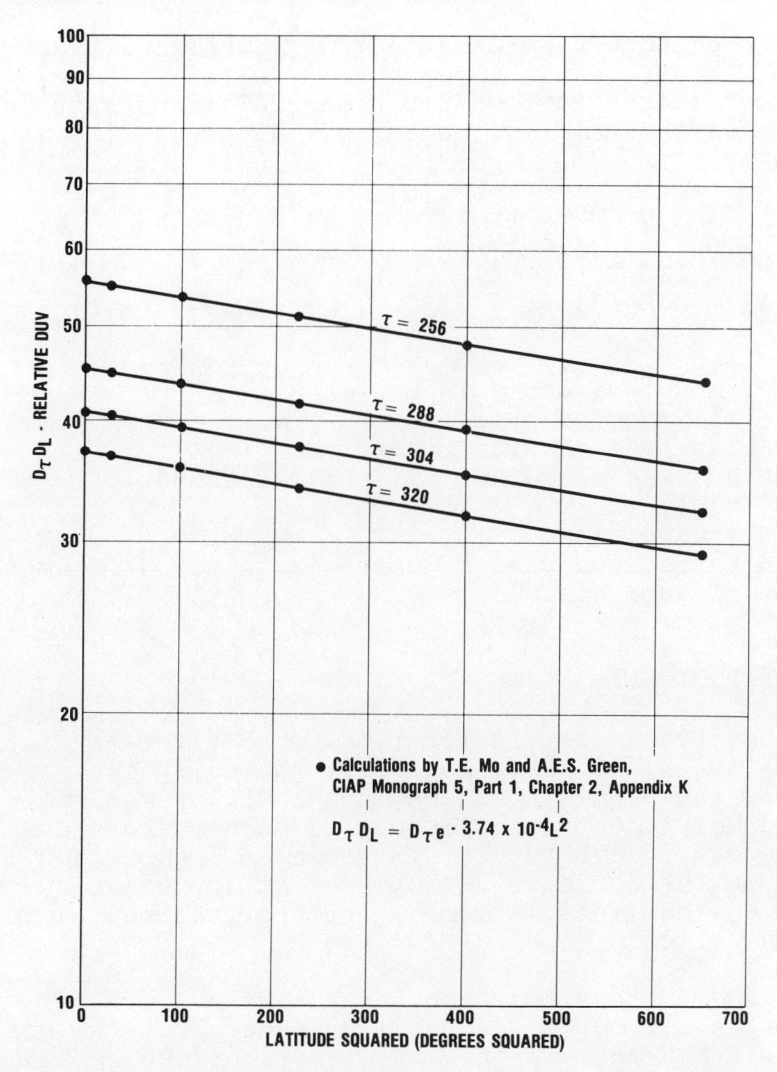

Figure 4. Relative DUV vs latitude squared

assuming equality of three modifying factors, i.e., cloudiness, ground albedo, and amount of aerosols.

Analogous formulas have been developed for altitude, solar zenith angle, cloudiness, ground albedo, and aerosols.

FORMULA FOR RELATIVE ANNUAL DUV AT TROPICAL SITES

Substituting various empirical relations into Eq. 1 gives, for the relative annual DUV at tropical sites, the formula

$$D = [1 - 0.00484 (\tau - 240)] (1 + 0.06 h) \quad (7)$$
$$\exp (-3.74 \times 10^4 L^2)$$
$$\times (1 - 0.50 C) (1 + 0.50 A) [1 - 0.093 (\beta - 1)]$$

where
- τ = amount of ozone in m atm-cm
- h = altitude of site in km
- L = latitude of site in degress
- C = average cloud amount (unity for complete overcast)
- A = ground albedo
- β = ratio of amount of aerosols to standard amount of aerosols.

MID-LATITUDE SITES

To derive a formula for relative DUV for mid-latitude sites, it is first necessary to modify the equation for the ozone thickness factor D_τ. The latter covered only the tropical range of ozone values 240-280 m atm-cm. Since τ values can reach as high as 400 m atm-cm in the Northern Hemisphere, it is necessary to derive another formula for D_τ for application to mid-latitude sites.

Figure 5 shows the annual relative DUV at the equator as a function of ozone thickness τ. The solid line is drawn through the circles, points which are based on calculations of Green and Mo (1975). The relative DUV units in Fig. 5 are based on daily erythema doses on the 15th day of each month in $(\text{Joules/m}^2)_e$ and differ from those in Fig. 4 which were based on annual monthly sums in MED.

A very good parabolic approximation, obtained by fitting the calculated DUV values at 300, 350, and 400 m atm-cm, is given by

$$D_\tau = 3.80 \times 10^{-5} \tau^2 - 3.95 \times 10^{-2} \tau + 11.19 \quad (8)$$

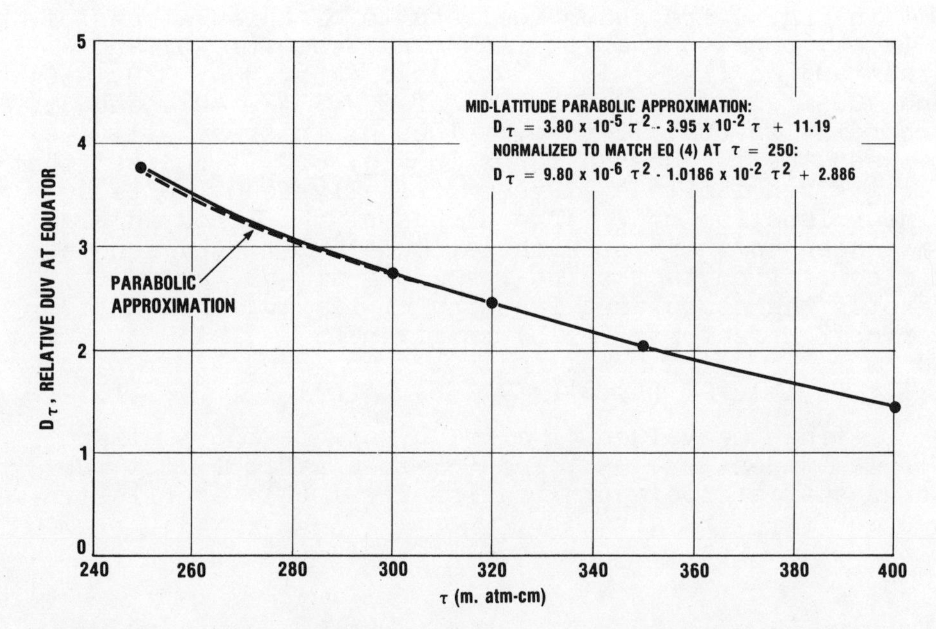

Figure 5. Relative DUV at equator vs ozone thickness.

In order that Eq. 8 be compatible with Eq. (4), the equations are matched at $\tau = 250$. According to Eq. (4), $D_\tau = 0.9516$ at $\tau = 250$, whereas, according to Eq. 8, $D_\tau = 3.69$. Thus, multiplying Eq. 8 by 0.9516/3.69 yields the normalized parabolic approximation

$$D_\tau = 9.80 \times 10^{-6}\ \tau^2 - 1.0186 \times 10^{-2}\ \tau + 2.886 \tag{9}$$

MID-LATITUDE CORRECTION FOR TROPICAL LATITUDE FACTOR

The tropical latitude exponential factor D_L in Eq. 5 must be modified for mid-latitude application. If D_{Lm} (τ, L) denotes the mid-latitude relative annual DUV for ozone thickness τ and latitude L, and α (τ, L) denotes the correction factor for the value of D_L as given by Eq. 5, then

$$D_{Lm}(\tau, L) = \left[1 - \alpha\ (\tau, L)\right] D_L \tag{10}$$
$$= \left[1 - \alpha\ (\tau, L)\right] \exp\ (-3.74 \times 10^{-4} L^2).$$

In Fig. 6 are shown the results of calculations of α (τ, L) based on Tables in Mo and Green (1975) and Green and Mo (1975) for τ values of 256, 300, 320, 350, and 400 m atm-cm and L values of 30°, 35°, 40°, and 55°. According to Fig. 1, the α values of interest will lie between the two dashed lines in Fig. 6. It is seen that for a given latitude, α is empirically found to be a linear function of τ. The slopes of the lines in Fig. 6 are seen to increase with latitude. If f (L) denotes the correction factor for τ = 256 m atm-cm, and S (L) denotes the slope as a function of latitude, then the correction factor α (τ, L) is given by

$$\alpha (\tau, L) = f (L) + (\tau - 256) S (L) \tag{11}$$

By fitting the values of α (τ, L) at τ = 256 m atm-cm for latitudes 30°, 45°, and 55° to a parabola, it is found that a good approximation for f (L), as illustrated in Fig. 7 is given by

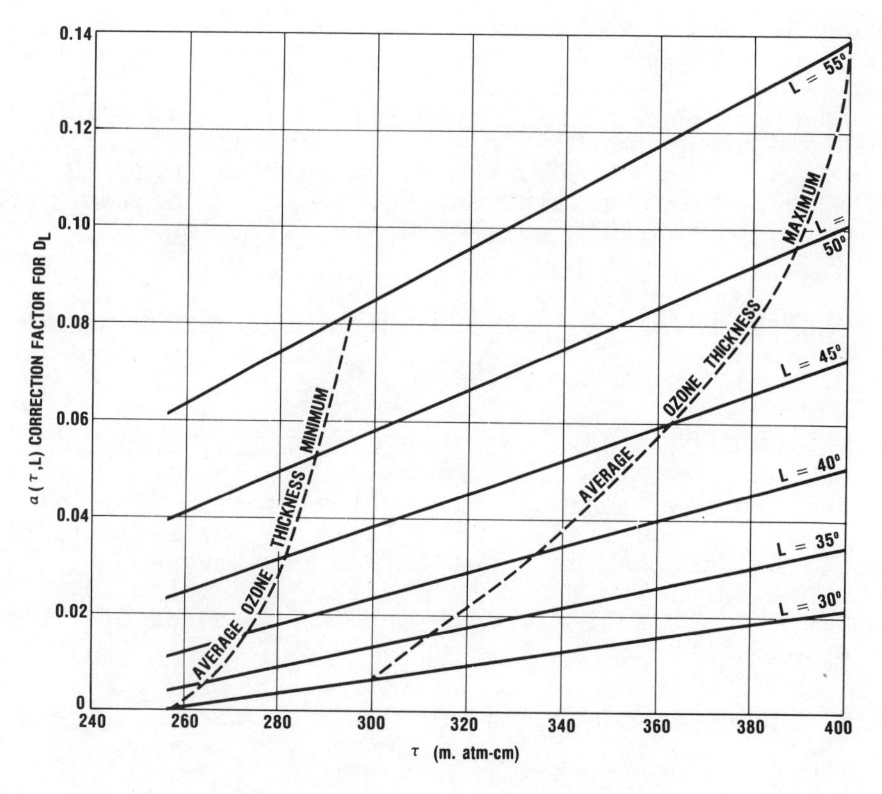

Figure 6. Correction factor for D_L.

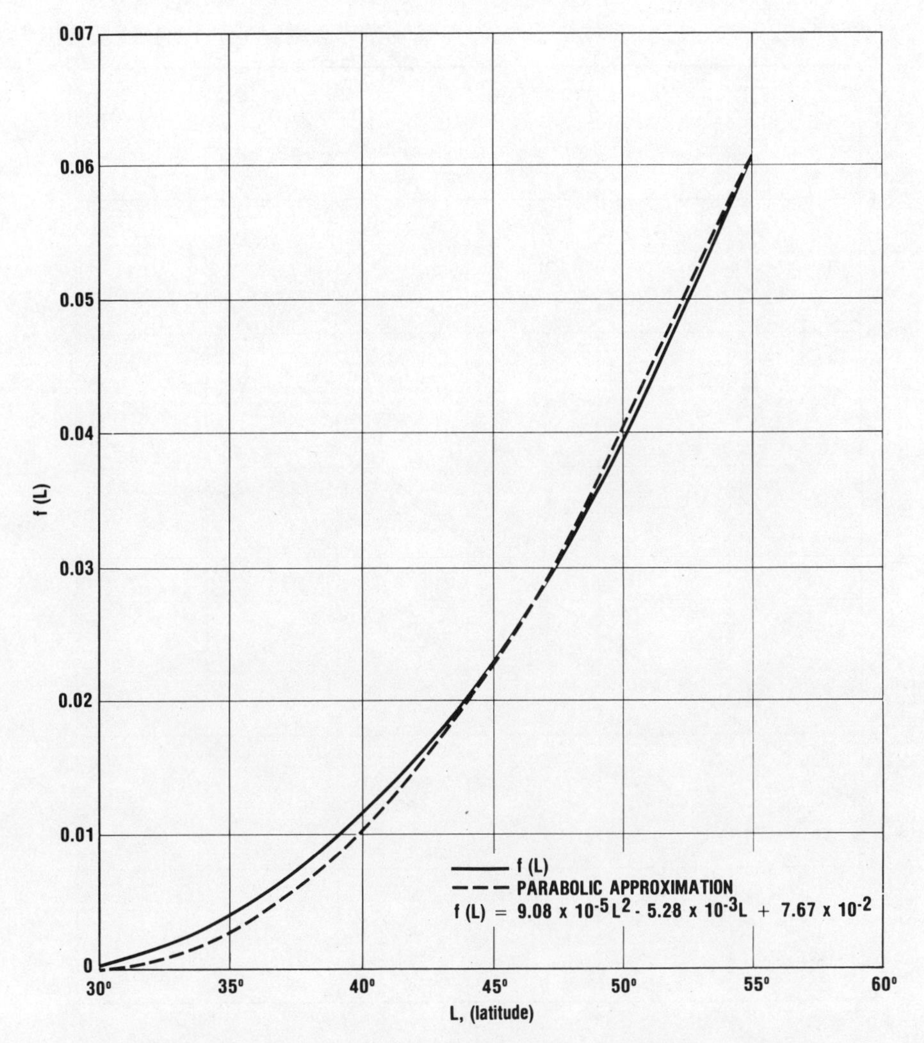

Figure 7. f(L) and a parabolic approximation.

$$f(L) = 9.08 \times 10^{-5} L^2 - 5.28 \times 10^{-3} L^2 + 7.67 \times 10^{-2} \quad (12)$$

The slope S (L) is found to be linear in the interval $30^o < L < 45^o$, as indicated in Fig. 8. The function S (L) may be approximated by the equation

$$S(L) = g(L) + \delta \mu (L) \quad (13)$$

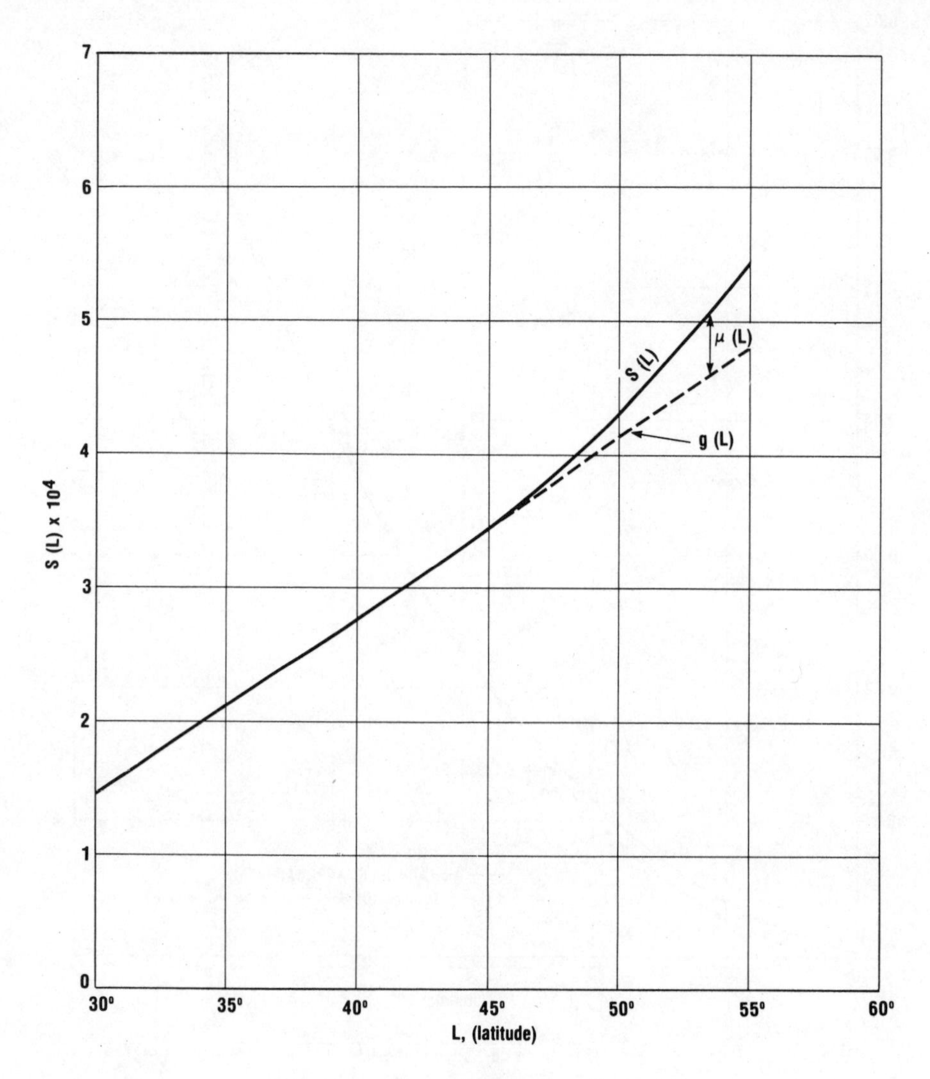

Figure 8. S(L) vs. L.

where $\delta = 0$ for the interval $30^0 < L < 40^0$ and $\delta = 1$ for $L > 45^0$; μ (L) is a slope correction factor for latitudes greater than 45^0. By assuming the slope correction factor μ (L) increases proportionally to the square of $(L - 45^0)$, the function $\bar{S}$ $(\bar{L})$ is well approximated by the equation

$$S\ (L) = 1.46 \times 10^{-4} + 1.34 \times 10^{-5}\ (L - 30^0) + 6.10 \times 10^{-7}\ \delta\ (L - 45)^2 \tag{14}$$

Substituting Eqs. 12 and 14 in Eq. 11, the mid-latitude correction factor α (τ, L) is therefore given by

$$\alpha\ (\tau, L) = 9.08 \times 10^{-5} L^2 - 5.28 \times 10^{-3} L^2 + 7.67 \times 10^{-2} + (\tau - 256) \left[1.46 \times 10^{-4} + 1.34 \times 10^{-5} (L - 30^o) + 6.10 \times 10^{-7} \delta (L - 45^o)^2\right] \quad (15)$$

where $\delta = 0$ for the interval $30^o < L < 45^o$ and $\delta = 1$ for $L > 45^o$.

CORRECTION FACTOR FOR SEASONAL OZONE VARIATION

If the average ozone thickness were a constant value independent of season, then the substitutions of Eq. 9 for the ozone thickness factor D_τ and Eqs. 10 and 15 for the latitude factor D_L in the tropical relative DUV formula would accurately estimate the relative DUV for a mid-latitude site. However, the ozone thickness varies significantsly from month to month at mid-latitude sites. This seasonal ozone variation introduces a significant fractional error ρ such that

$$D_s = D \left[1 + \rho(\tau, L)\right] \quad (16)$$

where D_s is the relative annual DUV as adjusted for seasonal ozone variation. It was shown that ρ is positive in the Northern Hemisphere and negative in the Southern Hemisphere, also that ρ is a function of the average annual ozone thickness, τ, as well as latitude, L, in the Northern Hemisphere, but appears to be essentially independent of τ in the Southern Hemisphere (Cutchis 1980).

$$\rho_s (L) = - 6.75 \times 10^{-5} L^2 + 3.75 \times 10^{-3} L - 6.66 \times 10^{-2} \quad (17)$$

$$\rho_n(\tau, L) = 0.010 + 1.20 \times 10^{-4} L^2 - 3.80 \times 10^{-3} L - (\tau - 197.8 - 2.46 L) (1.424 \times 10^{-3} + 1.955 \times 10^{-6} \times 10^{-6} L^2 - 9.25 \times 10^{-5} L) \quad (18)$$

FORMULA FOR RELATIVE ANNUAL DUV AT MID-LATITUDE SITES

Substituting Eq. 9 for the first term in Eq. 7, multiplying by the latitude correction $[1 - \alpha(\tau, L)]$ and the seasonal ozone correction factor ϱ gives, for the relative annual DUV at mid-latitude sites, the formula

$$D = (9.80 \times 10^{-6} \tau^2 - 1.0186 \times 10^{-2} \tau + 2.886$$

$$\times [1 - \alpha(\tau, L)] [\exp-(3.74 \times 10^{-4} L^2]$$

$$[1 + \varrho(\tau, L)] (1 + 0.06\ h)$$

$$\times (1 - 0.50C) (1 + 0.50\ A) [1 - 0.093 (\beta - 1)] \quad (19)$$

where $\alpha(\tau, L)$ is given in Eq. 15, and $\varrho(\tau, L)$ by Eq. 17 for the Southern Hemisphere and by Eq. 18 for the Northern Hemisphere. Other symbols are as previously defined.

CONCLUSIONS

It was shown (for details see Cutchis (1980) that it is possible to derive and apply, to the extent available ozone and meteorological and geographical data are available, a formula for a relatively quick determination of annual relative DUV dose for tropical and mid-latitude sites. The input parameters required are average annual amount of ozone, latitude, altitude, average cloud amount, ground albedo, and amount of aerosols. Ozone, latitude, and altitude information is readily available. Cloudiness information to the accuracy required is, unfortunately, not readily available on a worldwide basis. Input data on ground albedo and aerosol content are also not readily available, but, in general, can be expected to play a less significant role than the other four parameters.

The derived formulas for tropical and mid-latitude sites can be used to provide a fundamental input parameter, relative annual DUV dose, in models designed to investigate the effects of solar ultraviolet radiation and its possible increase resulting from stratospheric ozone depletion, on ecological systems on land and in the oceans of the world, and on the incidence of skin cancer in human populations.

Because of the sparsity of the data available and possible interdependency effects in the six multiplicative factors derived, the formula remains to be validated by further field study or calculations.

ACKNOWLEDGEMENTS

This paper was modified from a report prepared for the High Altitude Pollution Program of the Federal Aviation Administration under Contract No. DOT-FA77WA-3965.

The author is indebted to Dr. Ernest Bauer, Mr. Henry Hidalgo, and Dr. R. C. Oliver of IDA for their critical reviews of this paper. He is especially grateful to Dr. R. C. Oliver who suggested extending the tropical relative DUV formula to mid-latitude.

REFERENCES

Cutchis, P. 1974. Stratospheric ozone depletion and solar ultraviolet radiation on earth. Science 184:13.

Cutchis, P. 1980. A formula comparing annual damaging ultraviolet (DUV) radiation doses at tropical and mid-latitudes. Final Report High Altitude Pollution Program. U.S. Dept. of Transportation. IDA log # HQ 80-22403.

Green, A.E.S. and T. Mo. 1975. Erythema Radiation Doses, CIAP Monogr. 5, Part 1, Chapter 2, Appendix I, Department of Transportation Climatic Assessment Program.

London, J., R. D. Bojkov, S. Oltmans, and J. I. Kelley. January, 1976. Atlas of the Global Distribution of Total Ozone July 1957-June 1967, National Center for Atmospheric Research, NCAR/TN/113 + STR.

Mo, T. and A.E.S. Green. 1975. Systematics of Climatic Variables and Implications - Local Erythema Dose, CIAP Monogr. 5, Part 1, "Ultraviolet Radiation Effects," Chapter 2, Appendix K, Department of Transportation Climatic Assessment Program. Sept. 1975.

Robertson, D. F. September 1975. Calculated Sunburn Responses, CIAP Monogr. 5, Part 1, Chapter 2, Appendix J, Department of Transportation Climatic Assessment Program.

Scotto, J., T. R. Fears, and G. B. Gori, Measurements of Ultraviolet Radiation in the United States and Comparison with Skin Cancer Data, National Cancer Institute, DHEW No. (NIH) 76-1029.

Venkateswaren, S. V., R. J. Breeding, J. J. DeLuisi, J. Gille, A.E.S. Green, R. Greenstone, B. M. Herman, H. Hidalgo, F. M. Luther, E. P. Shettle, N. Sundararaman, September 1975. Radiation in the Natural and Perturbed Troposphere, CIAP Monogr. 4, Chapter 5, Department of Transportation Climatic Assessment Program.

PREFACE TO SECTION II - THE HYDROSPHERE

A column of water of variable thickness is interposed between the solar UV reaching the water surface and the aquatic organisms. While the nature of the chemical and the optical events occurring in the suspending water are known to be just as complex and controversial as the ways in which the atmosphere modifies solar UV radiation, it would appear that for ecological purposes the optical oceanographers need to answer only one encompassing question: How much solar UV reaches organisms at various depths in natural waters? In more rigorous terms: What are the attenuation coefficients of natural waters?

But closer examination reveals problems of great complexity: The questions of proper biological weighting arise again since the attenuation of solar UV in water is wavelength dependent. The attenuation processes, scattering and true absorption, combine in complex ways to produce the effective attenuation. Pure water is much more transparent to UV-B wavelengths than was previously believed; if not water itself, then what are the most significant attenuators, substances dissolved or suspended in natural waters? Could the attenuation characteristics of natural waters be deduced as the summation of contribution from measurable components present in the waters?

Aquatic systems have an additional problem not present in determination of exposure to solar radiation for terrestrial ecosystems, in general, land plants and animals receive either full sunlight or very little exposure. On the other hand, aquatic biota may undergo the complete possible range of exposure as they move about in the water column. Many larger aquatic organisms have full control of their position, but very critical parts of the aquatic ecosystems are carried about by the water movement. To fully evaluate solar UV in aquatic systems one must also inquire into the nature and magnitude of the mixing processes which may move critical organisms up and down in the water column.

While there is a tendency to concentrate on the direct photochemical actions of solar UV on living organisms, it should also be recognized that solar UV radiation is a potent agent for inducing chemical reactions in various materials which may be found in natural waters. Chemicals can be rendered far more toxic or injurious through the action of sunlight; on the other hand, dangerous pollutants in the water may be transformed into harmless substances through photochemical reactions. It is required that questions regarding chemical reactions in natural waters be investigated: Where do the photochemical reactions occur? How important is sunlight for toxin production or detoxification in natural waters?

While most work on marine optics focuses on either the visible or ultraviolet part of the solar spectrum, organisms in the seas are, in fact, exposed to the two radiation bands simultaneously. Through position, aquatic organisms can control either the UV or visible light incident over a substantial range but one must also ask: How much UV/visible exposure will certain positioning behavior produce and how could the ratio of visible (beneficial radiation) to UV (harmful radiation) be made optimum?

To begin Section II, Baker and Smith (Chapter 20) provide a further elaboration of a method previously proposed for the calculation of the penetration of solar UV into marine waters. The diffuse attenuation coefficient is considered to be the sum of contributions from 1) the water itself 2) the plants in the water (measured by the chlorophyll) and 3) the humic or dissolved organic material in the water. New data substantially reduce the UV absorption coefficients previously reported for waters of high productivity.

The aquatic ecologist may have little need to understand details of the optical phenomena occurring in the water, but requires some way of estimating the attenuation coefficient for biologically active UV in particular waters. Calkins (Chapter 21) proposes an alternative method for obtaining the diffuse absorption coefficient for natural waters. It was observed that attenuation of the biologically effective solar UV was related to the Secchi disc transparency for many natural waters. Formulas are provided to relate UV attenuation to observations of the attenuation of visible light in the actual water in question. It is suggested that this approach to determination of attenuation of solar UV could be very useful to aquatic biologists who may have only limited equipment and no access to alternative ways of estimating penetration of solar UV in the waters which are under study.

Water, free of dissolved or suspended materials, transmits solar UV wavelengths quite well. The component of natural waters which is most effective in absorbing solar UV is the "yellow substance", a material of great importance in quantitating solar UV effects, which is considered in detail by Højerslev (Chapter 22). Perspective regarding the UV problems in marine waters arises from considerations of the source of yellow substance. Højerslev investigates the possible sources of yellow substance which appears in the oceans.

Ways of estimating movements in natural waters and the origin and nature of these movements are discussed by Kullenberg (Chapter 23). Specific instances of photochemical reactions in natural waters are presented by Zepp in Chapter 24. Calkins and Thordardottir (Chapter 25) investigate the relationship of visible radiation (in particular, green light) and solar UV-B measured simultaneously at a number of locations off Iceland.

As in the presentations in Section I, studies of the role of solar UV in the hydrosphere are at various levels of development. Ideas about the magnitude and origin of absorption of UV in natural waters have evolved and changed radically over the last 30 years. Solar UV penetrates much more deeply in many waters than was widely believed. It remains difficult to estimate penetration in natural waters, especially freshwaters. Problems such as movements which circulate the biota in waters and the activation or decomposition of toxic chemicals by sunlight are clearly enormously complex problems which we are only beginning to address.

SPECTRAL IRRADIANCE PENETRATION IN NATURAL WATERS

Karen S. Baker and Raymond C. Smith

Scripps Institution of Oceanography
University of California, San Diego
La Jolla, California 92093

ABSTRACT

In order to investigate the influence of ultraviolet radiant energy on marine organisms, it is necessary to quantitatively describe the spectral irradiance, $E(z,\lambda)$, present in the UV-B region of the spectrum. Models have been developed which provide for the descriptive organization of spectral irradiance data for a wide range of atmospheric conditions (Baker et al., 1980, hereafter referred to as BSG; Baker et al., this volume) and for a wide range of ocean water types (Smith and Baker, 1978b, hereafter referred to as SBI; Baker and Smith, in press, hereafter referred to as BSII). In the following we summarize this previous work and discuss its implications in terms of the influence of underwater UV-B radiation on marine organisms.

ATMOSPHERIC MODEL

A BSG atmospheric model can be used to calculate the spectral irradiance reaching the surface of the ocean in the 280 to 380 nm region, i.e.

$$E(0+,\theta,\lambda) = E_{sun}(0+,\theta,\lambda) + E_{diff}(0+,\theta,\lambda) \quad (1)$$

where $E(0+,\theta,\lambda)$ is the total downward global flux just above the ocean surface (0+), $E_{sun}(0+,\theta,\lambda)$ is the direct (sun) component and $E_{diff}(0+,\theta,\lambda)$ is the diffuse component. This model permits as input parameters the time, location, and atmospheric conditions (including ozone thickness). The total downward irradiance just below the surface (0-) is calculated by

$$E(0-,\theta,\lambda) = t(\theta)\cdot E_{sun}(0+,\theta,\lambda) + t_d\cdot E_{diff}(0^+,\theta,\lambda) \quad (2)$$

where $t(\theta)$ is the transmittance of the air-sea interface as calculated using Fresnel's equation and t_d = 0.94 is the transmittance of the air-sea interface for a uniform radiance distribution (Preisendorfer, 1976; Austin, 1974).

OCEANOGRAPHIC DATA

For the BSII model, downwelling spectral irradiances have been measured as a function of depth using

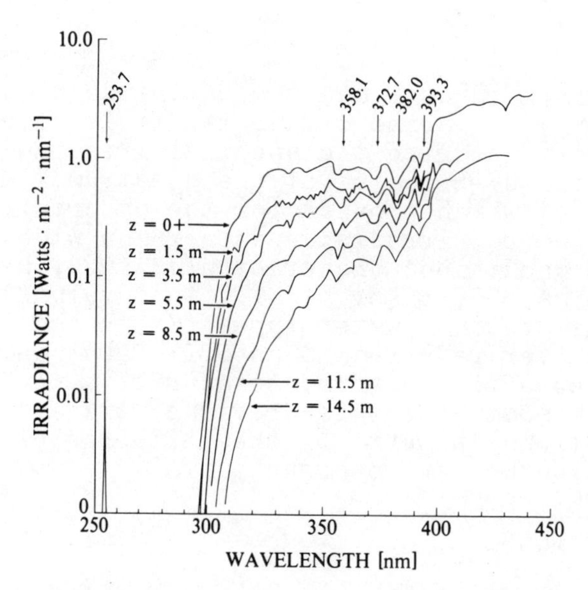

Figure 1. Downwelling spectral irradiance, $E[\text{watts}\cdot m^{-2}\cdot nm^{-1}]$, as a function of wavelength, λ[nm], measured using the UV-B underwater spectroradiometer at the depths 0, 1.5, 3.5, 5.5, 8.5, 11.5 and 14.5 meters. The arrows indicate the sun's Fraunhofer lines except for the 253.7 nm line which is from an internal mercury lamp. All of these lines aid in the wavelength calibration of the instrument.

the submersible Scripps spectroradiometer (Tyler and Smith, 1966, 1970) in the wavelength region from 350 nm to 700 nm and the UV submersible spectroradiometer (Smith et al. 1979) in the wavelength region from 280 nm to 700 nm. An example of data from the UV-B region of the spectrum is illustrated in Fig. 1. Although these data are limited, they comprise the first set of spectral irradiance data to span a range of biogeneous water types for the spectral region from 300 nm to 400 nm. Since the data do not include waters highly influenced by terrigenous material, the model does not simulate such waters.

The diffuse attenuation coefficient for irradiance, $K_T(\lambda)$, is the optical property which relates irradiance just beneath the ocean surface, $E(\theta^-,\lambda)$, to irradiance at depth, $E(z,\lambda)$:

$$E(z,\lambda) = E(0^-,\lambda) \exp -K_T(\lambda) \cdot Z \qquad (3)$$

The rationale for choosing $K_T(\lambda)$ as the optical property for modeling has been discussed elsewhere (Smith and Baker, 1978a).

From $E(z,\lambda)$ data the spectral diffuse attenuation coefficient can be directly determined for the water column between the depths z_1 and z_2:

$$K_T(\lambda) = -\frac{1}{z_2 - z_1} \ln \frac{E(z_2,\lambda)}{E(z_1,\lambda)} \qquad (4)$$

However, for UV-B data $K_T(\lambda)$ is very sensitive to small errors in the depth determination because $E(z,\lambda)$ is measured near the surface and because (z_2-z_1) is relatively small. Consequently, a more accurate data taking mode, when large depth intervals are impractical, is to fix the wavelength and to measure irradiance as a function of depth. The depth data for a single wavelength is then least-squares fit to a straight line on a semi-log plot, the slope of which gives

$$K_T(\lambda) = -\frac{1}{E(z,\lambda)} \frac{d(\ln E(z,\lambda))}{dz} \qquad (5)$$

This method has the advantage of averaging out variations in depth due to rough seas. For this analysis it is assumed that the water column is optically homogeneous to the depth of interest.

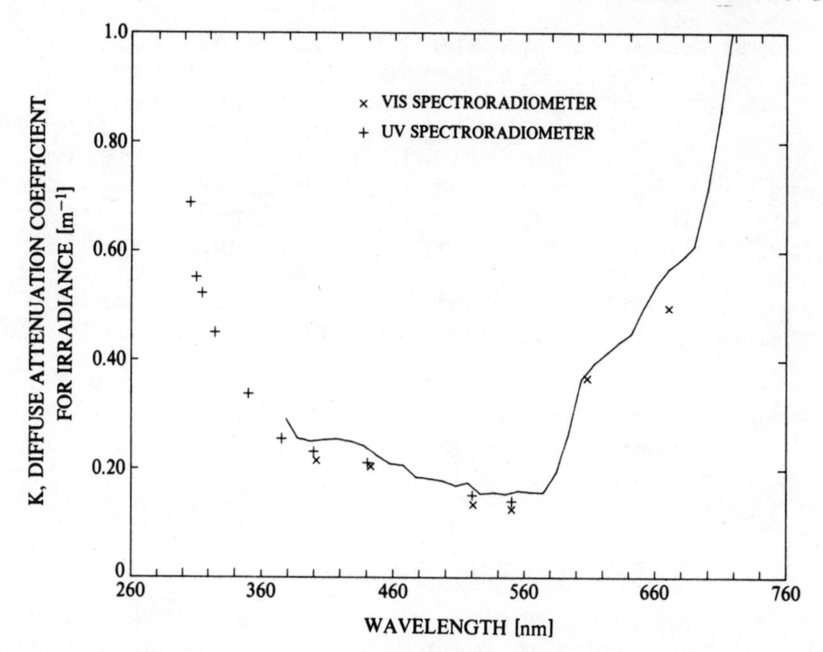

Figure 2. The diffuse attenuation coefficient for irradiance, $K_T[m^{-1}]$, as a function of wavelength, λ[nm], measured using three different methods (1) the solid line determined using the Scripps Spectroradiometer at two depths, Eq. (4); (2) the crosses using the Scripps Spectroradiometer at a fixed wavelength and varying the depth, Eq. (5); (3) the pluses from the UV submersible spectroradiometer in a similar manner. Eq. (5).

Figure 2 presents values of $K_T(\lambda)$ derived in these two ways: the solid line from $E(z,\lambda)$ data determined using the Scripps Spectroradiometer at two depths, Eq. (4); the crosses from $E(z,\lambda)$ data using the Scripps Spectroradiometer at a fixed wavelength and varying the depth Eq. (5); the pluses from $E(z,\lambda)$ data from the UV Submersible Spectroradiometer in a similar manner, Eq. (5). These data indicate that the results using the two instruments and the two different techniques for determining $K_T(\lambda)$ are consistent within experimental error and environmental change.

OCEANOGRAPHIC MODEL

In the BSII optical classification model, $K_T(\lambda)$ is composed of three terms:

$$K_T(\lambda) = K_W(\lambda) + K_C(\lambda) + K_D(\lambda) \tag{6}$$

The clear water component, $K_W(\lambda)$, is known and based upon an accurate and consistent set of data (Smith and Baker, 1981). The chlorophyll component, $K_C(\lambda)$, accounts for all "chlorophyll-like" pigments that co-vary with chlorophyll. The dissolved organic material component, $K_D(\lambda)$, accounts for that biogeneous material variously referred to as dissolved organic material (DOM), gelbstoff, yellow substance, or gilvin.

The diffuse attenuation coefficient for the clearest natural waters, $K_W(\lambda)$, has been determined with emphasis on the spectral region from 300 to 400 nm (Smith and Baker, 1981). It has been shown that $K_W(\lambda)$ can be related to the inherent optical properties of pure water, in particular the total absorption coefficient $a_w(\lambda)$ and the molecular scattering coefficient $b_m(\lambda)$, by means of equations derived from radiative transfer theory. Thus limiting values of $K_W(\lambda)$ can be estimated from $a_w(\lambda)$ and vice versa. A comparative analysis, using published $a_w(\lambda)$ and our own $K_W(\lambda)$ data, allows a consistent and accurate set of optical properties for the clearest natural waters and for pure fresh water and salt water to be estimated from 300 to 800 nm (See Table 1, Smith and Baker, 1981). We estimate the accuracy of the selected data in the UV-B to be within +25 and -5% between 300 and 400 nm.

$K_W(\lambda)$ values represent the lowest attenuation coefficient values for natural waters and are unique in that they represent a natural limit. Thus, these values can be used to estimate maximum penetration of radiant energy into natural waters. That is, when $K_C(\lambda) \approx K_D(\lambda) \approx 0$, the penetration of UV-B and biologically effective dose-rates (Smith and Baker, 1979) will be maximum.

The chlorophyll component of the diffuse attenuation coefficient of irradiance, $K_C(\lambda)$, was determined (BSII) using a larger data base than previous modeling efforts (SBI) which permitted the model to be extended further into the UV-B region. The behaviour of the nonlinear K_C term is shown by a plot of $(K_T - K_W)/C$ versus log C, as in Fig. 3. The Gaussian form of K_C is illustrated by this figure where the peak is centered about C = 0.5mg pigment$\cdot$m^{-3}, instead of C = 1.0mg pigment$\cdot$m^{-3} as was found in SBI. The expanded data set and our procedure for separating the nonlinear part of the analytic fit suggest that the co-varying material

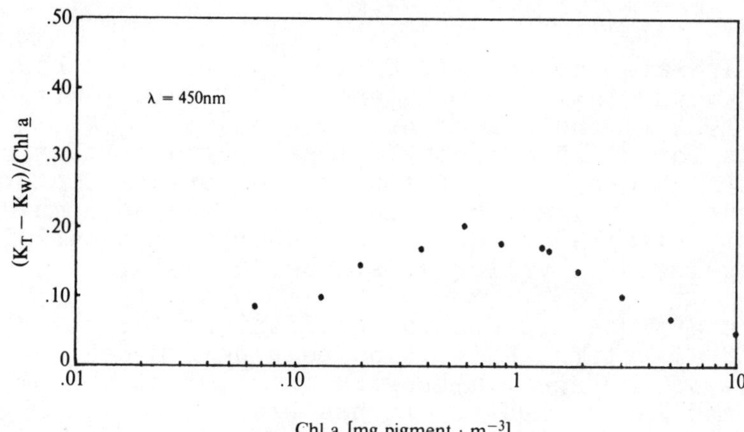

Figure 3. The chlorophyll component of the diffuse attenuation coefficient for irradiance divided by the chlorophyll concentration, $K_C/C = (K_T - K_W/C$, versus chlorophyll concentration, C[mg pigment $\cdot m^{-3}$] for wavelength 450 nm for those waters with negligible DOM.

changes in a smooth fashion as the environment changes from low ($C \ll 0.5$mg pigment $\cdot m^{-3}$) to medium ($C \approx 0.5$mg pigment$\cdot m^{-3}$) to high ($C \gg 0.5$mg pigment $\cdot m^{-3}$) chlorophyll concentrations. The nonlinearity of our analysis is the subject of further investigations with respect to its interpretation.

Since many of the early UV-B calculations have been based on SBI, it is important to realize the changes brought about through use of BSII. In general, the earlier model led to an underestimate of the penetration of MUV into water. A comparison of the two models is given in Fig. 4.

The attenuation due to DOM, $K_D(\lambda)$, has been found to have an exponential form as discussed in BSII. When experimentally determined K_T (λ)data are compared with values of $K_T(\lambda)$ calculated using the model and known values of C and D (Fig. 7 SBII), the agreement between model and experimental data is good. However, where data has been obtained at wavelengths near 300 nm, the comparison seems to indicate that a steeper exponential would be more representative of the measured data. There has been some debate as to whether the two slopes (Coastal versus Sargasso Sea) found by Steurmer (1975)

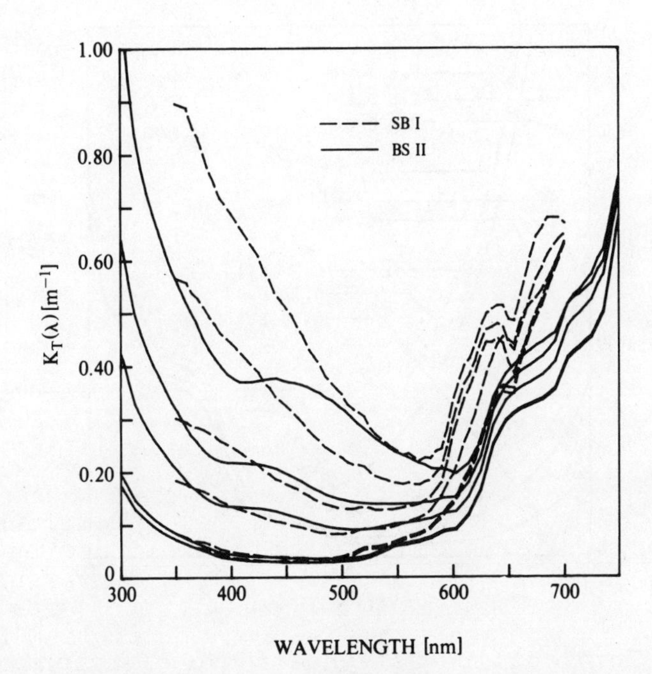

Figure 4. The diffuse attenuation coefficient of irradiance, $K_T[m^{-1}]$, versus wavelength, λ[nm], calculated for the range of chlorophyll values 0.05, 0.1, 0.5, 1.0, 5.0 mg pigment $\cdot m^{-3}$ using the SBI model (broken lines) and the BSII model (solid lines).

are generally applicable to the world's oceans. SBII used the slope of Hojerslev (1980) to calculate the K_D component which is very similar to the coastal Steurmer value. Steurmer's Sargasso Sea value may be the more appropriate for this data set which was not as heavily influenced by coastal effects as the Hojerslev data may have been.

Given the above observations of the BSII model's accuracy in the UV-B region, the model allows a full range of biogeneous water types to be analytically modeled. This is illustrated by Fig. 5 which uses the model to calculate $K_T(\lambda)$ for a range of C and D

BIOLOGICAL EFFECTIVE DOSE

Smith and Baker (1979, this volume) have discussed the penetration of UV-B and biologically effective dose-rates in natural waters. From a knowledge of the relative biological efficiency, or generalized action

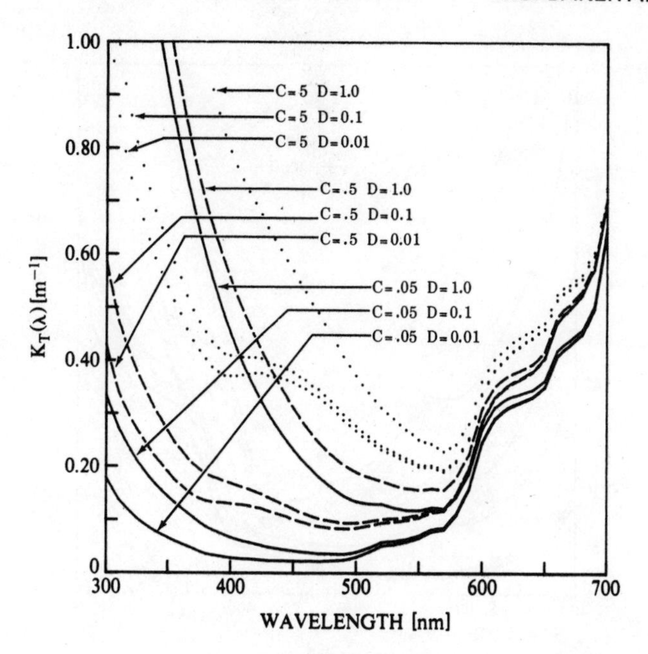

Figure 5. Total diffuse attenuation coefficient of irradiance, $K_T[m^{-1}]$, versus wavelength, λ[nm], as calculated using the BSII model for a range of chlorophyll and DOM concentrations.

spectrum, $\epsilon(\lambda)$, for the biological effect under study, it is possible to calculate the biologically effective radiation at depth z by

$$E_B(z,\theta)[W\cdot m^{-2}]_{\epsilon(\lambda)} = \int E(z,\theta,\lambda)[W\cdot m^{-2}\cdot nm^{-1}] \cdot \epsilon(\lambda) \cdot d\lambda[nm] \quad (7)$$

The total daily biological effective dose at depth z can be calculated by integrating $E_B(z,\theta)$ for all angles over the course of a day, i.e.

$$E_{TB}(Z)\ [J\cdot m^{-2} day^{-1}]_{\epsilon(\lambda)} = \int E_B(z,\theta)[J\cdot s^{-1}\cdot m^{-2}]_{\epsilon(\lambda)} \cdot dt[s] \quad (8)$$

We use the notation $[W\cdot m^{-2}]_{\epsilon(\lambda)}$ to indicate a physically measured (or measurable) absolute irradiance weighted by a relative weighting function with an arbitrary normalization wavelength. These units, while arbitrary, are precisely defined and are useful for a comparison with dose-rates in the same units obtained for different situations, e.g. different depths, different ozone thicknesses, etc. Also, different investigators, having agreed upon a weighting function and wavelength normalization, can quantitatively compare results.

A series of $\epsilon(\lambda)$ pertinent to the marine environment are shown in Fig. 6. As can be seen, these weighting functions are heavily weighted in the UV-B region of the spectrum. A key problem in the assessment of the potential influence of enhanced UV-B on marine organisms is an understanding of the photoprocess involved and the relevant biological weighting functions involved.

DISCUSSION

The bio-optical classification models outlined here provide: (1) systematic order to a wide range of experi-

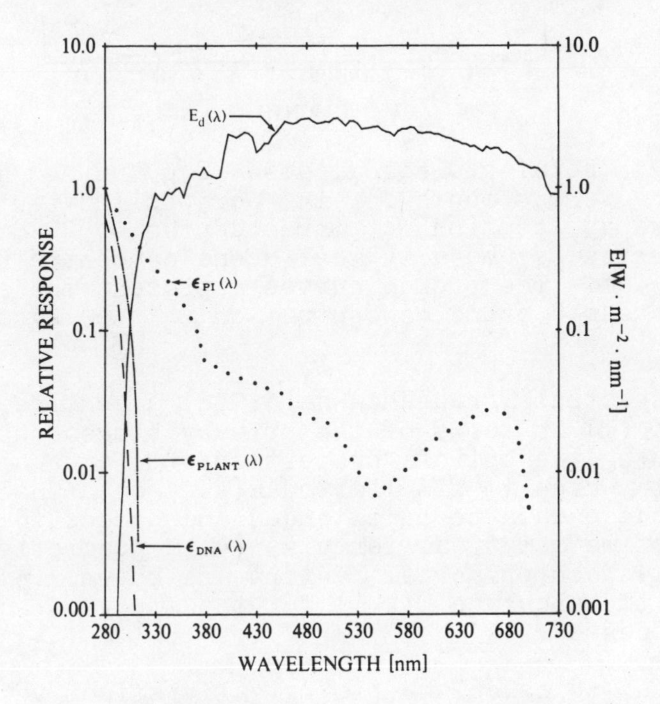

Figure 6. Relative biological efficiencies $\epsilon(\lambda)$ (left hand ordinate): for Setlow's (1974) average action spectrum for biological effects involving DNA (---); for Caldwell's (1971) generalized action spectrum for plants (- . -); and for Jones and Kok's (1966) action spectrum for photoinhibition of chloroplasts (---). The right hand ordinate (corresponding to the solid curve) gives the noon spectral irradiance for wavelength comparison of $\epsilon(\lambda)$ and $E_d(0^+,90,\lambda)$.

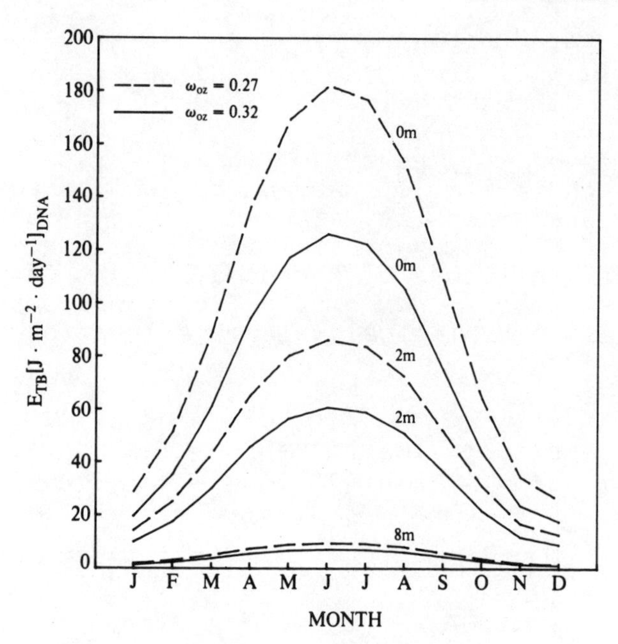

Figure 7. Total biological dose, E_{TB}, using ϵ_{DNA} weighting versus month for the Scripps Institution of Oceanography location for three depths 0, 2, 8m. The solid curves assume a typical ozone concentration ω_{OZ}= 0.32 atm-cm. The broken curves represent the situation of a decreased ozone concentration ω_{OZ} = 0.27 atm-cm.

mental spectral irradiance data; (2) a continuous index of $K_T(\lambda)$ in terms of the primary biogeneous factors influencing the optical properties; (3) an analytic form with which to fit limited $E(z,\lambda)$ data that then allows these data to be extended and used with the facility of mathematical formula; (4) a predictive model for $K_T(\lambda)$ when principal factors are known; (5) the ability to calculate $E(z,\lambda)$ for the study of aquatic photoprocesses.

Thus, to calculate biologically effective doses in natural waters we: (1) utilize the BSG marine atmosphere model to obtain $E(0^+,\theta,\lambda)$ for the atmospheric parameters of interest; (2) calculate $E(0^-,\theta,\lambda)$ using Eq. (2); (3) choose a diffuse attenuation coefficient for the water type of interest using our bio-optical classification model (BSII); (4) select a biological weighting function for the aquatic photoprocess under study; (5) determine $E(z,\lambda)$ using Eq. (3); (6) calculate the

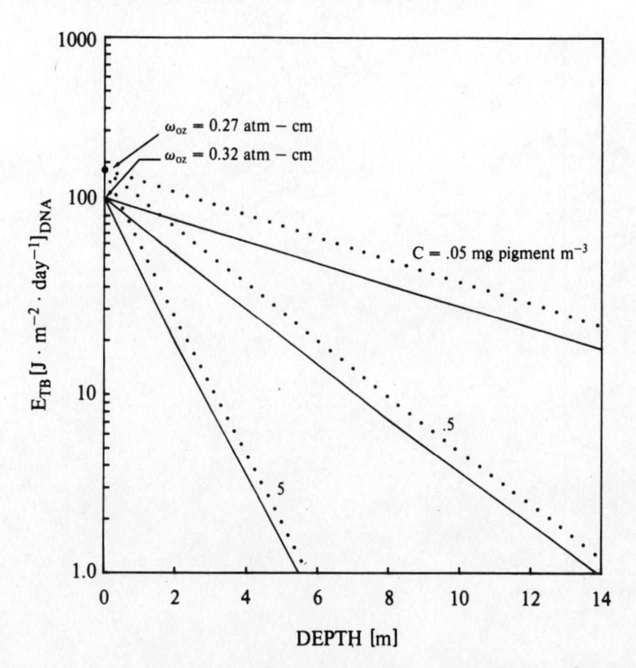

Figure 8. Total biological dose, $E_{TB}[J \cdot m^{-2}]_{DNA}$, using ϵ_{DNA} weighting versus depth, z[m], as calculated using the BSII model (Tables 1 and 2) with C = 0.05, 0.5, 5.0 mg pigment$\cdot m^{-3}$. The solid lines represent ω_{oz} = 0.32 atm-cm and the broken lines represent ω_{oz} = 0.27 atm-cm.

biological effective dose at depth z using Eq. (7); (7) determine the total daily biological dose using Eq. (8).

Figure 7 illustrates use of these calculations. The E_{TB} is plotted as a function of season by month. For the BSG model, our San Diego Scripps Institution of Oceanography (SIO) location was chosen along with the appropriate known ozone concentration (ω_{oz} = 0.32atm-cm) (step 1). Values of Chl=0.5mg pigment $\cdot m^{-3}$ and D=0.0mg DOM$\cdot l^{-1}$ were chosen for input to the BSII model for the calculation of $K_T(\lambda)$ (step 3). For this example ϵ_{DNA} was chosen (step 4) and then $E(z,\lambda)$, E_B, and E_{TB} were calculated (steps 5, 6, and 7) for three depths 0, 2, and 8 meters. The calculations were repeated assuming a 16% decrease in the atmospheric ozone thickness (National Research Council, 1979) to produce the comparative curves. At 2 meters, for example, these curves

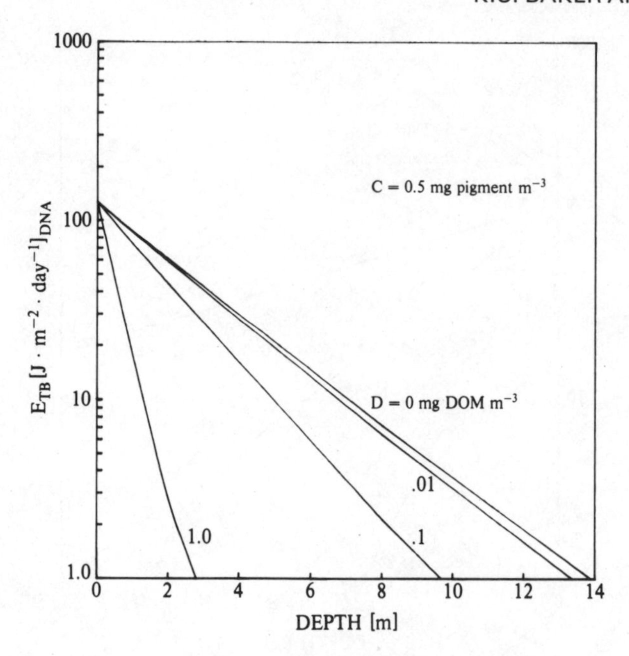

Figure 9. Total biological dose, $E_{TB}[J \cdot m^{-2} day^{-1}]_{DNA}$, using ϵ_{DNA} weighting versus depth, z[m], as calculated using the BSII model (Tables 1 and 2) with C = 0.5 mg pigment·m^{-3} and D = 0.0, 0.01, 0.1, and 1.0 mg DOM·l^{-1} assuming an ω_{oz} = 0.32 atm-cm.

show that the maximum daily dose of about 60[J·m^{-2}·day^{-1}] occurs only for a few days in June for the normal ozone thickness of ω_{oz} = 0.32atm-cm, whereas this dose is exceeded for nearly five months for an ozone thickness of ω_{oz} = 0.27atm-cm.

Figure 8 illustrates calculated values of E_{TB} as a function of depth for June in the coastal region near San Diego SIO for a range of chlorophyll values. Once again, this calculation has been carried out for two ozone values, ω_{oz} = 0.32atm-cm and ω_{oz} = 0.27atm-cm. Figure 9 shows a similar calculation with a fixed C=0.5mg pigment·m^{-3} and with a range of D values in order to show the rapid attenuation caused by DOM. If the maximum acceptable dose for a particular species is known, then such graphs illustrate the minimum depth at which the species can exist under the given circumstances. Hunter et al.,(this volume),have used our technique and model in their studies on the influence of UV-B on the Northern Anchovy.

The combination of the BSG atmospheric model and the BSII model provides a useful tool for calculating the spectral irradiance at any depth in a water column and allows the quantitative modeling of aquatic photo-processes.

ACKNOWLEDGEMENTS

This work was supported by the United States Environmental Protection Agency Stratospheric Impact and Assessment Program, Grant No. R 806489010.

REFERENCES

Austin, R. W. 1974. In: Ocean Color Analysis. SIO Ref. 74-10: 2.1-2.2.

Baker, K. S., R. C. Smith and A.E.S. Green. 1980. Middle ultraviolet radiation reaching the ocean surface Photochem. Photobiol. 32: 367-374.

Baker, K. S. and R. C. Smith. Bio-Optical classification and model of natural waters II. Limnol. Oceanogr. (In press).

Caldwell, M. M. 1971. In Photophysiology. A.C. Giese, [ed.] Academic Press.

Hojerslev, N. K. 1980. On the origin of yellow substance in the marine environment. Univ. Copenhagen Inst. Phys. Oceanogr. R Rep. 42.

Jones, L. W. and Kok, B. 1966. Photoinhibition of chloroplast reactions. Plant Physiol. 41: 1037-1043.

National Research Council. 1979. Protection against depletion of stratospheric ozone by chlorofluorocarbons. National Academy of Sciences. Washington,D.C.

Preisendorfer, R. W. 1976. Hydrologic Optics Vol. VI. U. S. Dept. of Commerce.

Setlow, R.B. 1974. The wavelengths in sunlight effective in producing skin cancer: A theoretical analysis. Proc. Natl. Acad. Sci. U.S. 71, 9: 3363-3366.

Smith, R. C. and K.S. Baker. 1978a. The bio-optical state of ocean waters and remote sensing. Limnol. Oceanogr. 23: 247-259.

Smith, R.C. and K.S. Baker. 1978b. Optical classification of natural waters. Limnol. Oceanogr. 23: 260-267.

Smith, R. C. and K. S. Baker. 1979. Penetration of UV-B and biologically effective dose-rates in natural waters. Photochem. Photobiol. 29: 311-323.

Smith, R.C., R.L. Ensminger, R.W. Austin, J.D. Bailey, G.D. Edwards. 1979. Ultraviolet submersible spectroradiometer. Ocean Optics VI, Soc. Photooptical Inst. Engs. 208: 27-140.

Smith, R. C. and K. S. Baker. 1981. Optical properties of the clearest natural waters (200-800 nm). App. Opt. 20: 177-184.

Stuermer, D. H. 1975. The characterization of humic substances in sea water. Ph.D. Thesis, Mass. Inst. Technol. Woods Hole Oceanogr. Inst.

Tyler, J. E. and R. C. Smith. 1966. Submersible spectroradiometer. J. Opt. Soc. Am. 56: 1390-1396.

Tyler, J. E. and R. C. Smith. 1970. Measurement of spectral irradiance underwater. Gordon and Breach, N. Y.

A METHOD FOR THE ESTIMATION OF THE PENETRATION OF BIOLOGICALLY INJURIOUS SOLAR ULTRAVIOLET RADIATION INTO NATURAL WATERS

John Calkins

Department of Radiation Medicine (1) and
School of Biological Sciences (2)
University of Kentucky, Lexington, KY 40536

ABSTRACT

In spite of the critical importance of attenuation data for the biologically potent short wavelength solar UV (primarily UV-B 280-320 nm) there are very few published data on the penetration of these wavelengths in freshwaters and only limited observations in marine waters, Smith and Tyler (1976). In the absence of extensive quantitative data, very erroneous impressions of the shielding of aquatic organisms by the water have arisen. Although superior methods of measuring or calculating attenuation are needed, and will doubtless be developed, certain regularities in the optical properties of natural waters have been observed and the use of these characteristics is proposed. The attenuation of solar UV radiation as weighted by the Robertson sensor (UV-RB) in a wide variety of natural waters has been measured. The waters surveyed included very clear oceanic and freshwaters, marine locations of high productivity, bays and estuaries, large and small lakes, reservoirs and streams. Various other measurements were made at the time of the UV-RB attenuation measurements.

Exponential attenuation of the biologically weighted solar UV was observed in the vast majority of the stations and a broad beam attenuation coefficient (K) was computed from the observations. It was found that the UV-RB attenuation coefficient maintained a relatively high correlation with the Secchi disc measurements if the waters were subdivided into two groups: "average oceanic", and a group encompassing average waters with a freshwater component, including "coastal, bays, and estuaries" and "average" lake. Two conditions require special consideration: 1) if the waters are highly productive and 2) if there are large amounts of "humic" substance evident.

Quantitative estimation of UV action in aquatic systems requires knowledge of the incident radiation, attenuation in reaching the organisms under study and the biological response to be expected from a given UV dose. It is necessary to use compatible units in evaluating the action of solar UV radiation. Large numbers of observations of incident radiation and biological actions of solar UV expressed in terms of the Robertson sensor units are becoming available. These measurements combined with estimates of attenuation of UV derived from the observations reported here will permit a much more extensive computation of the impact of solar UV radiation in aquatic ecosystems.

METHODS

Measurements of solar UV-RB have been made using the Robertson sensor, the sensing element of the "sunburning meter". Smith and Calkins (1976) consider the theory of the use of this device for UV-B attenuation measurements. The operating principal and other technical considerations are described in detail in Robertson (1969) and Berger (1976). In brief, the Robertson sensor uses the UV fluorescent characteristics of magnesium tungstate to simulate typical UV biological response. Fluorescence is large at 300 nm where biological action is high and declines with increasing wavelength in general correspondence to the reduction of biological effectiveness. The relative output of the Robertson sensor (per photon) is reduced by a factor of more than 1000 at 350 nm relative to 300 nm. It should be especially noted that the Robertson sensor,as used,does not measure the attenuation of a particular wavelength but the reduction of the biological potency of solar UV with depth in water. Since shorter wavelengths are attenuated more by the water column than longer UV, the peak of the action curve (See Figure 2 Calkins Chapter 10,this volume) will shift to longer wavelengths with depth. A similar shift occurs at the water surface each day; morning and afternoon solar action will peak at longer wavelengths than the solar noon action.

The University of Kentucky Robertson sensor was sealed and lowered into a wide variety of natural waters (note Table I). Solar UV-RB immediately below the surface and at appropriate depth intervals were noted. The Secchi disc transparency (Z_{sd}) was also determined at the same location (Holms, 1970; Tyler, 1968). The effective attenuation coefficient (K) for UV-RB was computed from the measured irradiance at depth (Z) relative to the irradiance just under the surface.

TABLE I

WATERS SURVEYED

SITE	NUMBER OF STATIONS SURVEYED
"Typical" Oceanic	
Off Puerto Rico	3
Off San Diego	14
Off Iceland	9
"Productive" Oceanic	
Off Iceland	12
Coastal, Bays, Estuaries and Harbors	
Puerto Rico	3
Delaware Bay	6
Chesapeake Bay-Patuxent River	10
Iceland	2
"Typical" Freshwater	
Lake Erie	6
Lake Huron	3
Lake Michigan	4
Lake Herrington, KY	11
Water Reservoir, KY	2
Wastewater Lagoon, KY	8
"Humic" Freshwaters	
Lake Superior	1
Douglas Lake, MI	2
Lake Herrington, KY	6
Atypical Sites	
Lake Superior	3
Lake Thingvalier, Iceland	1
Iceland Marine	4

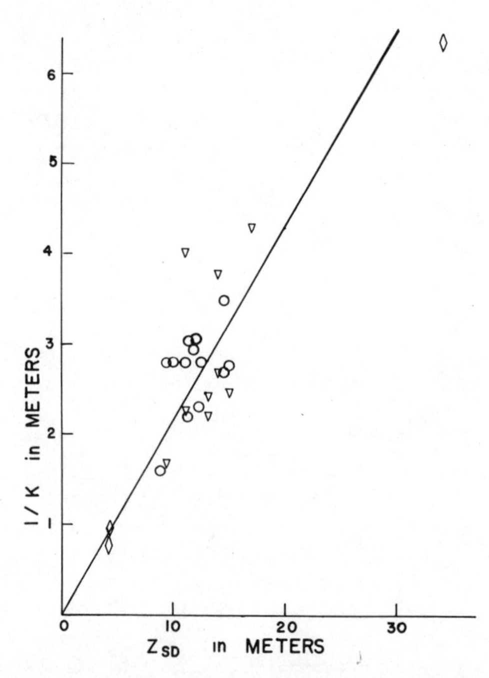

Figure 1. The relationship of the broad beam UV-RB attenuation coefficient to the Secchi disc depth for "typical oceanic" waters. The reciprocal of K, i.e., 1/K = depth attenuating the UV-RB irradiance by 1/e.≈ 37%,has been plotted against the depth where a 20 cm Secchi disc disappears from sight (Z_{sd}). Data from the waters off western Puerto Rico are indicated by ◊ , off San Diego, CA by ∇ and from waters off Iceland by O. Although productivity was measured only for the Icelandic data, the character of the water off Puerto Rico and San Diego would suggest low productivity at the time of measurement. The computer program used to fit the data points required the linear fit to pass through the point 0,0. The line is the computed least-square fit of the data points and has a slope of 0.215 with a standard error of .00866. The computed line fits the data points with a squared-multiple-correlation-coefficient (R-squared) equal to 0.961.

It was repeatedly observed that solar UV-RB irradiace as measured by the Robertson sensor declines in an exponential manner as the sensor is lowered into natural waters, i.e.:

$$I = I_o e^{-KZ} \quad (1)$$

where I_o is the UV-RB measured at the surface, I the UV-RB at depth Z, and K is a constant.

OBSERVATIONS

Figure 1 shows the values of 1/K and Z_{sd} simultaneously measured at the indicated marine stations. Figure 1 shows these values plotted and the slope of the least-square fit (required to pass through the origin) is indicated. The "average" oceanic waters seem to be waters of low productivity. The primary productivity of the Icelandic waters was measured and the optical properties of the other waters suggested a low productivity at the time of measurement. Figure 2 shows a similar plot of Z_{sd} and 1/K for high productivity waters off Iceland. It should be especially noted that when productivity is high, then the penetration of UV-RB is much greater relative to the penetration of visible light (as indicated by the Secchi disc) than when there is little productivity.

The properties of bays, estuaries and (near shore) coastal waters are plotted in Figure 3. Again, there is an apparently linear relation of 1/K and Z_{sd}; the solid line indicates the least-square fit of the data. Two types of freshwater observations have been included. If the water showed an intense brown coloration indicative of humic substances, it was grouped with the "humic lakes" (Fig. 4). If humic coloration was not noticeable, the lake observations were plotted as average lake" (Fig. 5). The slope of the coastal, bays, and estuaries (also indicated in Fig. 5) is almost identical to the relationship for "average" lake.

The observations are summarized in Fig. 6. Four oceanic and four lake observations which do not fit the classification scheme are indicated on Fig. 6 and are considered in the discussion section.

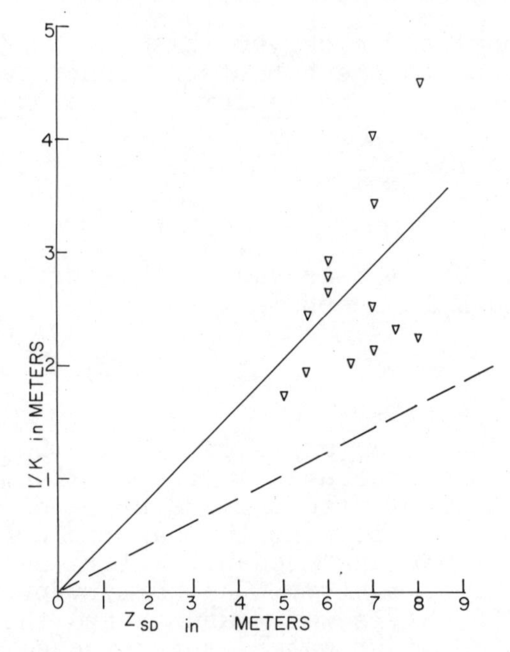

Figure 2. The relationship of Secchi disc depth (Z_{sd}) to the broad beam UV-RB attenuation coefficient (K) plotted as in Figure 1 for productive oceanic waters. All data points were locations off Iceland. Stations termed productive fixed 5-25 mg C/M^3/hr. Icelandic stations not considered productive and included in the "Typical Oceanic" data (Fig. 1) fixed less than 2.5 mg C/M^3/hr. Chlorophyll data paralleled productivity measurements. The slope of the computer fitted line (solid line) is 0.413 with a standard error of 0.028 and in R squared value of 0.942. The slope of "typical oceanic" waters (Figure 1) is indicated by the dotted line.

Clear shallow waters where the bottom is visible precludes the possibility of measuring Z_{sd}, therefore, measurements of Z_{sd} have been made in impoundments of various kinds rather than in the shallow local streams. The UV-B absorption of streams has been found to be quite similar to the lake or ponds from which the streams arise (unpublished observations).

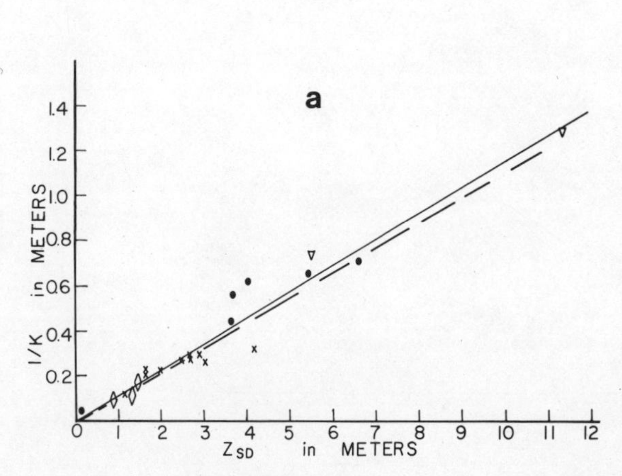

Figure 3a(top). The relation of Secchi disc depth (Z_{sd}) and the broad beam UV-RB attenuation coefficient (K) plotted as in Fig. 1 for various near coastal waters, bays, estuaries and harbors. Data from Iceland is indicated by ∇, Delaware Bay 0. Chesapeake Bay-Patuxent River X, Puerto Rico ◇ . The computer fitted line (solid line) has a slope of 0.116 with a standard error of 0.00345 and the R squared value is .982. The dotted line represents "average" freshwater.

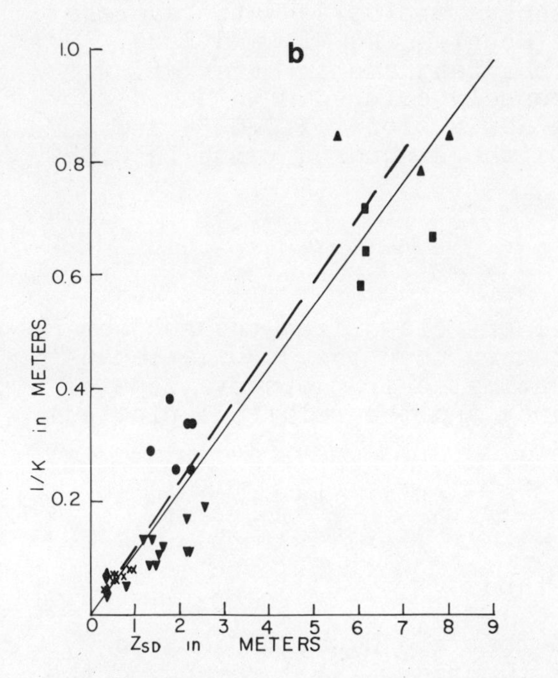

Figure 3b(bottom) The relation of Z_{sd} to K for "average" freshwaters. Data points include Lake Huron Δ , Lake Michigan □ , Lake Erie 0, Lake Herrington, Ky. ∇, a local (Ky.) water reservoir ◇, and a local (Ky.) wastewater treatment lagoon system X. The fitted computer line (solid line) has a slope 0.109 with a standard error of 0.00706 and an R squared value of 0.878. The dotted line represents fitting of coastal waters (Fig. 3a).

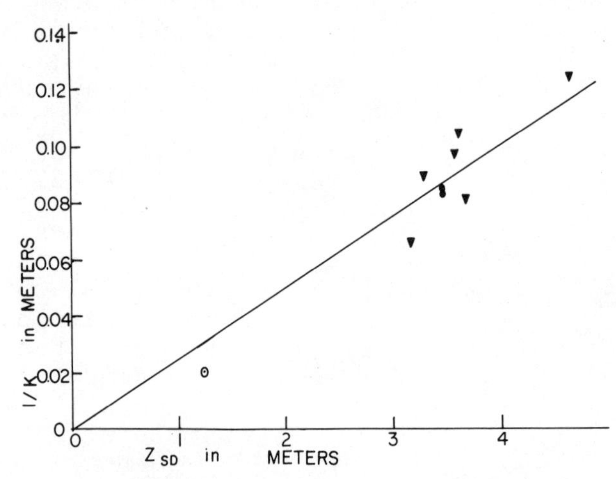

Figure 4. The relation of Z_{sd} to K for lakes showing humic coloration, plotted as in Figures 1-3. There were a number of measurements in Lake Herrington, Kentucky symbolized by ∇; Lake Herrington showed humic coloration in late fall and winter after local leaf fall. At other times of the year the lake was not obviously brown colored and the observations are plotted with "average lakes" in Figure 5. Lake Douglas, O, MI, and Lake Superior at the mouth of the Tahquamenon River MI, ʘ, were waters obviously intensely colored with humic substances. The fitted line has a slope of 0.0254 and a standard error of 0.00090; the R squared value is 0.989.

PROPOSED METHOD

It is proposed that if one classifies the subject water into the proper category then the UV-RB attenuation coefficient can be estimated from observations of the Secchi disc transparency and the relations plotted in Figures 1-5.

DISCUSSION

Rationalization of the method

It is widely accepted that the equation derived by Duntley and Prieserdorfer (Duntley, 1953) regarding the visibility of underwater objects is applicable to the Secchi disc observations (Holms, 1970; Tyler, 1968). The Duntley-Priesendorfer equation for a vertically

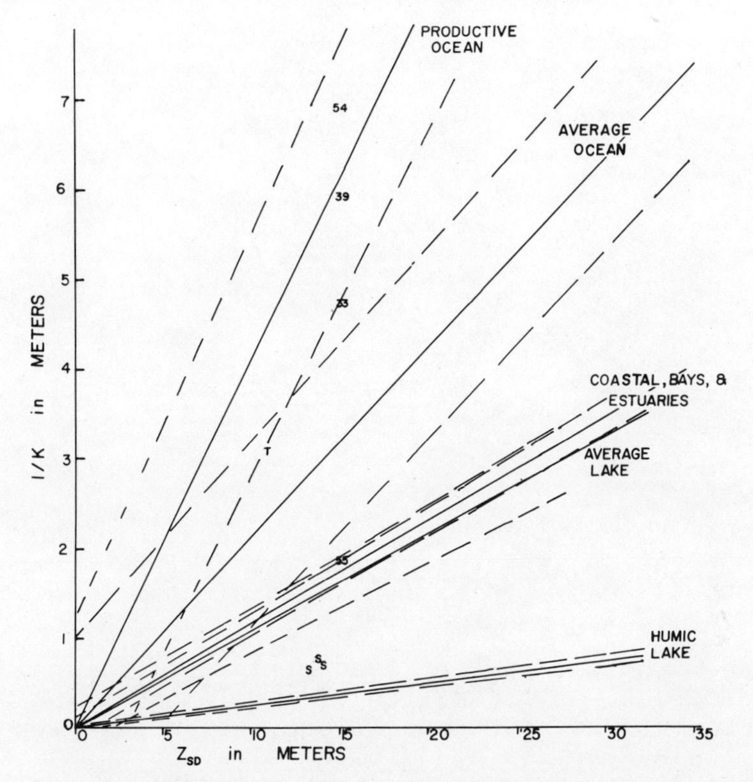

Figure 5. A summary of the relation of Z_{sd} to 1/K for the various waters. The slopes are indicated by the solid lines. Dotted lines indicate the computed 90% confidence limits for 1/K. The anomalous Icelandic marine observations are represented by station numbers plotted at the observed values of Z_{sd} and /K. The data from lakes Thingvalier (T) and Superior (S) are similarly plotted.

viewed object is:

$$C_z = C_o \exp -(\alpha + k)Z \tag{2}$$

where C_z is the apparent contrast to the background at depth Z, C_o is the inherent contrast against the background, and α and k are the narrow and broad beam attenuation coefficients of the water, respectively (of visible light). Equation 2 predicts that visibility of a Secchi disc declines exponentially with depth. Since the limiting visual contrast of the human eye is relatively constant, Z_{sd} will bear a constant relation

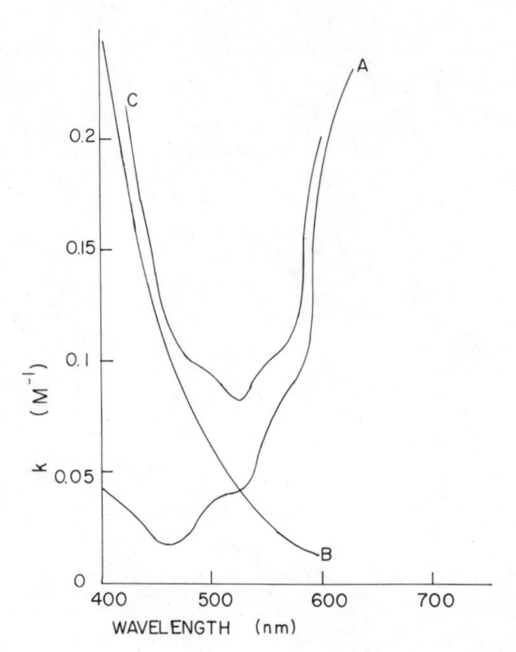

Figure 6. The narrow band absorption coefficient of humic water, k, as a function of wavelength (line C). Line A is the absorption coefficient of water alone and line B the absorption coefficient of humic substances, both lines A and B are replotted from data in Jerlov (1968); line C the absorption coefficient of humic substance dissolved in water would be the summation of points on lines A and B for each wavelength. Line C indicates that transmission of visible light in humic waters tends to be primarily in the 500-600 nm regions and that the narrow band absorption coefficient for the most penetrating wavelength would be approximately 0.09.

to the optical parameters of the water (α and k). i.e.:

$$Z_{sd} = A\exp-(\alpha + k) \qquad (3)$$

where A is a constant related to the optical properties of the Secchi disc and the human eye or:

$$\ln \frac{A}{Z_{sd}} = \alpha + k \qquad (3a)$$

Holms (1970) shows that:

$$Z_{sd}(\alpha+k) \cong 9.4 \quad (4)$$

Since the visibility of an object in water (Eq. 2) and the attenuation will depend on light in the 500-600 nm range. Assuming the effective value of k in this band to be 0.09, and λ = k and the corresponding value of K (for 310 nm) = .8, then the ratio of $1/K$ and Z_{sd} can be computed. The computed ratio of $1/K$ to Z_{sd} equals 0.0239 in excellent agreement with "humic water data (Fig. 4) where this ratio equals 0.0254.

Although K, α, and k cannot be predicted independently except for the humic waters (or other waters where true absorption >> scattering), typical freshwaters and coastal waters (Fig. 6) have a similar relation of Z_{sd} and $1/K$. A possible explanation for the optical similarity of very diverse waters can be suggesed. If the principal attenuating factor in these diverse waters is the non-specific scattering of light by small suspended particles (clay, organic debris, etc.) the overall result might be quite similar if Z_{sd} and $1/K$ are both linearly dependent on the concentration of scattering substances in the water. When freshwater reaches the oceans the high salt content precipitates much of the suspended material and an optically different "typical" ocean water results.

The most surprising observation is the fact that high productivity leads to deep penetration of UV-RB relative to visible light. The absolute penetration of UV-RB (indicated by $1/K$ is not greater than that in sterile oceanic areas, so the most likely interpretation is that there is a reduction of penetration of visible light in highly productive waters greater than the reduction of penetration of UV-RB. Proteins and nucleic acids are not strong absorbers at UV-RB wavelengths and it is possible that there is a strong selective absorption of visible light for photosynthesis but much less absorption of UV-RB. Regardless of the origin of the strange optical behavior of productive waters, it is clear that UV-RB is a greater problem to photosynthetic organisms in highly productive waters than in the oceans in general (Note Calkins and Thordardottir, Chapter 25, this volume).

ANOMALOUS OBSERVATIONS

There were some anomalous observations which were not included in the water group in which they would appear to belong (see Fig. 6). Three Icelandic stations of low productivity showed very deep penetration of UV-B (stations 33, 39, and 54). All three were at the extreme western (Greenland) end of the line of stations and could represent a type of water differing from the large category termed "typical ocean". One Icelandic station, 55, was at the edge of the Greenland ice pack and showed a very low penetration of UV-B compared with visible. The water contained minute ice crystals and the selective scattering of UV-RB relative to visible light might account for the anomaly.

The ratio of 1/K to Z_{sd} for the one Icelandic freshwater observation (Lake Thingvalier) was more like oceanic values than like other lakes. Lake Thingvalier is fed by fossil water; melted snow is filtered through volvanic material and requires many years to reach the lake. The input water is extraordinarily clear and Lake Thingvalier may better represent the behavior of some of the very clear freshwaters than the "average lake" observations.

The penetration of UV-RB to Z_{sd} in Lake Superior was much lower than would be expected from the other lake data. The observations in Lake Superior were a few miles off Whitefish Bay. A very humic river (the Tahquamenon) was discharging coffee colored water into the bay. It seems possible that the humic substances could lower the penetration of UV-B for some distance out into Lake Superior and thus explain the high UV-RB absorption of otherwise clean water. It is, of course, evident that there must be transition situations between the "productive" or "humic" conditions and "average" transmission characteristics of natural waters.

NOTES ON THE USE OF THIS METHOD

The relationships reported here will permit the estimation of penetration of UV-RB from Z_{sd}, a very simple measurement requiring an absolute minimum of equipment. More readily available estimates of UV-RB penetration in natural waters should greatly improve our comprehension of the impact of solar UV-B radiation on aquatic systems.

There are a number of caveats regarding the potential usefulness of this method which should be clearly understood. It is evident from the 90% confidence limits in Figure 6 that "true" values of $1/K$ could vary considerably from values estimated from Z_{sd}. Some uses of the broad beam attenuation coefficients (K) do not demand great accuracy while other applications may require very accurate determinations of K.

Observations of biological responses to solar UV-B indicate attenuation of incident UV-RB by a factor of 10 to 100 will in general render it biologically insignificant. Thus, one may often wish to calculate the depth of water column required to attenuate the incident radiation to 1% or 10%. From Equation 1 it will be noted that the depth required to produce a given attenuation depends linearly on K and thus a 10% error in K would produce a 10% error in the calculated depth receiving 1% of the surface UV-RB. Also, one may wish to calculate the average UV-RB exposure in a volume of water above a known depth (Z_b). Calkins et al. (1976) provide an appropriate formula:

$$I_{av} = \frac{I_o}{KZ_b} \left[1-\exp-(KZ_b)\right] \qquad (5)$$

where I_{av} is the average irradiance, I_o the irradiance at the surface, Z_b the lower limit of the volume, and K the broad beam attenuation coefficient. Unless Z_b is very small or the water very transparent $\exp-KZ_b$ will be small and I_{av} will be approximately linearly dependent on K, thus a 10% error in K would produce $\approx$ a 10% error in I_{av}.

If one desires to compute the irradiance (I) at a given depth (Z) from the surface irradiance (Eq. 1), then errors in K become greatly magnified as Z becomes large. One should use the limits plotted in Figure 6 to estimate the possible errors in calculated values of I and make a proper allowance in interpretations.

It is evident from Figure 6 that the proposed method will, in general, provide better estimates of K for fresh and coastal waters than for marine waters. The scatter in the marine points and anomalous points noted in Figure 6 should suggest that some care and common sense should be used in classification of waters; in natural waters the limits of penetration of UV-RB arise primarily from the materials in the water and not from a strong absorption by water itself. Exceptionally clear freshwaters should transmit UV-RB as well as marine waters. It is evident that there must be a graded transition from "non-productive" to "productive" marine waters and also for "humic" lake to "average" lake; only two classifications are proposed for lack of sufficient data.

The method proposed only provides an estimate of a very important and neglected property of natural waters. It is hoped that better methods of determining solar UV-RB intensities in aquatic ecosystems will soon become widely available. However, the method proposed here is presently available to anyone who can make a 20 cm disc and far superior to any alternative method known to the author short of possession of a submersible radiometer.

ACKNOWLEDGEMENTS

I wish to thank the following individuals,and institutions and staff for their generous assistance: The Lexington Water Company; The Lexington Division of Parks and Recreation; The Lexington Division of Sanitation and Public Works; Dr. K. W. Watters and the Ecology Division, Puerto Rico Nuclear Center, USAEC; Dr. D. A. Fleemer and the University of Maryland, Natural Resources Institute, Chesapeake Biological Laboratory; Dr. K. Price and the University of Delaware, College of Marine Studies Field Station; Dr. W. Carey and the Ohio State University, Center for Lake Erie Research; Dr. D. M. Gates, the University of Michigan, Biological Station Douglas Lake; Dr. Raymond Smith, Scripps Institute of Oceanography; and Drs. T. Thordardottir and U. Stefansson and the Marine Research Institute, Reykjavik, Iceland.

I gratefully acknowledge the excellent assistance received from the project technical staff at the University of Kentucky; Mr. S. Lewandowski, Ms. A. Knuckles, Mr. J. Buckles, Ms. A. Williams, Ms. J. Meadows, and Dr. J. Barcelo; and I thank Mr. Steve Thompson of the University of Kentucky Computing Center for assistance in computer fitting of the field data.

This work has been supported in part by the Climatic Impact Assessment Program, Office of the Secretary, U. S. Department of Transportation, and in part by the Office of Water Research and Technology, U. S. Department of Interior under the provisions of Public Law 88-379.

REFERENCES

Berger, D. S. 1976. The sunburning ultraviolet meter: design and performance. Photochem. Photobiol. 24: 587-593.

Calkins, J., J. D. Buckles and J. R. Moeller. 1976. The role of solar ultraviolet radiation in 'natural' water purification. Photochem. Photobiol. 24: 49-57.

Duntley, S. Q. 1963. Light in the sea. J. Opt. Soc. Amer. 53: 214-233.

Holms, R. W. 1970. The Secchi disc in turbid coastal waters. Limnol. Oceanogr. 15: 688-694.

Jerlov, N. G. 1968. Optical Oceanography, Elsevier, New York.

Robertson, D. F. 1969. Long-term field measurements of erythemally effective natural ultraviolet radiation. In Biological Effects of Ultraviolet Radiation [F. Urbach ed.] Pergamon Press, Oxford. 432-436.

Smith, R. C. and J. Calkins. 1976. The use of the Robertson meter to measure the penetration of solar middle-ultraviolet radiation (UV-B) into natural waters. Limnol. Oceanogr. 21: 746-749.

Smith, R. C. and J. E. Tyler. 1976. Transmission of solar radiation into natural waters. In Photochemical and Photobiological Reviews [K. C. Smith, ed.] Plenum Publishing Co., New York. 117-155.

Tyler, J. 1968. The Secchi disc. Limnol. Oceanogr. 13: 1-6.

YELLOW SUBSTANCE IN THE SEA

N. K. Højerslev

Institute of Physical Oceanography
University of Copenhagen
Haraldsgade 6, 2200 Copenhagen N., Denmark

ABSTRACT

The break-down processes of plankton and detritus in the marine environment cause no appreciable _in situ_ production of yellow substance. The amount of yellow substance is small and approximately the same in oligotrophic and eutrophic open oceanic areas.

In coastal areas, where it is dependent on river run-off and salinity, yellow substance is highly variable.

For both oceanic and coastal waters, yellow substance is not correlated with suspended organic matter and temperature.

Yellow substance and its fluorescent components are stable within certain salinity intervals. For this reason they can be used together with conventional temperature and salinity (T, S) analysis for determination of the mixing between different coastal water types. At salinities in the vicinity of 35 o/oo the amount of yellow substance and the fluorescent components fall abruptly with increasing salinity.

INTRODUCTION

Yellow substance is a mixture of different organic compounds dissolved in natural water. These compounds fall into two main groups: i) the phenolhumic acids which are light to dark brown, and ii) the carbohydrate - humic acids or melanoides which are light to golden yellow and more stable than the phenol - humic acids (Kalle, 1961).

Yellow substance in the marine environment absorbs, practically speaking, only ultraviolet and blue light in almost the same spectral manner. This is valid for a variety of water masses with salinities exceeding 7.5 o/oo (e.g. Højerslev, 1974a; Lundgren, 1976 and Bricaud et al., 1979).

Experimental results demonstrate that

$$a_y(\lambda) = a_y(\lambda_o) \exp\left[-0.0140(\lambda-\lambda_o)\right] \qquad (1)$$

in which $a_y(\lambda_o)$ (m^{-1}) is the light absorption coefficient for yellow substance at any fixed wavelength λ_o between 280 - 450 nm. The exponent in Eq. (1) can vary by ±0.0025 (one standard deviation). According to Nyquist (1979) $a_y(\lambda_o = 450 \text{ nm}) = 0.212 \text{ m}^{-1}$ corresponds to one mg/l of yellow substance. The concentration of yellow substance C_y(mg/l) can then be expressed as

$$C_y = 4.72\, a_y(\lambda) \exp\left[0.0140(\lambda-450)\right] \text{ (mg/l)} \qquad (2)$$

A solution of yellow substance fluoresces in the blue-green part of the spectrum around 490 nm when excited by ultraviolet light (e.g. 367 nm). Kalle (1937) suggested that the fluorescent matter was due to some by-products formed during breakdown processes of yellow substance. This is substantiated by the findings in Fig. 1 showing an increase of fluorescence but a decrease of yellow substance with depth.

Thus, a direct proportionality between fluorescent matter and yellow substance in the Baltic as postulated later by Kalle (1949) is not observed. However, for waters where mixing is pronounced, the above mentioned direct proportionality can be found (e.g. Malmberg, 1964; Højerslev, 1971, and Nyquist, 1979).

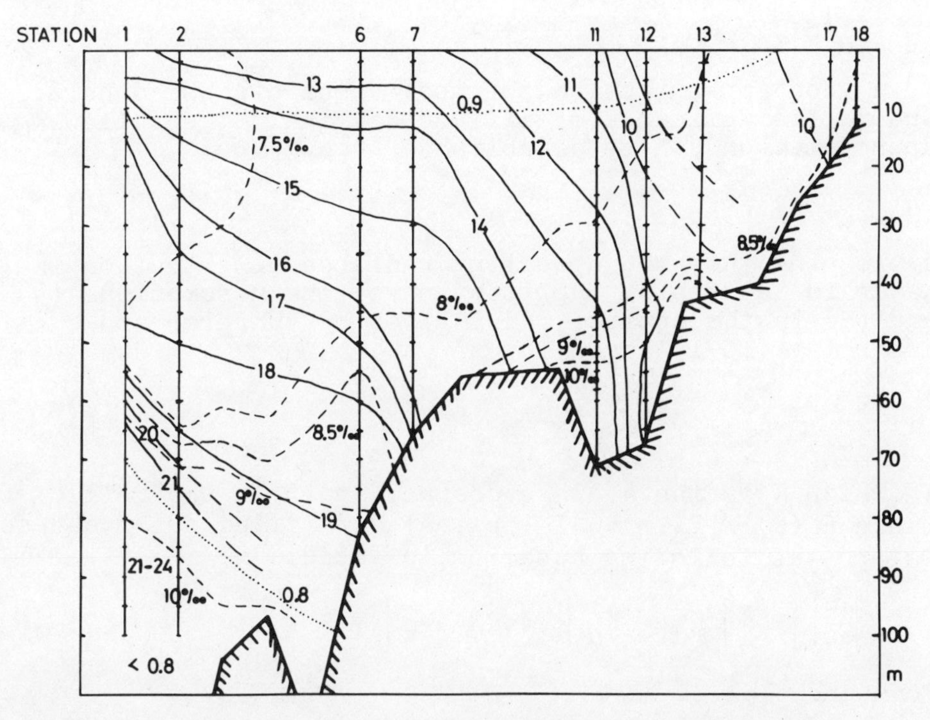

Figure 1. The vertical distribution of salinity (---), yellow substance (...), and fluorescence (——) in a section going NNE-SSW from Gotland to Bornholm in the Baltic proper.

The optical model for yellow substance

The optical notation used in this paper is the one recommended in IAPSO Standard Terminology on the Optics of the Sea (Anon., 1964).

The downward irradiance above the sea surface at 310 nm is defined as

$$E_d(0+) = \int_{2\pi} L \cos\theta \, d\omega \approx \pi L_{sky} + L_{sun} \cdot \cos\theta_{sun} \cdot \Delta\omega_{sun} \qquad (3)$$

in which L_{sky}, the radiance from the sky, is assumed to be independent of directions. The sky radiance term in Eq. (3) is always several times larger than the sun radiance term (e.g. Robinson, 1966), in particular at higher latitudes or with cloud cover. Accordingly,

$$E_d(0+) = \pi L_{sky} \quad (4)$$

is a good approximation for most cases discussed here. Due to the dominance of diffuse daylight, the following approximation can be established:

$$E_u(z) = \pi L_u \quad (5)$$

expressing that the upwelling radiance distribution in water is isotropic. Application of the assumptions (4) and (5) in the optical model given by Lundgren and Højerslev (1971) leads to

$$\frac{E}{E_o} = \bar{\mu} = 0.83 \frac{1-R}{1+4.9R} \quad (6)$$

in which $\bar{\mu}$ is the average cosine, and R is the reflectance $E_u(z) / E_d(z)$. In optical stratified or homogeneous waters the following equation is valid:

$$\left|\frac{dR}{dz}\right| = R \left| K_d - K_u \right| << R \left| a-c \right| = R \cdot b. \quad (7)$$

Consequently,

$$\frac{dR}{dz} \approx 0 \quad (8)$$

so R varies only little with depth. Moreover, it has been demonstrated that R can be given in terms of the absorption and the backscattering coefficient as:

$$R = \text{const} \frac{b_b}{a} \quad (9)$$

R is virtually independent of the surface light conditions (e.g. Gordon et al., 1975). Application of Eqs. (8) and (9) in Eq. (6) suggests that $\bar{\mu}$ at 310 nm should be independent of depth and solar elevation. Field experiments in Kattegat 1979 verified that $\bar{\mu}$ actually was constant in the solar elevation range $4^o - 52^o$. The vertical attenuation coefficient of the vector irradiance K_E was measured throughout the day and no significant variation with the solar elevation could be observed. Thus, from Gershun's equation (1936) formulated as

$$a = \bar{\mu} K_E \quad (10)$$

the independency of $\bar{\mu}$ on the solar elevation is seen. K_E can be expressed as

$$K_E = K_d + \frac{1}{1-R}\frac{dR}{dz} \cong K_d \qquad (11)$$

in which K_d is the vertical attenuation coefficient for the downward irradiance. By means of Eqs. (2), (6), (10) and (11), the concentration of yellow substance C_y (mg/l) can be calculated:

$$C_y = 0.55 \frac{1-R}{1+4.9R} K_d - 0.061 - 0.665\, a_p \qquad (12a)$$

R and K_d are measured at 310 nm and K_p (310 nm) is assumed equal to $2.2/z_q(1\%)$. (See Eq. (14).)

This assumption ensures that the concentration of yellow substance becomes negligible in the most blue and oligotrophic waters. The following assumption

$$a_p = 0.83 \frac{1-R}{1+4.9R} \cdot \frac{2.2}{z_q(1\%)} \qquad (12b)$$

allows C_y in (12a) to be calculated.

The concentration of yellow substance is often given in terms of light absorption at 375 nm, i.e. a_y (375 nm). This parameter can be obtained by applying (1), (2), (12a) and (12b) (See Table 1).

Optical measurements

During the period 1970 - 1980 the author participated in a number of expeditions to oligotrophic and eutrophic waters poor and rich in yellow substance (see Table 1 and Fig. 6).

At all locations the following optical parameters were measured: 1) downward quanta irradiance (350 - 700 nm) for which the instrumental details are given by Jerlov and Nygård (1969), ii) spectral scattering either *in vitro* or *in situ* for which the instrumental details are given by Højerslev (1971) and Jerlov (1961), respectively, and iii) spectral attenuation *in situ* for which the instrumental details are given by Højerslev (1974a, 1977b). Off Iceland, in the North Sea, the Baltic and the Western Mediterranean, spectral scalar and vector irradiances (372 - 633 nm) were measured *in situ*. At most locations the spectral reflectance R was calculated from measurements of the downward and upward irradiance (371 - 674 nm) near the water surface.

TABLE 1

Vertical attenuation coefficient for the downward UV-B irradiance K_d(310 nm) and the depth of the euphotic zone z_q(1%) measured in different waters

Area	K_d(310nm) (m^{-1})	z_q(1%) (m)	a_y(375nm) (m^{-1})
Baltic	3.0 - 3.5	16 - 20	0.89-1.04
Kattegat	1.2 - 2.4	17 - 20	0.30-0.68
Skagerrak	0.59-1.20	20 - 26	0.12-0.31
German Bight	0.53-1.77	12	0.07-0.47
" "	1.20	24.5	0.31
" "	4.7 - 5.0	7.6	1.37-1.49
" "	5.0	3.5	1.36
North Sea	0.48-0.67*	23 - 30	0.09-0.15
" "	0.37-0.65*	35 - 50	0.07-0.15
Orkney - Shetland	0.39	16 - 17	0.04
Shetland - Faroe	0.19	32	≈0
Faroe	0.17-0.19	35 -40	≈0
Iceland	0.30*	25 - 32	0.04
Western Greenland	0.19-0.21	46 - 57	0.01
Svalbard	0.20-0.25	50 - 56	0.01-0.02
Western Mediterranean	0.43*	35 - 40	0.08
" "	0.22*	56 - 69	0.02
" "	0.16*	70 - 75	≈0
Sargasso Sea	0.15	112	≈0

* Calculated from irradiance measurements at 372 nm and Eq. (1).

Usually, the applied instrument was a 12-channel irradiance meter equipped with double interference filters (half width ≈ 15 nm) constructed at our institute. Fluorescence was measured *in situ* in the Baltic, the Sound, Kattegat and off Iceland. *In vitro* fluorescence was measured in the Sound, Kattegat, Skagerrak and in the North Sea. The instrumental details are given by Kullenberg and Nygård (1971) and Højerslev (1971), respectively. Downward and upward irradiance at 310 nm (half width ≈ 2 nm) was measured at most locations except for off Iceland, the North Sea and the Western Mediterranean by means of a UV-B meter constructed at our institute. It records continuously the irradiance versus depth. Standard hydrography was always performed together with the optical measurements.

The data set from the Baltic, Danish coastal waters and the North Sea (Fladen Ground)is especially comprehensive. The total expedition period in these areas is of the order of one year.

RESULTS

The concentration of yellow substance decreases about 20% per 100 m downwards and about 10% per 100 km outwards the Baltic (Fig. 1). The fluorescence generally increases about 50% per 100 m downwards, but decreases about 25% per 100 km outwards the Baltic. The concentration of suspended matter, expressed in terms of light scattering at 655 nm, varies between 0.10 - 0.40m^{-1}. The amount of suspended matter attains its maximum in the bottom nephloid layers. Above the thermocline, situated around 30 m, the amount of suspended matter is typically twice as large as below. The smallest amounts are encountered in the central part of the basin where the influence from all solid boundaries, capable of acting as particle sources, is reduced (Fig. 2).

Yellow substance and fluorescence are not correlated in the Baltic where the horizontal advection is sluggish and the vertical mixing small, but for the Danish coastal waters with salinities above 15 o/oo they are correlated (e.g. Jerlov, 1976; Højerslev, 1971 and Nyquist, 1979). Fluorescence and salinity are not correlated in the Baltic (Fig. 1 and Højerslev, 1974a), but they are in certain salinity intervals in the Danish coastal waters (Fig. 4) where the vertical mixing is much more pronounced than in the Baltic. The proportionality

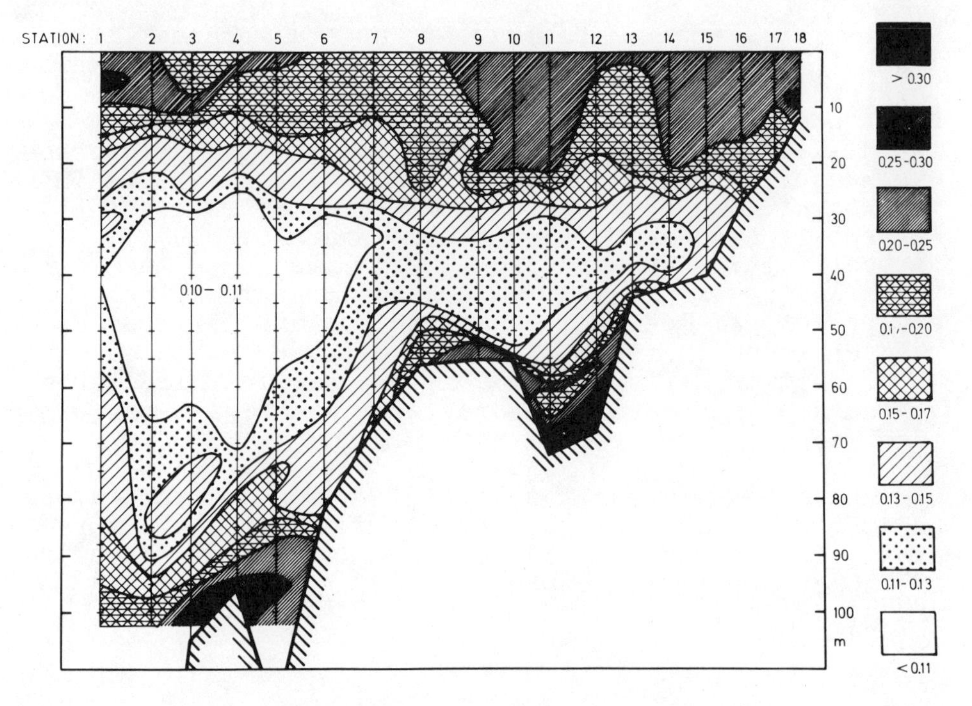

Figure 2. The vertical distribution of suspended matter for the same section as in Fig. 1.

between either fluorescence or yellow substance and salinity can be used as a means for the classification of coastal water types and their mixing (e.g. Jerlov, 1955 or Nyquist, 1979).

For coastal waters this method is more promising than the conventional (T. S.) analysis because the temperatures in the coastal water column are more subject to annual changes than the fluorescence (compare Figs. 3 and 4). Fig. 4 shows that: i) water in the Norwegian Coastal Current mixes with the underlying Atlantic water, which is characterized by intermediate high fluorescence values in the range 0.70 - 1.50 and salinities from 30‰ - 35‰ (another part of the Skagerrak water is formed through mixing with the water in the Jutland Current), ii) waters of Atlantic origin have fluorescence values below 0.70 and salinities above 35‰, and finally, iii) water from the Sound (fluorescence 2.30 and

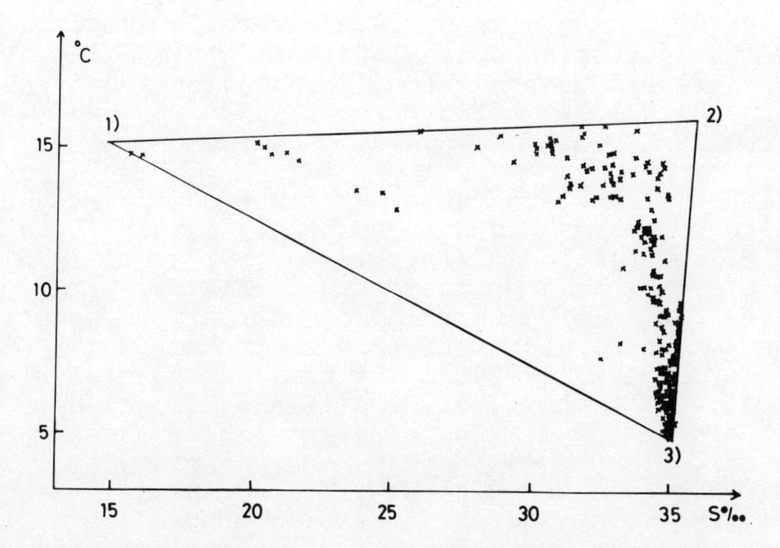

Figure 3. A (T, S) diagram for Danish water masses (the Sound, Kattegat, Skagerrack, the North Sea, and deep Atlantic water).

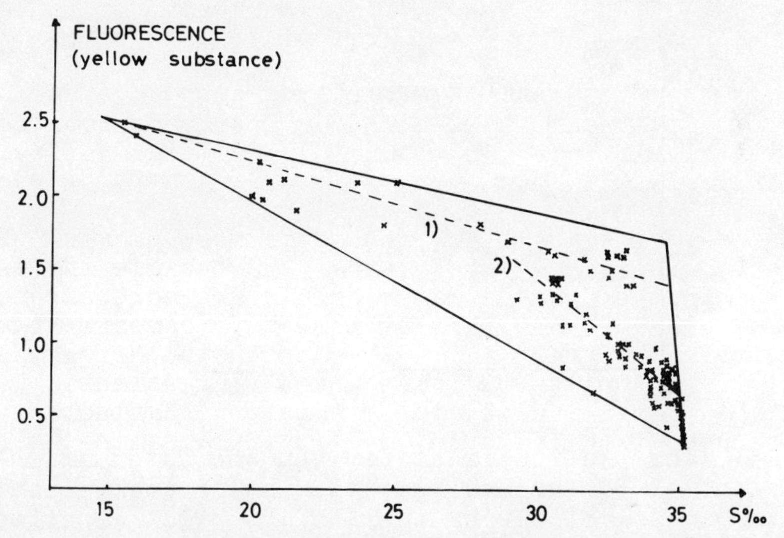

Figure 4. A (fluorescence, S) diagram for the same water masses as in Figure 3.

salinity 20‰) mixes with surface waters in Skagerrak outside the Norwegian Coastal Current - along line 1) - forming Kattegat water, which is characterized by high fluorescence values in the range 1.5o - 2.3o and salinities from 20‰ - 33.5‰.

Neither yellow substance nor fluorescence are observed to be correlated with suspended matter (Figs. 1, 2, and 5; see also Højerslev, 1974a, 1977a and 1978) at any of the locations presented in Fig. 6. Fig. 5 shows the result of a correlation analysis between the attenuation coefficient measured at 380 nm and 655 nm (Højerslev, 1977a, 1978) in the periods August-September, 1975 and March-May, 1976 at Fladen Ground down to a depth around 130 m. The straight lines crossing at (0.060 m^{-1}, 0.310 m^{-1}) indicate that no production of yellow substance took place in the course of a year (e.g. Duursma, 1974), neither from chemical breakdown of the plankton bloom in April-May, 1976, nor from settled organic matter at the sea floor. Spectral light absorption measurements give similar results (Table 1 and Fig. 6).

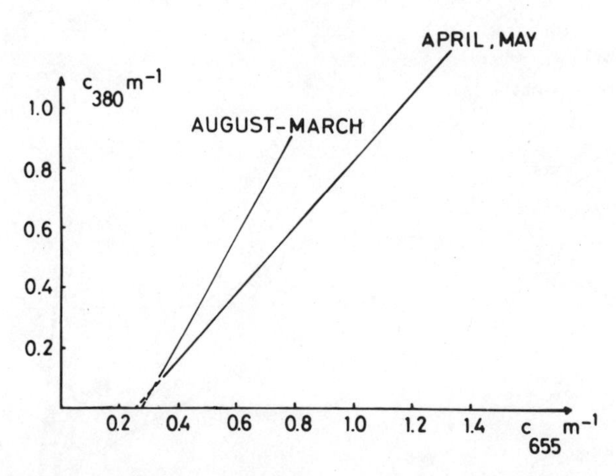

Figure 5. Correlation lines between the attenuation coefficient at 380 nm and the attenuation coefficient at 655 nm, measured during the FLEX 75 pilot experiment and the FLEX 76 experiment at Fladen Ground in the North Sea (around 2000 data points are included in the correlation analysis).

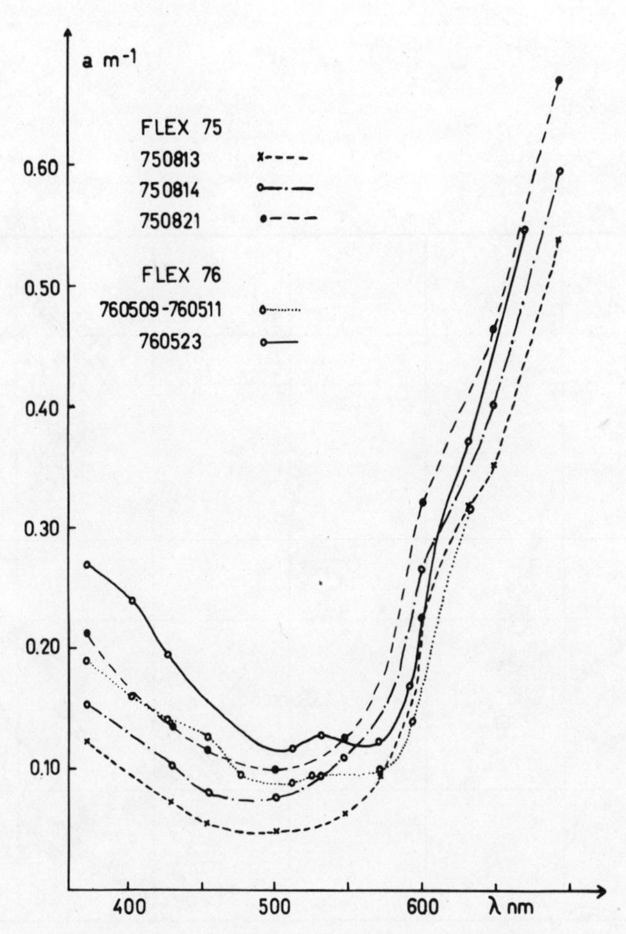

Figure 6. Spectral light absorption within the euphotic zone at Fladen Ground, before, during and after the occurrences of plankton blooms.

The annual temperature variation at Fladen Ground was 6°C to 18°C. Therefore, the temperature does not seem to be a dominating factor for the chemical degradation rate of organic matter into yellow substance. This is further supported by the quanta irradiance (350-700 nm) and the UV-B irradiance (310 nm) measurements made in the North Atlantic (Figs. 5, 7 and Table 1) in the temperature range -2°C to +12°C.

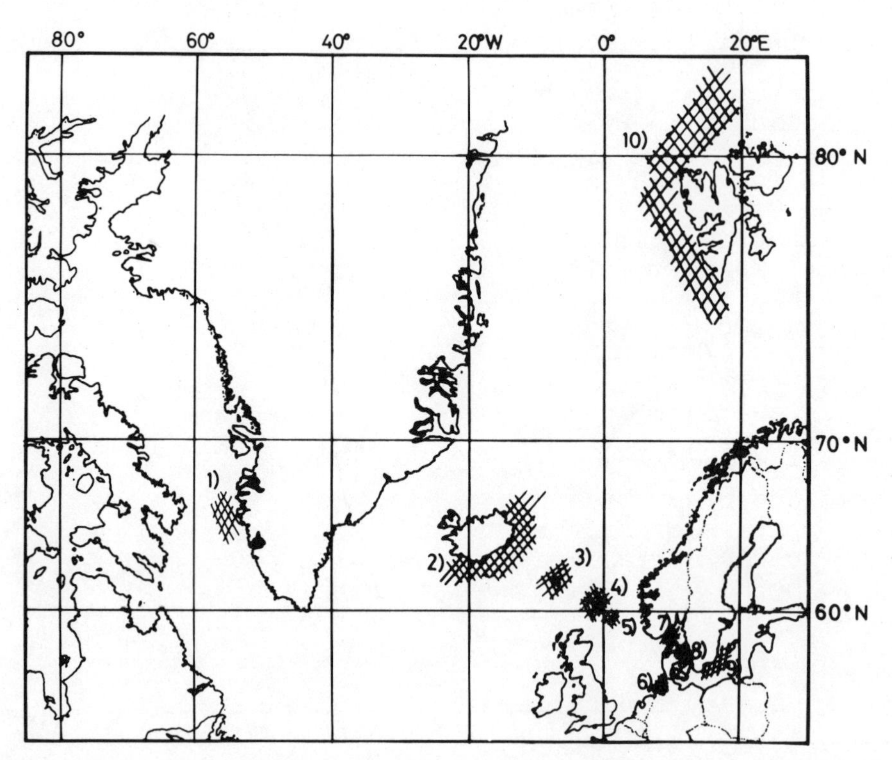

Figure 7. Sites for optical measurements made by the author during the period 1970 - 1980:
1) Off Holsteinsborg, Western Greenland, 1979, 2) off South East Iceland, 1973, 3) off the Faroe Islands, 1978, 4) off the Shetland Islands, 1978, 5) Fladen Ground (FLEX), 1975 and 1976, 6) the German Bight (MARSEN), 1979, 7) Skagerrack, 1970, 1976 and 1979, 8) Kattegat, 1970-1979, 9) The Baltic proper 1970 - 1979, 10) off Svalbard (NORSEX), 1979.

Low and almost constant values around 0.20 m^{-1} for K_d(310 nm), which is a measure for the concentration of yellow substance, correspond to depths of the euphotic zone z_q(1%) ranging from 30-60 m in open oceanic waters. In coastal waters the opposite effect is observed, i.e., almost constant values for z_q(1%) around 15 m is accompanied by variations of K_d(310 nm) from 0.5m^{-1} to 3.5 m^{-1}.

Except for the interior Danish sea waters, where the salinity is well below 35 o/oo, K_d(310 nm) was constant with depths down to 50 m or 10 optical depths, in the range 0.17 m^{-1} off the Faroe Islands to 5.0 m^{-1} in the German Bight. This suggests that the yellow substance is stable at salinities around 35 o/oo. Earlier measurements off Iceland of the fluorescence which was small and constant in the whole water column down to 100 m, leads to a similar conclusion (Højerslev, 1977b).

In the Sargasso Sea, where the concentration of yellow substance is assumed to be zero, K_d(310 nm) = 0.15 m^{-1}, z_q(1%) = 112 m, and R(310 nm) = 0.03. From these observed values, the average cosine $\bar{\mu}$ is calculated by means of Eq. (6) to be 0.70. The absorption coefficient $a = a_w + a_p + a_y$ becomes equal to 0.105 m^{-1} by using Eq. (10). a_w, a_p, a_y are the absorption coefficients due to pure sea water, particles and yellow substance, respectively.

At a very clear location in the Western Mediterranean, where $a_y = 0$ by assumption, K_d(310 nm) = 0.16 m^{-1}, z_q(1%) = 75 m, and R(310 nm) = 0.03. As before, calculations give $\bar{\mu}$ = 0.70 and a = 112 m^{-1}. Off the Faroe Islands K_d(310 nm) = 0.17 m^{-1}, z_q(1%) = 40 m, and R = 0.02. Accordingly, $\bar{\mu}$ = 0.74 and a = 0.125 m^{-1}. These values for a(310 nm) are almost 10 times smaller than those previously reported for pure sea water (e.g. Lenoble, 1956a,b; Armstrong and Boalch, 1961a,b) but are in excellent agreement with other _in situ_ findings (e.g. Calkins, 1975). It therefore seems pertinent to evaluate the absorbing and scattering properties of pure sea water in the ultraviolet part of the spectrum.

The particle absorption at 310 nm is assumed to be inversely proportional to the depth of the euphotic zone. For the Sargasso Sea and the Western Mediterranean, the following equation can be established, assuming $a_y = 0$:

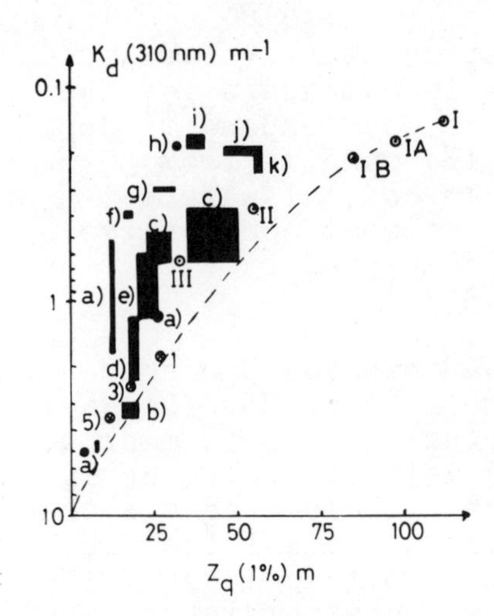

Figure 8. The vertical attenuation coefficient at 310 nm, K_d(310 nm), versus the depth of the euphotic zone, z_q(1%).

a) The German Bight, b) The Baltic proper, c) The Fladen Ground, d) Kattegat, e) Skagerrack, f) Shetland-Orkney Islands, g) off South East Iceland, h) Shetland-Faroe Islands, i) off the Faroe Islands, j) off Western Greenland, k) off Svalbard.

The Roman numbers I, IA, IB, II and III correspond to oceanic water masses according to Jerlov's optical classification, (1976).

$$(0.105 - a_w) \cdot 112 = (0.112 - a_w) \cdot 75 \qquad (13)$$

Consequently, $a_w = 0.091\ m^{-1}$ at 310 nm. If a similar relation as (13) is established between $K_d - K_w$ and $z_q(1\%)$, that is,

$$(0.15 - K_w) \cdot 112 = (0.16 - K_w) \cdot 75 \qquad (14)$$

then $K_w = 0.13\ m^{-1}$ where K_w is the vertical attenuation coefficient for pure sea water at 310 nm. The average cosine $\bar{\mu}_w$ and the reflectance R_w for pure sea water becomes equal to 0.70 and 0.03 as in the case of very clear blue ocean water.

For the waters off the Faroe Islands $(a - a_w) z_q(1\%) = (a_y + a_p) z_q(1\%) = 1.36$ i.e. $a_y + a_p = 0.034$ m^{-1} at 310 nm. By using Eq. 13 $a_p = 0.039$ m^{-1}, which is higher than $a_y + a_p$, obtained by direct calculation. This result suggests, however, that no in situ production of yellow substance takes place in the sea. At Svalbard and Greenland, where the salinities were in the range 32-34 o/oo, there was a slight indication of yellow substance, which is probably due to the freshwater input from Siberian rivers. At Iceland and off the Orkney and the Shetland Islands, the input of yellow substance from the neighbouring mainland was notable (Table 1).

DISCUSSION

The marine literature reports on 5 sources of yellow substance: i) river run-off, ii) rainfall, iii) melting of ice, iv) in situ production from plankton and detritus, and v) in situ production by brown algae close to the coast in temperate climates. Rainfall and melting of ice are not considered important. K_d(310 nm) was small off the Faroe Islands where the rainfall is considerable and also small off Western Greenland and Svalbard where melting of ice occurs. The in situ production of yellow substance in the open sea could not be observed at Fladen Ground throughout a year although there is a high primary and secondary production in the spring and early summer. The intense vertical mixing during the fall and the winter brings detritus from the sea floor up to the surface layers without leading to an increase of yellow substance. The in situ production yellow substance by brown algae does probably take place at salinities below 32 o/oo (e.g. Sieburth and Jensen, 1969). However, this production is completely dominated by the ever present high concentrations of yellow substance brought to coastal zones from land. The only important source of yellow substance is the river run-off. Yellow substance or its degradation products (fluorescence) behave in many ways as the clay sediments transported from the continents into the open sea. Flocculation of both yellow substance and fluorescence matter is observed at low salinities (e.g. Jerlov, 1955; Nyquist, 1979; and Fig. 1).

CONCLUSIONS

It is unlikely that significant amounts of yellow substance are formed in the sea by chemical break-down processes of either pelagic or benthic organic matter. Instead, the yellow substance is brought to the sea essentially by river discharge. On its way to the open sea parts of the yellow substance precipitates at rates given by certain salinity ranges. The yellow substance or the fluorescent matter in the open sea with salinities around 35 o/oo are fairly stable and can thus be considered as quasi-conservative passive properties. Consequently, the absorption due to yellow substance is approximately the same in both oligotrophic and eutrophic open oceanic areas. In the rare cases when the influence by river run-off is negligible, yellow substance cannot be observed in either oligotrophic or eutrophic open areas.

The amounts of yellow substance advected towards the open sea are of the order of 1 mg/l. The river run-off might be a sufficiently powerful source to cause the presence of yellow substance in the open sea, provided that it is very stable at oceanic hydrographical conditions which seems to be the case (e.g. Skopintsev, 1971). This demonstrates that *in situ* formation of yellow substance is not likely to take place in the sea.

In coastal areas the concentration of yellow substance is highly varying with location, depth and weather conditions. In addition, yellow substance is not correlated with the amount of suspended matter.

ACKNOWLEDGEMENTS

Due thanks are expressed to the Danish Space Agency (Rumudvalget), the Danish Ministry of Environment (Miljøministeriet), the Nordic Council for Physical Oceanography, and the University of Bergen for providing the necessary financial support.

I would also like to thank Dr. B. Lundgren and Mr. E. Buch for their competent assistance at sea. The captains and the crews of the following vessels are gratefully acknowledged for their valuable support: "Aranda", "Bjarni Saemundsson", Helland-Hansen", "Meteor", "Polarsirkel", "Tabasis" and "H. U. Sverdrup".

REFERENCES

Anon. 1964. Committee on Radiant Energy in the Sea. Standard Terminology on Optics of the Sea. [Ed. N. G. Jerlov, IAPSO (IAPO) publication office, Paris.

Armstrong, F.A. J. and G. T. Boalch. 1961. Ultraviolet absorption of sea water and its volatile components; In: Symposium on Radiant Energy in the Sea, [Ed. N. G. Jerlov. IUGG Monograph (10).

Bladh, J. O. and S. Bjørn-Rasmussen. 1978. Hydrografiska och växt-planktologiska undersökningar vid Skåne- och Blekingekusterna, 1970-75, resp. 1972-75. Medd. Havfiskelab. Lysekil, Gothenburg. 240.

Bricaud, A., A. Morel and L. Prieur. 1979. In: Symposium on Radiant Energy in the Sea. IUGG. IAPSO programme.

Calkins, J. 1975. Measurements of the penetration of solar UV-B into various natural waters. In: Climatic Impact Assessment Programme, Monograph 5, US Department of Transportation, Washington, D.C.

Duursma, E. K. 1974. The flourescence of dissolved organic matter in the sea. In: Optical Aspects of Oceanography: 237-256. [Eds. N. G. Jerlov and E. Steemann Nielsen, Academic Press, New York.

Gershun, A. 1936. O fotometri i mutnykk sredin. Tr. Gos Okeanogr. Inst. 11

Gordon, H. R., O. B. Brown and M. M. Jacobs. 1975. Computed relationships between the inherent and apparent optical properties of a flat homogeneous ocean. Appl. Opt. 14: 417-427.

Hill, H. W. 1973. Currents and water masses. In: The North Sea [Ed. E. D. Goldberg] North Sea Science. MIT Press: 18-42.

Højerslev, N. K. 1971. Tyndall and fluorescence measurements in Danish and Norwegian waters related to dynamical features. Rep. Inst. Phys. Oceanogr. University of Copenhagen. 16.

Højerslev, N. K. 1974a. Inherent and apparent optical properties of the Baltic. Rep. Inst. Phys. Oceanogr. University of Copenhagen. 23.

Højerslev, N. K. 1974b. Daylight measurements for photosynthetic studies in the Western Mediterranean. Rep. Inst. Phys. Oceanogr., University of Copenhagen. 26.

Højerslev, N. K. 1977a. Inherent and apparent optical properties of the North Sea, "Fladen Ground Experiment - FLEX 75", Rep. Inst. Phys. Oceanogr., University of Copenhagen. 32.

Højerslev, N. K. 1977b. Inherent and apparent optical properties of Icelandic waters. Bjarni Saemundsson Overflow 73, Rep. Inst. Phys. Oceanogr., University of Copenhagen. 33 (ICES Overflow 73 contr. no. 34).

Højerslev, N. K. 1978. Inherent and apparent optical properties of the North Sea. Fladen Ground Experiment FLEX 76. In: SFB [Ed.], FLEX-Atlas, Hamburg.

Jerlov, N. G. 1955. Factors influencing the transparency of the Baltic Waters. Medd. Oceanogr. Inst., Gothenburg. 25.

Jerlov, N. G. 1961. Optical measurements in the Eastern North Atlantic. Medd. Oceanogr. Inst., Gothenburg. 30: 1-40.

Jerlov, N. G. and K. Nygård. 1969. A quanta and energy meter for photosynthetic studies. Rep. Inst. Oceanogr. University of Copenhagen. 10.

Jerlov, N. G. 1976. Marine Optics. Elsevier Oceanogr. Series, Amsterdam, 2nd ed.

Kalle, K. 1937. Meereskundliche chemische Untersuchungen mit Hilfe des Zeisschen Pulfrich Photometers, VI, Mitt. Die Bestimmung des Nitrats und des Gelbstoffes. Ann. Hydr. u. Marit. Meteorol: 276-282.

Kalle, K. 1949. Fluoreszenz und Gelbstoff im Bottnischen und Finnischen Meeresbusen. Deutsch. Hydr. Z. 2: 117-124.

Kalle, K. 1961. What do we know about Gelbstoff. In: Symposium on Radiant Energy in the Sea. [Ed. N. G. Jerlov] IUGG Monograph (10)

Kullenberg, G. and K. Nygård. 1971. Fluorescence measurements in the sea. Rep. Inst. Phys. Oceanogr. University of Copenhagen. 15.

Lenoble, J. 1956a. L'absorption du rayonnement ultraviolet par les ions présents dans la mer. Rev. Opt. 35: 526.

Lenoble, J. 1956b. Sur le role des principaux sels dans l'absorption ultraviolet de l'eau de mer. Compt. Rend. 242: 806-808.

Lundgren, B. and N. K. Højerslev. 1971. Daylight measurements in the Sargasso Sea - Results from the "DANA" Expedition, January-April 1966. Rep. Inst. Phys. Oceanogr. University of Copenhagen. 14.

Lundgren, B. 1976. Spectral transmittance measurements in the Baltic. Rep. Inst. Phys. Oceanogr. University of Copenhagen. 30.

Malmberg, S. Aa. 1964. Transparency measurements in Skagerak. Medd. Oceanogr. Inst. Gothenburg. 31.

Nyquist, G. 1979. Investigation of some optical properties of sea water with special reference to lignin sulfonates and humic substances. Thesis Dep. Anal. Mar. Chem., Gothenburg.

Robinson, N. 1966. Solar Radiation. Elsevier, Amsterdam.

Sieburth, J. McN. and A. Jensen. 1969. Studies on algal substances in the sea. II. The formation of Gelbstoff (humic material) by exudates of phaephyta. J. Exp. Mar. Biol. Ecol. 3:275-289.

Skopintsev, B. A. 1971. Recent advances in the study of organic matter in oceans. Okeanologiya. II, No.6: 775-789.

NOTE ON THE ROLE OF VERTICAL MIXING IN RELATION TO EFFECTS OF UV RADIATION ON THE MARINE ENVIRONMENT

Gunnar Kullenberg

Institute of Physical Oceanography
University of Copenhagen
Haraldsgade 6, 2200 Copenhagen N., Denmark

ABSTRACT

Considering that the penetration depths of UV radiation into the sea are limited, and in particular as regards the levels at which UV radiation appears to be markedly harmful to marine organisms, the possible role of vertical mixing is discussed in relation to assessing harmful effects of UV radiation. The vertical mixing varies strongly in time and space, implying large variations of residence times of organisms in the near-surface layer. Periods of weak vertical mixing often coincide with periods of high biological activity.

PROBLEM STATEMENT

The ultraviolet light (UV) penetrates to very limited depths in the sea relative to parts of the visible light spectrum. The 1% level of the incoming UV light is found around 30 m depth in very clear (low productive) waters such as in the Sargasso Sea, around 20 m in moderately clear open ocean waters and at depths less than 3 m in most coastal waters. Smith and Baker (1979) found 'lethal doses' of UV radiation for depths less than about 1.5 m in the clearest waters, for depths less than about 0.7 m in moderately productive waters and for depths less than a few decimeters in highly productive waters, with much dissolved organic material. These numbers are of course all related to the action spectrum selected by Smith and Baker, but they nevertheless suggest that the penetration depths of potentially harmful doses of UV radiation are very small. It is also noted that the effective attenuation lengths for DNA are about 6 m in the clearest water and about 2.5 m in moderately productive water. The change of UV light conditions caused by a 25% reduction of the ozone layer thickness would imply an increase for the 1% depth ranges given above by a few decimeters to a few meters.

In order to evaluate the potential harm of the present UV radiation levels as well as an increase of these levels on organisms in the near-surface layers, it seems necessary to consider the influence of vertical mixing on the residence time of the organisms in the potentially harmful depth range. The possible significance of the vertical mixing in relation to harmful effects of the UV radiation is in turn related to the response time and recovery time of the organisms to the UV radiation. The response time does not seem to be well established for marine organisms, and could vary over a large range, from minutes to hours.

The purpose of this note is to discuss briefly the rate of vertical mixing in the near-surface layer and its dependence on physical conditions such as wind, heat input, and stratification in the water column.

RATES OF VERTICAL MIXING

The rate of vertical mixing in the surface layer can for instance be determined by means of tracer experiments or by following the seasonal heat penetration. For open sea areas the rate of mixing is as yet mostly quantified by means of an effective vertical mixing coefficient K_z, the so-called K-approximation. In Table 1 examples of observed mixing rates are given determined by various means (Kullenberg 1981). An intercomparison of the values is not directly required here. The results clearly show that the rate of mixing varies considerably. This variation is related to the conditions during the time of observation of the mixing as well as to the different conditions in the various areas, for instance as regards topography.

It should be noted that the K-approximation must be used only with great care. It will be indicated below

TABLE 1

SUMMARY OF RESULTS ON MAGNITUDE OF THE VERTICAL MIXING COEFFICIENT K_z

Method of determination	Region of value	K_z $cm^2 \cdot s^{-1}$
Circulation model	Oceanic tracer distribution	1 - 1.5
Mediterranean outflow	Atlantic intermediate water	0.35 - 0.7
Tritium, fallout	Sargasso Sea	0.2
Tritium, fallout	Greenland and Norwegian Seas	3
Double diffusion, Cox Number	Open Pacific	0.01
" " " "	California current)	
	)	10
" " " "	Gulf Stream)	
Saltfinger flux	Atlantic, central	0.05 - 0.8
Dye mixing (tracing)	Surface layer	1 - 200
Dye visual (divers)	Thermoclines	0.2
Dye diffusion (tracing)	Pycnoclines	0.02 - 0.4
Towed micro sensors	Pacific, coastal zone	0.06 - 3
" " "	Pacific equatorial	0.5 - 27
Radon and radium profiles	Atlantic bottom boundary	10 - 50
Dissipation measurements	Atlantic Equatorial undercurrent	0.02-3
Dissipation measurements	Atlantic near the Azores	0.3-5

that the mixing in the sea is not a continuous process but tends to occur in events related to various dynamical features, such as storms, surface cooling, tidal oscillations. Most of the mixing in the sea appears to be related to events at the boundaries: surface, sides and bottom. This implies that a constant K_z cannot be used. The coefficient will have to be related to the conditions and it will vary in time and space.

The values of K_z may be used to estimate the time t required for vertical mixing to move particles over a given distance h, using the expression

$$h^2 = 2\,K_z\,t \quad (1)$$

Some illustrative examples are given in Table 2. It is noted that for low mixing rates a particle subject to vertical mixing only will remain in the top 5 m for about 4 weeks. It is quite clear that different mixing rates will have an implication for the potential harmful effects of UV radiation. Important questions are: when does low mixing rates occur and for how long periods, and does significant biological activities take place during such periods?

TABLE 2

Times t for mixing across a given layer thickness h with a given mixing coefficient K_z.

h (cm)	K_z ($cm^2 \cdot s^{-1}$)	t (hr)	t (s)
50	0.01	35	-
50	10	-	126
50	100	-	12
100	0.1	14	-
100	10	0.4	-
100	100	-	50

It should in this context be noted that plankton organisms have a certain sinking rate. This varies with the type of organism and also depends on the conditions of the organisms. The sinking rates are usually of the order of a meter per day which would be equivalent to a K_z of about 0.1 $cm^2 s^{-1}$.

FACTORS INFLUENCING VERTICAL MIXING

Periods of very low mixing rates very near the surface may occur essentially during the warm season. The heat input will then tend to stably stratify the surface layer. Mechanical mixing generated by energy input from the wind will act in the opposite direction, tending to distribute the heat input over a certain depth. Low mixing rates will then occur during calm or low wind conditions. The heat input from a clear sky (sun) can prevent mixing completely up to wind velocities of 2-3 $m.s.^{-1}$ (Woods 1981). During calm conditions with a clear sun laminar flow conditions may prevail throughout the surface layer. This has been demonstrated by means of tracer studies (e.g. Woods 1968, 1981; Kullenberg 1974. An example of a dye layer observed during calm conditions in the open Baltic Sea is shown in Figure 1. The dye was traced by an *in situ* fluorometer using the technique developed by Kullenberg (1969, 1971). The sharp boundaries and structure of the layer suggest the presence of shear in the flow. However, such layers can be traced for several days up to a week at least in stably stratified conditions and calm weather (Kullenberg 1974).

It should also be noted that the distribution of some biological properties during such conditions are very similar to the tracer distributions (e.g. Holligan 1978, Kullenberg 1978, Pugh 1978, Owen 1980). This appears to be the case for chlorophyll *a* as observed by means of fluorescence (e.g. Pingree et al. 1975), as well as plankton sampled *in situ* with high vertical resolution.

In some shelf sea areas the mixing generated by the bottom friction of tidal currents will also act so as to prevent or delay the development of a stably stratified water column. In many coastal areas of which estuaries are very important examples, stable stratification is generated by freshwater run-off from land. This type of stratification may also vary with the season and it is also influenced both by tidal motion and wind effects.

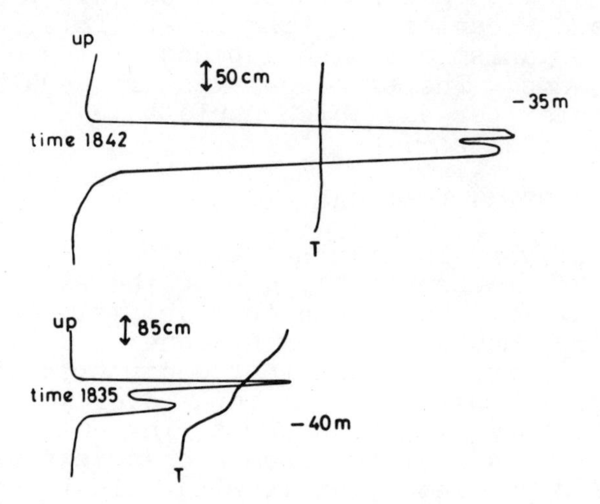

Figure 1. Layers of rhodamine and temperature profiles observed in the central Baltic Sea after 20 and 18 hours of tracing, using an in situ fluorometer with temperature sensor.

The importance of the wind for the vertical mixing in the surface layer can be demonstrated in many ways. The deepening of the mixed layer after the passage of a storm is often unambiguously related to the mechanical mixing generated by the energy input from the wind (Figure 2). The wind has to reach a certain strength with some duration in order to generate a surface mixed layer, the depth of which will depend on the balance between the effects of a stratification and wind. On the basis of results from dye mixing experiments, Kullenberg (1971, 1974a) found that the vertical mixing coefficient could be expressed in the form, for wind velocities above 4 $m \cdot s^{-1}$,

$$K_z = \text{constant} \cdot \frac{\overline{W^2}_{10}}{\overline{N^2}} \frac{dq}{dz} \qquad (2)$$

where W_{10} is the wind speed 10 m above the surface, dq/dz is the vertical gradient of the horizontal current vector q, and $N^2 = g/\varrho \cdot d\varrho/dz$ expresses the stability, with ϱ the density and g the acceleration of gravity. Alternatively, the rate of energy dissipation E_W per unit mass from the energy input by the wind over a layer thickness h may be given as

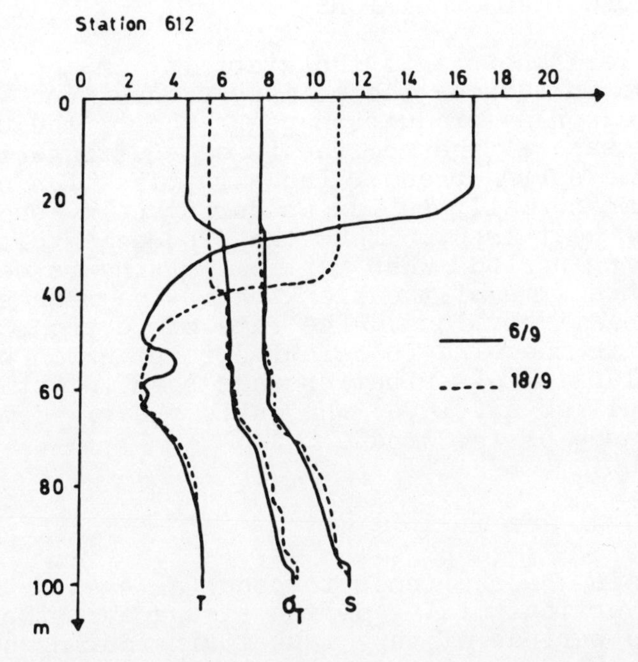

Figure 2. Profiles of salinity (s), temperature (T) and density (σ_T) before and after passage of severe storms over a five day period in the southern central Baltic Sea.

$$E_W = \frac{\rho_a \cdot C_d \cdot k \overline{W_{10}^3}}{\rho \cdot h} \quad (3)$$

where ρ_a is the density of air, c_d the surface drag coefficient and k the wind factor. The energy dissipation should be related to the rate of increase of the stratification from the surface heat input (see e.g. Pingree et al. 1978). These results demonstrate that the wind at a certain strength can counteract the surface heat input and generate a surface mixed layer. The increase of the thickness of this layer as a function of time t will vary depending upon the time scale, being proportional to $\sqrt[2]{t}$ and $\sqrt[3]{t}$.

Critical periods from the point of view of potential UV harm to organisms may therefore occur during clear sky days with weak winds. The duration of such periods varies with the meteorological conditions, often depending on latitude. They can vary from less than a day to weeks.

RELATED BIOLOGICAL CONDITIONS

It is well known that the stability in the water column also influences the primary production. In areas where spring production peaks occur, the development of a seasonal thermocline is often triggering the start of the spring bloom. The stratification of the water column normally builds up during calm conditions with strong heat input. Thus the production often starts during periods when vertical mixing is weak and the residence time of the plankton in the near-surface layer is relatively long. The rate of vertical mixing as related to the wind in relation to plankton productivity and plankton distributions has been investigated by Therriault et al.(1978) who quite clearly demonstrate the importance of the wind.

CONCLUSIONS

Mixing in the sea tends to occur as events. Periods of strong vertical mixing during strong atmospheric forcing and periods of very weak mixing during calm conditions with heat input occur in sequences. The great importance of this sequence of events for the biological system has been shown among others by Lasker (1978) and Walsh et al. (1978). During periods of heat loss the cooling and the wind will act together in generating vertical mixing and convection.

From this discussion it should be evident that the residence time of organisms in the near-surface layer will depend strongly upon the conditions, and can vary from minutes to days. This will have to be taken into account when assessing the effects of UV radiation on the marine biological system.

REFERENCES

Holligan, P. M. 1978. Patchiness in subsurface phytoplankton populations on the Northwest European continent shelf. In: Spatial Pattern in Plankton Communities [ed. J. H. Steele] Plenum Press, NATO Conf. Ser. IV, Vol. 3: 221-238.

Kullenberg, G. 1969. Measurements of horizontal and vertical diffusion in coastal waters. Kungl. Vetenskaps-och Vitterhets-samhället, Göteborg, ser. Geophysica 2.

Kullenberg, G. 1971. Vertical diffusion in shallow waters. Tellus 23 (2) 129-135.
Kullenberg, G. 1974. Investigation of small-scale vertical mixing in relation to the temperature structure in stably stratified waters. Adv. Geophys. 18A 339-351.
Kullenberg, G. 1979a. An experimental and theoretical investigation of the turbulent diffusion in the upper layer of the sea. Inst. for Physical Oceanography, University of Copenhagen, Rep. No. 25.
Kullenberg, G. 1978. Vertical processes and the vertical-horizontal coupling. In: Spatial Pattern in Plankton Communities [ed. J. H. Steele] Plenum Press, NATO Conf. Ser. IV. Vol. 3: 43-71.
Kullenberg, G. 1981. Physical processes. In: Pollutant Transfer and Transport in the Sea [ed. G. Kullenberg] CRC Press, (In press).
Lasker, R. 1978. The relation between oceanographic conditions and larval anchovy food in the California Current: identification of factors contributing to recruitment failure. Rapp. Proc. Vol. 173: 212-230.
McGowan, J. A. and T. L. Hayward. 1978. Mixing and oceanic productivity. Deap-Sea Res. 25: 771-794.
Owen, R. W. 1980. Patterning of flow and organisms in the larval anchovy environment. Report NOAA, National Mar. Fish. Serv. La Jolla.
Pingree, R. D., P. R. Pugh, P. M. Holligan and G. F. Foster. 1975. Summer phytoplankton blooms and red tides along tidal fronts in the approaches to the English Channel. Nature 258: 672-677.
Pingree, R. D., R. M. Holligan and G. T. Mardell. 1978. The effects of vertical stability on phytoplankton distributions in the summer on the northwest European shelf. Deep-Sea Res. 25: 1011-1028.
Pugh, P. R. 1978. The application of particle counting to an understanding of the small-scale distribution of plankton. In: Spatial Pattern in Plankton Communities [ed. J. H. Steele] Plenum Press, NATO Conf. Ser. IV, Vol. 3: 111-129.
Smith, R. C. and K. Baker. 1979. Penetration of UV-B and biologically effective dose-rates in natural waters. Photochem. Photobiol. 29: 311-323.
deSzoeke, R. D. and P. B. Rhines. 1976. Asymptotic regimes in mixed-layer deepening. J. Mar. Res. 34: 111-116.
Therriault, J. C., D. J. Lawrence and T. Platt. 1978. Spatial variability of phytoplankton turnover in relation to physical processes in a coastal environment. Limnol. Oceanogr. 23: 900-911.

Walsh, J. J., T. E. Whitledge, F. W. Barvenik, C. D. Wirick, S. O. Howe, W. E. Esaias, J. T. Scott. 1978. Wind events and food chain dynamics within the New York Bight. Brookhaven National Lab. 22783R.

Woods, J. D. 1968. Wave-induced shear instability in the summer thermocline. J. Fluid Mech. 32(4) 791-800.

Woods, J. D. 1981. Diurnal and seasonal variation of convection in the wind mixed layer of the ocean. Q. J. Roy. Met. Soc. (In Press).

PHOTOCHEMICAL TRANSFORMATIONS INDUCED BY SOLAR ULTRAVIOLET RADIATION IN MARINE ECOSYSTEMS

Richard G. Zepp
U. S. Environmental Protection Agency
Environmental Research Laboratory
College Station Road
Athens, Georgia 30613

INTRODUCTION

Various human activities during the past three decades have greatly increased the amounts of chemicals discharged into marine ecosystems. The dumping of hazardous chemicals into the ocean is likely to further increase in the near future as a response to mounting public concern over the use of landfills for such disposal. Oil spills and rainout of chemicals from the atmosphere represent other significant sources of chemical pollutants in the open ocean. To adequately assess the hazard of pollutants to aquatic organisms, toxicity data must be accompanied by analytical data concerning concentrations of pollutants or by some rational estimate of the concentrations in ecosystems that have not been analyzed. To make such estimates, computer models have been developed that utilize equations and data concerning rates and equilibria of various transport and transformation processes that affect the concentrations of chemicals. Processes considered in such models include microbial, thermal, and photochemical transformations as well as dilution, volatilization, and sorption to sediments and biota. Although these models have been developed specifically for discharges into freshwater systems, this approach could be applied to marine ecosystems with equivalent results.

Although little is known about marine photochemical processes, the high clarity of ocean water suggests that photochemical transformation may be an important degradation pathway for certain pollutants that enter the sea. Recent studies by Zafiriou and True (1979) and Zika (1977) have provided initial evidence that photochemical processes may play a significant role in the cycling of nitrogen and carbon-containing species in the sea. In this paper, approaches to forecasting rates of photoreactions in the sea are discussed.

EXPERIMENTAL

Experimental procedures

The quantum yield for direct photolysis of a chemical was computed from data concerning the concentration of the chemical, [P], as a function of time of exposure to monochromatic light of wavelength λ and irradiance $E_o^{av}(\lambda)$. Experiments were conducted in air-saturated water with [P] below the solubility limit of the chemical and sufficiently low that first order kinetics were observed. Under these conditions, a plot of ln[P] versus time was linear with a slope equal to $-k_{d,\lambda}$, the direct photolysis rate constant. The reaction quantum yield ϕ_r was computed using Eq. 1.

$$\phi_r = \frac{k_{d,\lambda}}{2.303 E_o^{av}(\lambda)\, \epsilon_\lambda} \qquad (1)$$

where ϵ_λ is the molar absorptivity of the chemical at the wavelength employed in the experiment. The irradiance was measured using chemical actinometers. A more detailed description of this procedure appears elsewhere (Zepp 1978).

Electronic absorption spectra were obtained on Perkin Elmer 602 or Perkin Elmer 356 spectrophotometers.

Action spectra for photosensitized reactions in natural water samples were obtained by a series of kinetic experiments that used various wavelengths of monochromatic light (Zepp et al. 1980). Natural water samples were obtained from the top meter of various water bodies and refrigerated at 5°C prior to use. Dark controls indicated no transformation of the chemical occurred by non-photochemical pathways during the time period required for the experiments. Response functions $X_{s,\lambda}$ were computed using Eq. 2, where $k_{s,\lambda}$ is the first order rate constant for photosensitized reaction.

$$X_{s,\lambda} = \frac{k_{s,\lambda}}{E_o^{av}(\lambda)} \quad (2)$$

Sunlight photolysis rate constants, k_p, were computed from photolysis experiments in sunlight using Eq. 3.

$$k_p = \frac{\ln \frac{[P_o]}{[P_t]}}{t} \quad (3)$$

where $[P_o]$ was the concentration prior to sunlight exposure and $[P_t]$ was the concentration after exposure to sunlight for time t. The photolysis halflife, which equals 0.693 k_p^{-1}, is the time required for one-half of the chemical in a system to be transformed.

Computation of photolysis rates

Values of k_p were calculated using a computer program described elsewhere (Zepp and Cline 1977). Most computations employed approximate levels of solar ultraviolet radiation that were calculated using empirical relations described by Bener (1972). Recent studies by Baker et al. (1980) have shown that Bener's relations underestimate the middle ultraviolet irradiance incident on the surface of open ocean water. Preliminary estimates indicate that rate constants computed by applying Bener's relations to the sea are no more than 50% too low, with the greatest error occurring at solar altitudes greater than 60°.

RESULTS AND DISCUSSION

The rate of any photochemical process, -d[P]/dt, depends upon the rate, I_a, at which light is absorbed by the species that initiates photoreaction and by the quantum yield of the process (Eq. 4). If all the absorbed light resulted in transformation, the photolysis rate would equal I_a. Usually, however, only a fraction of the absorbed light results in reaction--this fraction

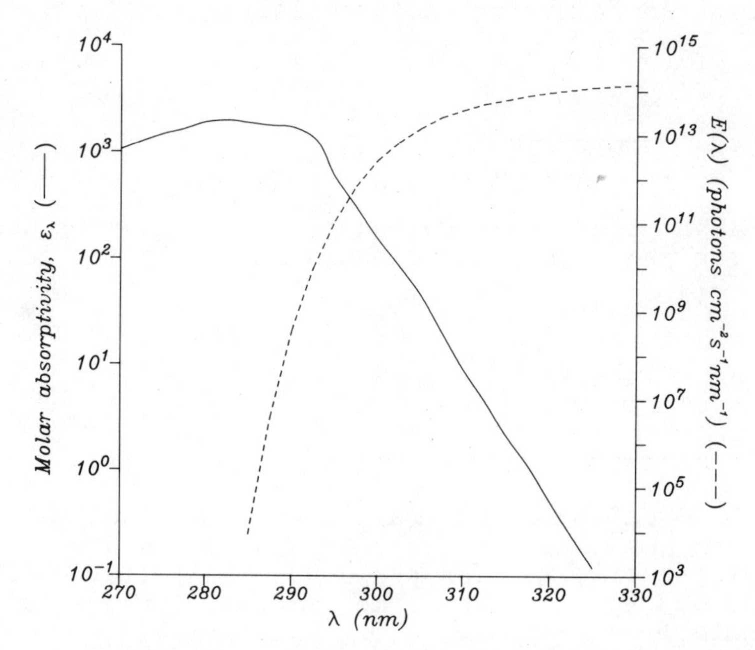

Figure 1. Comparison of electronic absorption spectrum of 2,4-D (Zepp et al. 1975) and solar spectral irradiance for midday during June, latitude 30°N, sea level (Baker et al. 1980).

is the quantum yield, ϕ

$$-\frac{d[P]}{dt} = I_a\phi \qquad (4)$$

In solution, quantum yields are usually wavelength independent, but I_a is always strongly wavelength dependent. Near and middle ultraviolet radiation from the sun are usually responsible for light-induced transformations of synthetic chemicals in natural waters. Two general types of transformations may occur: direct or photosensitized reactions. Examples of each reaction type are discussed below.

Direct photolysis in the sea

The simplest mechanism for photolysis involves direct absorption of light, yielding a molecule in its electronically excited state (Eq. 5). The excited molecule reacts (Eq. 6) or decays back to its ground state (Eq.7).

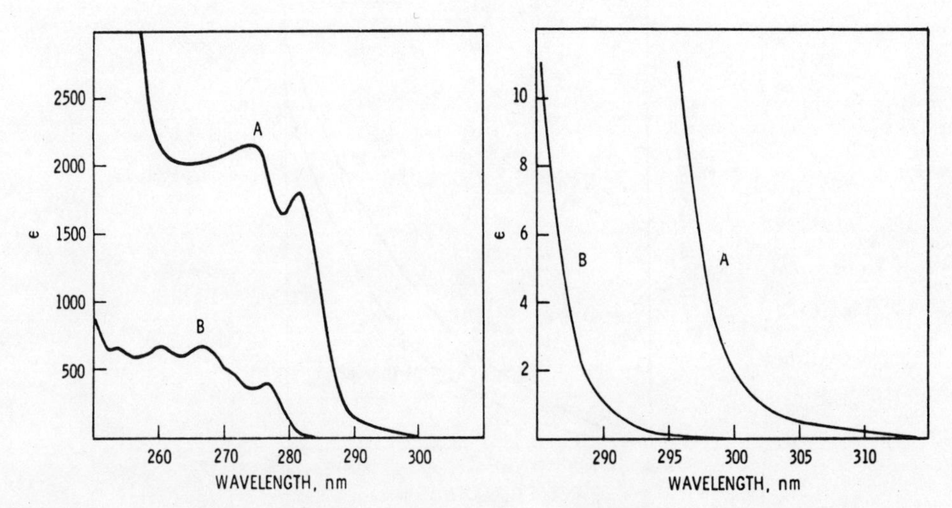

Figure 2. Electronic absorption spectra of the insecticides, methoxychlor (A) and DDT (B) (from Zepp et al. 1976).

$$P + \text{Light} \longrightarrow P^* \quad (5)$$

$$P^* \longrightarrow \text{Products} \quad (6)$$

$$P^* \longrightarrow P_o \quad (7)$$

Concentrations of pollutants in the sea are usually very low. Under these conditions, light attenuation by the pollutant is negligible compared to attenuation by natural substances in the water, and I_a is directly proportional to the pollutant concentration. The proportionality constant k_a, which is expressed in units of reciprocal time, is a measure of the spectral overlap between the solar spectral irradiance and the electronic absorption spectrum of the photoreactive chemical (Eq. 8)

$$\frac{I_{a,\lambda}}{[P]} = k_{a,\lambda} = 2.303\ E_o(\lambda)\ \epsilon_\lambda \quad (8)$$

where $k_{a,\lambda}$ is the specific absorption rate at wavelength λ,

$E_o(\lambda)$ is the irradiance at λ, and

ϵ_λ is the molar absorptivity of the chemical in liter mole $^{-1}$ cm^{-1}.

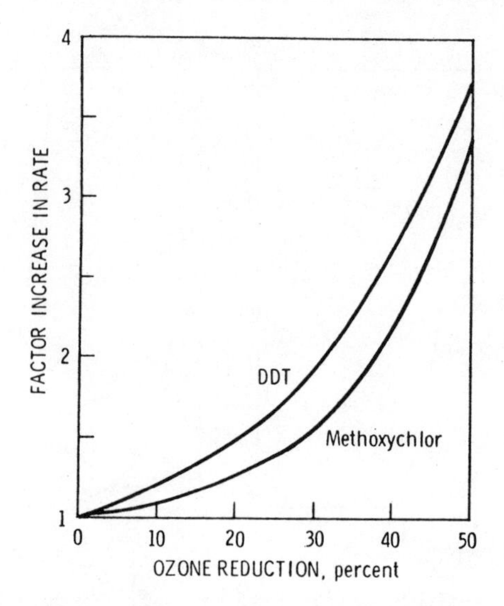

Figure 3. Computed effects of reduction of ozone layer on direct photolysis rates of DDT and methoxychlor. Assumptions: Solar altitude, 60°; average ozone layer thickness at no reduction, 0.319 cm (from Zepp et al. 1976).

In Eq. 8, $E_o(\lambda)$ is the scalar irradiance, expressed in units of millieinsteins cm^{-2} sec^{-1} to be compatible with ϵ_λ. Direct photolysis is described by a first order rate equation (Eq. 9) in which the rate constant k_d equals $\phi_r \int k_{a,\lambda} d\lambda$ where integration is over the effective wavelengths of sunlight. Reaction quantum yields are determined by laboratory experiments (Eq. 1).

$$-\frac{d[P]}{dt} = k_a \phi_r [P] = k_d [P] \quad (9)$$

Many commercially important chlorinated chemicals absorb sunlight most strongly in the middle ultraviolet region. The herbicide, 2,4-D(Figure 1) and the insecticides, DDT and methoxychlor (Figure 2), are examples of such chemicals. A decrease in the ozone layer thickness would increase the direct photolysis rates of these chemicals, as estimated for DDT and methoxychlor (Figure 3) employing computer-generated irradiances and Eq. 9 (Zepp et al. 1976).

Polycyclic aromatic hydrocarbons (PAH) represent another important class of pollutants that enter the sea. Benzo[a]pyrene is one important pollutant that absorbs solar radiation most strongly in the near ultraviolet region. Computations for benzo[a]pyrene (Figure 4) and other PAH indicate that photooxidation in sunlight is rapid with half-lives generally less than an hour, assuming clear skies (Smith et al. 1976; Zepp and Schlotzhauer 1979). Aromatic hydrocarbons absorb sunlight in spectral regions ranging from UV-B to visible, as shown for the chemicals, naphthalene, anthracene, and naphthacene (Figure 5). Computations employing

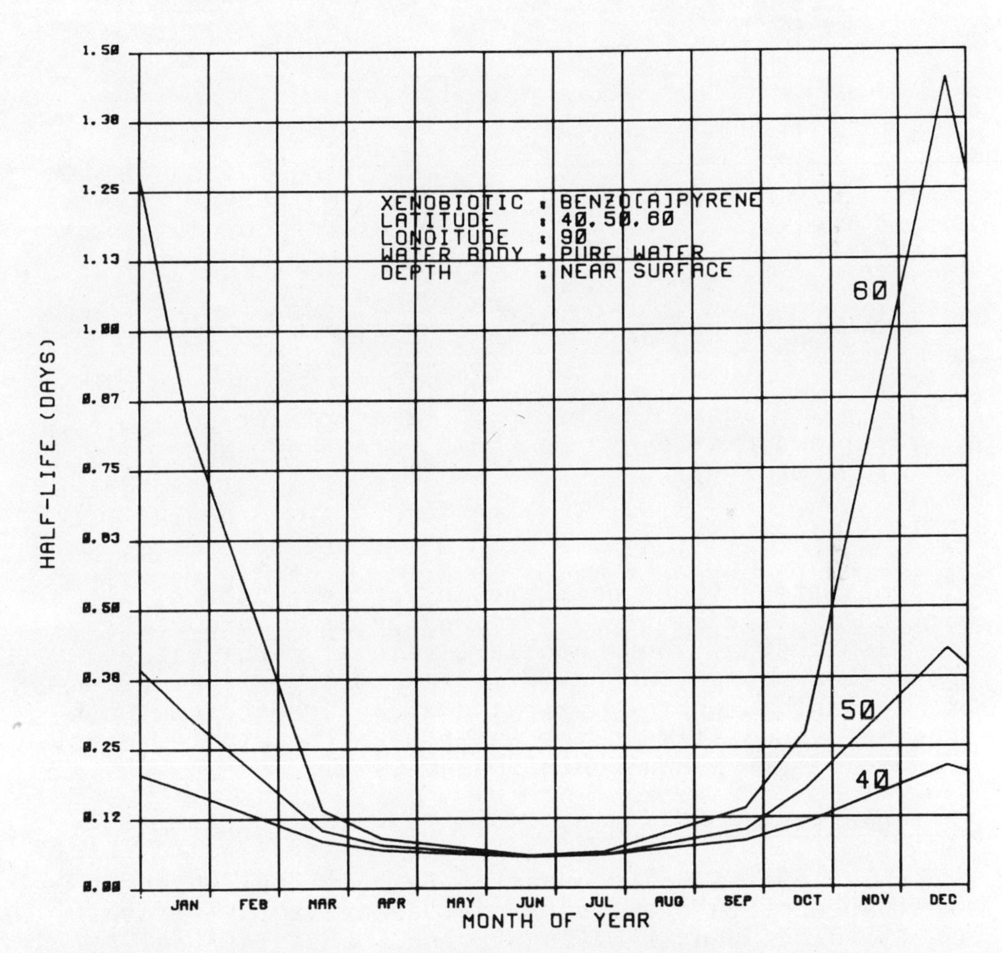

Figure 4. Computed annual variation of the near surface half-life for direct photolysis of benzo[a]pyrene at several Northern latitudes (from Zepp and Baughman 1978).

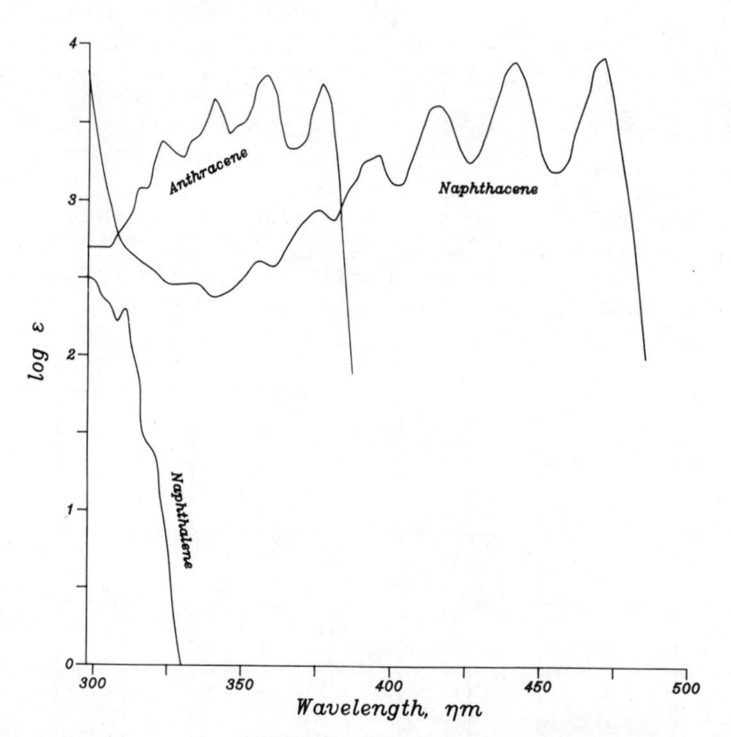

Figure 5. Electronic absorption spectra for several polycyclic aromatic hydrocarbons (from Zepp and Schlotzhauer 1979)

diffuse attenuation coefficients for sea water of various types recently determined by Smith and Baker (1979) show that the photolysis rate of naphthalene, the UV-B absorber, drops off more rapidly with increasing depth than photolysis rates of anthracene and naphthacene, especially in the water near the Florida coast (Figure 6) (Zepp and Schlotzhauer 1979).

Photosensitized Reactions

This type of transformation is initiated through light absorption by natural photosensitizers. Photosensitizers, upon light absorption, cause chemical changes that do not occur in their absence. A skeletonized scheme for possible photosensitized reactions that occur in natural waters is shown in Equations 10 through 15.

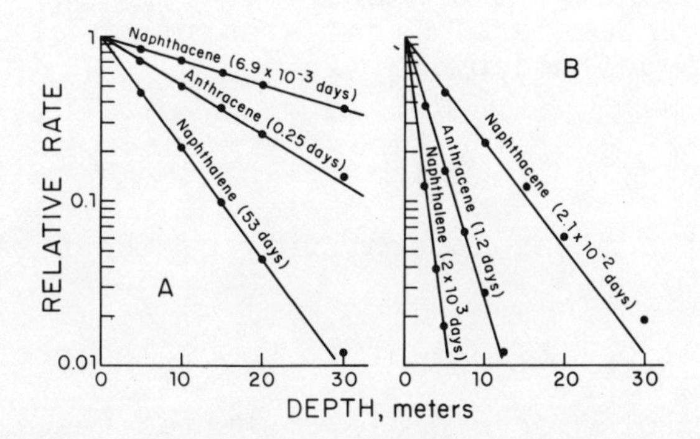

Figure 6. Computed depth dependence for direct photolysis of several polycyclic aromatic hydrocarbons during summer in Gulf of Mexico: A -Mid-Gulf; B- coastal Gulf near Tampa. Average half-lives in top 35 meters are in parentheses (from Zepp and Schlotzhauer 1979).

$$S + \text{Light} \longrightarrow {}^{1}S^{*} \quad {}^{3}S^{*} \qquad (10)$$

$${}^{3}S^{*} + O_2 \xrightarrow{k_{11}} S + {}^{1}O_2^{*} \qquad (11)$$

$${}^{3}S^{*} + P \xrightarrow{k_{12}} \text{Products} \qquad (12)$$

$${}^{3}S^{*} \xrightarrow{1/\tau} S \qquad (13)$$

$${}^{1}O_2^{*} + A \longrightarrow \text{Products} \qquad (14)$$

$${}^{1}O_2^{*} \longrightarrow O_2 \qquad (15)$$

Light absorption puts the sensitizer molecule into an excited state (Eq. 10) that can transfer energy to molecular oxygen (eg. 11), interact with another chemical P to form products (Eq. 12), or decay back to its ground state (Eq. 13). Singlet molecular oxygen, the product of reaction (11) can interact with certain types of chemicals to form oxidized products (Eq. 14), or decay back to ground state (Eq. 15). Rate constants have been published for many of the excited state processes shown above. Our goal has been to develop procedures that would permit usage of these published data to compute rates of photosensitized reactions in natural waters.

Equations that describe rates of photosensitized

reactions differ somewhat from those that describe direct photolysis, although the basic equation (Eq. 4) still is applicable. The rate of light absorption by a photosensitizer I_a^S is defined by Eq. 16.

$$I_a^S = 2.303\, E_o\, (\lambda)\; \epsilon_\lambda^S\, [S] \quad (16)$$

where λ^S is the molar absorptivity of the photosensitiz- at λ.

[S] is the molar concentration of the photosensitizer.

The quantum yield for a photosensitized reaction is defined by Eq. 17.

$$\phi_r^S = \phi_{IS} \frac{k_{12}\,[P]}{1/\tau\; k_{11}[O_2] + k_{12}[P]} \cong Q_\lambda\, [P] \text{ if } [P] \text{ very low.} \quad (17)$$

where ϕ_{IS} is the efficiency for formation of $^3S^*$,

k_{11}, k_{12}, and $1/\tau$ are rate constants for Equations 11-13,

$[O_2]$ is the molar concentration of molecular oxygen in water,

Q_λ is a constant equal to $\phi_{IS} k_{12} / (1/\tau + k_{11}[O_2])$.

Under typical conditions in the photic zone of the sea, [P] is very low, $k_{12}[P]$ is negligible compared to the other terms in the denominator of Eq. 17, and the quantum yield is approximately proportional to [P] (Eq. 17). When I_a^S and Q_λ are constant, the photosensitized reaction, like direct photolysis, obeys first order kinetics, and the rate expression is:

$$-\frac{d[P]}{dt} = 2.303\, E_o\, (\lambda)\, \epsilon_\lambda^S\, Q_\lambda\, [S][P] = k_{s,\lambda}[P] \quad (18)$$

By comparison of Eq. 2 and Eq. 18, it can be seen that the response function $X_{s,\lambda}$ equals $2.303 \epsilon_\lambda^S Q_\lambda$ [S]. The rate constant for a photosensitized reaction in sunlight k_s equals $\int E_o\, (\lambda)\, X_{s,\lambda}\, d\lambda$ where $E_o(\lambda)$ is the underwater solar irradiance.

To simplify our studies of sensitized photoreactions,

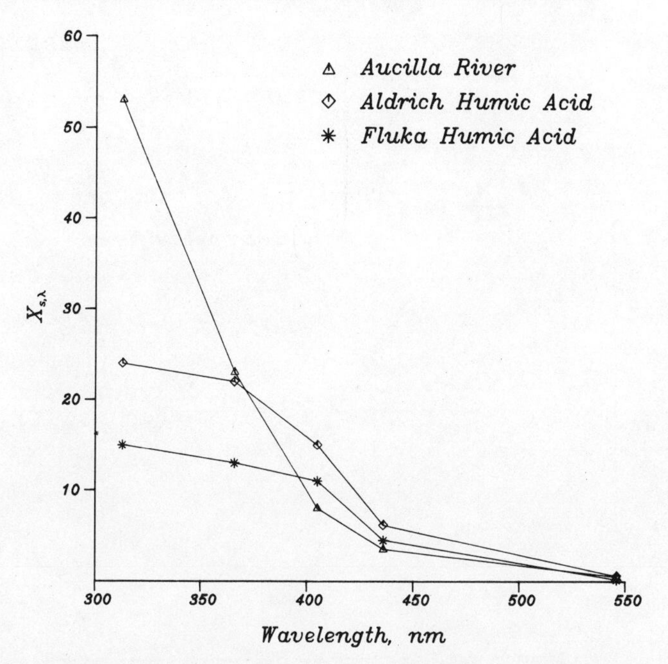

Figure 7. Action spectra for oxygenation of 2,5-dimethylfuran photosensitized by humic substances in water.

we have selected chemicals that do not undergo <u>direct</u> photolysis with exposure to light of wavelengths available in sunlight, i. e., the rate constant for sunlight photolysis, k_p, equals k_s. Quantitative kinetic data have been obtained for several different types of photosensitized reactions in natural waters (Zepp et al. 1980; Zepp et al. 1977). Results of those studies indicate that humic or "yellow" substances in natural waters act as photosensitizers. The sensitizing efficiencies of humic substances from a variety of locations were shown to be remarkably similar (Zepp et al. 1980). Plots of $X_{s,\lambda}$ versus wavelength, or action spectra, for the oxidation of 2,5-dimethylfuran photosensitized by soil-derived humic acids and the humic substances in a "blackwater" river indicate that ultraviolet and blue radiation are most effective (Fig. 7). This reaction has been shown to involve energy transfer from sensitizer to oxygen, forming singlet oxygen as an intermediate that oxidizes the dimethylfuran (Zepp et al. 1977). Other studies showed that the rate of the dimethylfuran reaction in the river water is independent of hydrogen ion activity in the pH 5 to 9 range

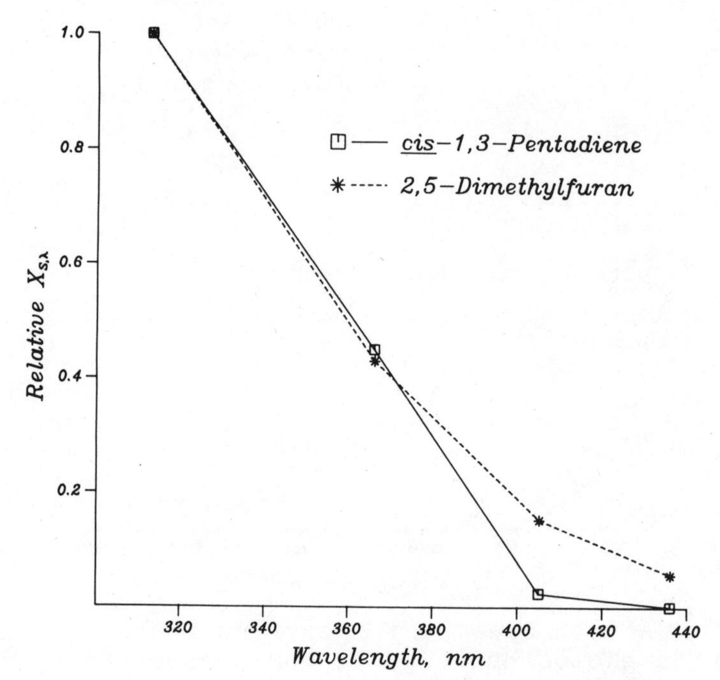

Figure 8. Comparison of action spectra for photosensitized reactions of cis-1,3-pentadiene and 2,5-dimethylfuran in a water sample from a blackwater river.

and that the response function at 366 nm is directly proportional to the concentration of humic substance, as predicted by theory (Zepp et al. 1980).

Another type of reaction sensitized by humic substances involves energy transfer from the sensitizer to the reactive chemical, cis-1,3-pentadiene, which then isomerizes to its trans isomer. The action spectrum for this reaction differs from that for the dimethylfuran reaction in that only ultraviolet light is effective (Fig. 8). The data in Figure 8 are qualitatively consistent with what is theoretically expected for these reactions. Energy transfer occurs efficiently only when the excited state energy of the photosensitizer exceeds that of the acceptor. Because oxygen has a lower excited state energy than pentadiene, it can accept energy from a wider range of sensitizers including some that absorb lower-energy visible light.

Other light-induced processes may involve humic substances. Mill and co-workers (1980) recently presented results that indicate that humic substances can initiate free radical oxidations of certain organic chemicals.

Photosensitization by humic substances is expected to be most important in coastal regions of the sea that receive riverine inputs of humic substances. It was recently demonstrated that photosensitized oxidations occur in coastal waters from the Gulf of Mexico (Zepp et al. 1977), and in coastal waters of Virginia and Massachusetts (Zafiriou 1977).

Inorganic or biological species in sea water may also initiate light induced reactions. Zafiriou and True (1979) have found that photolysis of nitrite in certain parts of the sea generates hydroxyl and other free radicals that are known to rapidly oxidize a wide variety of chemicals. The photolysis of nitrite is caused by solar ultraviolet radiation. Moreover, preliminary evidence exists that algae induce photochemical transformations of pesticides and aromatic hydrocarbons (Zepp 1980; Cerniglia et al. 1979).

CONCLUSION

In the above discussion, examples were selected to illustrate the approaches that are being used to assess the effects of photochemical transformations on concentrations of chemicals in natural waters. Clearly, these approaches closely parallel those being used in the computation of doses of solar ultraviolet radiation received by aquatic organisms. In particular, the reliability of the computed photolysis rates or doses is directly dependent on the accuracy of the solar spectral irradiance used in the calculations. Reliable procedures must be developed for estimating levels of solar ultraviolet radiation in unmeasured natural waters. In addition, the exposure to solar ultraviolet radiation also depends upon vertical mixing which moves chemicals or organisms from deeper, dark water up to the photic zone, and vice versa. Quantitative information concerning mixing rates in water bodies is required to model the photochemical behavior of extremely light sensitive chemicals such as polycyclic aromatic hydrocarbons.

REFERENCES

Baker, K., R. C. Smith, and A.E.S. Green. Middle ultraviolet radiation reaching the ocean surface. Photochem. Photobiol. (in press).

Bener, P. 1972. Approximate values of intensity of natural ultraviolet radiation. Technical Report, European Research Office, U. S. Army, London, Contract DAJA 37-68-C-1017.

Cerniglia, C. E. and D. T. Gibson. 1979. Algal oxidation of aromatic hydrocarbons: Formation of 1-naphthol from naphthalene by Agmellum quadruplicatem, strain PR-6. Biochem. Biophys. Res. Commun. 88:50.

Mill, T., D. G. Hendry, and H. Richardson. 1980. Free-radical oxidants in natural waters. Science. 207:886.

Smith, J. H., W. R. Mabey, N. Bohonos, B. R. Holt, S. S. Lee, T. W. Chou, D. C. Bomberger, and T. Mill. 1978. Environmental pathways of selected chemicals in freshwater systems. Part II. Laboratory studies. U.S. Environmental Protection Agency, Athens, GA, Report No. EPA-600/7-78-074.

Smith, R. C. and K. S. Baker. 1979. Penetration of UV-B and biologically-effective dose rates in natural waters. Photochem. Photobiol. 27:311.

Zafiriou, O. C. 1977. Marine Organic Photochemistry Previewed. Mar. Chem. 5:497.

Zafiriou, O. C. and M. B. True. 1979. Nitrite photolysis in seawater by sunlight. Mar. Chem. 8:9.

Zepp, R. G. 1978. Quantum yields for reaction of pollutants in dilute aqueous solution. Environ. Sci. Technol. 12:327.

Zepp, R. G. 1980. Assessing the photochemistry of organic pollutants in aquatic environments, pp. 69-110. In: R. Haque [ed.], Dynamics, exposure, and hazard assessment of toxic chemicals. Ann Arbor Science Publishers, Inc.

Zepp, R. G. and G. L. Baughman. 1978. Prediction of photochemical transformation of pollutants in the aquatic environment, p. 236-263. In: O. Hutzinger, I. van Lelyveld, and B. Zoeteman [eds.], Aquatic pollutants: Transformation and biological effects. Pergamon Press.

Zepp, R. G., G.L. Baughman, and P. F. Schlotzhauer. Comparison of photochemical behavior of various humic substances in water: Photosensitized oxygenations. Chemosphere (submitted for publication).

Zepp, R. G. and D. M. Cline. 1977. Rates of direct photolysis in aquatic environment. Environ. Sci. Technol. 11:359.

Zepp, R. G. and P. F. Schlotzhauer. 1979. Photoreactivity of selected aromatic hydrocarbons in water, p. 141-158. In: P. W. Jones and P. Leber [eds.], Polynuclear aromatic hydrocarbons. Ann Arbor Science Publishers, Inc.

Zepp, R. G., N. L. Wolfe, G. L. Baughman, and R. C. Hollis. 1977. Singlet oxygen in natural waters. Nature. 267:421.

Zepp, R. G., N. L. Wolfe, J. A. Gordon, and G. L. Baughman. 1975. Dynamics of 2,4-d esters in surface waters. Environ. Sci. Technol. 9:1144-1150.

Zepp, R. G., N. L. Wolfe, J. A. Gordon, and R. C. Fincher. 1976. Light-induced transformations of methoxychlor in aquatic systems. J. Agr. Food Chem. 24:727-733.

Zika, R. G. 1977. An investigation in marine photochemistry. Ph.D. Thesis. Dalhousie Univ.

PENETRATION OF SOLAR UV-B INTO WATERS OFF ICELAND

John Calkins (1) and Thorunn Thórdardóttir (2)

Department of Radiation Medicine
University of Kentucky
Lexington, Kentucky 40506 USA (1) and
Hafrannsóknastofnunin (Marine Research
Institute) Reykjavik, Iceland (2)

ABSTRACT

We have measured levels of solar UV-B incident on Iceland using the Robertson sensor, the relation of UV-B to total solar irradiance, and the levels at various depths in the oceans near Iceland. Typical penetration of UV-B and visible light into marine waters was observed. Penetration of UV-B is somewhat less variable than penetration of green light and may be relatively uniform over distances of hundreds of kilometers. The penetration of UV-B in relation to ^{14}C incorporation is discussed.

METHODS

We have measured the incidence and penetration of UV-B and visible sunlight in the waters off Iceland to facilitate modeling of the probable effects of ozone depletion on marine productivity.

The Robertson sensor (Robertson 1972, Berger et al. 1975) was used to observe the attenuation of solar UV-B at a number of locations in the waters off Iceland. The diffuse attenuation coefficient (K) has been calculated from the ratio of observed dose rate at various depths to the dose rate observed just under the surface. The use of the Robertson sensor for measurement of the diffuse attenuation coefficient (K) for irradiance in the UV-B region of the spectrum was considered in detail in Smith and Calkins (1976). Total incident radiation data was obtained from the Reykjavik Meteorlogical Institute.

At the stations surveyed for UV-B penetration in May-June, 1976 we also determined the penetration of visible (green) light, using a small submersible irradiance meter. The meter, constructed at the Science Institute of the University of Iceland, consisted of an opalizing disc mounted flush with the meter surface, a neutral density filter and a green filter (Figure 1), and a selenium photovoltaic cell with the recommended (50 ohm) load resistor which makes the current output linear with irradiance. The current output of the photovoltaic cell was read with a high gain DC amplifier. For computation of the diffuse attenuation coefficient for the green light we used the ratio of irradiance at various depths to the irradiance at a depth of 5 m. At 5 m reference

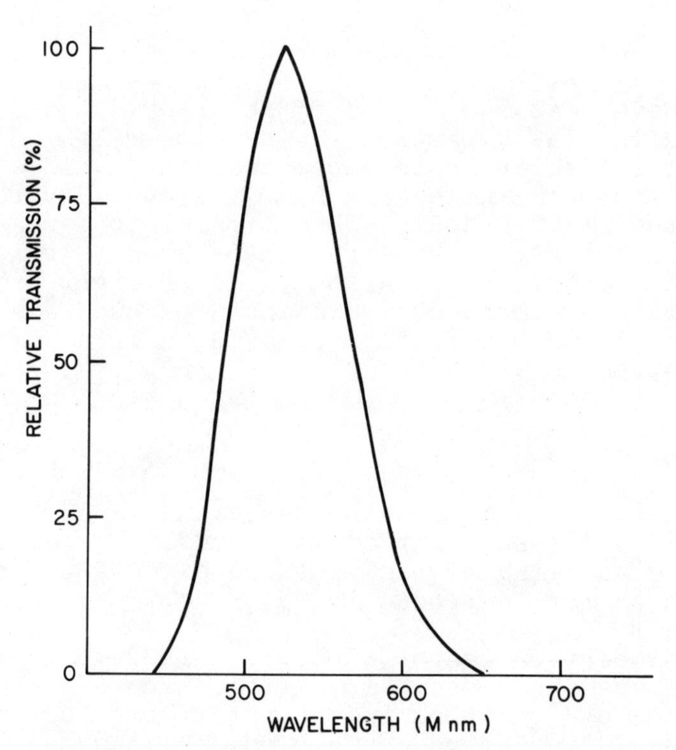

Figure 1. The transmission of the filter used for determination of penetration of visible light.

depth proved more stable and convenient than surface reference and avoids near surface optical phenomena. The attenuation was, with a few exceptions, exponential and the choice of reference depth would not be critical.

The location of the stations for UV-B penetration in 1975 and 1976 is shown in Figure 2. Secchi depth (transparency), primary productivity, chlorophyll content, phytoplankton composition, salinity, temperature and nutrient concentrations were determined at standard depths at the various stations (Thórdardóttir 1976).

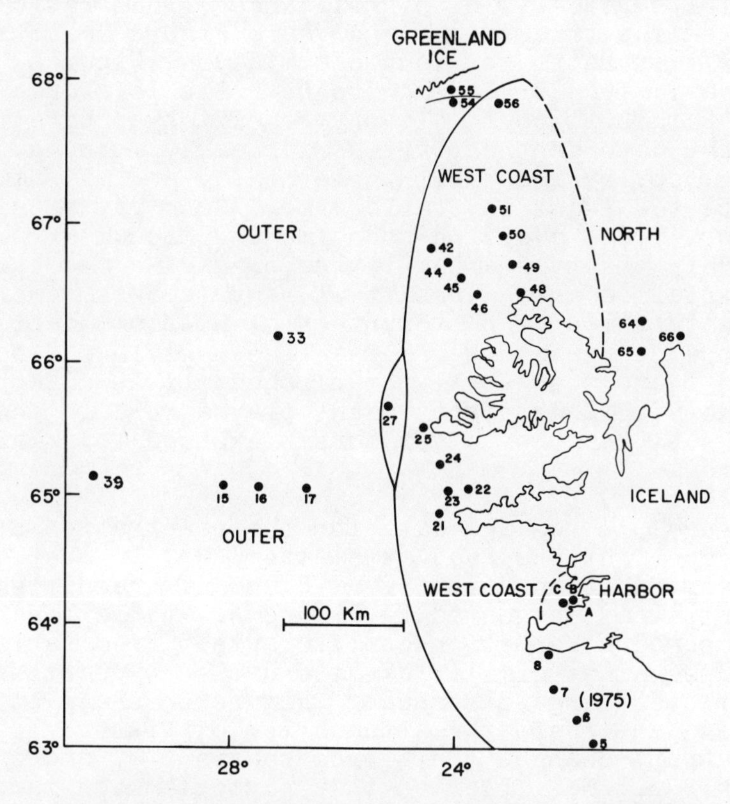

Figure 2. Location of stations surveyed for UV-B penetration and a general grouping of the stations according to UV-B optical properties is also indicated.

RESULTS AND DISCUSSION

Figure 3 is a plot of the daily variation of UV-B dose rate and the irradiance incident on Reykjavik, Iceland for selected days from May-October, 1976. The behavior of the relatively clear days resembles what would be anticipated from the nature of UV-B absorption in the ozone layer. The UV-B dose rate falls rapidly as the solar angle is decreased. Solar UV-B dose is high near midsummer noon (high solar angles). The total irradiance, which is only very weakly absorbed in the atmosphere (when there are no clouds), is less dependent on solar angle.

UV-B radiation is an injurious factor which light-requiring organisms, especially plants, must accept in order to receive light for photosynthesis, vision, etc. Thus, the ratio of total irradiance to the UV-B dose at the water surface is of significance. This ratio will depend on the time of day and year; Figure 4 shows a plot of this ratio for four clear days. Figures 3 and 4 show that, if an organism were near its tolerance limits for UV-B then it would clearly benefit by programming its light dependent activities for early morning, late evening, spring and fall; while midday and midsummer would be the least favorable time. Such patterns of behavior, i.e., avoiding noontime and midsummer sun, are widely observed among living organisms and have been explained as the result of many factors, including light saturation of photosynthesis, nutrient depletion, avoidance of predators and the like. While the non-UV related factors are well established, it is also possible that the UV-B sensitivity might play a role in the patterns of behavior of organisms exposed to solar radiation.

Figure 5 shows the data for the penetration of UV-B and green light observed at station 22. In most cases the observed attenuation of UV-B and of green light were exponential as suggested by Figure 5. Table 1 contains the measured attenuation coefficients. From Table 1 it can be seen that similar UV attenuation properties persist over large distances. Figure 6 is a plot of two instances where the attenuation of visible light shows abrupt changes (data from stations 27 and 65). At station 27 it was observed that phytoplankton samples from 30 meters showed a high capacity to incorporate ^{14}C while the surface and 10 meter samples were relatively unproductive. The 526 nm optical properties at

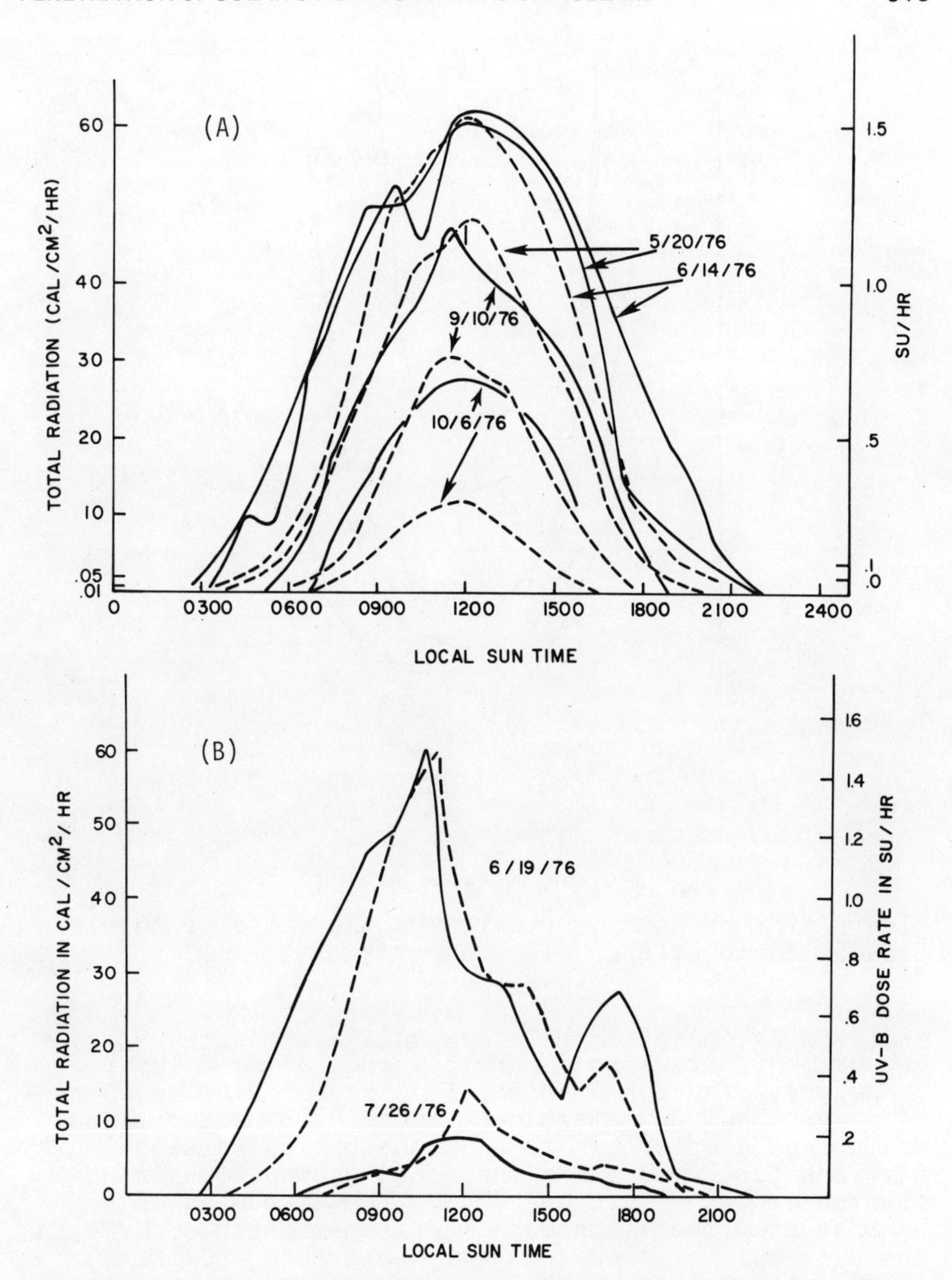

Figure 3A. The daily variation of UV-B intensity (dotted lines) on relatively clear days in Reykjavik and the variation of total irradiance (solid lines) for the same days. Figure 3B. Daily variations of UV-B (dotted) and total irradiance (solid) on partly cloudy and cloudy days.

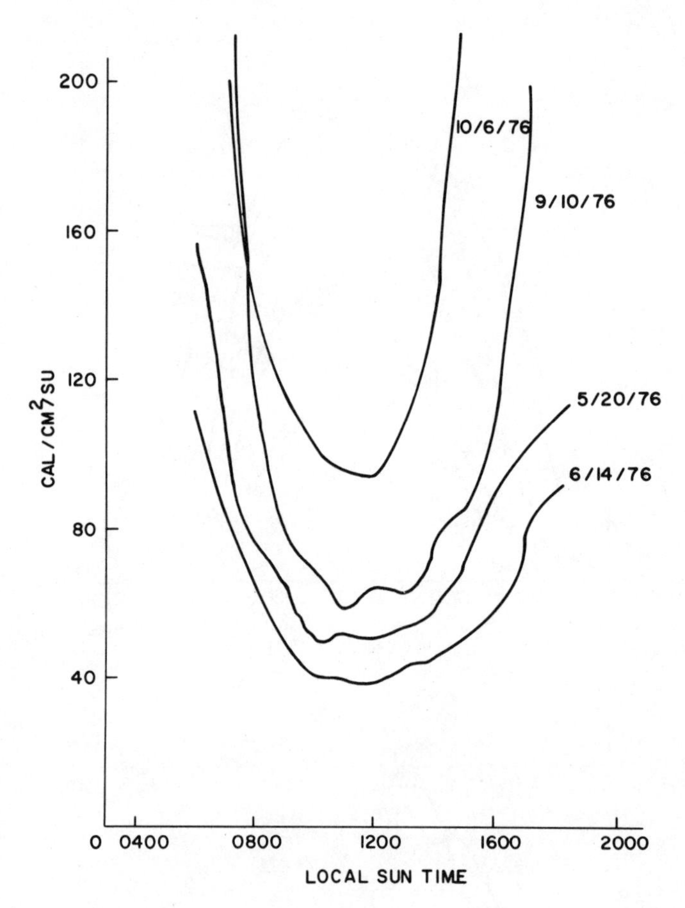

Figure 4. The daily variation of the ratio of total irradiance to UV-B on clear days in Reykjavik.

stations 27 and 65 correlate with the productivity measurements (bars in Figure 6), indicating a layer of dense phytoplankton overlain by a rather sterile layer of water about 20 meters deep. The UV-B measurements did not go deep enough to show the optical discontinuity. Stations furtherest from the coast showed high transmission of both green and UV-B. The transmission of UV-B at station 54 exceeded the transmission of UV-B in the extremely clear waters off Puerto Rico (Calkins, 1975), although the Secchi disc transparency was much inferior to the waters off Puerto Rico.

The stations nearer the coast showed a decreased transmission of both UV-B and green; the ratio of K_{UV-B}/K

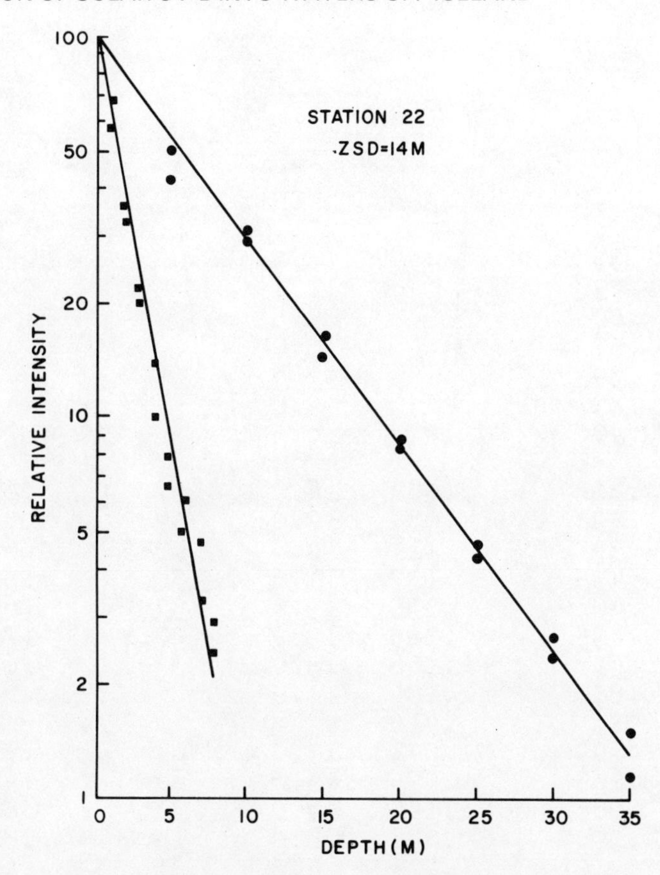

Figure 5. The attenuation of UV-B (Boxes) and green light (circles) in the waters off Iceland. The ratio of UV-B and green irradiance to incident values as a function of depth. Points indicate individual measurements, the lines are fitted "by eye".

green is similar in the two groups (Fig. 2). The one station at the edge of the Greenland ice pack (Station 55) showed very low penetration of UV-B relative to green and the stations in the "harbor" (surveyed in 1975) also showed low penetration of UV-B (as do most harbor measurements) considering the relatively large values of Z_{sd}.

The three measurements off the north coast of Iceland are distinctly different from other observations. The penetration of UV-B is quite high considering the relatively high attenuation of 526 and low values of Z_{sd}.

TABLE 1

Station	K_{UV-B}	K_{526}	Z_{SD}	Z_{EZ}*	UV-B dose to EZ(SU)**
A	1.35		5.5		
B	.77		11.2		
C	.60		9.5		
5	.57		5.0		
6	.39		7.0		
7	.29		7.0		
8	.49		6.5		
15	.262	.080	14.0	57	.61
16	.232	.084	17.0	59	.73
21	.39	.114	14.0	40.5	.63
22	.455	.127	11.0	38	.58
23	.40	.135	13.0	37	.68
24	.455	.166	13.0	27.8	.79
25	.416	.166	15.0	37	.64
27U	.25	.105	11.0		
27L		.416			
33	.209	.0725	15.0	64	.75
39	.169	.075	15.0	60	.99
42	.386	.147	7.0	31.4	.82
43	.59	.238	8.0	19.2	.88
45	.417	.21	7.5	21	1.14
46	.525	.308	5.0	30	1.26
48	.435	.127	8.0	36.4	.63
49	.455	.20	6.0	23.4	.94
50	.40	.217	5.0	21.4	1.16
54	.147	.067	15.0	69	.98
55	.55	.058	15.0	60	.30
56	.38	.159	6.0	28.8	.89
64	.238	.15	7.0	30.8	1.36
65U	.222	.11	8.0	31	1.45
65L					
66	.345	.204	6.0	23	1.25

*Depth of the euphotic zone is assumed to be the depth where the intensity of 526 nm light equals 1% of the surface value.

**Assuming 10 SU incident at the surface.

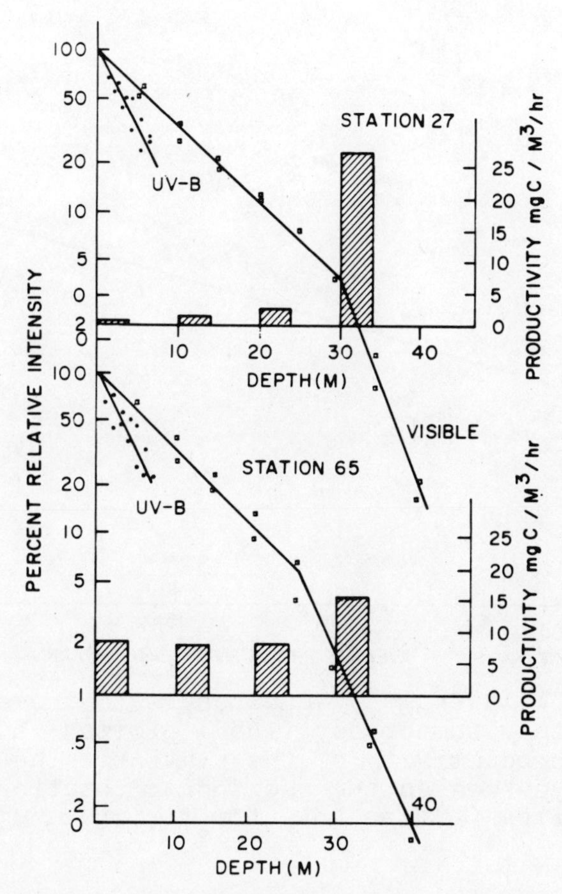

Figure 6. Plots similar to Figure 5 for two stations showing a discontinuity in optical properties. Productivity of samples from 0, 10, 20 and 30 m are indicated by crosshatched bars.

Primary production of waters from 0, 10, 20, and 30 m (^{14}C incorporation under standard illumination) was determined. Productivity correlates closely with the chlorophyll content of the samples which in our studies was determined for the 10 m sample only. The UV-B exposure that marine phytoplankton receive will be critically dependent on both the UV-B absorption characteristics and the absorption of photosynthetically active light. The band of visible light we have chosen to measure (488-570) (Fig. 1) penetrates well into the ocean water (Jerlov, 1968) and is efficient for photosynthesis. Our measurements of transmission of green

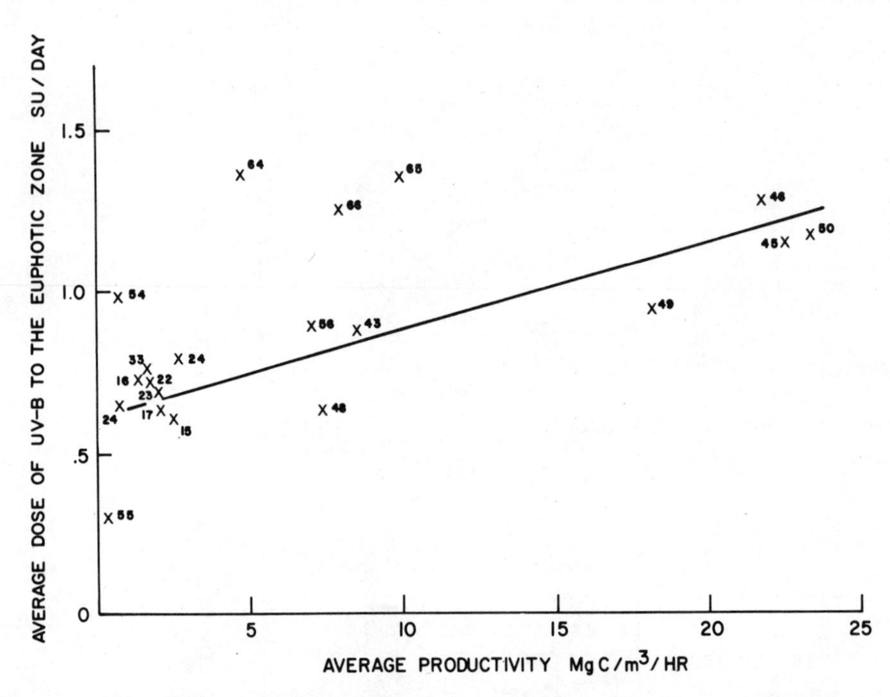

Figure 7. The average UV-B exposure to the euphotic zone on a sunny summer day (10SU) plotted as a function of average productivity of the station. The line is drawn "by eye" neglecting the "north" stations (64, 65 and 66). (From Calkins and Thordardottir, 1980).

light at the stations surveyed thus provides an estimate of the fraction of surface irradiance available for photosynthesis at various depths. If the ratio of K_{UV-B}/K green is the same for two stations, then the phytoplankton at the two stations receiving the same amount of photosynthetically active light will receive essentially the same UV-B dose. Organisms will of course receive a given amount of solar radiation at greater depths in clear water than in less transparent waters.

The average irradiance incident in a given volume of water can be calculated from the formula (Calkins et al. 1976):

$$Ia = \frac{Io(1-e^{-KZ_b})}{KZ_b} \qquad (1)$$

where Ia is the average irradiance, Io is the irradiance

incident just below the surface, K is the diffuse absorption coefficient and Z_b is the depth of the volume under consideration. If the common approximation is used, i.e., that the bottom of the euphotic zone (Z_{ez}) is the depth receiving 1% of the surface irradiance, then the average UV-B dosage to the euphotic zone (during a sunny summer day) can be computed; Figure 7 shows the average UV-B dosage to the euphotic zone plotted against productivity. We assume the depth of the euphotic zone to be the depth receiving 1% of 526 nm intensity just below the surface and 10 SU/day are incident at the surface. Although there is a large scatter, the trend is such that the more productive stations are subjected to greater UV-B exposure, an observation in contrast to one's intuitive assumption that high productivity in the water column would tend to protect the phytoplankton from UV-B exposure.

ACKNOWLEDGEMENTS

We wish to express our appreciation for the extensive assistance given this project by the Hafrannsoknastofnunin (Marine Research Institute), the kindness of its staff and the crew of the Bjarni Saemundsson.

REFERENCES

Berger, D., D. F. Robertson, and R. E. Davis. 1975. Field measurements of biologically effective UV radiation. In Climatic Impact Assessment Program Monogr. 5, Part 1, U.S. Dept. Transportation DOT-TST 75-55, National Technical Information Service, Springfield, VA. 233-264.

Calkins, J., J. D. Buckles, J. R. Moeller. 1976. The role of solar ultraviolet radiation in "natural" water purification. Photochem. Photobiol. 24: 49-57.

Jerlov, N.G. 1968. Optical Oceanography, Elsevier Publishing Company, Amsterdam.

Robertson, D. F. 1972. Solar ultraviolet radiation in relation to human sunburn and skin cancer. Ph.D Thesis, University Queensland, Australia.

Smith, R. C., and J. Calkins. 1976. The use of the Robertson meter to measure the penetration of solar middle-ultraviolet radiation (UV-B) into natural waters. Limnol. and Oceanogr. 21: 746-749.

Thordardóttir, T. 1976. Spring primary production in Icelandic waters 1970-1975. ICES C.M. 1976 1:31.

PREFACE TO SECTION III - THE BIOSPHERE

Sections I and II are directed toward methods which establish the amount of solar UV reaching organisms in natural waters; in this section we turn to the responses which the solar UV will induce in the aquatic biota. Obviously, if inquiries are made very specifically, then the biological responses will be as diverse and complex as the organisms which inhabit the seas. The variety of living organisms is so great that biologists have no choice but to generalize, to presume that the basic processes one observes in one or a few species are occurring in a more or less similar fashion in most or all living organisms. One should be very conscious that our biological information is gleaned from the sources presently available; the species most extensively studied are not organisms commonly found in the seas. On the other hand, it must be admitted that generalizing has for the most part worked well in biology; many mechanisms and processes are remarkably ubiquitous among living organisms.

With such caveats in mind we ask: How does solar UV kill or injure living organisms?, a question also posed by the terrestrial photobiologists. In spite of decades of study, there are still doubts as to the answer to the most fundamental question: "What are the primary photochemical lesions induced by solar UV?" Regardless of the nature of the initial lesion there is no serious doubt that living organisms possess very complex ways of coping with injurious agents in the environment, including solar UV. To understand the role of solar UV in the marine environment and especially for predictive purposes, one must inquire: What is the relative importance of the three major pathways that living organisms possess for mitigating the harmful effects of solar radiation, i.e., 1) generation of shielding substances, 2) avoidance of exposure and 3) reversal of the photochemical lesions or their injurious effects through repair processes?

We know much more about the mechanisms of repair processes than we know of pigment induction or avoidance of solar UV. Repair mechanisms have been extensively studied using 254 nm UV radiation and it has been found that the repair systems which function on 254 nm induced lesions also function (but often with very different efficiency) on solar UV induced damage. Perhaps the most startling inquiry one can make in UV photobiology is: What is the role of solar UV in pigmentation and avoidance of sunlight? Both photoresponses have long been explained without reference to the UV component of sunlight, yet if one compares tolerance to sunlight with exposure, it must be concluded that the observed avoidance responses and pigments are vital to surviving solar UV exposure, judging from the few cases where adequate data are available.

Turning from the characteristics which are common to living organisms in general to specific solar UV induced responses of animals and plants, one finds a rather limited data base, especially if only marine organisms are considered. Estuarian ecosystems; fish which have great economic implications and plants in general, a system where photoeffects are the natural focus, have received the most extensive specific study.

One must ultimately ask: How does solar UV impinge on the large scale processes which shape life on the earth? What is the role of solar UV in mutation, in the success or failure of differing species in the ecosystem, in the history and future of life on the earth, in evolution? Compared to the other questions considered in this Monograph, the insight regarding these questions which we now possess is much more cloudy. Some facts can be established beyond doubt. Biochemical lesions can be identified, repair pathways can be established, mutants defective in repair or pigmentation can be produced, movements in response to UV radiation can be observed, but still when one turns to the real world many doubts linger. Is an organism really the sum of its biochemical pathway; have we truly represented the various forces pushing and pulling the individual, the species, the population, the ecosystem?

Despite the problems, again a coherent pattern emerges; solar UV impinges on all levels of life. One sees subtle responses to UV in innumerable characteristics of living organisms; yet, the little attention that this most ubiquitous agent has been given in traditional biological training is bound to produce both reasonable and unreasonable questioning of the impact of solar UV.

In particular, our coverage of this Monograph was designed to respond to the various aspects of the biological action of solar UV with contributions which are general in scope and detailed upon the more important points. Peak and Peak (Chapter 26) present an overview of the present status of the molecular biology of solar UV damage and the various repair processes. Hariharan and Blazek and also Zigman continue the development at the molecular biology level with a description of novel forms of near UV induced lesions.

Photoreactivation, a reversal of far UV lesions requiring near UV or visible light has been clearly established both regarding lesions and repair mechanisms. But the nature of photoreactivation, implied in the contributions of Peak and Kubitschek, Coohill, and Teramura becomes much less clear and straightforward when one considers that solar UV encompasses wavelengths which both produce and reactivate lesions. In addition to repair through constitutive systems, Coohill presents evidence, in Chapter 32, that solar UV lesions may induce repair as is well established for far UV. Another facet of solar UV and its repair is the dependence on life cycle phase, discussed by Eisenstark.

Contributions by Coohill and by Zigman address the nature and parameters of response to sunlight by pigment formation and dynamics. Photomovements induced by solar UV in a number of organisms are considered by Barcelo; Häder describes photomovement in blue-green algae; Damkaer and Dey find larval crustaceans do not protect themselves by movement.

Turning to more specific research on solar UV effects, an extensive review of the literature focusing on marine organisms is given by Worrest in Chapter 39. Hunter, Kaup and Taylor describe a number of their studies which lead to a quantitative estimate of the potential action of solar UV on fish larvae. Adult fish are also found to be subject to solar UV injury as described by Bullock in Chapter 41.

Photosynthesis, in the oceans as in terrestrial ecosystems, supports all the living organisms. The actions of solar UV on the photosynthetic biota are the most direct pathways for possible damage to the marine ecosystem. Smith and Baker model the potential loss of productivity based on the optical properties of typical waters; Calkins introduces the concept that phytoplankton may be controlling their average vertical position to accommodate their tolerance limits for solar UV-B and through this postulate reaches another estimate of the possible reduction of photosynthetic productivity for a given increase in UV-B.

Observations of the effects of UV-B on diatoms, littoral plants, and coastal phytoplankton are reported by Geiger et al, Wells and Nachtwey, and by Modert et al. The efficiency of various UV components of sunlight for the depression of photosynthesis is reported by Bogenrieder and Klein who also report a very important observation, namely that the way rates of photosynthesis are determined may include a significant error arising from the lack of UV transparency of the measuring container.

Factors which bear on the long-term or large scale distribution of the earth's biota are considered in the final chapters of the Monograph. Bogenrieder and Klein present data regarding the relation of solar UV radiation to the competitive efficiency and special adaptive characteristics of higher plants. Tyrrell examines some of the special properties of sunlight as a mutagenic agent. Calkins offers a model for quantitating the tolerance of a species to a cyclic hazard such as solar UV. Caldwell evaluates the role of solar UV in the evolution of plant life. Going back even to the abiotic earth, Stolarski offers some concepts on the evolution of the atmosphere. Calkins proposes some attributes which may be useful for establishing the importance of an agent such as sunlight on ecosystems or in the evolution of successful species. The evolution of prokaryotic organisms and the possible role of solar radiation is examined by Yentsch and Yentsch, while Damkaer gives similar consideration to solar UV in the evolution of zooplankton.

One need only look at the biota of the ocean bottom or in caves to see how drastically solar radiation modifies the form and function of living organisms. The UV component is by far the most potent portion of sunlight. It is clearly impossible to know all the biological effects of solar UV in detail but one cannot doubt that the biological actions of solar UV deserve a much greater place in current biological education and research than they presently receive.

LETHAL EFFECTS ON BIOLOGICAL SYSTEMS CAUSED BY SOLAR ULTRAVIOLET LIGHT: MOLECULAR CONSIDERATIONS

Meyrick J. Peak and Jennifer G. Peak

Mutagenesis Group, Division of Biological and Medical Research, Argonne National Laboratory
Argonne, Illinois 60439

INTRODUCTION

The most closely examined organisms for the measurement of lethal effects of ultraviolet light are the bacteria, and the information we have regarding molecular events leading to cell death is almost entirely derived from work using these microbial systems. Speculations as to ultraviolet light effects upon higher, eukaryotic cells and multicellular organisms are largely extrapolations. A further limitation in our knowledge of ultraviolet effects is that most work has concentrated upon the actions of the most biologically efficient wavelengths below 300 nm (far ultraviolet light, FUV) which are not present in solar-UV reaching the surface of the earth. Further, ready availability has made 254 nm the most widely explored wavelength. Comparatively little emphasis has been placed upon the biologically inefficient, but ecologically important longer wavelengths above 300 nm (near ultraviolet light, NUV). These wavelengths are present in the solar ultraviolet reaching the surface of the earth and penetrating its waters. The following briefly summarizes some of the recent findings relating to the effects of NUV, especially upon bacterial cells and also transforming DNA, and compares these effects with the effects of FUV.

ULTRAVIOLET LIGHT BELOW 300 nm

A. The role of the pyrimidine dimer

It has been well established that the biological effects of FUV are mediated predominantly by cyclobutane pyrimidine dimers in DNA. Many studies show that action spectra for lethality and for mutagenesis by FUV correlate closely with the absorption spectrum of DNA (see, for instance, J. Setlow, 1962; J. Setlow and M. Boling, 1970; J. Setlow, 1967; R. Setlow, 1974). Further, photoreactivation (PR) sectors of lethality as high as 0.8 firmly establish the role of the dimer as the predominant lethal lesion caused by FUV. Thus, the lethal effects are caused by direct photochemical alterations of the genetic material itself.

There is a large literature which deals with the excision repair or the bypass repair of FUV-induced dimers; see Hanawalt et al. (1978) for a recent comprehensive collection of the literature. Repair of DNA damage is ubiquitous in healthy cells, and the details of molecular processes of various repair mechanisms (PR, polA or recA-dependent excision-repair, postreplication repair, base excision repair (N-glycosidase), and various dimer bypass processes) are quite well known (Smith, 1978).

B. The role of the single-strand break

The low frequency of single-strand breaks (SSB's) in DNA makes these lesions unlikely candidates for FUV lethality. At 254 nm, they are induced at a frequency approximately 10^3 less than dimers, or about one per F_{37} (fluence for 37% survival, Tyrrell et al., 1974).

ULTRAVIOLET LIGHT ABOVE 300 nm (SOLAR UV)

A. The decreasing role of the pyrimidine dimer

The Solar ultraviolet radiations (NUV), as FUV, do produce photorepairable pyrimidine dimers, possibly through direct absorption of the radiation above 300 nm by DNA. However, in normal, repair-proficient bacteria, these dimers become increasingly less important as lethal lesions with increasing wavelength. Dimers are induced in DNA at 365 nm, as detected chemically, but they are produced with approximately 10^3 less efficiency at this wavelength than at 260 nm, the most efficient wavelength for their production (Tyrrell, 1973). Only in repair-deficient cells is the dimer the prime cause of lethality at 365 nm. Above 300 nm, other lethal effects induced by NUV also differ increasingly from those induced by FUV. See R. B. Webb (1977) for a recent review. Action spectra for lethality apparently deviate strongly from the absorption spectrum of DNA, especially above 320 nm, and it seems probable that these lethal events are not due to direct absorption of the NUV by DNA (Webb and Brown, 1976). Photorepair and dark-repair pathways become increasingly ineffective at restoring viability at longer wavelengths. At 365 nm, for instance, PR of lethality was not observed in repair-proficient strains of _Escherichia coli_ (_uvr_A, wild type) (Webb et al., 1976), and oxygen dependence for lethality becomes important. Further, a series of chemicals (histidine, 2-aminoethylisothiouronium bromide hydrobromide (AET), 1,4-diazabicyclo[2.2.2] octane (DABCO), azide, and glycerol)(see section on molecular mechanisms for solar UV lethality below) that modify the lethal effects of NUV but not those of FUV has been discovered. These discoveries have been experimentally facilitated by the use of purified, genetically active DNA instead of whole cells (see section on molecular mechanisms for solar UV lethality) so that physical (permeability) barriers and chemical barriers (due to extragenomic reactions) cannot obscure the reactions between UV, DNA, and the modifier molecule.

B. Molecular targets of NUV

NUV above 320 nm affects molecular organization at all levels of cellular activity studied in bacterial cells. Protein, RNA, and DNA have all been shown to be susceptible to the effects of NUV. For example, at the outer boundary of cells, plasma membranes have been shown to be NUV targets. Damage is due both to nonspecific generalized permeability alterations and specific inhibition of substrate uptake (permease) systems. Robb and Peak (1979) quoted references for bacterial systems. The contribution of membrane damage to NUV-induced cell lethality is not clear, however. It is possible that genetic damage (see below) could be enhanced by metabolic disorders due to some kinds of membrane damage. Such damage could lead to loss of metabolites and those essential to repair processes would be especially important for cell survival.

Nonmembrane bound enzymes have also been shown to be inactivated by NUV. For instance, Coetzee and

Pollard (1975) showed that tryptophanase and tryptophan synthetase are sensitive to both broad band NUV and 365 nm light. Especially important is the probability that enzymes involved in DNA repair are themselves sensitive to NUV. Concomitant damage to DNA and inactivation of the enzymes that repair the DNA damage may well have combined lethal actions greater than either lesion alone. For example, Tyrrell et al. (1973) and Tyrrell (1976) showed inactivation of bacterial PR enzyme by NUV *in vivo* and *in vitro*, with a marked oxygen-dependent sector. Also, there is indirect biological evidence that both excision repair and recombination repair processes are inhibited by NUV, since 365 nm sensitizes cells to 254 nm. This is accompanied by reduced excision of dimers (Tyrrell and Webb, 1973; Webb et al., 1978; Webb, 1977), which might be due to inhibition of induced protein synthesis.

Damage by NUV to RNA has been shown in *E. coli*. Favre et al. (1969) demonstrated a specific photoreaction in 4-thiouridine of tRNA, with maximal effect at about 340 nm, whereby a covalent intramolecular cross-link is formed between positions 8 and 13 in tRNA. This molecular lesion is responsible for NUV-induced growth delay through inhibition of specific amino acid acylation reactions (Thomas and Favre, 1975; Ramabhadran and Jagger, 1976). Since fluences of NUV an order of magnitude greater than those which cause growth delay are required for lethal effects, it is unlikely that this RNA cross-link is important for lethality.

It is clear that with this variety of possible chromophores at a variety of levels of cellular organization (protein, RNA, DNA), elucidation of specific lethal lesions for any particular broad band spectrum or monochromatic NUV radiation is highly complex. However, there is little doubt that DNA is at least one of the major molecular targets for NUV lethality in microbial systems, despite its apparent lack of absorption at these wavelengths. Evidence for this conclusion, discussed in detail by Webb (1977), is as follows:

1. Purified, genetically active DNA is sensitive to NUV (Cabrera-Juarez, 1964; Peak et al., 1973a; Cabrera-Juarez et al., 1976).

2. Oxygen similarly sensitizes whole cells and purified DNA to NUV, but not to FUV (Peak et al., 1973c; Peak et al., 1981).

3. Glycerol protects both DNA and whole cells against NUV but not FUV (Peak and Peak, 1980 and unpublished data).

4. Mutants sensitive to FUV damage show the same sequence of relative sensitivities to NUV (Webb and Brown, 1976).

5. Specific molecular lesions have been observed in DNA after NUV irradiation (dimers, breaks, glycols), see section Ultraviolet Light Above 300 nm (Solar UV). C. 2 below.

6. Genetic mutants with changed sensitivity to NUV have been isolated (Tuveson, 1980; Tuveson and March, 1980). These mutants map at sites remote from those for FUV repair systems, and are not altered markedly with respect to FUV sensitivity.

7. Mutants can be induced in significant yields by NUV (Tyrrell, 1980; Turner and Webb; 1981; Webb and Turner, 1981).

The combined evidence suggests, therefore, that in DNA-repair-proficient cells NUV causes death by effects upon DNA, although the lesions are probably not dimers.

C. Lesions in DNA caused by NUV

Despite the evidence cited above, the specific lesions induced by solar UV in DNA are not yet clearly identified. Pyrimidine dimer induction by NUV and the role of repair in bacterial cells were discussed above. The induction of other types of lesions by NUV has been studied very little and would thus appear to be a profitable research area.

1. Single-strand breaks. NUV induces SSB's in bacterial DNA *in vivo* (Tyrrell *et* al., 1974; Ley et al., 1978), and also in vivo in mammalian cells (Elkind and Han, 1978). In bacterial cells, the *polA* function repairs a portion of the NUV-induced SSB's. Evidence that these NUV-induced SSB's are due to the direct action of NUV, and not caused by enzymatic nicking of DNA in attempts to repair other lesions, was obtained by measurement of SSB induction in DNA *in vitro* (Peak and Peak,

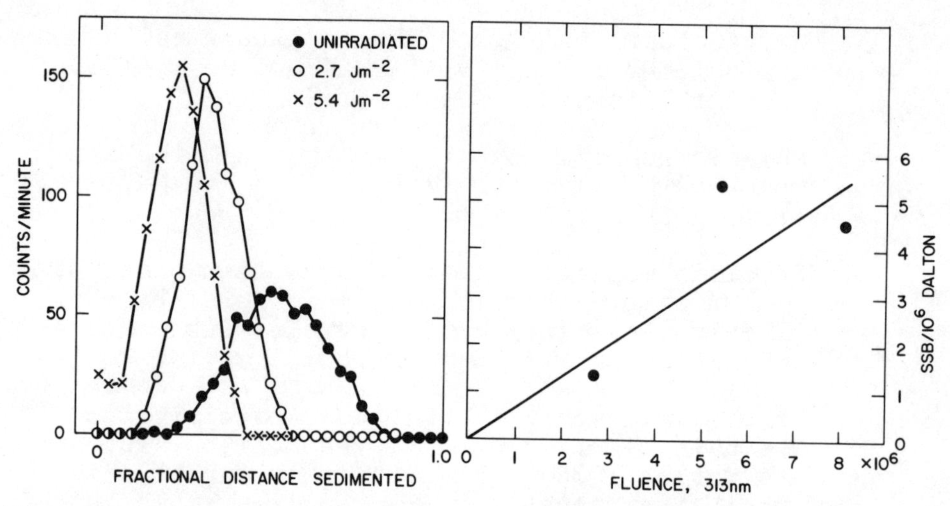

Figure 1. Induction of single-strand breaks in purified DNA by monochromatic 313 nm UV. DNA was purified from B. subtilis, and sedimented through alkaline sucrose gradients after the method of McGrath and Williams (1966). Left hand panel: alkaline sucrose gradients of control DNA, and DNA irradiated with two fluences of 313 nm. Right hand panel: induction of SSB's by 313 nm monochromatic light.

in preparation). Figure 1 shows the linear kinetics of the induction of SSB's in B. subtilis in vitro by 313 nm monochromatic light. The rate of accumulation at 365 nm was close to that obtained in vivo by Ley et al. in lambda phage (1978).

The ratio of SSB's to dimers increases sharply with wavelength, from about 1 x 10^{-3} at 254 nm to about 2 at 365 nm (Tyrrell et al., 1974) and to over 102 at 405 nm. There are several lines of evidence suggesting that NUV-induced SSB's in DNA may be important NUV lethal lesions. First, agents that modify lethality also modify SSB induction. For instance, there is a large oxygen-dependent sector for SSB induction by 365 nm. Glycerol (Peak and Peak, in preparation) and AET (Tyrrell et al., 1974) both mitigate NUV-induced SSB's, in parallel with lethal events. Second, the polA function repairs 365 nm induced SSB's, and similarly the polA mutation sensitizes E. coli to 365 nm lethality (Webb and Turner, submitted for publication). The dose modifying factors defined by Jagger (1967) for lethality caused by glycerol (0.5) and for polA mediated repair of SSB's (0.3) are not dissimilar.

2. Other lesions. Little information is available pertaining to DNA lesions other than pyrimidine dimers and SSB's caused by NUV. Rahn (1978) itemizes a number of nondimer lesions caused in DNA by UV light, mainly FUV, but the role that any of these play in NUV lethal events has not been seriously addressed, and remains an open and fruitful area for investigation. Recently, a NUV-induced photoproduct other than pyrimidine dimer or spore photoproduct has been isolated from irradiated naked DNA (Cabrera-Juarez and Setlow, 1980). This product, not yet identified, may be a 5,6 dihydrodihydroxythymine photoproduct. Hariharan and Cerutti (1977) found that irradiation of HeLa cells by 313 nm induced saturated thymine-type products more efficiently than pyrimidine dimers. They did not investigate longer wavelengths. Analysis of the melting and annealing of DNA after high fluences of 365 nm UV did not reveal the existence of DNA cross-links (Peak et al., 1973c).

MOLECULAR MECHANISMS FOR SOLAR UV LETHALITY

Histidine (Peak et al., 1973b), AET (Peak and Peak, 1975) DABCO (Peak et al., 1981) azide (Peak et al., in preparation) all protect transforming DNA against NUV (but not FUV), and oxygen (Peak et al., 1973c, Peak et al., 1981) sensitizes DNA to NUV, but not FUV. These observations lead to our model for a mechanism of the lethal action of NUV (Fig. 2), and the possible role played by the modifiers described above. Radiation is absorbed by the endogenous sensitizer forming the triplet state photosensitizer via the singlet state. At this point, either Type I or Type II reactions could occur (see Foote, 1978, for a recent discussion). The triplet state sensitizer may react with ground-state oxygen forming singlet oxygen. This could react with DNA, or preferentially with AET, DABCO, azide, or histidine, if present (Type I reaction).

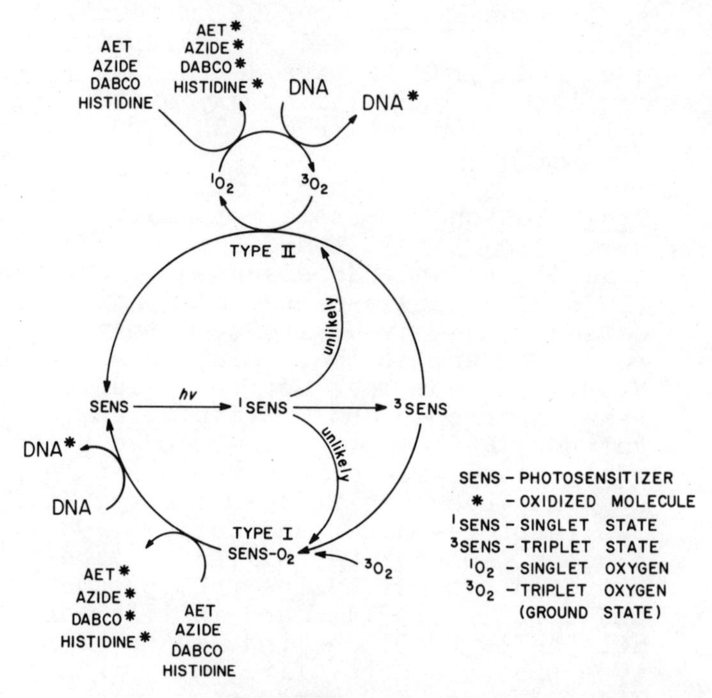

PROTECTION

Figure 2. Proposed model for the roles of oxygen and protective agents in the inactivation of DNA by NUV (Peak et al., in preparation).

Alternatively, an oxidized sensitizer may be formed which reacts with DNA or the other compounds, if present. (See Peak et al., 1981 for a recent discussion of the possible role of singlet oxygen in the effects of NUV upon DNA.)

CONCLUSIONS

Near-UV present in solar ultraviolet wavelengths reaching the surface of the earth kills and mutates cells. Although pyrimidine dimers are formed in biologically significant yields, these are effectively repaired and the critical lethal lesions remain as yet unidentified. One strong candidate may be breakage of DNA strands. Whatever the actual lesion, molecular oxygen forms part of the reaction, and singlet oxygen might be one intermediate.

ACKNOWLEDGMENTS

This work was supported by the U. S. Department of Energy under contract No. W-31-109-ENG-38, and by the Council for Scientific and Industrial Research, Pretoria, South Africa. The U.S. Government retains a nonexclusive royalty-free license to publish or reproduce the published form of this contribution, or allow others to do so, for U.S. Government purposes.

We thank Drs. R. B. Webb and H. E. Kubitschek for helpful comments.

REFERENCES

Cabrera-Juarez, E. 1964. "Black light" inactivation of transforming deoxyribonucleic acid from Haemophilus influenzae. J. Bacteriol. 87:771-778.

Cabrera-Juarez, E. and J. K. Setlow. 1980. Different chromatographic behavior of a thymine-containing new ultraviolet photoproduct and spore photoproduct. Photochem. Photobiol. 31:603-606.

Cabrera-Juarez, E., J. K. Setlow, P. A. Swenson, and M. J. Peak. 1976. Oxygen-independent inactivation of Haemophilus influenzae transforming DNA by monochromatic radiation: Action spectrum, effect of histidine and repair, Photochem. Photobiol. 23: 309-314.

Coetzee, W. F., and E. C. Pollard. 1975. Near ultraviolet inactivation studies on Escherichia coli tryptophanase and tryptophan synthetase. Photochem. Photobiol. 22: 29-32.

Elkind, M. M., and A. Han. 1978. DNA single strand lesions due to 'sunlight' and UV light: a comparison of their induction in Chinese hamster and human cells, and their fate in Chinese hamster cells. Photochem. Photobiol. 27: 717-724.

Favre, A., M. Yaniv, and A. M. Mickelson. 1969. The photochemistry of 4-thiouridine in Escherichia coli t-RNA^{val}, Biochem. Biophys. Res. Commun. 37: 266-271.

Foote, C. S. 1978. Mechanisms of photooxidation. In Singlet Oxygen-Reactions with Organic Compounds and Polymers [B. Ranby and J. F. Rabeck, eds.] 135-146. John Wiley and Sons.

Hanawalt, P. C., E. C. Friedberg, and C. F. Fox. 1978. DNA Repair Mechanisms, Academic Press, New York.

Hariharan, P. V., and P. A. Cerutti. 1977. Formation of products of the 5,6-dihydroxydihydrothymine type by ultraviolet light in HeLa cells. Biochemistry 16: 2791-2795.

Jagger, J. 1967. Introduction to Research in Ultraviolet Photobiology, Prentice-Hall, Englewood Cliffs, N. J.

Ley, R. D., B. A. Sedita, and E. Boye. 1978. DNA polymerase I-mediated repair of 365 nm-induced single strand breaks in the DNA of Escherichia coli. Photochem. Photobiol. 27:323-328.

McGrath, R. A. and R. W. Williams. 1966. Reconstruction in vivo of irradiated E. coli deoxyribonucleic acid, the rejoining of broken pieces. Nature, London 212:534-535.

Peak, M. J., J. G. Peak, and R. B. Webb. 1973a. Inactivation of transforming DNA by ultraviolet light. I. Near-UV action spectrum for marker inactivation. Mutat. Res. 20:129-135.

Peak, M. J., J. G. Peak, and R. B. Webb. 1973b. Inactivation of transforming DNA by ultraviolet light. II. Protection by histidine against near-UV radiation: Action spectrum. Mutat. Res. 20: 137-141.

Peak, M. J., J. G. Peak, and R. B. Webb. 1973c. Inactivation of transforming DNA by ultraviolet light.III. Further observations on the effects of 365 nm radiation. Mutat. Res. 20: 143-148.

Peak, M. J., and J. G. Peak. 1975. Protection by AET against inactivation of transforming DNA by near-ultraviolet light: action spectrum. Photochem. Photobiol. 22: 147-148.

Peak, J. G., M. J. Peak, and C. S. Foote. 1981. Protection by DABCO against inactivation of transforming DNA by near-ultraviolet light: action spectra and implications for involvement of singlet oxygen. Photochem. Photobiol. 34: 45-49.

Peak, M. J., and J. G. Peak. 1980. Protection by glycerol against the biological actions of near-ultraviolet light. Radiat. Res. 83: 553-558.

Rahn, R. O. 1978. Non-dimer damage in DNA caused by UV radiation [K. C. Smith, ed.]. Photobiological Reviews 4.

Ramabhadran, T.V., and J. Jagger. 1976. Mechanism of growth delay induced in Escherichia coli by near ultraviolet radiation. Proc.Nat. Acad. Sci. USA 73: 59-69.

Robb, F. T., and M. J. Peak. 1979. Inactivation of the lactase permease of Escherichia coli by monochromatic ultraviolet light. Photochem. Photobiol. 30: 379-383.

Setlow, J. K. 1962. Evidence for a particular base alteration in two T4 bacteriophage mutants. Nature 194: 664-666.
Setlow, J. K. 1967. The effects of ultraviolet radiation and photoreactivation. Comprehensive Biochem. 27: 157-209.
Setlow, J. K. and M. E. Boling. 1970. Ultraviolet action spectra for mutation in Escherichia coli. Mutat. Res. 9: 437-442.
Setlow, R. B. 1974. The wavelengths in sunlight effective in producing skin cancer: A theoretical analysis. Proc. Nat. Acad. Sci. USA 71: 3363-3366.
Smith, K. C. 1978. Multiple pathways of DNA repair in bacteria and their roles in mutagenesis. Photochem. Photobiol. 28: 121-130.
Thomas, G. and A. Favre. 1975. 4-thiouridine as the target for near-ultraviolet light induced growth delay in Escherichia coli, Biochem. Biophys. Res. Commun. 66: 1454-1461.
Turner, M. A. and R. B. Webb. 1981. Comparative mutagenesis and responses to near-UV (313-405 nm) and far UV (254 nm) radiation in strains of E. coli with differing repair capabilities. J. Bacteriol. 147: 00-00.
Tuveson, R. W. 1980. Genetic control of near-UV sensitivity independent of excision deficiency (uvrA6) in E. coli K12, Photochem. Photobiol. 32: 703-705.
Tuveson, R. W., and M. E. March. 1980. Photodynamic and sunlight inactivation of Escherichia coli strains differing in near-UV sensitivity and recombination proficiency. Photochem. Photobiol. 31: 287-289.
Tyrrell, R. M. 1973. Induction of pyrimidine dimers in bacterial DNA by 365 nm radiation. Photochem. Photobiol. 17: 69-73.
Tyrrell, R. M. 1976. Rec A+-dependent synergism between 365 nm and ionizing radiation in log phase Escherichia coli: A model for oxygen dependent near UV inactivation by disruption of DNA repair, Photochem. Photobiol. 23: 345-352.
Tyrrell, R. M. 1980. Mutation induction by and mutational interaction between monochromatic radiations in the near-ultraviolet and visible ranges. Photochem. Photobiol. 31: 37-46.
Tyrrell, R. M., R. D. Ley, and R. B. Webb. 1974. Induction of single-strand breaks (alkali-labile bonds) in bacterial and phage DNA by near-UV (365 nm) radiation. Photochem. Photobiol. 20: 395-398.
Tyrrell, R. M. and R. B. Webb. 1973. Reduced dimer excision following near ultraviolet (365 nm) radiation. Mutat. Res. 19: 361-365.

Tyrrell, R. M., R. B. Webb, and M. S. Brown. 1973. Destruction of photoreactivating enzyme by 365 nm radiation. Photochem. Photobiol. 18: 249-254.

Webb, R. B. 1977. Lethal and mutagenic effects of near ultraviolet radiation, In Photochemical and Photobiological Reviews [K. C. Smith, ed.], Vol. 2, 169-261. Plenum Press, New York.

Webb, R. B., and M. S. Brown. 1976. Sensitivity of strains of Escherichia coli differing in repair capability to far UV, near UV and visible radiations. Photochem. Photobiol. 24: 425-532.

Webb, R. B. and M. A. Turner. 1981. Mutation induction by monochromatic 254 nm and 365 nm radiation in strains of Escherichia coli that differ in repair capability. Mutat. Res. (In press).

Webb, R. B., M. S. Brown, and R. M. Tyrrell. 1976. Lethal effects of pyrimidine dimers induced at 365 nm in strains of E. coli differing in repair capability. Mutat. Res. 37: 163-172.

Webb, R. B., M. S. Brown, and R. M. Tyrrell. 1978. Synergism between 365 and 254 nm radiations for inactivation of Escherichia coli. Radiat. Res. 74: 298-311.

DNA DAMAGE OF THE 5,6-DIHYDROXYDIHYDROTHYMINE TYPE INDUCED BY SOLAR UV-B IN HUMAN CELLS

P. V. Hariharan and E. R. Blazek

Department of Biophysics
Roswell Park Memorial Institute
Buffalo, New York 14263

ABSTRACT

The magnitude of an acute skin dose of ionizing radiation which produces an abundance of 5,6-dihydroxydihydrothymine (t') equal to that produced by a plausible daily exposure to solar UV (313±6 nm) is calculated to range from 0.27 krad (winter) to 5.65 krad (summer) in temperate latitudes. Considering the evidence for the possible biological importance of the t' lesion, and the carcinogenicity of a 1.6 krad acute skin dose of ionizing radiation in one strain of rat, these t' damage equivalent doses emphasize the magnitude of the challenge presented by solar UV-B.

The correlation between skin cancer and solar UV exposure (Blum 1959, Urbach 1969) exists not only for squamous and most basal cell carcinoma (Urbach 1971) but also for the *lentigo maligna* type of melanoma (Kopf et al. 1977). Setlow (1974) has suggested an action spectrum which is a composite of those for various types of DNA damage. Erythemal action spectra weight longer wavelengths more heavily than does Setlow's action spectrum (Berger et al. 1968). In the present work, the 313±6 nm band will be uniformly weighted and the other wavelengths will be ignored for the technical reason that t' damage product formation rates have been measured only for this band. When Setlow's action spectrum is applied to a representative solar UV fluence, approximately 30% of the carcinogenic dose so determined originates in the 313±6 nm band.

We assume that the molecular basis of UV carcinogenesis is modification of the DNA, which falls into the categories of strand breakage, crosslinking, or base damage. Each of these types is heterogeneous; in particular, base damage may be further classified as dimeric or monomeric. It should be emphasized that the spectrum of all UV photoproducts exhibits dependence upon both dose and wavelength. This dependence has been most thoroughly studied for pyrimidine dimers (Unrau et al. 1972, Mang and Hariharan 1980, Carrier et al. 1980), which have been directly implicated in UV carcinogenesis (Hart and Setlow 1975).

Monomeric base damage due to UV frequently involves the 5,6 double bond of pyrimidines (Hariharan and Cerutti 1977). The product 5,6-dihydroxydihydrothymine (abbreviated t') is specifically measured by the alkali-acid degradation assay of Hariharan and Cerutti (1974), or, with greater sensitivity, by the alkali degradation assay of Hariharan (1980). It is the only DNA damage product for which, in human cells, rates of formation by both UV and ionizing radiation have been measured and repair kinetics have been studied.

Formation of t' is mediated by the OH radical. This has been established by studies with radical quenching agents (Remsen and Roti Roti 1977) and by evidence that only the OH radical, among the products of radiolysis, reacts significantly with 2'-deoxypyrimidines (Cadet and Teoule 1978). It has become standard practice to attempt to interpret abnormal sensitivity to any DNA damaging agent as a genetic deficiency in the repair enzyme system. The fact that some agents act through radical or excited molecular mediators implies that conditions should be sought which influence availability of the mediator. For example, acatalasia would increase the intracellular concentration of H_2O_2, which is believed to damage DNA via the OH radical in the presence of long wavelength UV (Hahn and Wang 1974) and in the dark (Hoffman and Meneghini 1979). The OH radical population derived from the excess H_2O_2 would increase the rate of formation of t'. Thus cells which experience the same exposure to UV-B do not necessarily incur the same burden of DNA damage, even if their repair systems are equally competent.

The biological significance of t' is uncertain, but consistent with its possible importance are several lines of evidence: an ionizing radiation sensitive strain of *Micrococcus radiodurans* is defective in removal of t' (Targovnik and Hariharan 1980); oxygen enhancement effects have been observed in the inactivation of *Bacillus subtilis* transforming DNA by long wavelength UV (Peak 1981); UV-B, but not 254 nm UV, mimics the effect of ionizing radiation upon survival and recombination in *Saccharomyces cerevisiae* (Hannan and Calkins 1981); and some human cell lines afflicted with ataxia telangectasia, an ionizing radiation sensitivity syndrome, are also abnormally sensitive to 313 nm UV, although their sensitivity to 254 nm UV is the same as that of normal cell lines (Smith and Paterson 1981).

To calculate a t' damage equivalent dose of ionizing radiation for solar UV-B exposure, it is first necessary to determine the clear-day irradiance in the 313±6 nm band which reaches the target cells at a specified latitude, time of day, and day of year. The present calculations are a simplified version of those performed by Green and coworkers (1974a, 1974b, 1979; Mo and Green 1974 , Johnson et al. 1976, Baker et al. 1980). The global (total) solar irradiance $G(\lambda,Z)$ was measured as a function of wavelength λ and solar zenith angle Z by Bener (1972), and the measurements were fit analytically by Green and colleagues (1974a, 1979). $G(\lambda,Z)$ is numerically integrated over the wavelengths of the 313 ±6 nm band to yield $G_{313}(Z)$. It can be shown that:

$$Z \cong \arccos\,(\sin L \cos\theta + \cos L \sin\theta \cos\,(0.2618\,t)) \qquad (1)$$

where: L = the (north) latitude in degrees
θ = (90 + 23.5 sin (0.9863 s)), where s is the number of days after Sept. 23
t = the number of hours away from local noon, either am or pm

Eq. 1 allows the Z-dependence of G_{313} to be expressed as time-dependence. $G_{313}(t)$ is numerically integrated over the exposure time for the specified day of the year to give the fluence F_{313}.

The second problem is to calculate the instantaneous abundance of the t' lesicn from what is known about its rates of formation and repair. We define the t' abundance $A(t) = \frac{t'}{T}$. The time rate of change of $A(t)$ is given by:

$$\frac{dA}{dt} \cong - \alpha A + \beta k G_{313}(t) \quad (2)$$

where: α = the fractional decrement in A per hour due to repair, β = the fractional increment in A per unit of fluence in the absence of repair, and k = the fraction of the irradiance incident upon the skin surface which reaches the target cells for carcinogenesis. The solution of this differential equation is:

$$A(t) = k \beta \exp -\alpha(t-t_o) \int_{t_o}^{t} G_{313}(\tau) \exp (\alpha\tau) d\tau, \quad (3)$$

with exposure beginning at t_o.

The repair parameter α may be estimated from the observation of Mattern et al. (1975) that WI-38 human lung fibroblasts and CHO cells excise at least 70% of the t' residues produced by γ- or x-radiation in 15 minutes. If this rate were continued for one hour, it would imply a fractional decrement of 0.99 hr^{-1}. The repair rate decreases after 15 minutes, however, and we prefer to assign the value 0.95 hr^{-1} to α,

Hariharan and Cerutti (1977) found that in HeLa S3 cells the t' abundance was $A = 1.6 \times 10^{-4}$, after a dose of 5 kJm^{-2} delivered at low temperature to prevent repair. Thus:

$$\beta = \frac{1.6 \times 10^{-4}}{5.0 \text{ kJm}^{-2}} = 3.2 \times 10^{-5} (\text{kJm}^{-2})^{-1}.$$

To estimate the transmission factor k, the target cells of the skin must be located. Although most experimental skin cancers in animals are of dermal origin, most human tumors arise in the epidermis (Kirby-Smith et al. 1942). Assuming that a typical target cell is covered by half the optical depth of the epidermis, Fig. 4 of the preceeding reference indicates that $k \cong 0.10$ is a reasonable estimate for Caucasians with slight tanning only.

Except for cases of occupational exposure, where tanning will cause k to be less than 0.10, it is unlikely that one area of skin would be exposed to direct sunlight for an entire day. A plausible recreational exposure might be for the two hours from 2 pm until 4 pm, and this is the exposure period chosen for all subsequent numerical evaluations.

We define the ultraviolet abundance increment

$\Delta A_{313} = \frac{A}{F_{313}}$, which measures the effectiveness of a unit of fluence in producing t' in the presence of repair. The abundances A at 4 pm, the fluences F_{313} delivered from 2 pm until 4 pm, and the abundance increments ΔA_{313} for representative latitudes and days of the year are given in Table 1 below.

The corresponding problem for an acute dose of ionizing radiation is simplified by the fact that repair during the short irradiation period is negligible. The time rate of change of A(t) is therefore:

$$\frac{dA}{dt} \cong - \alpha A, \text{ which is solved by:} \qquad (4)$$

$$A(t) = A_o \exp [- \alpha(t - t_o)] \qquad (5)$$

where A_o is the initial abundance. We define the ionizing abundance increment $\Delta A_{ion} = \frac{A}{D}$. Then $A_o = \Delta A_{ion} D(t_o)$, where $D(t_o)$ is the acute dose delivered at time t_o.

Mattern et al. (1975) measured the t' abundance following a 250 krad dose of ^{137}Cs γ-radiation delivered at 10 krad/min but with repair prevented by low temperature. Three cell lines studied had values of ΔA_{ion} differing by less than a factor of two. The value of $\Delta A_{ion} = 2.04 \times 10^{-6}$ found for WI-38 fibroblasts will be used in all subsequent calculations.

The final calculational problem is to meaningfully compare t' abundances with different time dependencies. The abundance averaged over time should be a better measure of biologically significant damage than the abundance at some chosen moment. Comparison of time-

TABLE 1

Latitudinal and Seasonal Dependence of Abundance, Fluence, and Abundance Increment

Latitude	Day of Year	A(4pm) x 10^6	$F_{313}(kJm^{-2})$	$\Delta A_{313} \times 10^6 \ (kJm^{-2})^{-1}$
30 N	Mar. 21, Sept. 23	6.40	5.51	1.162
30 N	December 22	1.58	1.48	1.068
30 N	June 22	11.50	9.37	1.227
42 N	Mar.21, Sept.23	4.07	3.57	1.140
42 N	December 22	0.54	0.52	1.038
42 N	June 22	11.53	9.20	1.253

TABLE 2

t' Damage Equivalents: Dose Ratio (E) and Dose (E x F_{313})

Latitude	Day of Year	E (krad/kJm^{-2})	F_{313}(kJm^{-2})	E x F_{313} (krad)
30 N	Mar. 21,Sept.23	0.570	5.51	3.14
30 N	December 22	0.524	1.48	0.77
30 N	June 22	0.601	9.37	5.63
42 N	Mar.21,Sept.23	0.559	3.57	1.99
42 N	December 22	0.509	0.52	0.27
42 N	June 22	0.614	9.20	5.65

averaged abundances would enhance the importance of any regimen which prolongs the abundance--in this case, the UV. Here, for simplicity, the t' abundance at the end of exposure to UV (4 pm) will be compared to that due to ionizing radiation immediately after its delivery.

We define the t' damage equivalent dose ratio E to be the acute dose of ionizing radiation needed to produce a t' abundance equal to that produced by 1 kJm^{-2} of 313±6 nm UV immediately after the specified exposure:

$$E \times \Delta A_{ion} = 1\ kJm^{-2} \times \Delta A_{313}, \text{ or } E = \frac{1\ kJm^{-2} \times \Delta A_{313}}{\Delta A_{ion}}.$$

Table 2 below contains E, the fluence F_{313} from 2 pm until 4 pm, and the product E x F_{313}, which is the damage equivalent dose of ionizing radiation to the skin for the entire specified solar UV exposure.

These results show that an afternoon exposure to 313±6 nm solar UV can produce as much t' damage product as 0.27-5.65 krad rapid dose of ionizing radiation. McGregor (1976) has shown that 1.6 krad of β-radiation from a flat ^{90}Sr source is carcinogenic for outbred albino rats of the Charles River CD strain, with nearly half of the irradiated animals bearing tumors at 32 weeks. The dose rate of 0.7 krad/min and the peak electron energy of 0.546 MeV are consistent with our postulated acute dose to the skin only. Although the extent to which t' participates in carcinogenesis is unknown, it is unquestionable that solar UV-B presents a major challenge to the DNA repair systems of epidermal cells.

ACKNOWLEDGEMENTS

E. R. B. is a graduate student trainee supported by PHS grant ES-07057-03.

REFERENCES

Baker, K. S., R. C. Smith, and A.E.S. Green. 1980. Middle ultraviolet reaching the ocean surface. Photochem. Photobiol. 23: 179-188.

Bener, P. 1972. Technical report (unpublished). European Research Office, U. S. Army (London). Contract No. DAJA 37-68-C-1017.

Berger, D., F. Urbach, and R. E. Davies. 1968. The action spectrum of erythema induced by UV radiation, p. 1112-1117. In Proc. 13th Int. Congress of Dermatology, vol. 2. Springer-Verlag.

Blum, H. F. 1959. Carcinogenesis by ultraviolet light. Princeton University.

Cadet, J. and R. Teoule. 1978. Comparative study of oxidation of nucleic acid components by hydroxyl radicals, singlet oxygen, and superoxide anion radicals. Photochem. Photobiol. 28: 661-667.

Carrier, W. L., R. D. Snyder, and J. D. Regan. 1980. Pyrimidine dimer excision repair in human cells and the effects of inhibitors. Abstract P321 bis of the VIII International Congress of Photobiol. Strasbourg.

Green, A.E.S., T. Sawada and E. P. Shettle. 1974a. The middle ultraviolet reaching the ground. Photochem. Photobiol. 19: 251-259.

Green, A.E.S., T. Mo, and J. H. Miller. 1974b. A study of solar erythema radiation doses. Photochem. Photobiol. 20: 473-482.

Green, A.E.S., K. R. Cross, and L. A. Smith. 1979. Improved analytic characterization of ultraviolet skylight. Photochem. Photobiol. 31: 59-65.

Hahn, B. and S. Y. Wang. 1974. Study of radiation chemistry of thymine and thymidine through their photolysis in the presence of hydrogen peroxide. Abstract A-13-3 of 5th Int. Cong. of Radiation Research. Radiat. Res. 59.

Hannan, M. A. and J. Calkins. 1981. Personal communication.

Hariharan, P. V. and P. A. Cerutti. 1974. Excision of damaged thymine residues from gamma-irradiated poly (dA-dT) by crude extracts of E. coli. Proc. Natl. Acad. Sci. U. S. A. 71: 3532-3536.

Hariharan, P. V. and P. A. Cerutti. 1977. Formation of products of the 5,6-dihydroxydihydrothymine type by ultraviolet light in HeLa cells. Biochemistry 16: 2791-2795.

Hariharan, P. V. 1980. Determination of thymine ring saturation products of the 5,6-dihydroxydihydrothymine type by the alkali degradation assay. Radiat. Res. 81: 496-498.

Hart, R. W. and R. B. Setlow. 1975. Direct evidence that pyrimidine dimers in DNA result in neoplastic transformation, 719-724 In P. C. Hanawalt and R. B. Setlow.[eds.] Molecular Mechanisms for the repair of DNA, part B. Plenum Press.

Hoffmann, M. E. and R. Meneghini. 1979. Action of hydrogen peroxide on human fibroblasts in culture. Photochem. Photobiol. 30: 151-155.

Kirby-Smith, J. S., H. F. Blum, and H. G. Grady. 1942. Penetration of ultraviolet radiation into skin, as a factor in carcinogenesis. J. Natl. Cancer Inst. 2: 403-412.
Kopf, A. W., R. S. Bart, and R. S. Rodriguez-Sains. 1977. Malignant melanoma: a review. J. Dermatol. Surg. Oncol. 3: 41-125
Johnson, F. S., T. Mo and A.E.S. Green. 1976. Average latitudinal variation in ultraviolet radiation at the earth's surface. Photochem. Photobiol. 23: 179-188.
Mang, T. S. and P. V. Hariharan. 1980. Production of cyclobutane type pyrimidine dimers in the DNA of Chinese hamster lung fibroblasts (V-79) exposed to UV-B light. Int. J. Radiat. Biol. 38: 123-125.
Mattern, M. R., P. V. Hariharan, and P. A. Cerutti. 1975. Selective excision of gamma ray damaged thymine from the DNA of cultured mammalian cells. Biochimica et Biophysica Acta 395: 48-55.
McGregor, J. F. 1976. Tumor-promoting activity of cigarette tar in rat skin exposed to radiation. J. Natl. Cancer Inst. 56: 429-430.
Mo, T. and A.E.S. Green. 1974. A climatology of solar erythema dose. Photochem. Photobiol. 20: 483-496.
Peak, J. 1981. Personal communication.
Remsen, J. F. and J. L. Roti Roti. 1977. Formation of 5,6-dihydroxydihydrothymine-type products in DNA by hydroxyl radicals. Int. J. Radiat. Biol. 32: 191-194.
Setlow, R. B. 1974. The wavelengths in sunlight effective in producing skin cancer: a theoretical analysis. Proc. Natl. Acad. Sci. U. S. A. 71: 3363-3366.
Smith, P. J. and M. C. Paterson. 1981. Abnormal response to mid-ultraviolet light of cultured fibroblasts from patients with disorders featuring sunlight sensitivity. Cancer Res. 41: 511-518.
Targovnik, H. S. and P. V. Hariharan. 1980. Excision repair of 5,6 dihydroxydihydrothymine from the DNA of Micrococcus radiodurans. Radiat. Res. 83: 360-363.
Unrau, P., R. Wheatcroft and B. S. Cox. 1972. Methods for the assay of ultraviolet light-induced pyrimidine dimers in Saccharomyces cerevisiae. Biochimica et Biophysica Acta 269: 311-321.
Urbach, F. 1969. The biologic effects of ultraviolet radiation. Pergamon.
Urbach, F. 1971. Geographic distribution of skin cancer. J. Surg. Oncol. 3: 219-234.

MECHANISMS OF ACTIONS OF LONGWAVE-UV ON MARINE ORGANISMS

Seymour Zigman

Departments of Opthalmology and Biochemistry
University of Rochester School of Medicine and Dentistry, 601 Elmwood Avenue (Box 314)
Rochester, New York 14642

INTRODUCTION

As a result of the interaction between sunlight UV and sensitive molecular species in the internal and external environments of marine organisms, photoproducts are formed that interfere with metabolic events, so as to inhibit survival and growth. This discussion deals with events not directly involving the genetic material or the synthesis of macromolecules, although these may be influenced secondarily by such events. It does deal with the central role of tryptophan in near-UV induced cellular photosensitization.

Two levels of action are to be considered. These are the production of toxic photoproducts, using tryptophan as the photosensitive molecule, and the influence of these products on enzymes essential for the livelihood of the cells. We already demonstrated the inhibition of mitosis in sea urchin eggs due to their exposure to exogeneous tryptophan near-UV photoproducts. This treatment inhibited the synthesis of all macromolecules in the eggs. Interference with energy metabolism has been postulated as a mechanism for this result (Zigman and Hare 1976 , Zigman et al. 1978).

Both organic peroxides (Sun and Zigman 1979) and H_2O_2 (McCormick et al. 1976) have been shown to be photoproducts of tryptophan hydrolysis by near-UV light. The toxicity of these photoproducts would be manifested if the usual enzymatic defense mechanisms (i.e., catalase, SOD) were overcome by high levels of peroxides and superoxide radicals. A near-UV photoproduct hydroxypyrroloindole (see Figure 1), was found to be a sensitizer for several enzymes in the oxidation: reduction system utilized by cells to protect them against the action of strong oxidants. These have included catalase and gluthathione reductase. Superoxide dismutase (SOD) was not found to be sensitive to pyrolloindole plus near-UV light. The end result of such a mechanism would be the loss of activity of SH dependent enzymes and oxidation of structural components of cells.

Not demonstrated previously is the finding that photosensitizing cyanine dyes used to transport electrons in nonbiological systems, were also cell growth and mitosis inhibitors of sea urchin eggs and dinoflagellates. In the case of dinoflagellates, these dyes induced cytolysis in short order (i.e. minutes). A sharp decrease of ATP levels in the effected cells indicated greatly reduced energy availability for supporting osmotic pumping as well as respiration. Interesting to note is the fact that several of these dyes are quickly bleached by sunlight and are thus rendered inactive.

FINDINGS AND DISCUSSION

Appreciable effects on enzymes result from their exposure to near-UV light in the presence of tryptophan or some of its oxidation products described above. Presently there is a list of enzymes that have been tested for their sensitivity to near-UV light plus tryptophan photoproducts, either purified or without purification.

The first of this series is catalase (see Figure 2). As shown by Zigman et al.1976) exposure to near-UV light in the presence of tryptophan leads to an inactivation of this enzyme that occurs because both the prosthetic group and the protein chain of the catalase are drastically altered by this treatment. There were also losses of soret and ultraviolet absorption properties, enhancement of non-tryptophan fluorescence, and altered electrophoretic mobility of the protein. In this case tryptophan photoproduct was firmly bound to the enzyme proteins.

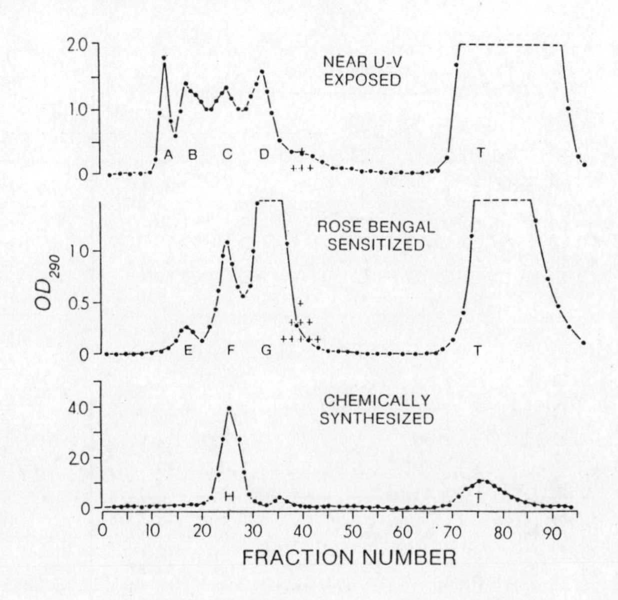

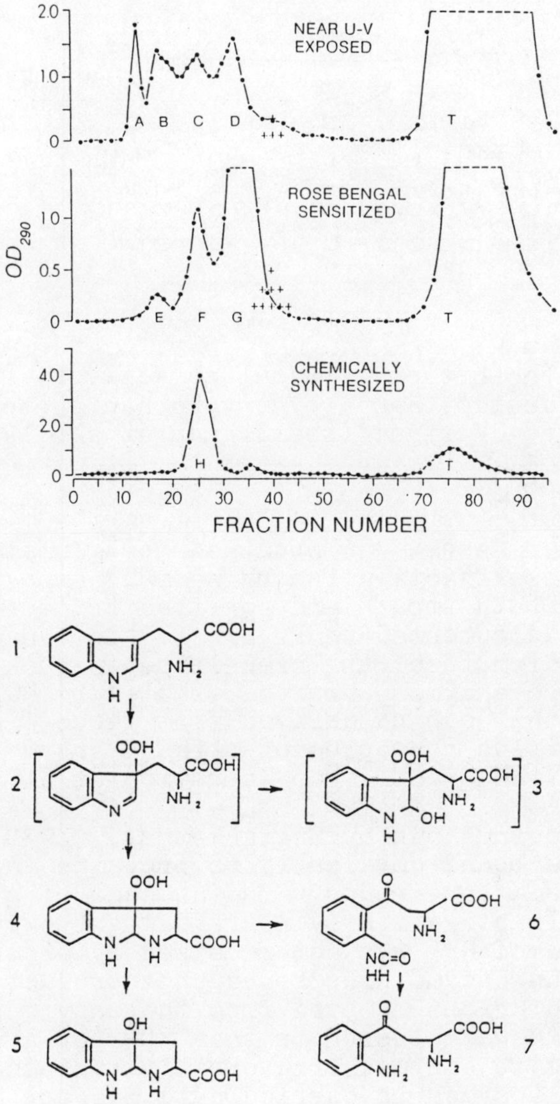

Figure 1. a) Induction of tryptophan photooxidation products by longwave-UV light (see Sun and Zigman, 1979 for details)

1 = TRP; 4 = peroxypyrolloindole (PPI)
5 = Hydroxypyrollindole (HPI)
6 = N-formylkynurenine (NFK); and
7 = Kynurenine (KY)

b) Preparation of purified tryptophan photoproducts by longwave-UV exposure.

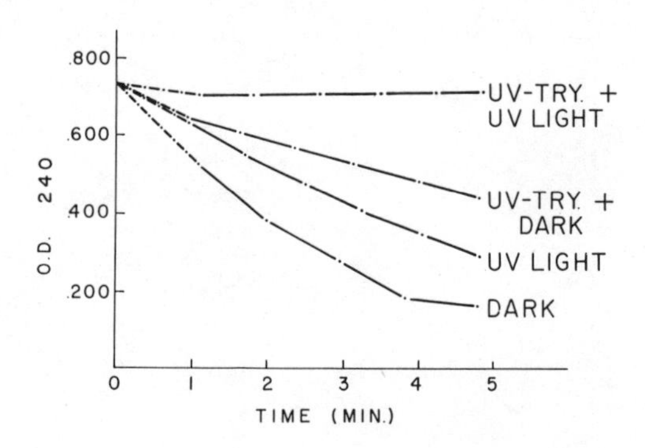

Figure 2. Inhibition of Catalase Activity (i.e. its ability to destroy H_2O_2) by tryptophan photoproducts and longwave -UV light (see Zigman et al. 1976).

A similar result was obtained for glutathione reductase, whose inactivation by near-UV exposure in the presence of both impure and purified tryptophan photoproducts is documented in Figure 3. On the other hand, superoxide dismutase was insensitive to such treatment. Both of these enzymes were inactivated by PCMBS. A conclusion that can be derived from these findings is that the UV plus photoproduct site of enzyme inactivation is not -SH groups (see Kalustian et al. 1981).

It has been shown by Zigman et al. (1973) that photoproducts are bound covalently to proteins and polypeptides via amino groups. In the case of glutathione reductase, inactivation by near-UV light exposure in the presence of HPI was not reversed by extensively dialyzing the enzyme plus bound photoproduct, nor could ^{14}C-labelled HPI be removed from the enzyme protein dialysis. The conclusion is that HPI (or a photoproduct of it) binds to enzymes irreversibly (unless harsh treatment of the enzymes is carried out) thereby altering chemical and physical properties leading to their inactivation.

The three enzymes discussed above are components of the oxidation reduction system of cells, and their interrelationship is illustrated in Figure 4. Both glutathione reductase and catalase activity in cells are known to diminish with aging, and both of these enzymes are now known to be inhibited by near-UV light in the presence of tryptophan photoproducts. While superoxide dismutase activity also diminishes with aging, it is

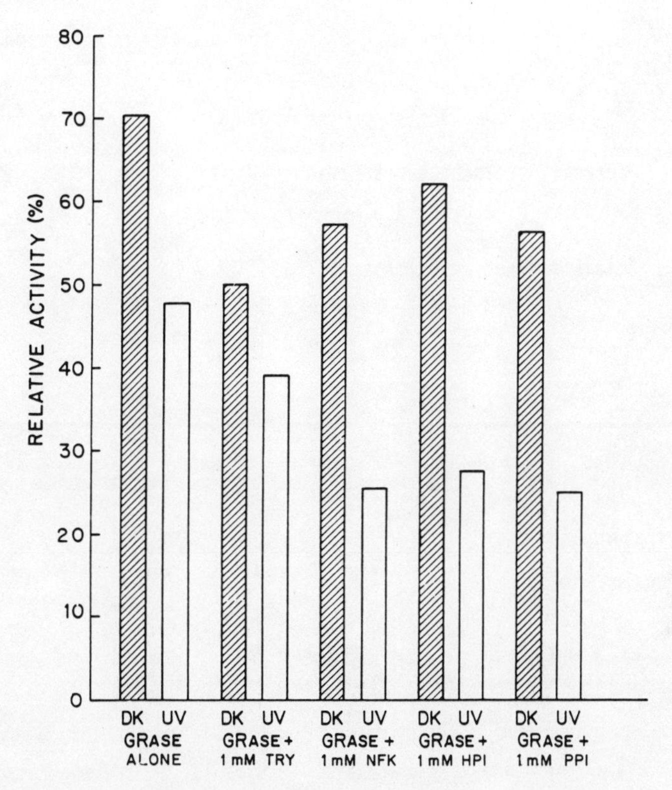

Figure 3. Inhibition of Glutathione Reductase activity (i.e. its ability to reduce oxidized glutathione) by tryptophan photoproducts and longwave-UV light (see Kalustian et al. 1981).

not sensitive to near-UV light action. As a result of environmental near-UV exposure, four major defects in cellular function would participate in causing extensive irreversible cell damage. These include inability to destroy superoxide radicals generated in metabolism or by radiation, inability to destroy peroxides, extensive oxidation of both GSH and protein sulfhydril-groups, and loss of energy production from oxidative metabolism by inability to regenerate NADP from NADPH.

Other enzymes essential for the normal functioning of many cells and tissues that have shown sensitivity to near-UV light and tryptophan photoproducts include ATPase, xanthine oxidase and cytochrome oxidase. The importance of ATPase in controlling salt and water balance is well documented. If ATPase is inactivated

ENZYMATIC, CHEMICAL OR PHOTOCHEMICAL UNIVALENT REDUCTION OF O_2

$$O_2 + e^- \longrightarrow O_2^{-\cdot}$$

SUPEROXIDE DISMUTASES REACTION

$$2\,O_2^{-\cdot} + 2\,H^+ \longrightarrow H_2O_2 + O_2$$

CATALASES AND PEROXIDASES

$$2\,H_2O_2 \longrightarrow 2\,H_2O + O_2$$
$$H_2O_2 + H_2R \longrightarrow 2\,H_2O + R$$

a

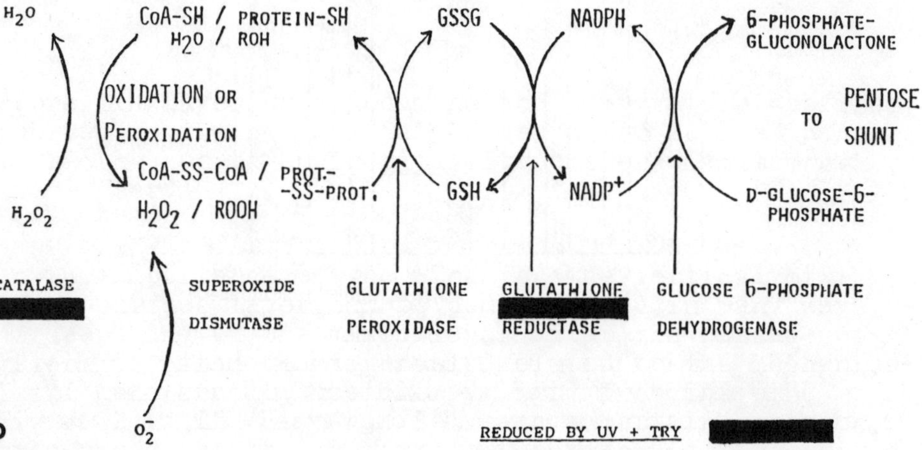

H_2O | CoA-SH / PROTEIN-SH | H_2O / ROH | GSSG | NADPH | 6-PHOSPHATE-GLUCONOLACTONE

OXIDATION OR PEROXIDATION | PENTOSE TO SHUNT

H_2O_2 | CoA-SS-CoA / PROT.-SS-PROT. | H_2O_2 / ROOH | GSH | $NADP^+$ | D-GLUCOSE-6-PHOSPHATE

CATALASE | SUPEROXIDE DISMUTASE | GLUTATHIONE PEROXIDASE | GLUTATHIONE REDUCTASE | GLUCOSE 6-PHOSPHATE DEHYDROGENASE

O_2^-

b

REDUCED BY UV + TRY

Figure 4. a) Cellular defense against superoxide radical and hydrogen peroxide. b) Scheme of Cellular oxidation: reduction reactions involving catalase, superoxide dismutase, glutathione reductase, glutathione, and hydrogen peroxide, and the inhibitory influence of tryptophan photoproducts plus longwave-UV light.

by near-UV plus tryptophan oxidation products, this defect would again add to the osmotic problems of cells to maintain their integrity. A summary of the photosensitized inhibition of enzymes studied so far is presented in Table I.

Not extensively discussed as a mechanism for damage to organisms exposed to sunlight ultraviolet wavelengths is the production of free radicals. Tryptophan again occupies a central role in this

TABLE I

Enzymes sensitive to tryptophan photoproducts plus longwave-UV light

Enzyme	Sensitivity
Catalase	yes
Superoxide dismutase	no
ATP-ase	yes
DNA polymerase I	no
Glutatione reductase	yes
Xanthine oxidase	yes
Prostaglandin dehydrogenase	yes
Cytochrome Oxidase	yes

mechanism. Stable free radicals were produced by Zigman and Knispel (1980) merely by exposing aqueous solutions of tryptophan to longwave UV light with maximum output at 365 nm with a range of 300 to 400 nm and with total irradiance of 5 mW/cm^2 for several hours. The characteristics of these radicals measured at 295°K are as follows: signal peak of 3215G; g-value of 2.0057; line width of 10G; frequency of 9.1; power of 2 mW; UV induced lifetime of 100 hours.

It was also found that such radical species have great affinity for proteins, and that when they combine with proteins the excited state characteristics remain with the protein. Protein aggregation also results from this reaction, but it appears that a quenching of the radical signal intensity occurs at the same time. These findings suggest quinone-like radicals. Many workers now agree that the most usual means whereby these free radicals are produced involves the photoexcitation of tryptophan. Antioxidants and free radical scavengers would serve to partially protect cells from the damage resulting from UV radiation due to free radical production.

To further observe the role of oxidation-reduction in cell viability, series of light-sensitive cyanine dyes, with reduction potentials (E_R) ranging from -1.30 to -0.20 V, were tested for their ability to inhibit growth and mitosis, using fertilized sea urchin eggs as a test system (Zigman and Gilman 1980). Dyes with E_R values more negative than -1.0 V were generally growth inhibitors, whereas those with positive E_R values were not. In this study the active dyes penetrated into the cells, permeated to all subcellular compartments, and were bound to numerous macromolecules. As a result, they inhibited cell division and growth. These findings suggest that the dyes interfere with electron transfer processes in cells. Indeed, preliminary results have shown that these same dyes do interfere with ATP production, which is quite dependent upon electron transfer reactions.

Energy reduction is further emphasized by studies of the action of these dyes on dinoflagellates (Martin et al. 1980). Representative cyanine dyes were tested for cytopathology against the unarmored dinoflagellate *Gymnodinium breve* (*Ptychodiscus brevis*). The active dyes caused cytolysis. Cell survival followed a zero-order kinetic pattern. In addition, cytolysis occurred to the same extent during the 24-hour test whether the dye and cells were illuminated or not, which suggests that the active dye was not acting as a sensitizer for singlet oxygen production. Since these organisms require much ATP to support both their rapid locomotion and osmotic balance with the sea water, interference with electrochemical events that have a role in energy generation are catastrophic.

CONCLUSIONS

This presentation shows that interference with oxidation-reduction reactions of the cells of several marine organisms that results from photosensitized reactions is a very important means whereby the UV wavelengths in sunlight damage them. Tryptophan photoproducts and free radicals appear to be the effectors that bind to enzyme proteins so as to inhibit their catalytic role in the control of cellular oxidation. Since tryptophan is only present in the vast expanses of the oceans at low concentrations, living cells in areas rich in organic matter from decaying plant or animals would most likely be influences by this mechanism. However, the enhanced levels of solar UV radiation reaching to greater depths of the ocean would increase the importance of the above-described process by creating greater levels of photoproducts and free radicals not only in the water, but also internal to unprotected organisms.

REFERENCES

Kalustian, A., S. Zigman, and M. Sun. 1981. Sensitized near-UV light inactivation of glutathione reductase in ocular tissues. (Chapter in book Photochemical Mechanisms in Cataract Formation [ed. G. Duncan] Academic Press).

Martin, B. B., D. F. Martin, S. Zigman, and B. Antonellis, 1980. Cytolysis of the Florida red tide organism *Ptychodiscus brevis* by redox dyes. *Microbios Letters* *14*: 77-80.

McCormick, J. P., J. R. Fischer, J. P. Pachlatko, and A. Eisenstark. 1976. Characterization of a cell-lethal product from the photooxidation of tryptophan: hydrogen peroxide. *Science* *191*: 468-469.

Sun, M., and S. Zigman. 1979. Isolation and identification of tryptophan photoproducts from aqueous solutions of tryptophan exposed to near-UV light. *Photochem*. *Photobiol*. *29*: 893-897.

Zigman, S., and P. Gilman, Jr. 1980. Inhibition of cell division and growth by a redox series of cyanine dyes. *Science* *208*: 188-191.

Zigman, S., and J. D. Hare. 1976. Cell growth inhibition by near ultraviolet light photoproducts of tryptophan. *Mol*. *Cell*. *Biochem*. *1*: 131-135.

Zigman, S., J. D. Hare, T. Yulo, and D. Ennist. 1978. Differential effects of tryptophan photoproducts and H_2O_2 on the growth of mouse embryonic fibroblasts. *Photochem*. *Photobiol*. *27*: 281-284.

Zigman, S. and Knispel, R. 1979. Stable free radicals in human lenses relative to aging, pigmentation and cataract. (chapter in book Photochemical Mechanisms in Cataract Formation [ed. G. Duncan 1981. Academic Press.

Zigman, S., J. Schultz., T. Yulo. and G. A. Griess.

1973. The binding of photo-oxidized tryptophan to a lens gamma crystallin. Exp. Eye Res. 209-217.
Zigman, S., T. Yulo. and G. A. Griess. 1976. Inactivation of catalase by near-UV light and tryptophan photoproducts. Mol. Cell. Biochem. 11: 149-154.

SURVIVAL OF THE BRINE SHRIMP, ARTEMIA SALINA, AFTER EXPOSURE TO 290-NM ULTRAVIOLET RADIATION, WITH AND WITHOUT MAXIMUM PHOTOREACTIVATION

M.J. Peak (1) (2) and H.E. Kubitschek (2)

Rhodes University, Grahamstown, South Africa (1) and Mutagenesis Group, Division of Biological and Medical Research, Argonne National Laboratory, Argonne, Illinois 60439 (2)

It is possible that increased exposure of planktonic organisms to far-ultraviolet light with biologically efficient wavelengths might occur as a result of attenuation of our stratospheric ozone shield due to man's activities. It is thus important to assess experimentally the effects this light might have upon our marine ecosystems. Ultraviolet light (UV) at 290 nm is one potentially important wavelength whose fluence at the surface of the earth might increase with a reduction in ozone (Baker, et al. 1980). For this reason, we report here studies of the lethal effects of 290 nm UV upon larvae of the salt water shrimp, *Artemia salina*. These experiments allowed for full expression of dark and photorepair, enabling assessment concerning the possible impact of attenuated atmospheric ozone shielding upon this marine organism.

MATERIALS AND METHODS

Animals

Artemia eggs were obtained commercially from a local South African pet shop. One batch of eggs, viability 0.24, was used for this entire study. Eggs were hatched in sterile natural sea water. Three days later, the viable larvae were separated from the egg cases and non-viable eggs by allowing them to swim through an aperture towards a beam of light. They were then concentrated to about 15 larvae/ml for irradiation in 1-ml batches. After irradiation, each batch of larvae was transferred with a broad aperture pipette to a 1 x 6 inch test tube. Five ml of fresh sea water was added, and the animals were held in the dark for 3 additional days. The larvae were then checked for "viability" (normal swimming movements) using a dissecting microscope. Appropriate numbers, up to as many as 3000 individuals, were checked to obtain any one survival point. No food was supplied during the six days of the experiment, and apart from photoreactivation (PR) all operations were carried out under weak yellow light at 25°C. All pipettings were performed extremely gently. Survival of nonirradiated control larvae was 100% using these conditions.

Radiations

During irradiation the larvae were held in a quartz glass vessel and stirred by a gentle stream of moist air bubbles. UV radiation was obtained from a 2.5-kW Hg-Xe Hanovia lamp coupled to a 500 mm Bausch and Lomb monochromator [equipment described by Robb and Peak (1980)]. At 290 nm the light was filtered by Corning filter 0-56, and the entrance and exit slits of the monochromator were set at 5 mm. Photoreactivation (PR) was performed at 405 nm, close to the optimal wavelength. In this case the slits were set at 10 mm and Corning filter 3-75 was used to eliminate stray light of shorter wavelengths. Radiation rates were 1 Jm^{-2}/sec. (290 nm) and 380/sec Jm^{-2} (405 nm).

RESULTS

Figure 1 shows the effects of 290 nm UV upon the normal swimming activity of *Artemia*. In the complete absence of PR (405 nm) a dose of about 50 Jm^{-2} inactivated the animals by two orders of magnitude. This may be compared with the similar survivals of *E. coli* WP2, which is inactivated to the same level by 50 Jm^{-2} using 254 nm UV (Webb and Lorenz 1970), and *E. coli* B/r, which requires 180 Jm^{-2} at 290 nm UV (extrapolated from Webb and Brown 1976 for this inactivation). There is a shoulder on the *Artemia* 290 nm survival curve which may be a region of dark repair. However, it is evident from Fig. 1 that PR of UV damage is highly important in the life of *Artemia*. Essentially complete PR by 405 nm light was obtained after 30 Jm^{-2} of 290 nm irradiation. Even after five orders of magnitude inactivation (120 Jm^{-2} 290 nm; exponential

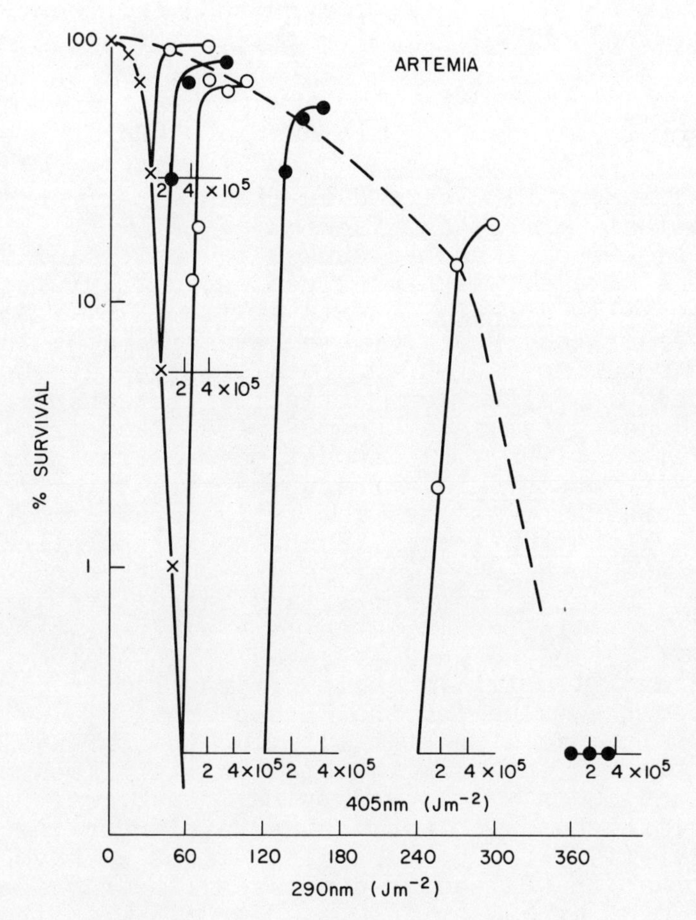

Figure 1. Survival of <u>Artemia salina</u> after exposure to monochromatic 290 nm UV light, with (broken curve) and without (continuous curve) maximum photoreactivation curves fitted by eye. The inserted abscissa show kinetics of PR (405 nm) after the various fluences of 290 nm. After 360 Jm^{-2} of 290 nm, no PR was observed. X-X, survival after 290 nm in the absence of PR; o-o, kinetics of PR after 30, 60, and 240 Jm^{-2} of 290 nm UV; •-• kinetics of PR after 45, 120, and 360 Jm^{-2} of 290 nm UV. (In no case did any of the filled or open circles overlap).

extrapolation) PR restores swimming activity to about 80% of the original individuals. However, after 360 Jm^{-2} 290 nm UV (16 orders of magnitude, exponential extrapolation), PR of swimming activity was not observed.

DISCUSSION

The efficient PR observed in these experiments is evidence that inhibition of swimming activity in *Artemia* by 290 nm light is caused by photoreactivable pyrimidine dimers. At high fluences, ultraviolet light at 290 nm kills *Artemia* despite maximum PR. From the results described in Fig. 1, we have estimated the effect of daily solar near-UV upon *Artemia* larvae at the surface of the ocean. Values for terrestrial irradiance on an average June day at latitude 30° were taken from Johnson et al.(1976). As no action spectrum for lethality exists for *Artemia*, we assumed that this spectrum has the same shape as that for fully repair proficient wild type *Escherichia coli* (Such an extrapolation may be valid, since both inactivations are DNA phenomena, as shown by the large PR sectors, and lethal events follow the DNA absorption spectrum in this region (Webb and Brown 1976)), and we used averages of the values for *E. coli* WP2 and *E. coli* K12 AB1189 (Tyrrell 1976). We calculated the value for the PR sector at 290 nm to be 0.85, from the data in Fig. 1, using the survival values observed for dark irradiation with fluences of 30, 40, and 60 Jm^{-2} followed by PR. In *E. coli* the PR sector diminishes with increasing wavelength (Ascenzi and Jagger (1979). For our calculations we have also assumed that the PR sector decreases linearly from the value of 0.85 at 290 nm to 0 at 340 nm. With these assumptions, our calculations indicate the number of lethal hits per day for the wavelength range 290 to 340 nm (incident fluence per day/37% inactivation fluence) for *Artemia* is 14.4.

Clearly, *Artemia* larvae could not survive June daily surface exposures to terrestrial solar UV radiation at latitude 30° without employing tactics other than repair. It is known that these larvae migrate daily to deeper levels of the ocean, which would protect from near-UV radiation.

We have also calculated the effect upon survival of *Artemia* at the surface of the ocean following a 10 percent decrease in the ozone layer, using recent data for extraterrestrial and terrestrial irradiance distributions (Kostkowski et al. 1980 and Green and Schippnick this volume). The fluence is increased by more than 30 percent, and survival is decreased by more than a factor of 100. The increase in lethal hits/day under these conditions would be 5.0 hits/day. If the animals failed to adjust their behavior patterns to compensate, the cumulative effect of 5 extra lethal events/day would exterminate the population.

From these calculations, we conclude that _Artemia_, and probably other lightly pigmented life forms as well, would be exterminated even at current ozone levels if they did not employ protective maneuvers against solar radiation. Under an ozone layer diminished by 10%, the calculated decrease in survival by a factor of more than 100-fold would appear to make survival unlikely even under current migrational behavior. This conclusion is strengthened by the fact that we used three-day swimming activity as our only end point. We consider that long-term effects of 290 nm UV upon inactivation of genetic end points might show more sensitivity compared with three-day swimming activity. Such genetic damage would decrease fecundity.

ACKNOWLEDGEMENTS

This work was supported in part by the C.S.I.R., Pretoria, South Africa, and in part by the U. S. Department of Energy under contract No. W-31-109-ENG-38. The submitted manuscript has been authorized by a contractor of the U. S. Government under contract No. W-31-109-ENG-38. Accordingly, the U.S. Government retains a nonexclusive, royalty-free license to publish or reproduce the published form of this contribution or allow others to do so, for U. S. Government purposes.

REFERENCES

Ascenzi, J. M. and J. Jagger. 1979 . Ultraviolet action spectrum (238-405 nm) for inhibition of glycine uptake in _E. coli_. _Photochem. Photobiol._ _30_: 661-666.

Baker, K. S., R. C. Smith and A.E.S. Green. 1980. Middle ultraviolet reaching the ocean surface. _Photochem. Photobiol._ _32_:367-374.

Johnson, F. S., I. Mo and A.E.S. Green. 1976. Average latitudinal variation in ultraviolet radiation at the earth's surface. Photochem. Photobiol. 23: 179-188.

Kostkowski, H. J., R. D. Saunders, J. F. Ward, C. H. Popenoe and A.E.S. Green. 1980. New state of the art in solar terrestrial spectroradiometry below 300 nm. Optical Radiation News, No. 33:1.

Robb, F. T. and M. J. Peak. 1980. Inactivation of the lactose permease of Esherichia coli by monochromatic light. Photochem. Photobiol. 30:379-383.

Tyrrell, R. M. 1976. Synergistic lethal action of ultra-violet-violet radiations and mild heat in Escherichia coli. Photochem. Photobiol. 24:345-351.

Webb, R. B. and J. R. Lorenz. 1970. Oxygen dependence and repair of lethal effects of near ultraviolet and visible light. Photochem. Photobiol. 12:283-289.

Webb, R. B. and M. S. Brown. 1976. Sensitivity of strains of Escherichia coli differing in repair capability to far UV, near UV and visible radiations. Photochem. Photobiol. 24:425-432.

PHOTOREACTIVATION: MAMMALIAN CELLS

Thomas P. Coohill

Biophysics Program
Western Kentucky University
Bowling Green, Kentucky 42101

The phenomenon known as photoreactivation (PR) was discovered in _Streptomyces griseus_ by Kelner in 1949. Thereafter, much of the initial work involving the characterization of this light effect was carried out using other prokaryotes (Jagger, 1967). I will attempt to briefly review the pertinent data accumulated for PR in mammalian cells.

PR is widespread in the biological kingdom (Cook and McGrath, 1967). Until the 1970's it was generally believed that mammalian cells were not capable of photoreactivating ultraviolet (UV) radiation damage (Cook, 1971). No convincing theory - outside of a vague evolutionary one - was proposed to account for this apparent cellular defect. However, the available evidence showed that attempts to demonstrate PR in mammalian cells were not successful. Although a general interest existed in why mammals, especially humans, lacked such an important repair system, these negative results seemed unequivocal.

The pioneering work of B. M. Sutherland, first published in 1974, rekindled an interest in looking for PR in mammalian cells. Her results showed that, under the appropriate conditions, PR could be demonstrated in a mammalian cell strain (Sutherland, 1974; Sutherland et al., 1974). For several years these results were questioned by other investigators, but by 1977 the controversy was decided in favor of PR by a series of experiments in other cell lines (Harm, 1978; Mortelmans et al., 1976). It has now been shown that the demonstration of PR in mammalian cells is dependent upon certain conditions, a most important one being the composition of the media in which the cells are grown. In general, cells in "rich" (e.g. DMEM) media exhibit PR while those in "poor" (e.g. IXMEM) media do not (Sutherland and Oliver, 1976; Mortelmans et al. 1977).

PR in mammalian cells is similar in most aspects to that in bacteria. The action spectrum for PR in mammalian cells is somewhat broader than it is in bacteria, and extends further into the visible light range (Sutherland et al., 1974). The PR enzyme from bacteria (e.g. *E. coli*) can be placed into mammalian cells and shown to exhibit biological activity there (Sutherland and Hausrath, 1979). PR enzyme can be measured in most mammalian cells (Sutherland, B. M., personal communication).

In human cell lines the PR situation is interesting. "Normal" cell lines have the PR enzyme in higher amounts (Sutherland et al., 1976) than do certain repair deficient cell lines [e.g. xeroderma pigmentosum (XP) cells]. However, PR is much more prevalent in repair deficient cells - presumably because it has less competition from excision repair enzymes (Wagner et al. 1975; Henderson, 1978). That is, "the inability to detect PR routinely in normal cells may be because PR is "masked" by normal amounts of excision repair. Alternatively, PR processes may be repressed in excision-proficient cells and derepressed in cells deficient in excision repair" (Henderson, 1978). Thus, it has now been reported that PR can repair up to 90% of the biological damage inflicted by UV light on certain cells (Henderson, 1978; Harm, 1980b). In addition, "*in vitro* tests have indicated that sunlight > 375 nm can cause photorepairable DNA lesions which are virtually fully repaired by the same light" (Harm, 1980a).

Interestingly the degree of PR seems to be dependent upon the cell culture "age", as defined by passage number. Normal human fibroblasts (KD) in their 48th passage can repair about 50% of the available UV lesions, while those in their 9th passage show no evidence of PR ability (Harm, 1980b). This effect is also true for the human skin fibroblast line HH, wherein cells at the 21st passage show PR ability of about 80% while those in their 3rd passage show much less (Harm, 1980b). Whether this effect is linked with other cell "age" effects on repair systems mentioned previously at this meeting (Moore and Coohill, 1979) is unknown at this time.

Finally, it now seems most interesting to determine the existence and extent of PR in the human body itself. This can be viewed as looking at the "wildtype cell in nature". Some important, as yet unanswered, questions seem to be:

(1) Is PR demonstrable *in vivo* in humans (Sutherland et al., 1980)?

(2) To what extent does PR ameliorate the damaging effects of UV radiation?

(3) Is there another role for the PR enzyme in humans beyond the repair of UV damage alone?

Thus, PR in humans is an open book again. The next chapter should be both informative and interesting.

REFERENCES

Cook, J. S. and J. R. McGrath. 1967. Photoreactivating enzyme activity in metazoa. Proc. Natl. Acad. Sci. (USA) 58: 1359.

Cook, J. S. 1971. Photoreactivation in animal cells/ virus UV lesions. *Photophysiol*. *3*: 191.

Harm, H. 1978. Damage and repair in mammalian cells after exposure to non-ionizing radiations: 1. ultraviolet and visible light irradiation of cells of the rat kangaroo and determination of photorepairable damage *in vitro* . *Mutat*. *Res*. *50*: 353-366.

Harm, H. 1980a. Damage and repair in mammalian cells after exposure to nonionizing radiations. *Mutat*. *Res*. *69*: 157-165.

Harm, H. 1980b). Damage and repair in mammalian cells after exposure to nonionizing radiations: 3. UV and visible light irradiation of cells of placental mammals including humans, and determination of photorepairable damage *in vitro*. *Mutat*. *Res*. *69*: 167-176.

Henderson, E. 1978. Host cell reactivation of Epstein-Barr virus in normal and repair defective leukocytes. *Cancer Res*. *38*: 3256-3263.

Jagger, J. 1967. Introduction to research in ultraviolet photobiology. Prentice-Hall, Englewood Cliffs, N. J.

Kelner, A. 1949. Effect of visible light on the recovery of Streptomyces griseus conidia from ultraviolet irradiation injury. Proc. Natl. Acad. Sci. (USA) 35: 73-79.

Moore, S. P. and T. P. Coohill. 1979. An effect of cell-culture passage on ultraviolet enhanced viral reactivation by mammalian cells. Mutat. Res. 62: 417-423.

Mortelmans, K., J. Cleaver, E. C. Friedberg, M. C. Paterson, B. P. Smith and G. H. Thomas. 1977. Photoreactivation of thymine dimers in UV irradiated human cells: unique dependence on culture conditions. Mutat. Res. 44: 433-445.

Sutherland, B. 1974. Photoreactivating enzyme from human leukocytes. Nature (London) 248: 109-112.

Sutherland, B., P. Runge and J. C. Sutherland. 1974. DNA photoreactivating enzyme from placental mammals. Origin and characteristics. Biochem. 13: 4710-4715.

Sutherland, B. M., M. Rice and B. K. Wagner. 1975. Xeroderma pigmentosum cells contain low levels of photoreactivating enzyme. Proc. Natl. Acad. Sci. (USA) 72: 108-107.

Sutherland, B., and R. Oliver. 1976. Culture conditions affect photoreactivating enzyme levels in human fibroblasts. Biochem. Biophys. Acta 442: 358-367.

Sutherland, B. M., R. Oliver, C. O. Fuselier and J. C. Sutherland. 1976. Photoreactivation of pyrimidine dimers in the DNA of normal and xeroderma pigmentosum cells. Biochem. 15: 402-406.

Sutherland, B. M. and S. G. Hausrath. 1979. Insertion of E. coli photoreactivating enzyme into mammalian cells. 7th Annual Meeting - American Society for Photobiology, Asilomar, Ca.

Sutherland, B. M., L. C. Harber and I. E. Kochevar. 1980. Pyrimidine dimer formation and repair in human skin. Cancer Res. 40. (In press).

Wagner, E. K., M. Rice and B. M. Sutherland. 1975. Photoreaction of herpes simplex virus in human fibroblasts. Nature (London) 254: 627-628.

THE AMELIORATION OF UV-B EFFECTS ON PRODUCTIVITY BY VISIBLE RADIATION

Alan H. Teramura

Department of Botany
University of Maryland
College Park, Maryland 20742

INTRODUCTION

Ultraviolet radiation is strongly absorbed by proteins and nucleic acids, and therefore has important photobiological consequences (Caldwell 1971; Klein 1978; NAS 1979). The principal attenuator of solar UV radiation passing through the earth's atmosphere is a thin layer of stratospheric ozone, which effectively absorbs short wavelength UV and sets the lower wavelength limit reaching the earth's surface at approximately 290 nm (Koller 1965). Therefore, the naturally occurring portion of the UV spectrum on the surface of the earth is in the UV-B (290-320 nm) and UV-A (320-380 nm) regions, and does not contain the highly actinic UV-C (200-290 nm) waveband. Although the middle portion of the UV spectrum accounts for only 3-5% of the total radiation penetrating the atmosphere, its energy level is sufficient to have a disproportionately large biological significance.

Natural stratospheric ozone concentrations vary substantially, producing significant variations in UV-B reaching the surface of the earth. For example, ozone concentrations vary by as much as 30% within the continental United States along latitudinal gradients. Furthermore, ozone concentrations can vary temporally: as much as 10% hourly variation with the passage of a frontal system, 25% seasonally, and 5% annually (NAS 1979). In addition to natural variations, recent attention has been directed toward anthropogenic decreases in ozone concentrations due to atmospheric pollutants. These include halogenated hydrocarbon from aerosol propellants and refrigeration systems (Cicerone et al. 1974; Molina and Rowland 1974) and solvents (McConnell and Schiff 1978). These compounds have long retention times in the stratosphere and catalytically destroy ozone molecules. A reduction in stratospheric ozone concentrations would undoubtedly increase the level of UV-B radiation penetrating the atmosphere. The current estimate of global ozone depletion due to man-made pollutants is approximately 16% (Hudson and Reed 1979; NAS, 1979).

Exposure to artificially enhanced UV-B radiation levels has been shown to have deleterious effects on a large number of unrelated organisms. The importance of UV-B radiation as a constraint on the evolution of terrestrial organisms has been recently reviewed by Caldwell (1980). As a result of these constraints, most biological organisms have developed UV protection mechanisms. These include repair mechanisms such as photoprotection and photoreactivation (Jagger 1964) as well as UV avoidance. Green plants generally receive a greater UV-B dose than do animals or aquatic microorganisms. This is primarily due to the sessile or fixed nature of most plants. Unlike motile organisms which can escape UV by behaviorial responses, higher plants can only avoid UV by the shielding effects of plant pigments, epidermal waxes or hairs, or multiple tissue layers. Because of these reasons, green plants are ideal systems to study the effects of enhanced levels of UV-B radiation on biological systems.

In UV research, a great deal of emphasis has been primarily focused upon UV-B irradiation lamp systems and their configuration, UV measurement, and weighting functions. This emphasis, coupled with facility limitations, often results in experimental conditions having little or no ecological equivalence in natural situations. For example, nearly all rigorously controlled experiments are conducted in controlled environment chambers or in greenhouses with artificially supplied UV-B irradiances. One potential disadvantage inherent with such growth facilities is that longer wavelength irradiances (such as photosynthetically active radiation between 400 and 700 nm) are typically quite low, ranging from one-tenth to one-third that found in natural sunlight. It is well known that many of the damaging effects of UV-B irradiation are reversible in the presence of longer wavelength radiation (Caldwell

1971). Therefore, if naturally occurring levels of visible radiation have an ameliorating effect on UV-B radiation-induced damage, it would be important to consider this effect prior to extrapolating experimental data into the field. Otherwise we might substantially overestimate the consequences of a reduction in global ozone concentrations. Therefore an experiment was specifically designed to test the ameliorating effects of photosynthetic photon flux densities (PPFD'S) on UV-B-induced damage.

MATERIALS AND METHODS

Plant materials and growth conditions

Hardee soybean (*Glycine max* (L.) Merr.) seeds were planted into 250 cm^3 of 1:1 mixture (v:v) of coarse sand and vermiculite in a controlled environment greenhouse at the Duke University Phytotron, North Carolina, U.S.A. Day and night temperatures in the greenhouse were 26°C and 20°C, respectively and the natural photoperiod was extended to 16 hrs by using incandescent floodlamps. Seedlings were thinned to two per pot soon after germination. Nine replicate pots were grown under each of four UV-B irradiation levels and four PPFD'S in a factorial design (N= 288).

Ultraviolet-B irradiation was supplied by four Westinghouse FS-40 sunlamps in each treatment. Irradiance from the lamps was stabilized by pre-burning each lamp for 100 hrs prior to use, since previous studies indicated that total lamp irradiance changed less than 5% between approximately 100 and 600 hrs of use. Radiation from the lamps was filtered with either 1) 0.076 mm cellulose acetate (CA), which transmitted UV-B radiation only down to 292 nm since shorter wavelengths do not naturally occur at the earth's surface, or 2) Mylar Type S which completely absorbs radiation below 320 nm. Since CA filter degradation was exponential, filters were routinely changed every 3 days to maintain the desired spectral transmission. The four UV-B irradiances were achieved by varying the lamp-plant distance and by using Mylar filters as a control. Spectral UV-B flux densities were measured with a high resolution, Gamma Scientific Corp., grating spectroradiometer (Model 2900 photometer and 700-31 monochromator). Ultraviolet-B irradiation was weighted with two action spectra for comparative purposes. One was based on a generalized plant response function to biologically effective radiation (Caldwell 1971) and the other on a DNA action spectrum (Setlow 1974). Irradiances used in this study were equivalent to 0 (Mylar control), 17.5, 35.0, and 70.0 effective mW m^{-2} biologically effective radiation (UV-B_{BE}), or 0, 2.39, 4.78, and 9.56 mW m^{-2} DNA effective radiation. The spectral distribution of these irradiances is presented in Figure 1. On a daily dose basis, the highest irradiance used in this study (1512 effective Jm^{-2} d^{-1}) was similar to that normally incident at Gainesville, Florida (29°36' latitude) at the summer solstice during clear sky conditions when calculated as biologically effective UV-B radiation.

The four PPFD'S were obtained by shading the plants with neutral density filters, achieving 0 (unshaded),

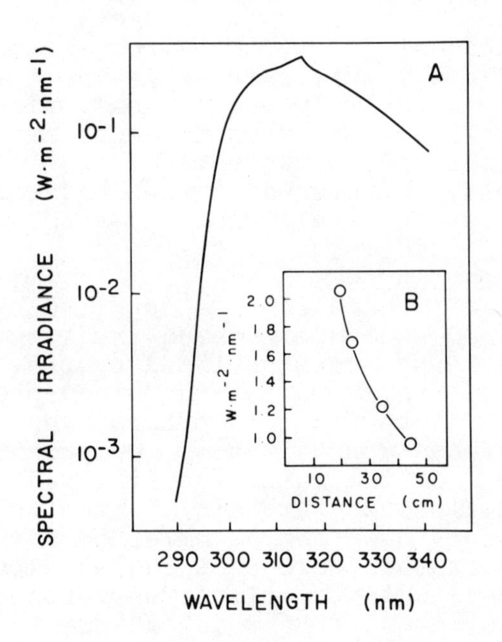

Figure 1. A: Spectrum of UV-B irradiance from four Westinghouse FS-40 lamps filtered by 0.076 mm cellulose acetate. Distance between the lamp and spectroradiometer was adjusted to 35.0 mW m^{-2} of biologically effective radiation (UV-B_{BE}). B: Relationship of total UV-B irradiance and lamp-to-plant distance. This distance was adjusted to produce UV integrated effective irradiances which were equivalent to 70.0, 35.0, and 17.5 effective mW m^{-2} UV-B_{BE}. Control doses were filtered with Mylar, which transmitted no UV-B_{BE}.

33, 55 and 88% shade. These correspond to an average daily maximum PPFD of 1600, 1408, 880, and 528 $\mu E\ m^{-2}s^{-1}$, respectively. PPFD'S were measured with a Lambda Instruments Corp. LI-190S quantum sensor at plant height. Leaf temperatures remained within ± 3°C of ambient temperatures under all PPFD'S.

Gas exchange measurements

The equilibrium flux of CO_2 and water vapor was simultaneously measured on single, attached leaves after two levels of UV-B accumulation. Measurements were made after 14 and 49 days of a 6-hr/day UV-B radiation exposure (84 and 294 hrs of accumulated UV-B radiation). Carbon dioxide was measured with a Beckman 215B infrared gas analyzer and water vapor with a Cambridge Systems EG&G model 880 dewpoint hydrometer. All gas exchange measurements were done at a leaf temperature of 30°C, 320 µl/l CO_2 concentration, a vapor pressure deficit of 1 kPa, and saturating irradiances of 1300 $\mu E\ m^{-2}s^{-1}$. Diffusive resistances to CO_2 and water vapor were calculated similar to Gaastra (1959, using 1.56 as the coefficient relating the diffusivities of water vapor to CO_2. Non-stomatal (mesophyll) resistances were calculated as a residual term.

Plant harvest measurements

At the end of 7 weeks, overall height was recorded to the nearest mm and plants were harvested. Leaf areas were determined with a Lambda Instruments Corp. LI-3000 area meter. Plants were dried in a forced-draft oven at 60°C and weighted to the nearest mg.

RESULTS

A summary of the effects of a 2-week exposure on net photosynthesis and the diffusive resistances to CO_2 are presented in Table 1. When UV-B irradiation was simultaneously applied with a high PPFD (1600 $\mu E\ m^{-2}s^{-1}$) no depression in photosynthesis was observed. Instead there was a mild stimulation of photosynthesis (not significant) in plants exposed to low UV-B fluences. However, under reduced PPFD'S, 70 effective mW m^{-2} UV-B_{BE} produced significant ($P<0.05$) reductions in leaf photosynthesis. These reductions were primarily the result of increases in mesophyll resistances to CO_2 (Table 1). Stomatal resistances also contributed toward reductions in net photosynthesis in leaves irradiated with 70 effective mW m^{-2} UV-B_{BE}, but was variable at

TABLE 1. Mean effects of a 2-week exposure to four UV-B irradiances and four photosynthetically active photon flux densities on net photosynthesis and diffusive resistances in soybean and a 7-week exposure on several plant growth characteristics. Values in columns for each PPFD followed by same letter are not statistically different at 5% level.

PPFD	$UV\text{-}B_{BE}$	Net Photosynthesis	Mesophyll Resistance	Stomatal Resistance	Plant Height	Leaf area	Total Leaf no.
$\mu Em^{-2}s^{-1}$	mWm^{-2}	mg $CO_2 dm^{-2}h^{-1}$	$s\ cm^{-1}$		cm	cm^2	
1600	0	15.9a	13.1a	1.2a	48d	1034a	21a
	17.5	17.2a	13.8a	1.0b	73a	1170a	20a
	35.0	15.7a	12.5a	1.7ab	59c	1146a	21a
	70.0	15.0a	12.2a	2.4a	69b	997a	20a
1408	0	12.1a	17.3b	1.7b	181a	1182a	20a
	17.5	15.6a	13.1b	1.0b	135b	1272a	21a
	35.0	14.6a	14.3b	1.7b	101c	829a	18b
	70.0	9.9b	20.2a	2.8a	93c	776b	17b
880	0	13.1a	16.9b	0.9ab	181a	1131a	19ab
	17.5	13.9a	15.6b	0.6ab	177a	807a	20a
	35.0	15.8a	14.1b	0.9ab	94b	763a	18b
	70.0	5.5b	47.1a	1.1a	97b	296b	13c
528	0	12.0a	16.5b	1.0b	138a	460a	18a
	17.5	11.3a	18.5b	1.1ab	109a	490a	13a
	35.0	11.0a	18.8b	1.7a	118a	386a	15a
	70.0	9.3b	23.2a	1.9a	78b	168b	10b

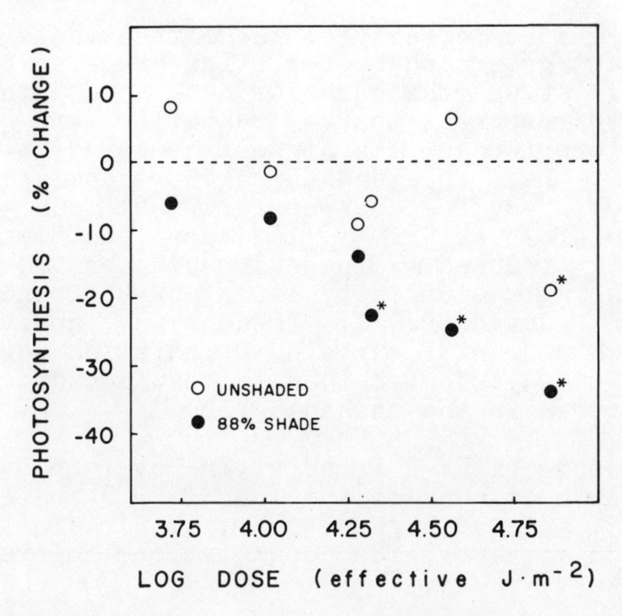

Figure 2. Effects of log UV-B radiation incident dose on percent change of net photosynthesis from controls in soybeans under two levels of shade. Each point represents the mean of four observations. Values above the dashed horizontal line indicate enhancements, those below it indicate reductions in net photosynthesis. Asterisks indicate values significantly ($P<0.05$) less than control.

lower fluences. Dark respiration, measured as CO_2 efflux in the dark, was unaffected by UV-B irradiation in all four PPFD'S (Teramura et al. 1980).

These observations were generally similar for plants exposed to 6 weeks of UV-B irradiation and therefore these 6-week data are not presented; however, some of these data were used in Figure 2 to calculate the effects of incident dose (UV-B flux density x exposure time) on net photosynthesis. For clarity, only two levels of shade are presented in Figure 2. Nevertheless, it strongly suggests that plants grown in the highest PPFD accumulated larger UV-B radiation doses than plants grown in the lowest PPFD prior to incurring an equivalent reduction in net photosynthesis.

The cumulative effects of these treatments on several plant growth characteristics are also presented on Table 1. As was the case for net photosynthesis, plant growth responses to UV-B radiation were qualitatively different under the highest compared to the lowest PPFD. Leaf area and number were unaffected by the range of UV-B fluences when grown at 1600 $\mu E\ m^{-2} s^{-1}$; however, under lower PPFD'S, leaf area and number were significantly reduced by these irradiances. Soybeans were increasingly stunted by larger UV-B fluences when grown under reduced PPFD'S. In contrast, however, plants which did not receive UV-B radiation were significantly ($P<0.05$) shorter than those which did, when both were grown in the unshaded PPFD.

The effects of UV-B irradiation level on overall soybean growth at each of the four PPFD'S are summarized in Figure 3. At the highest PPFD, UV-B radiation had no effect on total biomass accumulation; however, as

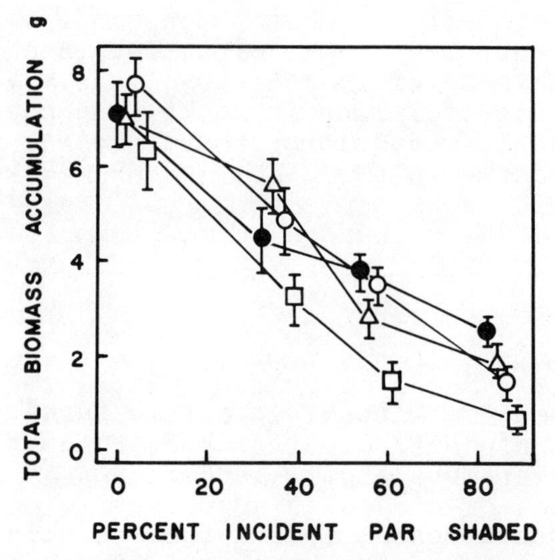

Figure 3. Effects of four UV-B and four levels of shade on total dry matter accumulation in soybean after seven weeks. Each symbol represents the mean of nine independent observations. Symbols and standard errors (vertical bars) were offset for ease of comparison. All plants were grown at 0, 33, 55, and 88% shade. Closed circles = Mylar controls; triangles = 17.5 mW m^{-2} UV-B_{BE}; open circles = 35.0 mW m^{-2} UV-B_{BE}; squares= 70.0 mW m^{-2} UV-B_{BE}.

PPFD'S were reduced, UV-B fluences resulted in greater biomass reductions. Except in the highest PPFD, 70 effective mW m^{-2} UV-B_{BE} produced significant ($P<0.05$) reductions in soybean dry weight accumulation.

DISCUSSION

It is evident from the photosynthesis and growth responses presented here that soybeans are sensitive to moderate levels of UV-B irradiation. More importantly, however, is the observation that the degree of sensitivity is highly dependent upon the flux density of longer wavelength radiation simultaneously incident upon the plant. For example, a 2 week exposure to 70 effective mW m^{-2} UV-B_{BE} and a PPFD of 1600 $\mu E\ m^{-2} s^{-1}$ had no effect on net photosynthesis, while the same UV-B irradiance at reduced PPFD'S resulted in at least 18% decrease in net photosynthesis. This ameliorating effect of high PPFD'S was also observed in total leaf area, number, and overall plant productivity (estimated from total biomass).

In addition to the quantitative differences resulting from UV-B irradiation under reduced PPFD'S, there were qualitative differences in some responses. Plants which received UV-B radiation and unshaded PPFD'S were taller than control plants. This increase in plant height was accompanied by a proportionately greater stem dry weight and was therefore not due simply to longer internodes, but rather represented a substantial shift in resource allocation (Teramura 1980).

The mechanism for the ameliorating effect of high PPFD'S on UV-B induced alteration of plant processes is presently unknown. It could possibly involve photoreactivation or photoprotection by the subsequent exposure to longer wavelength radiation. There are two different types of processes resulting in photoreactivation. One involves an _in situ_, photoenzyme-mediated repair of UV-induced cyclobutane pyrimidine dimers (direct photoreactivation), and the other is an intensification due to longer wavelength-induced physiological changes of enzyme-independent dark repair mechanisms (Rupert 1975). Photoprotection is the phenomenon whereby this enhancement of repair processes is ellicited by longer wavelength radiation supplied prior to the damaging UV irradiation. Since the UV-B radiation exposures used in this study were simultaneous to and bracketed by exposures to contrasting PPFD'S, it would be impossible to distinguish between photoprotection or photoreactivation. Nevertheless, it is clear that the effectiveness of the mechanism(s), whatever they may be operating in this soybean system diminished as PPFD'S were reduced.

TABLE 2
Examples of UV-B irradiation studies on terrestrial plants

Species	PPFD ($\mu Em^{-2}s^{-1}$)	Waveband	Reference
Pisum sativum L.	400 G	UV-B	Brandle et al.
	800 C	UV-B	(1977)
Pisum sativum L.	500 (?)G	UV-B	Van et al.
Brassica oleracea L.	500 G	UV-B	(1977)
Arachis hypogaea L.	500 G	UV-B	
Impatiens balsamina L.	100 C	UV-A	Klein et al.
Lycopersicum esculentum			(1965)
Mill.	100 C	UV-A	
Tagetes erecta L.	200 C	UV-A	
Pisum sativum L.	250 G	UV-B	Vu et al.
Zea mays L.	250 G	UV-B	(1978)
Glycine max (L.) Merr.	250 G	UV-B	
Coleus bluemei Benth.	400 G	UV-B	Semeniuk and Stewart (1979)
Chenopodium quinosa Willd.	170 G	UV-B	Semeniuk and Goth (1980)

TABLE 3

Percent change from Mylar controls in photosynthesis or growth by UV-B irradiation simultaneously applied with different photosynthetically active photon flux densities (PPFD)

Species	PPFD ($\mu E \cdot m^{-2} s^{-1}$)					
	2000	1600	1400	800-900	400-500	200
Rumex patientia L. [1]	-38			-30	-38	
Phaseolus vulgaris L. [2]	+5					
Zea mays L.	+7				-8	
Lycopersicum esculentum L.	-22				-24	
Pisum sativum L.					+4	-33
Glycine max (L.) Merr. [3]		-6	-18	-58	-23	
Glycine max (L.) Merr. [4]		-14	-29	-65	-59	-52
Pisum sativum L. [5]				-19		-35
Lycopersicum esculentum L.				-12		-33
Brassica oleracea var. acephala L.				-10		-27
Brassica oleracea var. capitata L.				-6		-33
Zea mays L.				+5		-16
Avena sativa L.				+5		-24

1) Sisson and Caldwell (1976)
2) Bartholic et al. (1974)
3) Teramura et al. (1980)
4) Teramura (1980)
5) Van et al. (1976)

TABLE 4

Examples of UV-B irradiation studies on marine organisms

Species	PPFD ($\mu Em^{-2}s^{-1}$)	Waveband	Reference
Dunaliella tertecolata	≈660 C	UV-A & UV-B	McLeod and Kanwisher (1962)
Phaeodactylum tricornutum	≈600 C	UV-A & UV-B	
Melosira nummuloides	≈200 C	UV-B	Thompson et al.(1980)
Thalassiosira pseudonana	≈ 80 C	UV-B	Worrest et al.(1980)
Acartia clausii	(?)	UV-B	Karanas et al. (1979)
Pandalus platyceros	(?)	UV-B	Damkaer et al. (1980)
P. Hypsinotus	(?)	UV-B	
P. danae	(?)	UV-B	
Cancer magister	(?)	UV-B	
C. oregonensis	(?)	UV-B	
Thysanoessa raschii	(?)	UV-B	
Engraulis mordax	(?)	UV-B	Hunter et al. (1979)
E. mordax	200 C	UV-B	
Scomber japonicus	(?)	UV-B	
S. japonicus	200 C	UV-B	

The universality of the general conclusions drawn from this study is presently unknown. It is significant, however, that nearly all the published studies examining the effects of enhanced UV-B radiation on plant growth and photosynthesis have been conducted in controlled environment chambers or in greenhouses, where PPFD'S are typically below the light-saturation level for photosynthesis. Some representative studies with terrestrial plants are presented in Table 2. In controlled environment chambers (indicated with a "C"), PPFD'S rarely are above 200-300 $\mu E\ m^{-2}s^{-1}$. This represents approximately 10-15% of mean maximum daily PPFD'S naturally occurring in the field. Due to various types of shading necessary for partial temperature control during the summer months, PPFD'S in greenhouse studies (indicated with a "G") are often less than 500-600 $\mu E\ m^{-2}s^{-1}$, or about 25-30% of daily maxima.

Table 3 summarizes the results of those few studies which examined the effects of UV-B radiation on plants under at least two contrasting PPFD'S. The data from Bartholic et al. (1974) represents the effects of ambient levels of UV-B radiation on total dry matter accumulation, while the other studies examined the effects of artificially supplied UV-B radiation (from FS-40 sunlamps) on net photosynthesis and growth. The study by Teramura et al. (1980) illustrates the ameliorating effects of PPFD'S on soybean net photosynthesis, while Teramura (1980) and Teramura et al. (1980) exemplifies this effect on overall productivity or dry matter accumulation. The data presented in this paper appears to be consistent with the general observation that higher PPFD'S have an ameliorating effect on UV-B-induced alterations of normal plant processes. One noteworthy exception is the response of *Rumex patientia* which is apparently PPFD independent. It is presently unknown if other plants also exist which have UV responses which are not modified by PPFD.

Table 4 summarizes the experimental conditions used in some selected UV-B studies on marine organisms. Like the studies with terrestrial plants, PPFD'S were characteristically supplied by fluorescent lamps and therefore quite low. Additionally, there were a large number of studies where longer wavelength radiation was largely ignored and unreported. If any of the ameliorating effects of higher PPFD'S found in terrestrial plant studies are operative in marine organisms, then we must be careful in any attempts to relate these studies to natural situations. In conclusion, this illustrates the complexity of the interactions between UV-B radiation and other ecological factors, specifically longer wavelength radiation. Furthermore, this study points out the need for field validation of conclusions drawn from laboratory studies.

REFERENCES

Bartholic, J. F., L. H. Halsey, and R. H. Biggs. 1974. Effects of UV radiation on agricultural productivity. Third Conference on CIAP (Feb. 1974), U.S. D.O.T., Wash., D. C. 498-504.

Brandle, J. R., W. F. Campbell, W. B. Sisson and M. M. Caldwell. 1977. Net photosynthesis, electron transport capacity, and ultrastructure of Pisum sativum L. exposed to ultraviolet-B radiation. Plant Physiol. 60: 165-169.

Caldwell, M. M. 1971. Solar UV irradiance and the growth and development of higher plants. In Photophysiology [Giese, A. C.ed], Vol. 6:131-177. Academic Press, NY.

Caldwell, M. M. 1980. Plant life and ultraviolet radiation: some perspective in the history of the earth's UV climate. BioScience 29: 520-525.

Cicerone, R. J., R. S. Stolarski, and S. Walters. 1974. Stratospheric ozone destruction by man-made chlorofluoromethanes. Science 185:1165-1167.

Damkaer, D. M., D. B. Dey, G. A. Heron, and E. F. Prentice 1980. Effects of UV-B radiation on near-surface zooplankton of Puget Sound. Oecologia (Berl.) 44: 149-158.

Gaastra, P. 1959. Photosynthesis of crop plants as influenced by light, carbon dioxide, temperature, and stomatal diffusion resistance. Med. Landbouwh. Wageningen 59: 1-68.

Hudson, R. D. and E. I. Reed. 1979. The stratosphere: present and future. Chlorofluoromethanes and the stratosphere. NASA Ref. Publ. 1049. Sci. and Techn. Info. Serv., Springfield, VA.

Hunter, J. R., J. H. Taylor, and H. G. Moser. 1979. Effect of ultraviolet irradiation on eggs and larvae of the northern anchovy, Engraulis mordax, and the pacific mackerel, Scomber japonicus, during the embryonic stage. Photochem. Photobiol. 29: 325-338.

Jagger, J. 1964. Photoreactivation and photoprotection. Photochem. Photobiol. 3: 451-461.

Karanas, J. J., H. VanDyke, and R. C. Worrest. 1979. Midultraviolet (UV-B) sensitivity of Arcartia clausii Giesbrecht (Copepoda). Limnol. Oceanogr. 24: 1104-1116.

Klein, R. M. 1978. Plants and near-ultraviolet radiation. Bot. Rev. 44: 1-127.

Klein, R. M., P. C. Edsall, and A. C. Gentile. 1965. Effects of near ultraviolet and green radiations on plant growth. Plant Physiol. 40: 903-906.
Koller, L. R. 1965. Ultraviolet Radiation. John Wiley and Sons Inc., N. Y.
McConnell, J. C. and H. I. Schiff. 1978. Methyl chloroform: Impact on stratospheric ozone. Science 199: 174-177.
McLeod, G. C. and J. Kanwisher. 1962. The quantum efficiency of photosynthesis in ultraviolet light. Physiol. Plant. 15: 581-586.
Molina, M. J. and F. S. Rowland. 1974. Stratospheric sink for chlorofluormethanes: chlorine atom-catalysed destruction of ozone. Nature 249: 810-812.
National Academy of Sciences. 1979. Protection Against Depletion of Stratospheric Ozone by Chlorofluorocarbons. J. W. Tukey and M. S. Peters, Chairmen, Office of Publ., NAS, Washington, D. C.
Rupert, C. S. 1975. Enzymatic photoreactivation: Overview. In P. C. Hanawalt and R. B. Setlow [eds.] Molecular Mechanisms for Repair of DNA. Basic Life Sciences, Vol. 5A, 73-87, Plenum Press, N. Y.
Semeniuk, P. and R. N. Stewart. 1979. Comparative sensitivity of cultivars of coleus to increased UV-B radiation. J. Amer. Soc. Hort. Sci. 104: 471-474.
Semeniuk, P. and R. W. Goth. 1980. Effects of ultraviolet irradiation on local lesion development of potato virus S on Chenopodium quinoa 'Valdivia' leaves. Environ. and Exp. Bot. 20: 95-98.
Setlow, R. B. 1974. The wavelengths in sunlight effective in producing skin cancer: a theoretical analysis. Proc. Nat. Acad. Sci. (USA) 71: 3363-3366.
Sisson, W. B. and M. M. Caldwell. 1976. Photosynthesis, dark respiration, and growth of Rumex patientia L. exposed to ultraviolet irradiance (288 to 315 nanometers) simulating a reduced atmospheric ozone column. Plant Physiol. 58: 563-568.
Teramura, A. H. 1980. Effects of ultraviolet-B irradiances on soybean. I. Importance of photosynthetically active radiation in evaluating ultraviolet-B irradiance effects on soybean and wheat growth. Physiol. Plant. 48: 333-339.
Teramura, A. H., H. Biggs, and S. Kossuth. 1980. Effects of ultraviolet-B irradiances on soybeans. II. Interactions between ultraviolet-B and photosynthetically active radiation on net photosynthesis, dark respiration, and transpiration. Plant Physiol. 65: 483-488.
Thomson, B. E., R. C. Worrest, and H. Van Dyke. 1980.

The growth response of an estuarine diatom (Melosira nummuloides [Dillw.] Ag.) to UV-B (290-320 nm) radiation. Estuaries 3: 69-72.

Van, T. K., L. A. Garrard, and S. H. West. 1976. Effects of UV-B radiation on net photosynthesis of some crop plants. Crop Sci. 16: 715-718.

Van, T. K., L. A. Garrard, and S. H. West. 1977. Effects of 298-nm radiation on photosynthetic reactions of leaf discs and chloroplast preparations of some crop species. Environ. and Exp. Bot.17: 107-112.

Vu, C. V. Allen, Jr., L. H. and L. A. Garrard. 1978. Effects of supplemental ultraviolet radiation (UV-B) on growth of some agronomic crop plants. Soil and Crop Sci. Soc. Fl. 38: 59-63.

Worrest, R. C., D. L. Brooker, and H. Van Dyke. 1980. Results of a primary productivity study as affected by the type of glass in the culture bottles. Limnol. Oceanogr. 25: 360-364.

ERROR PRONE REPAIR-EMPHASIS WEIGLE REACTIVATION

Thomas P. Coohill
Biophysics Program, Western Kentucky
University, Bowling Green, Kentucky 42101

There are several ways that living cells repair damage inflicted upon their genetic material (DNA) by ultraviolet light. The best understood, but by no means the only, type of photochemical damage is the dimerization of adjacent pyrimidine bases in the DNA chain, the most prevalent ones being thymine-thymine (TT) dimers. Almost certainly, we haven't discovered all of the repair mechanisms present in cells. However, at present, three distinct types of repair are demonstrable;

(1) dark repair

(2) photoreactivation

and (3) "error prone" repair

Error-prone repair (EPR) had, until the last eight years or so, been considered to be a "minor" repair pathway. The fact that it is error-prone now makes it exceedingly interesting for those of us working with human cells, in that these errors can lead to mutations, and mutation, perhaps to carcinogenesis.

EPR is initiated by the existence of an unrepaired lesion in the DNA (Devoret, 1978). Such an event could be due to the timing of repair by the two other pathways (dark repair, or photoreactivation) or by the existence of a dimer (for example) in the DNA strand near where the DNA copying mechanism is replicating new DNA

(Witkin, 1976; Gudas and Pardee, 1975), a lesion causing the DNA polymerase to "stall" with dire consequences for cell survival. One way around such a "stalled" replication site is for the polymerase machinery to "over-ride" or copy around the damaged site (Radman, 1975). DNA synthesis can thus continue again but at the expense of having added additional or incorrect base sequences into the progeny genetic material. Such error can give rise to cells that have survived but are mutated.

There are a series of cellular functions that appear to be induced by such a "stalled" DNA polymerase, EPR system. These include viral induction, mutagenesis, filament formation (in bacteria), and enhanced viral reactivation (Weigle reactivation). Radman (1975) has developed an SOS repair hypothesis to explain these cellular responses. Since these functions are assumed to be linked together by a common signal, a discussion of one may be sufficient to shed some light on all of them. Accordingly, I will limit my talk to a brief analysis of Weigle reactivation (WR).

In 1953 J.J. Weigle (1953) reported that ultraviolet damaged bacteriophage survived <u>better</u> if the host cells that were infected with these damaged viruses were pretreated with low levels of ultraviolet radiation. The extent of such "reactivation" of damaged virus could be as much as three hundred fold above the levels produced in control (unirradiated) cells. In addition, these reactivated viruses were highly mutated (Weigle, 1953). This phenomenon was clearly inducible since a delay between irradiating the cells and infecting them with damaged virus, gave higher levels of virus reactivation.

Weigle reactivation (WR) has been demonstrated in several phage bacterial systems (Rupert and Harm, 1966). An analogous phenomenon has been reported in mammalian cells (Bockstahler and Lytle, 1970) involving the reactivation of UV irradiated mammalian viruses grown in irradiated mammalian cells. The term "ultraviolet enhanced reactivation" (UVR) has been recommended to describe this phenomenon in mammalian cells since the molecular events that produce this reactivation may not be identical to those that produce Weigle reactivation in bacteria.

The implications of the phenomenon to the study of mutation, virus activation and carcinogenesis, have been recently stated by Devoret (1978) and Witkin (1976), among others.

We have determined the wavelength dependency for ultraviolet enhanced reactivation (UVR) in a mammalian cell-virus system. Substantial amounts of UVR can be demonstrated by irradiating mammalian cells at wavelengths from 238-302 nm. For wavelengths above 310 nm no UVR was evident within the limits of our radiating system (2000 Jm^{-2}) (Coohill, et al. 1978). If a photosensitizer (in our case 8-methoxypsoralen is added to the system, then UVR can be demonstrated for wavelengths as long as 370 nm. Above 370 nm no UVR was observed even in the presence of the photosensitizing drug (James and Coohill, 1979). The wavelengths for the most efficient induction of UVR occurred between 260 and 275 nm with or without an exogenous sensitizer.

UVR also appears to be "age" related. Moore and Coohill (1979) reported that the levels of UVR in mammalian cells increased as the cells "aged" in culture. "Age" being measured here in culture passage number from the host animal. However, although there was a monotonic increase in UVR from passage number 20 to 62, cell cultures at passages beyond 62 rapidly lost their ability to reactivate viruses.

Finally, the literature is contradictory with regard to the mutagenic effects of UVR in mammalian cells. Das Gupta and Summers (1978) report that it is, while Lytle, et al. (1980) report that it is not. We have attempted to demonstrate mutagenicity with UVR in our cell virus system but have not been able to show that UV irradiated cells are any more proficient in producing viral mutants than are unirradiated cells. This important point is in doubt. Weigle reactivation in bacteria is highly mutagenic; UVR in mammalian cells appears to be a similar phenomenon but may not be mutagenic. In fact, the proof of the existence of error prone repair processes in mammalian cells may await the outcome of this controversy.

It should be remembered that:

(1) essentially all living cells have their attendent viruses (A. Eisenstark, personnal communication). In addition, in the natural setting, both the viruses and their host cells will be exposed to radiation; thus, UVR and WR should be prevalent in most living systems.

(2) the reported studies show that UVR and WR occur at very low exposures to ultraviolet radiation (when compared to the exposures required to inactivate cells). This includes radiation in the UV-B region.

(3) whether UVR is identical to WR awaits further study. However, either phenomenon is of general interest in that they are important cellular responses to ultraviolet radiation damage.

REFERENCES

Bockstahler, L. E. and C. D. Lytle. 1970. UV enhanced reactivation of a mammalian virus. Biochem.Biophys. Res. Commun. 41:184-189.

Coohill, T. P., L. C. James and S. P. Moore. 1978. The wavelength dependence of ultraviolet enhanced reactivation in a mammalian cell-virus system. Photochem. Photobiol. 27:725-730.

Das Gupta, U. B. and W. C. Summers. 1978. UV reactivation of herpes simples virus is mutagenic and inducible in mammalian cells. Proc. Natl. Acad. Sci.USA 75:2378-2381.

Devoret, R. 1978. Inducible error prone repair: one of the cellular responses to DNA damage. Biochimie 60: 1135-1140.

Gudas, L. J. and A. B. Pardee. 1975. Model for regulation of E. coli DNA repair functions. Proc Natl. Acad. Sci. USA 72:2330-2334.

James, L. C. and T. P. Coohill. 1979. The wavelength dependence of 8-methoxypsoralen photosensitization of radiation enhanced reactivation in a mammalian cell-virus system. Mutat. Res. 62:407-415.

Lytle, C. D., J. G. Goddard and Chen-ho-lin. 1980. Repair and mutagenesis of herpes simplex virus in UV radiated monkey cells. Mutat. Res.(In press).

Moore, S. P. and T. P. Coohill. 1979. An effect of cell culture passage on ultraviolet enhanced viral reactivation by mammalian cells. Mutat. Res. 62: 417-423.

Radman, M. 1975. Molecular Mechanisms for Repair of DNA [ed. P. C. Hanawalt and R. B. Setlow], Plenum Press, part A, 355-368.

Rupert, C. S. and W. Harm. 1966. Advances in Radiation Biology, [ed. L. G. Augenstein and R. Maston]. 2, 1-81. Academic Press, N.Y.

Weigle, J.J. 1953. Induction of mutations in a bacterial virus. Proc. Natl. Acad. Sci. USA 39:628-636.

Witkin, E. M. 1976. Ultraviolet mutagenesis and inducible DNA repair in Escherichia coli. Bacteriol. Rev. 40: 869-907.

SENSITIVITY TO UV-B IRRADIATION AS RELATED TO BACTERIAL LIFE CYCLES

A. Eisenstark

Division of Biological Sciences
University of Missouri
Columbia, Missouri 65211

Although their life cycles may be relatively simple when compared to eukaryotic organisms, bacteria do have distinct stages of growth. Indeed, some bacterial (prokaryote) species have complex developmental life cycles (spore former, myxobacteria, cytophaga). However, even simple bacteria such as *Escherichia coli* have sharply different sensitivities to UV-B when they are in different stages of growth (Eisenstark 1970). The various stages in a bacterial growth cycle can be illustrated by the states of its chromosomal replication and cell division (Figure 1). In a resting state, the bacterial cell has a single chromosome, double stranded and circular, without any growing forks. The first stage of cell division is the initiation of DNA replication and the development of a chromosomal growing fork. DNA synthesis zippers along the chromosome until the replication cycle ends with two circular chromosomes, the formation of a cell wall partition and binary fission of the cell into two daughter cells. While this appears to be very simple, the total process is complex; it is highly orchestrated with a number of regulatory programs. In particular, cell division is coupled to chromosome replication. It is important to emphasize that there are vast differences in UV-B sensitivity and repair of UV-B damage depending on the stage of the cycle (Figure 2).

We have taken advantage of the different stages of UV-B susceptibility in designing experiments that will give us new information on DNA repair mechanisms

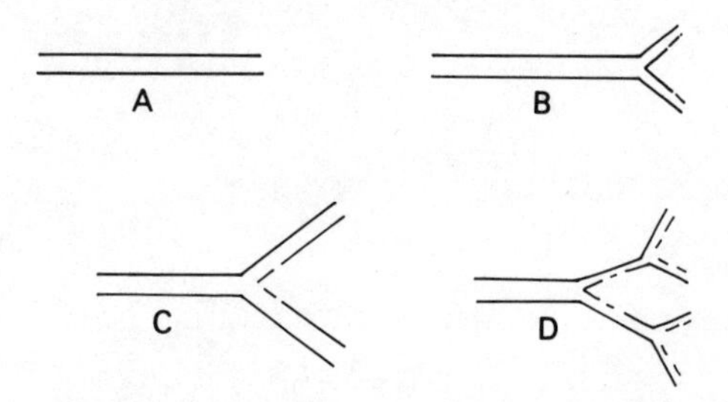

Figure 1. Bacterial chromosome in various stages of growth. (A) Stationary phase cell, without any DNA growing forks; (B) Initiation of DNA replication; (C) DNA in exponential phase cells; (D) In cells that are growing very rapidly, there may be multiple growing forks, and thus cells are more sensitive to damage.

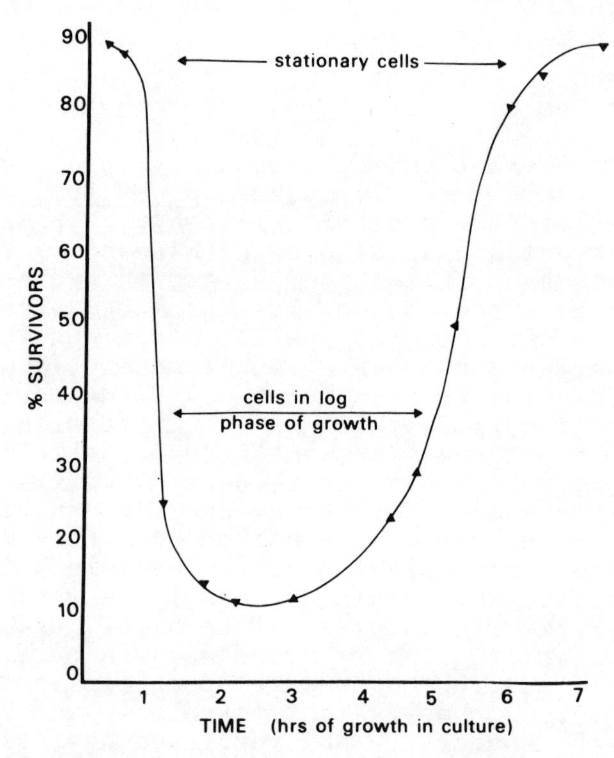

Figure 2. Differences in UV-B sensitivity during different stages of growth. Samples from a growing culture of Salmonella typhimurium were irradiated with a constant dose.

(Hartman and Eisenstark 1978). In particular, we have found that only at certain stages of growth does the *recA*+ product (a particular repair enzyme) operate to repair DNA damage (Figure 2). This knowledge of when in the growth cycle the *recA*+ repair is operative permits us to study some specific molecular events in UV-B damage and repair. In particular, we have recently assessed the role of double strand DNA breaks in the absence of recombinational (*recA*+) repair; we were able to determine this only because of the knowledge that *recA*+ enzyme operates in growth phase (but not stationary phase) cells (Hartman and Eisenstark 1980).

In summary, the state of the bacterial chromosome (whether it is in a resting state or is in a DNA replicative state) is extremely important with regard to sensitivity to UV-B. Whether one can extrapolate this information to higher organisms, of course, must be examined carefully. No doubt higher organisms have other factors influencing sensitivity in various stages of the life cycle, but the stage of chromosome replication could still be a major factor.

REFERENCES

Eisenstark, A. 1970. Sensitivity of *Salmonella typhimurium* recombinationless (*rec*) mutants to visible and near-visible light. *Mutat. Res.* *10*: 1-6.

Hartman, P. S. and A. Eisenstark. 1978. Synergistic killing of *Escherichia coli* by near-UV radiation and hydrogen peroxide: Distinction between *recA*-repairable and *rec-A*-nonrepairable damage. *J. of Bacteriol*:*133*: 769-774.

Hartman, P. S. and A. Eisenstark. 1980. Killing of *Escherichia coli* K-12 by near-ultraviolet radiation in the presence of hydrogen peroxide: Role of double-strand DNA breaks in absence of recombinational repair. *Mutat. Res.* *72*: 31-42.

PIGMENT DISPERSION BY LIGHT IN THE MELANOPHORES OF THE FIDDLER CRAB

Thomas P. Coohill

Biophysics Program, Western Kentucky University, Bowling Green, Kentucky 42101

Adaptive color changes in animals have been the subject of a large number of investigations. One genus that has been extensively studied is the fiddler crab, Uca (Fingerman, 1968). The effects of light, temperature, and hormones on the chromatophores of this animal have been successfully described and analyzed. However, incident radiation can elicit chromatophore responses in Uca in two ways (Brown and Sandeen, 1948). One is the direct action of light on the chromatophores (primary response). The other effect (secondary response) is an indirect one caused by light entering the eyes of the animal and ultimately resulting in secretion of neurohormones that affect the crab's chromatophores. Background responses of Uca are typical secondary responses (Brown and Hines 1952).

PRIMARY (Direct) RESPONSE

First I will discuss our results involving the primary (direct) response (Coohill et al. 1970). The direct responses of chromatophores to light stimuli have been observed in a wide diversity of organisms (Fingerman, 1968; Van der Lek, 1967). With very few exceptions, pigment dispersion occurs in bright light (visible and ultraviolet) and pigment concentration in darkness. For example, when melanophores of eyestalkless fiddler crabs Uca pugilator, were exposed to sunlight, pigment dispersion occurred as a direct response to the illumination (Brown and Sandeen 1948). However, the relative spectral efficiency of this response was not known.

Our objective was to compare the relative efficiencies of ultraviolet and visible light in inducing direct melanin dispersion in the fiddler crab, U. pugilator. Because we could not remove intact melanophores from the crab, we decided to determine the relative efficiencies of three broad spectral ranges of light in melanophores *in situ*.

For our light studies three different light sources were used. They were a GE15T8-15 watt Germicidal lamp (far ultraviolet) primary (86%) output at 254 mm, a GEF15T8-15 watt Blacklight (BLB) lamp (near ultraviolet), and a GEF15-CW-15 watt Cool White lamp (visible) 370-730 mm. The intensities of both UV (ultraviolet) lamps were measured with a Blak-Ray Ultraviolet intensity meter (Ultraviolet Products, San Gabriel, Calif.). The J-225 shortwave cell was used for the far-UV and the J-221 long-wave cell for the near-UV. Visible light was measured with a General Electric PR-2 exposure meter, and the readings converted to microwatts per square centimeter, using the known intensities at various band widths (General Electric Co., personal communication). The animals were irradiated while mounted on sponges. The eyestalks were removed before irradiation to eliminate the secondary (hormonal) response.

Our results showed that, near-UV light was most efficient in eliciting direct dispersion of the melanophoric pigment (Coohill, et al. 1970). If one corrects for absorption of light by the exoskeleton, near-UV was more than 200 times as effective as visible light in causing pigment dispersion. Far-UV light had no effect on pigment dispersion. We made no attempt to correct for absorption due to the epidermal layer between the cuticle and the melanophores. To do this would require knowing the exact position of each melanophores and all of its processes, which is not practical for melanophores *in situ*.

Because the sun is a good source of near-UV light (about 10% of its total energy (Koller 1965) it is interesting to compare the doses necessary to drive the melanophores to dispersion with the near-UV light source used by us with sunlight. In the former case, using our data one can estimate that about $5.2 \times 10^5 \mu w$ per square centimeter were needed. With sunlight approximately 4.5×10^6 μw per square centimeter are in the near-UV radiation range. Even though these calculations are approximate (e.g., estimating the response of the meter), it would appear that there is enough near-UV radiation in sunlight to elicit responses of the same magnitude as we have obtained with our near-UV radiation lamp. Therefore, it would appear that the near-UV radiation component of sunlight is of primary importance in eliciting the direct melanin-dispersing response. We could conclude from our data that melanophores of *Uca pugilator* respond most efficiently to light in the 300-400 nm range.

SECONDARY (Hormonal) RESPONSE

Our results involving the secondary (hormonal) response involved experiments similar to those mentioned above (Coohill and Fingerman, 1975). However the eyestalks were not removed from the animal and the chromatophores were shielded from the light source. Hence only the response due to eyestalk stimulation was measured. Our results showed that the secondary response of these melanophores is similar for each of the three spectral regions tested. Small differences in the initial and final responses were apparent, but no overall qualitative difference was noticeable. Low doses (5,000 μw/cm^2) of near UV elicited the strongest initial response, but as the dose was raised above this value no consistent difference was apparent. Thus, the broad spectral sensitivity of the chromophore(s) involved in eliciting the secondary response is not the same as that involved in the primary response.

It is not possible to explain the similar secondary response to all spectral regions by the presence of small components of light from other spectral regions present in each light source. Although the strongest initial response to low levels of radiation (5,000 μw/cm^2) is the near-UV regions, the other two sources emit only 3%-4% of their radiation in this region. This is a factor of 25x-33x below that produced by the near-UV source. For example, if only near UV were causing the secondary response, then a total dose of 150,000 μw/cm^2 of white light should give the same melanophore response at 4,500 μw/cm^2 (3%) of near-UV light. However, the response in the former case was twice as high as in the latter case.

In addition, reciprocity of time and dose rate did not hold for the secondary response under the current conditions. This implies that either the number of chromophores available to react with incident photons is limited for short time periods, or that the ability to produce and release melanin-dispersing hormone requires the continued presence of incident radiation. Reciprocity did hold for the primary response under similar conditions. No attempt was made to determine the absorption of light by the tissues of the eyestalk. It is possible that the radiation reaching the chromophores involved in the secondary response are differentially absorbed by the intervening tissue. Since the exact location of these chromophores is not known, any correction for this effect would be questionable.

These results suggest that in eyed fiddler crabs bright illumination causes the release of some melanin-dispersing hormone in addition to the amount of this hormone that is normally released in response to the shade of background and as a consequence of the circadian and circatidal rhythms of color change of these crabs. This response of the eyed crabs to the increased illumination was probably a response to the brightness of the illumination incident on the eyes directly from the source of illumination. An exposure of approximately 100 times more illumination is required to produce the same response in eyestalkless crabs than in intact crabs (Coohill and Fingerman, 1976). Even with an unchanging albedo, melanin dispersion increased in eyed crabs with increasing illumination when the exposure of illumination was too weak to evoke a primary response of the melanophores of *Uca pugilator* which has a much higher threshold than does the secondary response. The fact that these intact crabs still exhibit melanin dispersion when exposed to illumination while being held against a white background which fosters melanin concentration is further support for concluding there is a brightness component which is not part of the response to the albedo. Small changes in either the length of the staging period or the amount of illumination used can have a profound effect on the results obtained. These observations, and the fact that cytochalasin B inhibits both the primary and secondary responses (Coohill and Fingerman, 1976), suggest that the same pigment-dispersing mechanism (perhaps involving microfilaments) is responsible for pigment dispersion in the melanophores regardless of whether the dispersion is induced by a blood-borne factor or by direct irradiation of the pigment cells.

In summary:

(1) the direct response is most sensitive to near UV light (300-400mm).

(2) the hormonal response is more sensitive by about 100 fold than is the direct response.

(3) the hormonal response is approximately the same for each of the three wavelength regions tested.

(4) the method of presenting the radiation had a significant effect on the melanophore response. Animals exposed to the same total amount of light spread out over a longer period of time had a greater response. That is, reciprocity of time and dose rate does not hold for the hormonal response.

REFERENCES

Brown, F. A. and M. I. Sandeen. 1948. Responses of the chromatophores of the fiddler crab, Uca, to light and temperatures. Physiol. Zool. 21:361-371.

Brown, F. A. and M. N. Hines. 1952. Modifications in the diurnal pigmentary rhythm of Uca affected by continuous illumination. Physiol. Zool. 25:56-70.

Coohill, T.P., C. K. Bartell and M. Fingerman. 1970. Relative effectiveness of UV and visible light in eliciting pigment dispersion directly in melanophores of the fiddler crab, Uca pugilator. Physiol. Zool. 43:232-239.

Coohill, T. P. and M. Fingerman. 1975. Relative effectiveness of UV and visible light in eliciting pigment dispersion in melanophores of the fiddler crab, Uca pugilator, through the secondary response. Physiol. Zool. 48:57-63.

Coohill, T. P. and M. Fingerman. 1976. Comparison of the effects of illumination on the melanophores of intact and eyestalkless fiddler crabs, Uca pugilator, and inhibition of the primary response of cytochalasin B. Experientia 15:569-570.

Fingerman, M. 1968. Crustacean color changes with emphasis on the fiddler crab. Scientia 53

Koller, L.R. 1965. Ultraviolet Radiation. Wiley, New York.

Van der Lek, B. 1967. Photosensitive melanophores. Bronder-Offset, Rotterdam.

YELLOW LENS PIGMENT: AN ADAPTATION FOR ACTIVITY IN BRIGHT SUNLIGHT

Seymour Zigman

Departments of Ophthalmology and Biochemistry
University of Rochester School of Medicine and Dentistry, 601 Elmwood Avenue (Box 314)
Rochester, New York 14642

INTRODUCTION

Anterior eye light filters are present in many vertebrates which modify the quality and quantity of sunlight radiant energy that reaches the site of vision, the retina. In this way, nature seems to have provided a means to protect the retina from an overabundance of short wavelength visible and long wavelength UV light which could interfere with vision and cause damage to it. Any additional light of 300 to 400 nm reaching to the eye could upset a delicate balance that would be detrimental to vision.

Among all of the fishes, there is a strong positive relationship between diurnal activity and the presence of a pigment in the visual system anterior to the retina. Most often this pigment is yellow, and is present mainly in the ocular lens (see Table I). However, some of the fish (i.e. tautog) have yellow corneas instead, while others have colorless lenses, that are still fine absorbers of shortwave visible and longwave UV light. The following are important functions of a yellow lens pigment:

1) Reduce chromatic aberrations;

2) Sharpen vision; provide contrast;

TABLE I

INTRA-OCULAR COLOR FILTERS*
THE HABITS AND LENS PIGMENTATION OF VARIOUS FISHES

ANIMALS	HABITS	LENS PIGMENTS	COMMENTS
1. FISHES			
a) Hagfish	Blind	Colorless	
b) Lampreys			
1) Petromyzontidae	Diurnal	Yellow	
2) Geotriiade	Nocturnal	Colorless	
c) Elasmobranchs			
** 1) Dogfish (Squalus) (Mustelis)	Nocturnal	Colorless	
** 2) Sharks Tiger	Shallow-swimmers mixed diurnal and nocturnal	Yellow	
Nurse	Benthic Nocturnal	Colorless	
Brown	Deep swimmers	Colorless	
Dusky	Deep and Shallow swimmers	Yellow	
3) Skates Rays	Deep and Nocturnal	Colorless	
d) Chondrosteans	Benthic	Colorless	
e) Holosteans	Diurnal	Colorless	
f) Teleosts			
1) Esox	Diurnal	Colorless	
2) Flounder	Benthic	Colorless	
3) Scup	Various	Colorless	
4) Yellow Perch	Shallow swimmers	Yellow	
5) Others	Variable	Mainly Colorless	Yellow Cornea (Cyprinus)

* Gordon Lynn Walls, "The Vertebrate Eye and its Adaptive Radiation", Cranbrook Institute of Science, Bulletin No. 9 (August, 1942)

** Observed by Zigman

3) Reduce glare by scattered light at lower wavelengths;

4) Protect posterior eye from short wavelength photochemical damage.

This communication will describe some features of the lenses of sharks that inhabit the waters of Cape Cod, Massachusetts and the Gulf of Mexico near Sarasota, Florida, some of which have yellow pigmented lenses. Here again, the relationship between the presence of a yellow lens pigment and activity in bright sunlight is strong. Some chemical features of the pigment will also be described.

FINDINGS AND DISCUSSION

The degree of pigmentation of several sharks found in Florida Gulf Coast waters is shown in Figure 1 (Zigman and Gilbert 1978). The most voracious surface or shallow feeder of this series is the tiger shark, an animal with intense yellow pigment in the ocular lens. This filter is useful for improved vision, especially in bright sunlight entering the water from above, and for better contrast between prey and the blue sky background. Contrasting this situation is that of the entirely pigmentless lens of the bottom swimming nurse sharks, which spend much of their time in caves. Such pigment would be totally useless and even detrimental to the vision of these animals, since it would absorb a portion of the already dim shortwave visible light available. The dogfish lens also transmits most of the blue light and even half of the longwave UV light to the retina. These animals are deep swimmers, but seem to feed actively only in the evening. The variability of the spectral transmission of many fish was shown by Kennedy and Milkman (1956).

As in the lenses of higher animals, such as squirrels, monkeys, and humans, lens pigment in sharks is yellow, and exhibits longer wavelengths of fluorescence as compared to those of tryptophan, another abundant constituent of ocular lenses both in peptide linkage and free form. Table II provides data on the fluorescences and absorptions of the constituents of tiger shark lenses. A comparison of the degree of fluorescence in several sharks lenses is made in Table III. The findings are that the fluorescent material

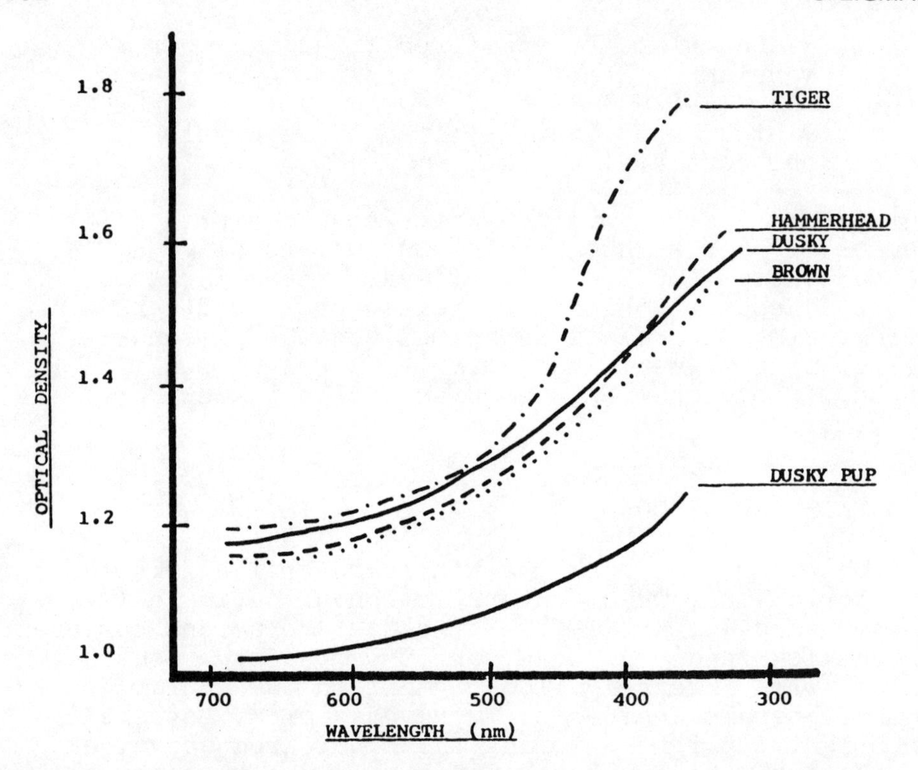

Figure 1. Absorption spectra of the lenses of Florida Gulf Coast sharks. The OD at 400 nm represents the yellow pigment. Enhancement of absorbance of the young dusky shark lens to the level of that of the adult represents increased thickness of the lens mainly and some slight increase in pigmentation. (data from Zigman and Gilbert 1978).

is present in the non-protein and protein phases extracted from these lenses. In dogfish lens, hardly any of the fluorescence is present other than in the insoluble or aggregated membrane-bound protein. In tiger sharks, there is greater fluorescence in the non-protein phase than in the others, but again, the major fluorescence is in the insoluble fraction.

While the nature of the pigment is not yet known, it is thought to represent a series of tryptophan oxidative products resulting from metabolism and photochemistry. It is of great interest however, that the

TABLE II
FLUORESCENCES OF TIGER SHARK LENSES*

Fraction	Absorption	Fluorescence	
		Excitation	Emission
Non Protein	265 nm	290 nm	340 nm
		340 nm	400 nm
All Protein Fractions**			
TSP	275 nm	290 nm	328 nm
INS			
HMW	282 nm	320 nm	395 nm

* Measured at neutral pH using a Perkin Elmer Spectrofluorometer.
** TSP = Total soluble protein. INS = Insoluble (in water) protein.
HMW = High molecular weight colloidal protein.

(see reference Zigman and Yulo (1979) for details of tissue fractionation).

TABLE III
NON-TRYPTOPHAN FLUORESCENCES OF LENS PROTEIN FRACTIONS

Species	NON-TRYPTOPHAN (360/440 nm) FLUORESCENCE*			
	Non-Protein	Soluble	High Mol.	Insoluble
Dogfish (Woods Hole) (*Mustelus canis*)	3	5	6	28
Brown Shark (Sarasota) (*Carcharhines milberti*)	15	17	8	40
Tiger Shark (Sarasota) (*Galeocerdo cuvieri*)	20	50	10	160
Human (Rochester) (*Homo sapiens*)	70	70	60	500

* Relative Fluorescence per mg protein. (See reference Zigman and Yulo (1979) for details of tissue fractionation).

yellow lens pigments observed are quite similar chemically to those of higher animals.

CONCLUSIONS

Consideration of the above-detailed information leads to the conclusion that the pigment present in the ocular lens of marine vertebrates represents an adaptation to diurnal activity. This pigment diminishes the short wavelength visible and long wavelength UV light of sunlight that reaches the retina. The benefits apply both to the improvement of vision and the protection of the retina from the lower damage thresholds for short wavelength light (see Ham et al. 1978). However, the yellow pigment is photosensitive and darkens when exposed to long wavelength UV light. This cumulative process eventually does lead to diminished visual acuity, and the chemical consequences can lead to brown cataracts (see Zigman and Gilbert 1978).

The question of whether a thinner ozone layer would present a challenge to the visual system due to the additional radiant energy in the 300 nm to 450 nm wavelength range that enters the eye remains to be answered.

REFERENCES

Ham, W. T., Jr., H. A. Mueller, J. J. Ruffolo, Jr. and A. M. Clarke. 1978. Sensitivity of the retina to radiation damage as a function of wavelength. _Photochem_. _Photobiol_. _29_: 735-744.

Kennedy, D. and R. D. Milkman. 1956. Selective light absorption by the lens of lower vertebrates and its influence on spectral sensitivity. _Biol_. _Bull_. _111_: 375-386.

Zigman, S. and P. W. Gilbert. 1978. Lens colour in sharks. _Exp_. _Eye_ _Res_. _26_: 227-231.

Zigman, S. and T. Yulo. 1979. Eye lens ageing in the dogfish (_Mustelus_ _Canis_). _Comp_. _Biochem_. _and_ _Physiol_. _63B_: 379-385.

PHOTOMOVEMENT OF AQUATIC ORGANISMS IN RESPONSE TO SOLAR UV

Jeanne A. Barcelo

University of Kentucky
School of Biological Sciences
Lexington, Kentucky

Photomovement in response to sunlight has been reported in many species of planktonic organisms. The literature includes numerous studies of vertical migration patterns of planktonic species, but relatively few studies examine the role of solar UV in influencing these movements. Studies by Cowles and Brambel (1936), Hasle (1950, 1954) and Barcelo and Calkins (1979) suggest that UV light may influence movements during the day, particularly avoidance of the surface at midday. These daily movement patterns were observed to change in varying climatic conditions e.g. the amount of cloudcover, intensity of irradiance, etc. changed the distribution pattern (Hasle 1950, 1954; Eppley et al. 1968; Goff 1974). Other examples of this phenomenon were reported by Cowles and Brambel (1936) in which they observed that *Gonyostomum* were distributed lower in a pond on bright clear mornings than on other somewhat less bright mornings, while on cloudy mornings the maximum counts were at the surface. Agamaliev and Bagirov (1975) also observed a similar pattern by ciliates in the Caspian Sea. Barcelo and Calkins (1979) examined various external and internal environmental factors (i.e. light, temperature, wind, etc.) to determine if it is indeed sunlight, solar UV, or some other factor which is most important in influencing the vertical distribution of certain aquatic species. In this particular study the distribution of the ciliate, *Coleps*, was examined and low correlation was found between the location of the animals and wind, temperature gradients, food gradients, etc. On the other hand, irradiance factors (both total solar irradiance and the UV-B component of sunlight) showed high correlations with the location of the protozoan, in good agreement with the aforementioned studies.

Because responses to UV light are difficult to separate from responses to other components of sunlight in natural settings, direct methods of examining behavioral responses to UV have been utilized in various laboratories. As early as 1926, Inman et al. demonstrated that an amoeba will withdraw its pseudopodia and immediately reverse direction when entering an area illuminated by ultraviolet light. Giese (1953, 1967) notes that protozoa will accumulate under a screen opaque to UV light when exposed to these deleterious wavelengths. He postulates that this negative reaction to UV might serve to protect them in the wild. Positive and negative phototaxis of the alga _Platymonas subcordiformis_ has been demonstrated using various UV wavelengths and intensities (Halldal 1961). Also Kimeldorf and Fontanini (1974) have shown that the amphibian, _Taricha granulosa_, can detect an exposure to near-ultraviolet radiation and will subsequently move to a shielded location. More recently Barcelo and Calkins (1978) and Barcelo (1980) have shown that various species of protozoans, crustaceans and planarians avoid UV-B exposure in correlation to their tolerance to these wavelengths (i.e. sensitive organisms avoid exposure by moving into sheltered areas whereas more resistant species show less pronounced avoidance). Furthermore, in the case of _Cyclops_, a crustacean, the rate of movement out of UV into a protected location was shown to be dose-dependent (Barcelo and Calkins, 1980).

The above studies suggest that UV-B is an important environmental factor since avoidance behavior is highly correlated with UV-B tolerance. They also indicate that these organisms must possess relatively complex sensing and analytical capabilities in order to sense exposure, find shelter, and move to shielded locations fast enough to avoid injury. These complex behavior mechanisms may be very important in allowing sensitive species to survive in sunlight exposed niches in nature.

REFERENCES

Agameliev, F.G. and R.M. Bagirov. 1975. Diurnal vertical migrations of microbenthic, planktonic and periphytonic ciliates of the Caspian Sea. _Acta Protozoologica_ _14_: 195.

Barcelo, J.A. 1980. Photomovement, pigmentation and UV-B sensitivity in planaria. Photochem. Photobiol. 32: 107.

Barcelo, J.A. and J. Calkins. 1978. Positioning of aquatic microorganisms in response to visible light and simulated solar UV-B irradiation. Photochem. Photobiol. 29: 75.

Barcelo, J.A. and J. Calkins. 1979. The relative importance of various environmental factors on the vertical distribution of the aquatic protozoan *Coleps spiralis*. Photochem. Photobiol. 31: 67.

Barcelo, J. A. and J. Calkins. 1980. The kinetics of avoidance of simulated solar UV radiation by two arthropods. Biophys. J. 32: 921.

Cowles, R.P. and C.E. Brambel. 1936. A study of the environmental conditions in a bog pond with special reference to the diurnal vertical distribution of *Gonyostomum semen*. Biol. Bull. 71: 286.

Eppley, R.W., O. Holm-Hansen and J.D.H. Strickland. 1968. Some observations on the vertical migration of dinoflagellates. J. Phycol. 4: 333.

Giese, A.C. 1953. Protozoa in photobiological research. Physiol. Zoo. 26:1.

Giese, A.C. 1967. Effects of radiation upon protozoa. Research in Protozoology [T.T. Chen, ed.] 2:267.

Goff, N.M. 1974. Observation on the vertical migration of *Gymnodinium nelsoni* Martin in an estuarine environment, Masters thesis, University of Maryland, College Park, Md.

Halldal, P. 1961. Ultraviolet action spectra of positive and negative phototaxis in *Platymonas subcordiformis*. Physiol. Plant. 14: 133.

Hasle, G. R. 1950. Phototactic vertical migrations in marine dinoflagellates. Oikos 2: 162.

Hasle, G.R. 1954. More on phototactic diurnal migration in marine dinoflagellates. Nytt. Mag. Bot. 2: 139.

Inman, O.L., W.T. Bovie and C.E. Barr. 1926. The reversal of physiological dominance in amoeba by ultraviolet light. J. Exp. Zool. 43: 475.

Kimeldorf, D.J. and D.F. Fontanini 1974. Avoidance of near-ultraviolet radiation exposures by amphibious vertebrae. Environ. Physiol. Biochem. 4: 40.

MOVEMENT REACTIONS OF BLUE-GREEN ALGAE IN RESPONSE TO THEIR PHOTOENVIRONMENT

Donat-Peter Häder

Philipps-Universität Marburg, Fachbereich Biologie-Botanik, Lahnberge D-3550 FRG

ABSTRACT

The photophobic response is the basic mechanism for the orientation of blue-green algae in a photoenvironment. The photoreceptor system is identical with the photosynthetic pigments, and the sensory transduction between photoreceptor and motor apparatus is mediated by light-induced potential changes.

A computer model has been developed to simulate the photophobic behavior under various conditions.

Orientation of microorganisms to light stimuli can be achieved by three basic mechanisms: 1) phototaxis, which is a directed movement with respect to the direction of the impinging light rays, 2) photokinesis, which is the dependence of velocity on the quantum flux density, and 3) the photophobic response which is a transient reaction caused by a sudden change in light intensity. Though the first two mechanisms have been described for blue-green algae (see reviews by Häder 1979 b; Nultsch and Häder 1980), the photophobic response is the main reaction of Cyanophyceae to actively search for suitable light conditions.

In the filamentous blue-green alga Phormidium uncinatum a photophobic response usually consists of a sudden reversal of movement caused by a decrease in the quantum flux density. These organisms populate the bottom of both marine and freshwater habits and since their main energy source is photosynthesis, they depend on a sufficient irradiation. The strategy to avoid obstruction of the solar irradiation by particles or other organisms in the body of water above them is based on a trial-and-error mechanism: The organisms continuously glide over the substrate until the front end encounters a decrease in light intensity, which may be caused by a shadow cast by an object above the filaments, upon which they reverse the direction of movement. A difference in light intensity as low as 4% is sufficient to trigger a response (Häder 1973).

The action spectrum indicates the participation of chlorophyll a and the accessory pigments C-phycoerythrin and C-phycocyanin in photoreception. Furthermore, the action spectrum extends well into the near UV caused by pigments not yet identified. The sensor for the primary photoevent has been identified as plastoquinone, a redox component in the linear electron transport chain between photosystem II and photosystem I. The absorption maxima of plastoquinone are 295 nm in its reduced form and 255 nm in its oxidised form (Amesz 1973). According to the electron pool hypothesis (Häder 1975, 1976) phobic responses are triggered by a sudden decrease of the flux through the pool. Plastiquinone is involved in generating an electrochemical gradient across the thylakoid membrane (Mitchell 1978). Sudden changes in the light intensity result in a membrane potential change, which can be detected by both external and internal electrodes (Häder 1978 a, b, 1979). These light-induced electrical potential changes are believed to be the means for sensory transduction in blue-green algae (Nultsch and Häder 1980).

Recently a computer model has been developed to simulate photophobic responses in blue-green filaments under various light conditions. In the model each cell is treated as a compartment which shows a distinct electrical potential. The total potential consists of a dark resting potential, a light-induced potential and a phobic potential caused by a stimulation. It is constantly modified by the changing light conditions in each cell and by the currents into or out of the neighboring compartments. The model assumes an electrical potential gradient between front and rear end. The resulting partial gradient across a single cell determines the direction of the motor force.

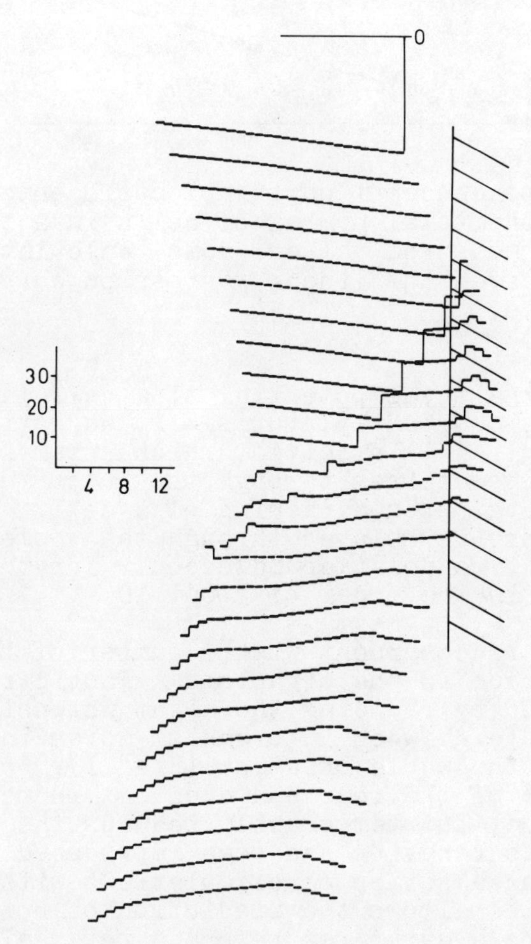

Figure 1. Computer simulation of a photophobic reversal at a light-dark border. Units on x-axis correspond to cell length. Y-axis, potential in mV. The zero potential is given by the horizontal bar in the top graph. The trichome's position and potential are drawn every third cycle with a vertical offset of 10 mV.

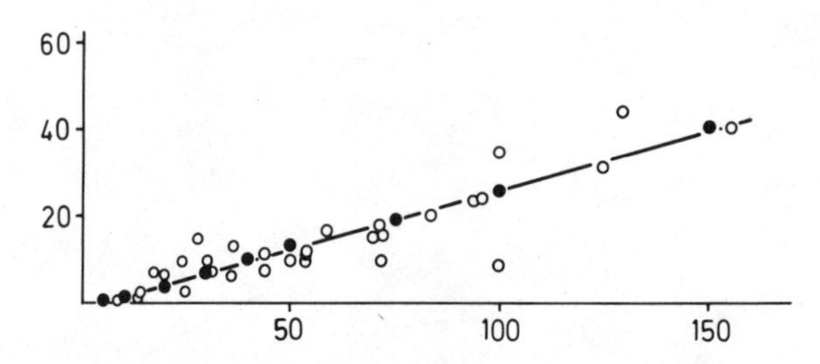

Figure 2. Dependence of number of cells entering the dark area on the total number of cells in a trichome. Open circles, measured values; dots, calculated values. The line indicates the linear regression curve for the measured values.

Whenever the total potential of a cell falls below the level of a constantly, but slowly adjusting sensor a phobic potential is elicited, which assumes the influx of positive charges through the outer membrane; this happens e.g. upon darkening of a cell. This primary response is relayed through the whole trichome. The resulting time delay could account for the observed response time in the order of about 10 s.

The model is congruent with a number of behavior patterns observed in the organisms. Phobic responses can be elicited by reducing the light intensity over the front end (e.g. when an organism moves into a dark area) or over the whole organism (Fig. 1). A shadow over up to 60% of the rear end does not induce a photophobic response, but temporarily reduces the speed of movement. This behavior has been reproduced by the model. Running a series of calculations with various filament lengths allows the prediction of how far an organism enters a dark area before a reversal of movement occurs (Fig. 2). A comparison of the predicted data with the observed values using a time lapse microvideographic system showed a congruence within about 1%.

The model can be used as well to simulate the situation in a light trap. This population method has been often used in analysis of photomovement (Nultsch 1975). A light field is focused into a homogeneous suspension of organisms. Filaments enter the light trap either by their random movement or by positive

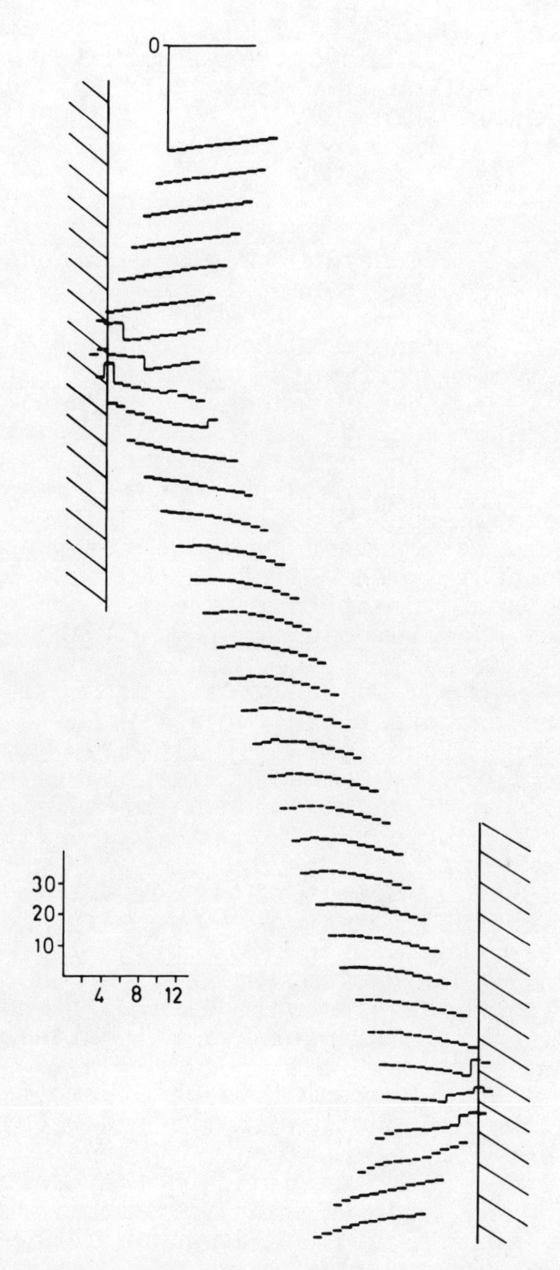

Figure 3. Computer simulation of repetitive photo-phobic reversal of a filament 12 cells long in a light trap 40 cells wide. Details as in Fig. 1.

phototaxis toward the light scattered from particles inside the field (Burkart and Häder, in press). Each time an alga tries to leave the illuminated area it reverses its direction of movement and is thus trapped in the light field (Fig. 3).

REFERENCES

Amesz, J. 1973. The function of plastoquinone in photosynthetic electron transport. Biochim. Biophys. Acta 301: 35-51.

Burkart, U., D.-P. Häder. 1980. Phototactic attraction in light trap experiments. J. Math. Biol. 10: 257-169

Häder, D.-P. 1973. Untersuchungen zur Photo-phobotaxis bei Phormidium uncinatum. Thesis Marburg.

Häder, D.-P. 1975. The effect of inhibitors on the electron flow triggering photophobic reactions in Cyanophyceae. Arch. Microbiol. 103: 169-174.

Häder, D.-P. 1976. Further evidence for the electron pool hypothesis. The effect of KCN and DSPD on the photophobic reaction in the filamentous blue-green alga Phormidium uncinatum. Arch. Microbiol. 110: 301-303.

Häder, D.-P. 1978. a. Evidence of electrical potential changes in photophobically reacting blue-green algae. Arch. Microbiol. 118: 115-119.

Häder, D.-P. 1978. b. Extracellular and intracellular determination of light-induced potential changes during photophobic reactions in blue-green algae. Arch. Microbiol. 119:75-79.

Häder, D.-P.1979 a. Effect of inhibitors and uncouplers on light-induced potential changes triggering photophobic responses. Arch. Microbiol. 120: 57-60.

Häder, D.-P. 1979 b. Photomovement. In: Haupt, W., M. Feinleib [eds.] Encyclopedia of plant physiology. N. S. VII Springer Berlin, Heidelberg: 268-309.

Mitchell, P. 1978. Protonmotive chemiosmotic mechanisms in oxidative and photosynthetic phosphorylation. TIBS 3: N58-N61.

Nultsch, W. 1975. Phototaxis and photokinesis. In Carlile, M. J. [ed.] Primitive sensory communication systems. Acad. Press, London, New York, San Francisco: 29-90.

Nultsch, W., D. -P. Häder. 1980. Light perception and sensory transduction in photosynthetic prokaryotes. In: Hemmerich, P. [ed.] Structure and bonding. Springer, Berlin, Heidelberg, New York: 111-139.

SHORT-TERM RESPONSES OF SOME PLANKTONIC CRUSTACEA EXPOSED TO ENHANCED UV-B RADIATION

David M. Damkaer (1),(2) and Douglas B. Dey (2)

University of Washington WB-10, Seattle,
WA 98195 (1) and
National Marine Fisheries Service/NOAA
Manchester, WA 98353 (2)

ABSTRACT

With shrimp and crab larvae, several-day exposures to UV-B radiation below lethal threshold levels of dose-rate and total dose had no significant effects on either activity or development (molting). Above those UV levels, activity, development, and survival rapidly declined. The specimens in those experiments were held in flow-through seawater containers of 8 cm depth. Perhaps these near-surface zooplankton in nature could increase their depth slightly, if they could sense fatal dose-rates or doses of UV-B. With such a short-term response they might completely avoid damaging solar UV.

Experiments in paired 2-liter glass cylinders (38 cm water depth) suggested that there is no difference in behavior between irradiated and non-irradiated shrimp larvae (*Pandalus platyceros*) or copepods (*Epilabidocera longipedata*). The irradiated specimens maintained their near-surface positions (as did the controls) until the time of decreased activity and death at about 4 days. The progress and outcome were similar in experiments using a 44-liter plastic cylinder (140 cm water depth).

The zoea stage of the shore crab *Hemigrapsus nudus* is extremely attracted to strong visible light. No differences in response-time (seconds) was noted between control larvae and larvae receiving lethal UV-B doses until decreased activity and death of the irradiated larvae at about 10 days. There was no apparent reluctance of the crab larvae to swim toward and to hold themselves within lethal doses of UV-B.

Investigations of the biological effects of UV-B radiation on aquatic organisms have included a variety of approaches, ranging from histological analyses to simple survival studies. Most of this work has been concerned with the direct effects of natural or enhanced UV-B radiation. The assumption implicit in some of the conclusions is that the organisms which live in the near-surface layer will be passive recipients of ambient UV-B radiation, regardless of its intensity.

Because some of these near-surface organisms appear to be living near their UV tolerance limits (Damkaer et al., 1980), it is possible they have evolved a sensitivity to fluctuations in UV-B radiation. In addition to photorepair mechanisms, they may possess an ability to avoid harmful levels of UV-B by appropriate defensive behavior. In moderately productive ocean waters (0.5 mg Chl a · m^{-3}), the DNA-weighted irradiance at 1 m is only about 40% of incident (Smith and Baker, 1979). Under these conditions, the capacity to sense harmful UV-B and to avoid it by simply swimming or sinking a meter below the surface would provide considerable protection.

Only recently have attempts been made to determine the behavioral response of aquatic organisms to UV-B. For small, primarily benthic crustaceans, there is some evidence that correlates horizontal positioning behavior with resistance to UV-B (Barcelo and Calkins, 1979). That is, sensitive organisms (shown by low survival) tended to avoid UV radiation more than did the more tolerant organisms.

However, even if sensory mechanisms exist to warn an organism of dangerous UV-B radiation, the particular stimulus which elicits the avoidance response may not be effective under a new spectral distribution caused by ozone depletion. With a reduction in the ozone layer the UV-B band would increase disproportionately to UV-A and visible light, and an avoidance response linked to high levels of visible irradiance, for example, might not be activated quickly enough. It seems likely that only a sensory system which directly detects irradiance in the UV-B range and does so before irreparable damage occurs will afford a surface-living organism a safety potential.

The present short-term response experiments supplement the previous observations of UV-effects on shrimp and crab larvae (Damkaer et al., 1980). The 8 cm depth of the flow-through water-table containers could have restricted vertical movement if test organisms attempted to avoid exposure to the UV-B irradiation. However, examination of the containers during exposures always showed the healthy shrimp and crab larvae near the surface until UV-B damage (when it occurred) curtailed their activity. Afforded a greater vertical range in the present study, these larval forms still behaved in a similar fashion.

METHODS

Experiments and measurements were conducted at the National Marine Fisheries Service/NOAA Experimental Laboratory at Clam Bay, Manchester, Washington, directly across Puget Sound from Seattle. Seawater for all experiments was pumped directly from the bay, sand-filtered, and partially sterilized with germicidal UV. Simulation of ambient and enhanced solar irradiance, measurements of spectral irradiance, and the water tables used in response-time studies were as described by Damkaer et al. (1980). Control organisms in all experiments were irradiated by one FS-40 sunlamp and one cool-white lamp, both filtered by a clear Mylar® plastic sheet (0.25 mm thickness). The duration of all light exposures was 3 h each day, centered around solar noon.

Adult copepods <u>Epilabidocera longipedata</u> were collected at midday from surface swarms in Clam Bay. Egg-bearing females of the shore crab <u>Hemigrapsus nudus</u> and the shrimp <u>Pandalus platyceros</u> were held in tanks until the eggs hatched. All shrimp and crab larvae were less than 24 h old at the beginning of experiments.

One set of experiments consisted of paired 2-liter glass cylinders each filled to a depth of 38 cm and containing 10 specimens. The cylinders were nearly submerged in an aquarium with continuously flowing seawater. The bottom, sides, and top of the aquarium were covered with opaque black plastic sheeting, except for an opening directly above each cylinder. The light sources were placed directly above the cylinders. A glass side of the tank could be uncovered at intervals during experiments, permitting direct observation of the cylinders with the test organisms. Visible markings divided the cylinders vertically into thirds, and the number of organisms within each division (upper, middle, lower) was recorded at five times: in relative darkness (before turning on lamps) and at 30, 60, 120, and 180 min after lamps had been turned on. Seawater in the cylinders was changed daily.

A dose-rate of 0.018 $Wm^{-2}_{[DNA]}$ was administered for 3 $h \cdot d^{-1}$ to shrimp larvae (*Pandalus platyceros*) and adult copepods (*Epilabidocera longipedata*) in the 2-liter glass cylinders. This dose-rate is comparable to the noon irradiance of a clear summer day and is well above irradiance levels shrimp larvae are likely to encounter during their season of surface occurrence in spring. Moreover, this dose-rate as well as the daily dose far exceed the threshold values established for shrimp larvae (Damkaer et al., 1981. The threshold values for *Epilabidocera longipedata* have yet to be determined. This copepod is often found at the surface in full spring and summer sunlight.

To further investigate behavioral responses to potentially damaging UV-B and to complement initial findings, a larger cylinder was constructed which allowed even greater freedom of movement of the test organisms. This clear plastic cylinder, 165 cm high and 20 cm in diameter, was filled to a depth of 140 cm, at which level it contained about 44 l of seawater. Because only one large cylinder was available, UV-B treatment and control phases of these experiments were consecutive, with different sets of organisms, rather than concurrent. The cylinder, containing 50 specimens, was in a darkened room where the only source of light was the UV or control lamp-combinations suspended above the opening at the top of the cylinder. The cylinder was marked at depths of 30 cm, 60 cm, and 90 cm, and the vertical distributions of test organisms were recorded before turning on the lamps and at 30, 60, 120, and 180 min after UV exposure began. Seawater in this cylinder was slowly drained and replenished once each day.

We also considered response-time as a possibly sensitive index of exposure to UV radiation. Newly hatched zoea larvae of the shore crab *Hemigrapsus* nudus were placed in 1,000 ml beakers which were held in plastic baskets in the water table. Ten specimens were placed in each of ten beakers, with five beakers receiving UV-B and five beakers as controls receiving no UV-B. Daily 3 h light exposures were administered while the animals were in the water table. Each day specimens were removed from the beakers and placed in a small petri dish (6 cm diameter) which was covered to eliminate light for 5 minutes. In a darkened room, the cover was removed, a 60-W incandescent light 30 cm to one side of the dish was turned on and a stop-watch activated. When 60% of the specimens had actively responded by swimming to the light edge of the dish, the watch was stopped and the elapsed time recorded. The animals were returned to the beakers which had been cleaned and refilled with seawater, and the beakers were again placed in the water-table baskets.

UV-B irradiance (285-315 nm) is reported in DNA-weighted effective doses. These values were calculated using the analytical representation of Green and Miller (1975) which is based on the DNA action spectrum (Setlow, 1974).

RESULTS

Throughout the first day's exposure the control copepods and the copepods receiving a high daily dose of UV-B (194 $Jm^{-2}{}_{[DNA]}$) at a high dose-rate (0.018 $Wm^{-2}{}_{[DNA]}$) tended to swarm at the surface of the small cylinders (Fig. 1). During the next two days the general condition of the copepods receiving the UV-B deteriorated (including mortalities), although specimens which remained active continued to move toward the surface during the UV-B (with visible light) exposure. All control organisms were active and 40-50% were always in the top third of the cylinder during light treatments. Even control organisms which were at times in the bottom third of the cylinder remained active and mobile. During the final light treatment on the fourth day, the effects of the UV-B radiation were apparent. All of the irradiated copepods were either moribund (50%) or dead (50%) and thus were on the cylinder bottom. The control copepods, however, continued to be active and 30-50% were within the upper two-thirds of the cylinder at all times (Fig. 1).

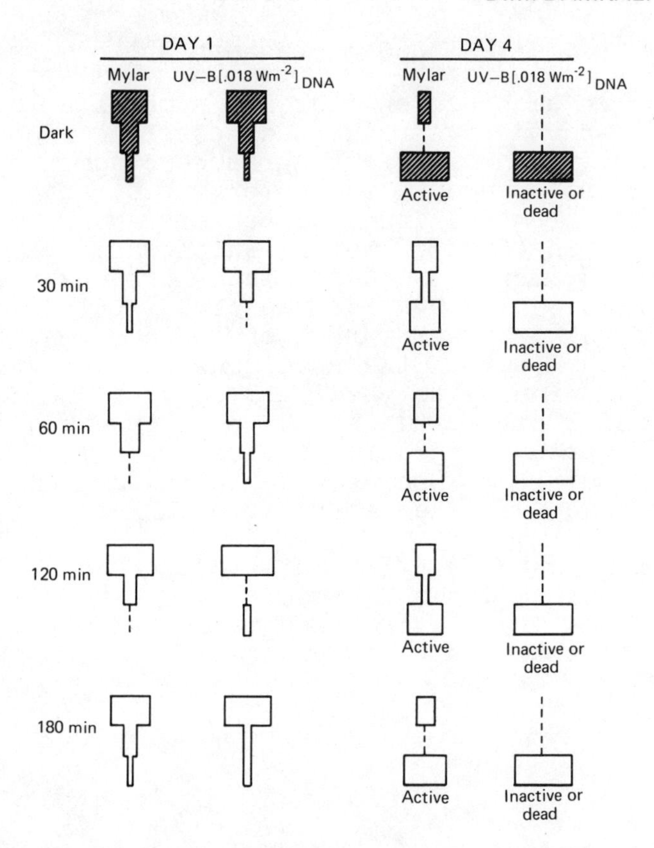

Figure 1. Vertical distributions of the copepod Epilabidocera longipedata in paired glass cylinders (water depth 38 cm), during 3-h exposures with and without UV-B radiation.

An experiment of similar design with Pandalus platyceros led to equivalent results. Before and during the first day's 3 h exposure in the small cylinders there was no apparent difference between the vertical positioning of the shrimp larvae receiving damaging doses of UV-B irradiance and the response of the control shrimp. The vertical distributions were almost uniform with depth in both cylinders before the lights were turned on. After 30 min of exposure and when observations were made at 60, 120, and 180 min the greatest number of shrimp larvae were consistently found in the upper third of each cylinder.

Before the second day's 3 h exposure all shrimp larvae were in the bottom third of each cylinder. When the lamps were turned on some of the larvae in each cylinder immediately swam to the surface. Throughout the remainder of that light exposure the vertical distributions of the larvae in the two treatments were similar, with the larvae most often congregating in the top third of the cylinders.

By the third day, shrimp larvae receiving the high doses of UV-B became far less active than the control organisms, with at least 80% of the UV-B irradiated larvae inactive and on the bottom of the cylinder. In contrast, some control larvae were always in the upper third, with at least 50% within the upper two-thirds during exposure. During the fourth and last day of exposure, the UV-B irradiated larvae were lying moribund on the bottom of the cylinder. In the control cylinder, all larvae were still active and 40-50% were in the upper two-thirds of the cylinder during light treatment.

In the first of two experiments with the large cylinder, shrimp larvae were irradiated with 0.0047 Wm^{-2} $[DNA]$ for 3 h. This dose-rate is more than twice the dose-rate threshold for shrimp larvae (Damkaer et al., 1981).

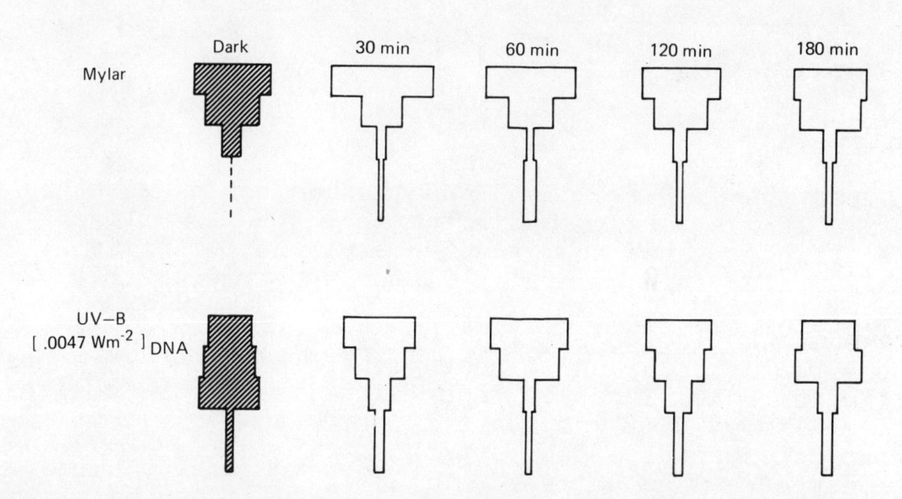

Figure 2. Vertical distributions of shrimp larvae *Pandalus platyceros* in a plastic cylinder (water depth 140 cm), during the first day's 3 h exposure with and without UV-B radiation.

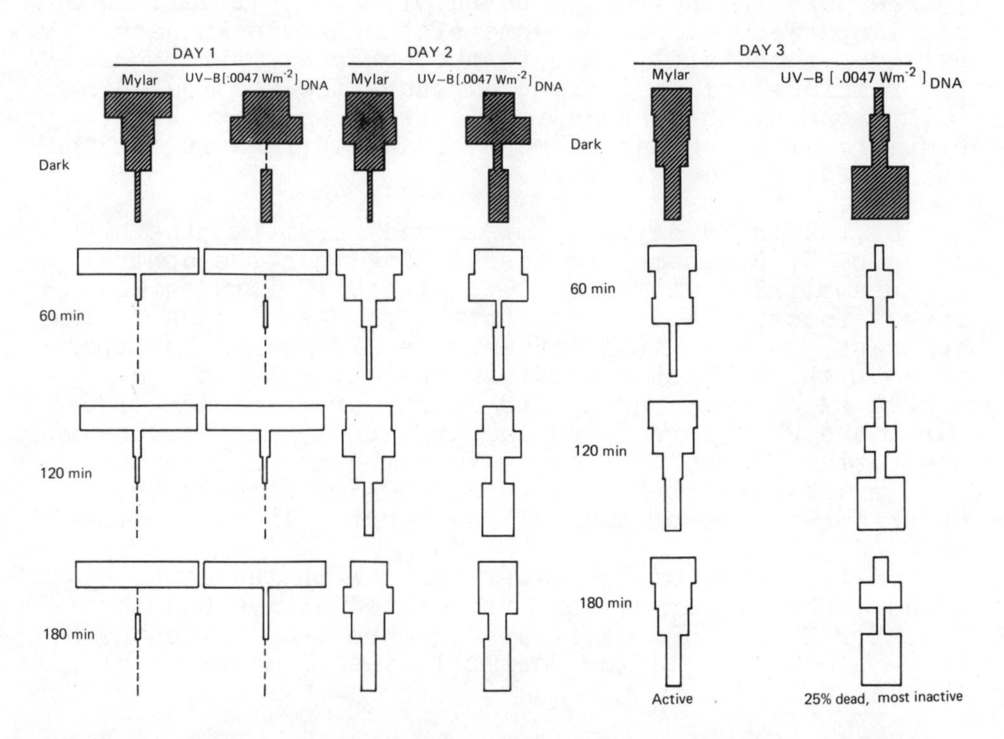

Figure 3. Vertical distributions of the copepod Epilabidocera longipedata in a plastic cylinder (water depth 140 cm), during 3-h exposures with and without UV-B radiation.

In both the UV-B treatment and the control phases the shrimp larvae were concentrated in the upper half of the cylinder throughout the 3-h exposure, with the maximum abundance generally in the upper 30 cm (Fig. 2).

Adult Epilabidocera longipedata were also placed in the large cylinder and exposed to UV-B (0.0047 Wm^{-2}[DNA]). As in the other large-cylinder experiments, the control phase without UV-B was run afterwards with a second group of copepods. The response during the first day's 3-h exposure was nearly identical for each phase (Fig. 3). No avoidance of this harmful level of UV-B irradiance was observed. During the second 3-h exposure (Day 2), the vertical distributions for the UV-B treatment and the control were again similar, and while the concentration

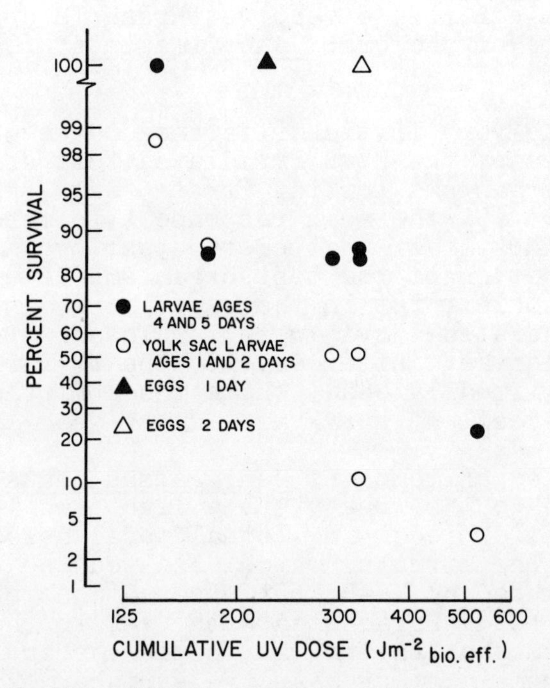

Figure 4. Visible-light response time in seconds for zoea larvae of the shore crab <u>Hemigrapsus nudus</u>; closed circles are UV-B irradiated groups, open circles are controls without UV-B exposure; large circles are means, small circles are ranges.

of copepods in the upper 30 cm of the cylinder was somewhat reduced from the previous day, the majority of the organisms remained in the upper half of the cylinder throughout the 3-h period. On Day 3 the damaging effect of UV-B became evident. While the vertical distribution of the control animals was similar to that of the previous day and all of the control animals were active, the copepods receiving a third exposure of UV-B were now concentrated and inactive in the bottom half of the cylinder.

The zoea stage of the shore crab <u>Hemigrapsus nudus</u> was selected as an experimental organism because of its rapid positive response to strong visible light. Zoeae were first timed within hours of hatching and then before each day's exposure. The test organisms received 0.0073 Wm^{-2} [DNA] for 3 h per day, which is more than twice the biologically effective threshold dose-rate for the zoea stage of the crab Cancer magister (Damkaer et al., 1981).

There was very little difference between the mean response times of the UV-B irradiated test organisms and the control organisms for the first 5 days (Fig. 4). However, by Day 9, the mean response time for the control groups was almost half that of the test organisms. The physical condition of the test organisms deteriorated rapidly from this point in the experiment, and by Day 12 the UV-B irradiated larvae were far less responsive than the control larvae. After Day 12, the UV-B irradiated larvae were virtually dead, while the controls remained active with excellent survival.

In this experiment the Hemigrapsus nudus larvae continued to seek out strong visible light (while being timed) long after receiving lethal total doses of UV-B irradiance. The stimulus which elicited the timed response included no UV-B, but observations of the larvae while being irradiated with UV-B in the water table indicate that as long as the larvae are active they mostly remain at or near the surface.

DISCUSSION

Each of these experiments suggests that there were no differences in behavior between UV-B irradiated organisms and the control organisms until lethal doses of radiation reduced their activity. This is not considered to be an active response to UV, and it would have no value in avoidance. The apparent inability to perceive potential danger from UV-B occurred at dose-rates well above established laboratory thresholds and at exposures to total doses which, in most experiments, surpassed lethal total dose thresholds for similar animals (Damkaer et al., 1981). If the animals tested in this study possess a behavioral mechanism for protection from dangerous levels of irradiation it seems unlikely that it could be based on the direct sensing of UV-B intensity. The shrimp and crab larvae and the adult copepods tested here seemed to be attracted to wavelengths longer than in the UV-B range, and the additional exposure to high levels of UV-B irradiance did not alter their short-term behavior. Within the limits examined, these animals generally positioned themselves as near the light source as possible. That they continue to seek out a strong light source even while doomed from past UV-B exposures demonstrates not only the strength of this photo-positive response but also, probably, their inability to independently discriminate between safe and dangerous levels of UV-B irradiance.

REFERENCES

Barcelo, J.A., and J. Calkins. 1979. Positioning of aquatic microorganisms in response to visible light and simulated solar UV-B irradiation. Photochem. Photobiol. 29: 75-83.

Damkaer, D.M., D.B. Dey, and G.A. Heron. 1981. Dose/dose-rate responses of shrimp larvae to UV-B radiation. Oecologia (Berl.) 48: 178-182.

Damkaer, D.M., D.B. Dey, G.A. Heron, and E.F. Prentice. 1980. Effects of UV-B radiation on near-surface zooplankton of Puget Sound. Oecologia (Berl.) 44: 149-158.

Green, A.E.S., and J. H. Miller. 1975. Measures of biologically effective radiation in the 280-340 nm region. Impacts of Climatic Change on the Biosphere, CIAP Monogr. 5 1(2):60-70.

Setlow, R.B. 1974. The wavelengths in sunlight effective in producing skin cancer: a theoretical analysis. Proc. Nat. Acad. Sci. 71: 3363-3366.

Smith, R.C., and K.S. Baker. 1979. Penetration of UV-B and biologically effective dose-rates in natural waters. Photochem. Photobiol. 29:311-323.

REVIEW OF LITERATURE CONCERNING THE IMPACT OF UV-B RADIATION UPON MARINE ORGANISMS

Robert C. Worrest

Department of General Science
Oregon State University
Corvallis, Oregon 97331

INTRODUCTION

UV-B radiation (e.g. 310 nm) penetrates approximately the upper 10% of the coastal marine euphotic zone before it is reduced to 1% of its surface irradiance (Jerlov, 1976). There is good evidence that current levels of solar UV radiation depress near-surface primary production in marine waters (e.g., Steemann Nielsen, 1964; Jitts et al., 1976). Marine animals may tolerate current levels of solar UV-B radiation by means of protective screens, avoidance behavior, and repair processes which reverse much of the potential damage inflicted by the radiation. However, as early as 1925, scientific investigators have documented the damage inflicted on marine animals by exposure to sunlight, especially the UV component of the sunlight (Huntsman, 1925; Klugh, 1929, 1930; Harvey, 1930). Utilizing currently available laboratory and field data the present contribution will review the literature relating to the impact of UV-B radiation upon marine organisms.

PHOTOSYNTHESIS

In 1962 Steemann Nielsen (1964) noted that the UV component of sunlight was capable of depressing both the rate of light-saturated photosynthesis and the initial slope of the *P* vs. *I* curves for water samples from Friday Harbor, Washington. Following exposure to full sunlight for 90 minutes the rate of photosynthesis for a surface sample was depressed about 35% in comparison to a sample covered with a 3-mm thickness of clear glass in full sunlight. In another experiment described in the report it was demonstrated that those phytoplankton adapted to low light and/or UV fluences (light levels found in the lower region of the euphotic zone) are significantly more sensitive to UV exposure than phytoplankton found in surface samples. This was discussed in some detail at the Second Conference on Marine Biology, Princeton, New Jersey (Halldal, 1966). Steemann Nielsen (1964) suggested that solar UV radiation should be considered in simulated in situ productivity experiments. Jitts et al. (1976) also found that surface samples of marine waters underwent greater primary production when covered with clear glass than when exposed to full sunlight. The results showed that a 2-mm thickness of glass filtering the solar radiation would increase primary productivity by about 50%. The authors speculated that most reports for primary production in surface waters are overestimated by about 40% due to the common use of thick-walled glass bottles in the studies.

More recently Lorenzen (1979) found that radiocarbon uptake by phytoplankton exposed to full sunlight in Vycor containers (fairly UV transparent) was significantly less than for samples contained in Vycor vessels covered with a 0.1-mm thickness of Mylar (a UV-B cut-off filter). This effect was noted in the upper one-third to one-half of the euphotic zone. Although the difference in production levels for the surface samples was about 50%, the difference for the total euphotic zone was calculated to be 2%. The decrease in overall impact results from the attenuation of UV-B radiation at depth in the productive water column. The effect of using different types of bottles for productivity studies was further quantified in a laboratory investigation by Worrest et al. (1980). Three types of bottles were considered: standard soda-lime, commercial Pyrex, and quartz. Transmittance within the UV-B waveband varied significantly (quartz>Pyrex>soda-lime). Measurements of radiocarbon uptake by cultures of Thalassiosira pseudonana in the three bottle types under white fluorescent lamps alone were not significantly different, but when combined with a sunlamp/filter system (UV-B enhancement) there was a significant difference in productivity. Radiocarbon uptake within the Pyrex bottles was 17% less than within the soda-lime bottles (reasonably UV-B opaque), and uptake within the quartz bottles (UV-B transparent) was 29% less than in the soda-lime bottles. Further evidence of the impact of UV-B radiation was reported by Smith et al. (1980). The authors noted, as did the previous investigators, that when UV-B radiation is excluded from ambient sunlight, primary production in oceanic samples (as determined by static bottle *in situ* ^{14}C- uptake) is enhanced. In addition, the authors simulated solar irradiance at depth and noted that enhanced levels of UV-B radiation depressed productivity. For the various water-types investigated the impact of reduced or enhanced levels of UV-B radiation was limited to the upper 1.5 attenuation lengths (the product of the diffuse attenuation coefficient and the depth).

Other laboratory studies have demonstrated a similar impact of UV-B radiation on primary production within unialgal cultures (Wolniakowski, 1980; Worrest et al., 1981a). In the study by Wolniakowski (1980) cultures of *Dunaliella tertiolecta* were exposed to UV-B radiation for up to 10 days. There was a significant depression of ^{14}C-uptake by the algae during this time period. The series of experiments by Worrest et al. (1981a) demonstrated similar results with seven species of marine phytoplankton (Figure 1).

The results of studies by Van Dyke and Thomson (1975a) and Worrest et al. (1981b) regarding photosynthetic rates within marine microcosms exposed to deficient and enhanced levels of UV-B radiation paralleled the results of Smith et al. (1980). The work by Van Dyke and Thomson (1975a) showed that exposure of static marine microcosms to radiation from sunlamp/filter systems depressed oxygen evolution by the autotrophic component of the ecosystems. Worrest et al. (1981b) utilized naturally recruited phytoplankton within flow-through microcosms exposed to natural levels of UV-A and visible radiation, and demonstrated that enhanced levels of UV-B radiation from sunlamp/filter systems depressed radiocarbon uptake (Figure 2). The exposure levels in the 290-320 nm waveband for the four treatment groups in this experiment were 1.92×10^2, 6.58×10^3, 7.02×10^3, and 7.78×10^3 $J \cdot m^{-2} \cdot d^{-1}$. This study corroborated many of the results obtained from an earlier investigation regarding the impact of UV-B radiation upon an attached community of marine algae (Worrest et al., 1978). Among other results from the earlier work it

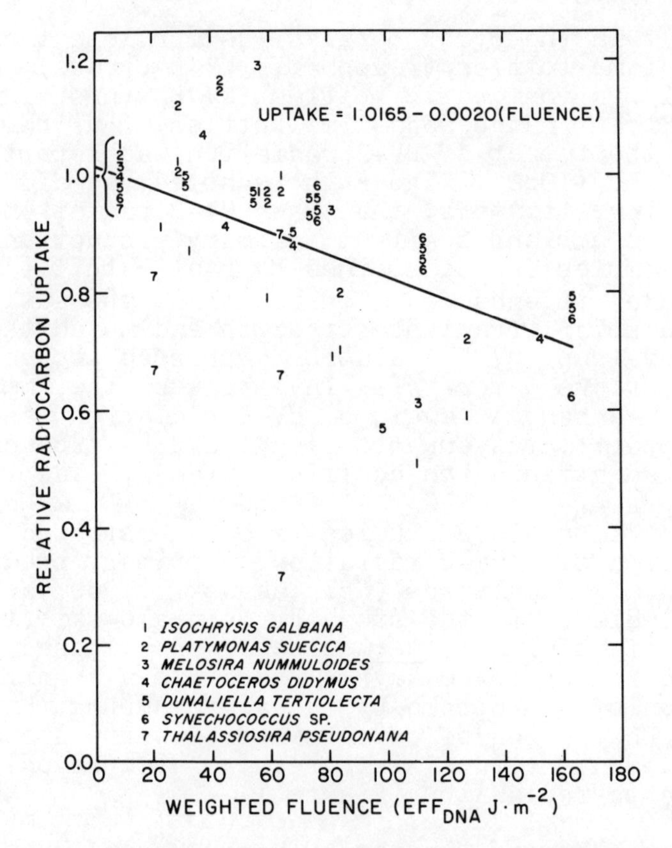

Figure 1. Regression curve representing the effect of UV-B enhanced photosynthetically active radiation upon the radiocarbon estimate of primary production by seven species of marine phytoplankton. The UV-B fluence was weighted at each wavelength in the 290-320 nm waveband by a relative DNA action spectrum (Setlow, 1974). R^2 = 0.34 (n=67). Slope is significant at the p=0.001 level. (From Worrest et al., 1981a.)

was found that UV-B radiation depresses the chlorophyll a content of the autotrophic community. Harris (1978), in his review regarding primary productivity, cites the work of several other authors who have found that current levels of solar UV radiation depress photosynthetic processes in algae.

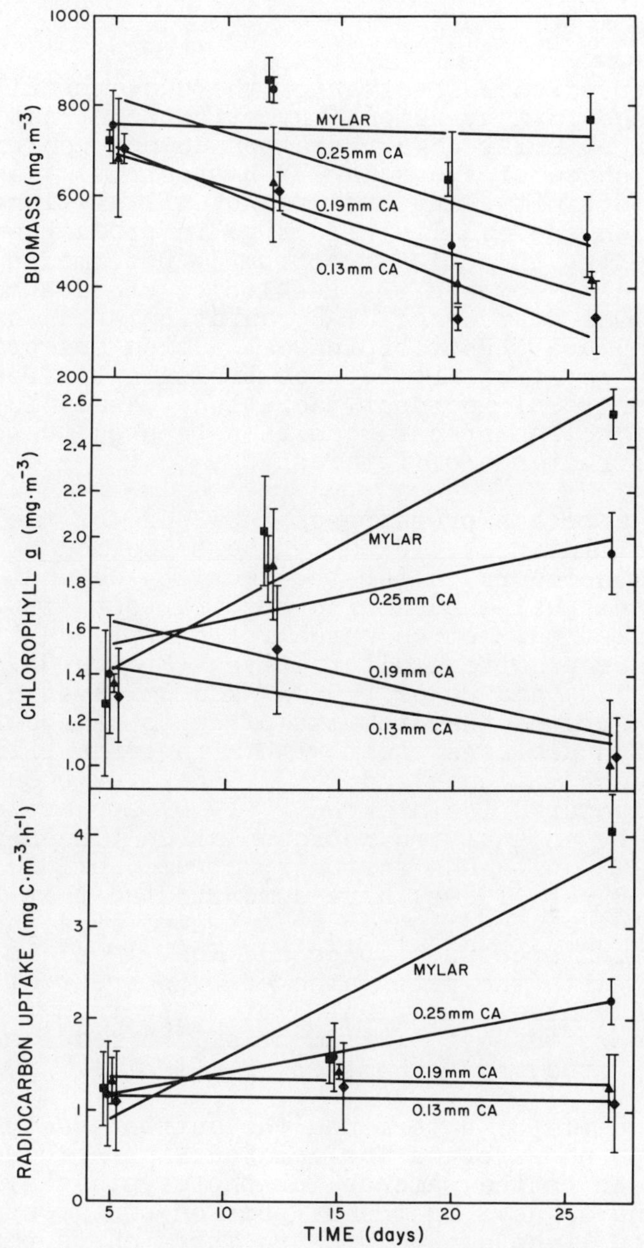

Figure 2. Least-squares regression analysis of biomass (ash-free dry weight), chlorophyll a concentration, and radiocarbon uptake following exposure to UV-B enhanced solar radiation. Filters used were a 0.18 mm thickness of Mylar 'D' and three different thicknesses of cellulose acetate (CA). Error bars represent 95% confidence intervals. (From Worrest et al., 1981b.)

ACTION SPECTRA

The irradiance spectra for the sunlamp/filter systems used in many recent investigations that concern the impact of solar UV-B radiation do not conform exactly to the shape of the solar UV-B spectrum. Due to this lack of conformity, and because not all wavelengths of UV radiation are equally effective in producing biological effects, an action spectrum is determined to make it possible to compare the results of studies making use of artificial sources of UV-B radiation with what might occur in nature. Photobiological action spectra demonstrate the relationship between biological effectiveness and wavelength of incident radiation. Analytic representations of action spectra can then be used to weight the spectral irradiance of interest.

This approach presupposes knowledge of the action spectrum of interest. Usually the selection of a weighting function occurs 'after-the-fact' by determining which of the published action spectra best 'fits' the data in question; not an unrealistic approach if there are no other practical alternatives (Figure 3). For example, Smith and Baker (1980a) and Smith et al. (1980) have demonstrated that results of acute radiocarbon uptake studies are consistent with a photoinhibition action spectrum (Jones and Kok, 1966), but results of chronic exposures (Wolniakowski, 1980) appear to be best described by an analytic representation (Green and Miller, 1975) of a DNA action spectrum (Setlow, 1974). Smith and Baker (1980b) have demonstrated that the biologically effective fluence as weighted by a photoinhibition action spectrum (Jones and Kok, 1966) is fairly consistent with the data gathered by Worrest et al. (1980). A generalized plant action spectrum (Caldwell, 1968) appears to best 'fit' most of the results of a subsequent study by Worrest et al. (1981b).

McLeod (1958) determined the action spectrum in the 250-750 nm waveband for delayed light production (presumed to be the same as for photosynthesis) by four marine algae. In the UV portion of the spectrum the delayed light action spectra for a chrysophyte, *Monochrysis* sp.; a green alga, *Dunaliella euchlora*, and two diatoms, *Phaeodactylum tricornutum* and *Coscinodiscus* sp., showed a gentle shoulder from 440 nm down to about 280 nm, indicating some photosynthetic potential for radiation in the UV-B waveband. In no case was inhibition observed. In a later study by

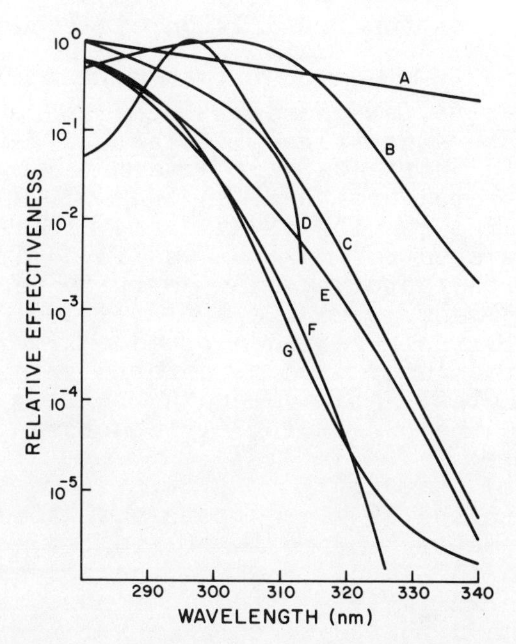

Figure 3. Plots of several action spectra used as weighting functions for assessing the biological impact of UV-B radiation.

A. Photoinhibition action spectrum.
B. Robertson-Berger Meter.
C. Erythema.
D. Caldwell generalized plant action spectrum.
E. AΣ21, used at the University of Florida.
F. AΣ9, used by the USDA Agricultural Research Center, Beltsville, Maryland.
G. Setlow generalized DNA action spectrum.

(Adapted from Nachtwey and Rundel, 1981.)

McLeod and Kanwisher (1962) the authors presented information on the quantum efficiency of photosynthesis in the UV-A and UV-B wavebands for the flagellated, green alga Dunaliella tertecolata. Utilizing the polarographic method for determining oxygen evolution it was shown that the quantum efficiency between 435 nm and 350 nm was relatively constant (about 0.067 oxygen molecules per quantum), and then dropped sharply at wavelengths shorter than 350 nm. It was not possible to measure any net oxygen evolution below 270 nm. The time scale for determination of these action spectra is important because, even though the shorter wavelengths are utilized in the photosynthetic process, damage to the cell resulting from the irradiation can occur quite rapidly (Halldal, 1966).

Action spectra for photosynthesis and photosynthetic inhibition in the green alga Ulva lactuca and the red alga Trailliella intricata were determined by Halldal (1964). The spectral response in the 400-700 nm waveband coincided with the results of Haxo and Blinks (1950). In the UV region a spectral maximum was found at about 375 nm for Ulva, with photosynthetic activity decreasing from there to 302 nm. The spectral response in the UV region for Trailliella showed a broad maximum at about 375 nm, which then decreased more rapidly than Ulva at shorter wavelengths. At wavelengths shorter than 313 nm for Trailliella and 300 nm for Ulva photosynthetic inhibition occurred. The inhibition curves increased rapidly at shorter wavelengths down to 225 nm. Halldal suggests that in the UV region both activated photosynthesis and photosynthetic inhibition may occur simultaneously, with the latter masking the former (Figure 4). As stated by Halldal (1979), the action spectrum for photosynthesis and photosynthetic inhibition in Ulva lactuca "is one of the most detailed analyses performed on a green alga and, with small adjustments, may be representative for several algal groups (diatoms, dinoflagellates, unicellular green algae)." Halldal has found that the response in the UV-B waveband for green, blue-green, and red algae and dinoflagellates is quite similar. Halldal (1967) and Halldal and Taube (1972)

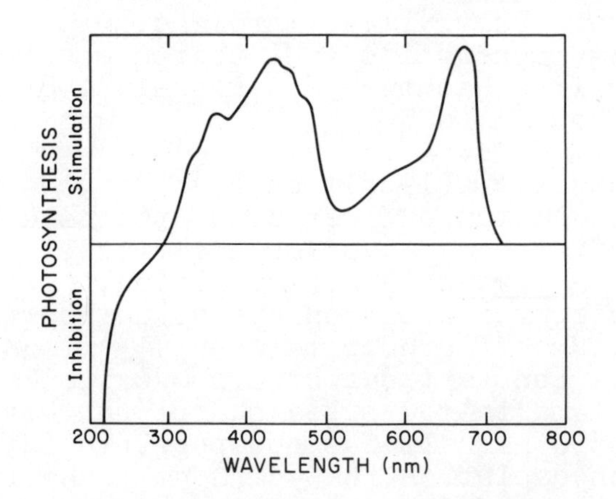

Figure 4. Action spectrum for photosynthesis and photosynthetic inhibition in the green alga Ulva lactuca. (Adapted from Halldal, 1979.)

present reviews of the determination of action spectra by many of the previous authors. More recent evidence of the photosynthetic activity generated by UV-B radiation is presented by Halldal (1968). The zooxanthellae and Siphonales which inhabit many corals and clams evolved oxygen in response to radiation between 300 nm and 720 nm. The action spectra decreased rapidly at wavelengths shorter than 340 nm. At wavelengths shorter than 300 nm photooxidation occurred. Haxo (1970) also presents data indicating the photosynthetic potential of UV-B radiation. Action spectra for oxygen evolution were presented for three dinoflagellates, _Gonyaulax polyedra_, _Amphidinium carterae_ and _Cachonina niei_; two diatoms, _Chaetoceros_ sp. and _Thalassiosira fluviatilis_; a coccolithophorid, _Cricosphaera carterae_, and a cyanophyte, _Dermocarpa violacea_. In addition action spectra were shown for three samples of mixed phytoplankton concentrated from seawater collections obtained near the Scripps Institute of Oceanography pier. In almost all cases in this study there was net photosynthesis at the shortest wavelengths studied (about 320 nm). An action spectrum of photosynthesis for _Calothrix crustaceae_ was also determined and was detectable down to about 300 nm, below which photooxidation was noted.

UV-B IMPACT UPON ECOSYSTEM DYNAMICS

As cited by Steemann Nielsen (1964), Gessner and Diehl (1951) investigated the destruction of chlorophyll in phytoplankton by solar UV radiation and found that there was a differential response among the species. Species of _Scenedesmus_ and _Ankistrodesmus_ were found to be more sensitive than one of _Chlorella_. Other laboratories have also found differential sensitivities of marine algae to UV-B radiation. In the work of Calkins and Thordardottir (1980) six species of diatoms isolated from the North Atlantic were irradiated with simulated solar UV-B radiation. Based on survival _Thalassiosira gravida_ and _T. polychorda_ were relatively resistant in comparison to four other species: _Chaetoceros debilis_, _C. decipiens_, _Skeletonema_ sp. and _Nitzschia_ sp. In this study and in a report by Calkins and Nachtwey (1975) utilizing sunlamp/filter systems as a source of simulated solar UV-B radiation the authors suggest that a few hours exposure to the summer sun at current levels of UV-B irradiance can be lethal to a variety of marine diatoms. They approach the assessment of the impact of UV-B radiation through the replacement limiting model (Calkins and Thordardottir, 1980). Two parameters in this model are

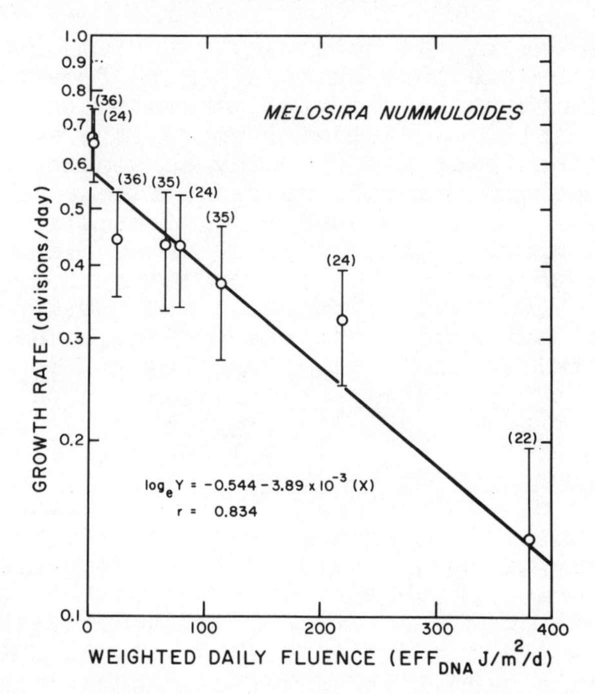

Figure 5. Regression curve representing the effect of UV-B radiation upon the growth rate of Melosira nummuloides. The UV-B fluence was weighted at each wavelength in the 290-320 nm waveband by a relative DNA action spectrum (Setlow, 1974). Mean growth rates (±st. dev.) are indicated at each experimental daily fluence. Figures in parentheses represent total number of chains at each fluence. (From Thomson et al., 1980a.)

the dose-response relations for growth delay and growth rate. In a study of growth rates Thomson et al. (1980a) irradiated short chains of the littoral diatom Melosira nummuloides to UV-B radiation from a sunlamp/filter system. Least squares regression analysis indicated a significant depression in the growth of this species by UV-B radiation (Figure 5). Exposure of Dunaliella tertiolecta to UV radiation emitted by a sunlamp/filter system caused a short-term (24 h) reduction in the growth rate of the cultures; however, following the initial depression the growth rate returned to previous levels (Wolniakowski, 1980). Continued exposure to UV-B radiation did not significantly affect the growth rates during a subsequent 10-day period.

Worrest et al. (1981a) described the impact of acute exposure to UV-B radiation upon the radiocarbon uptake by seven species of marine phytoplankton (Figure 1). The phytoplankton were obtained from established stock cultures which had been reared under UV-B deficient conditions. The *in vitro* response of the seven species indicated that each possessed a unique sensitivity to UV-B radiation (Table 1). The diatom *Thalassiosira pseudonana*, in this study, was the most sensitive of the group; whereas the flagellated, green alga *Dunaliella tertiolecta* was the most resistant member. A blue-green alga, *Synechococcus* sp. was similar to *D. tertiolecta* in resistance to UV-B radiation. Shibata (1969) extracted a UV-absorbing pigment, S-320, from a marine, blue-green alga. S-320 has an extremely strong absorbance maximum at 322 nm. This pigment may be very similar in chemical structure to UV-absorbing pigments found in other tropical forms and, therefore, it may function as a UV screen to protect the tropical organisms from solar UV radiation. Another pigment, a UV-absorbing substance with an absorbance maximum at 337 nm, has been isolated from red alga, *Porphyra yezoensis* (Yoshida and Sivalingham, 1970). Further evidence consistent with the UV-protection hypothesis is found in the paper by Sivalingham et al. (1974). A UV-absorbing pigment with an absorbance peak in the 320-340 nm range was extracted from seventy different species of marine algae. The data indicate that the concentration of the pigment decreases in species found at increasing depths; therefore, in species with reduced exposure to solar UV radiation.

Evidence of differential sensitivity of marine algae is also found in laboratory studies by Van Dyke and Thomson (1975b) and Worrest et al. (1978). Both of these studies exposed simulated ecosystems to UV-B stress and noted shifts in the community composition. Diatom species constituted the dominant autotrophic component of the ecosystems prior to the supplemental UV-B enhancement. In the study by Van Dyke and Thomson (1975b) an immediate decline in the number of diatoms was noted following exposure to UV-B radiation in comparison to other algae. At the time of peak gross photosynthesis for the developing ecosystems the dominant primary producers were filamentous, blue-green algae. In UV-deficient ecosystems serving as controls diatom species continued as the dominant primary producers until the end of the experiment, when several species of green, filamentous algae became evident. In the study by Worrest et al. (1978) the dominant

TABLE 1

Sensitivity of seven species of marine phytoplankton based on least-squares linear regression of relative radiocarbon uptake versus biologically weighted UV-B exposure. Regression equation of the form: [RELATIVE UPTAKE] = $B_o - B_1$ [FLUENCE($EFF_{DNA} J \cdot m^{-2}$)].

Species	n	$B_o \pm SD$	$B_1 \pm SD$	Significance Level of Regression
Dunaliella tertiolecta	19	1.017±0.029	0.001543±0.000357	0.001
Synechococcus sp.	8	1.040±0.043	0.001868±0.000405	0.004
Chaetoceros didymus	5	1.042±0.049	0.002082±0.000618	0.044
Platymonas suecica	11	1.114±0.053	0.002591±0.000883	0.017
Melosira nummuloides	4	1.136±0.211	0.003246±0.002859	0.374
Isochrysis galbana	12	1.081±0.057	0.004096±0.000832	0.001
Thalassiosira pseudonana	8	0.966±0.099	0.005667±0.002473	0.062

autotrophic component was comprised of diatoms throughout the experiment in both the experimental and control chambers. However, there was a temporal shift in community composition; the diversity of the community exposed to enhanced UV-B radiation being depressed in comparison to the community exposed to control levels of UV-B radiation. The difference in diversity between the two exposure regimes increased with time.

The results of the two previously described laboratory studies were, in part, corroborated by an investigation by Worrest et al. (1981b). In this study naturally recruited organisms were exposed, in flow-through microcosms, to nearly natural levels of visible and UV-A radiation in conjunction with the experimental levels of UV-B radiation. As was noted in the previous studies which utilized relatively low-level, artificial light sources there was a significant depression of community diversity following exposure to supplemental UV-B radiation (Table 2).

Utilizing a mathematical model Worrest et al. (1981a) have simulated the impact of UV-B radiation upon a phytoplankton community. The model assumed that the reproductive rate of the plankters was a direct function of the rate of carbon fixation following UV-B exposure, a rate determined in the study. The model [$C_{it} = C_{io}2\ (R_i - k_iC_{io})^t$, where C_{it} represents the percentage composition of species i within a community at time t (days), C_{io} is the percentage composition of species i at some previous reference time, R_i is the doubling rate of species i (day^{-1}), and k_i is a constant of proportionality (grazing or 'preference factor') for species i] yielded dramatic changes in community composition following a simulated 50% increase in UV-B exposure.

The interpretation of changes in community diversity is subject to speculation, but one can say that, if the species composition of the primary producers is altered, the quality and possibly the quantity of food present for primary consumption could also be altered. Therefore, organic carbon exchange between trophic levels could be affected. The impact would be significant if the organisms selected for by enhanced UV-B radiation were of lower nutritional value (i.e. if they impacted growth and fecundity of the consumers). Another effect might be to alter the size distribution of the component producers in the ecosystem. Decreasing the size of the representative producers upon which

TABLE 2

Analysis of community composition. The seminatural ecosystems were allowed to develop four weeks, with samples collected at the beginning and end of the experiment. The filters used in the exposure apparatus were a 0.18 mm thickness of Mylar 'D' and three different thicknesses of cellulose acetate (CA). N is the number of diatoms in the sample, S is the number of species represented in the sample, H" is the estimator for the uncertainty index, and R' is a measure of redundancy. Estimate of sample variance for H" in all cases was in the 0.003-0.004 range.

	N		S		H" (nats)		R'	
Filter	Initial	Final	Initial	Final	Initial	Final	Initial	Final
Mylar	504	503	17	37	1.619	2.831	0.466	0.311
0.25mm CA	502	502	23	22	1.691	1.653	0.512	0.563
0.19mm CA	500	503	17	19	1.381	1.645	0.558	0.554
0.13mm CA	508	500	21	11	1.568	1.224	0.535	0.595

consumers graze can significantly increase the energy allotment required for consumption; thereby reducing the feeding efficiency of the consumer. On the other hand, changes in the phytoplankton species composition might be beneficial to primary producer-consumer trophodynamics. As stated by Worrest et al. (1981b), the actual impact of altering the community composition of an ecosystem requires further investigation.

Finfish

Information exists regarding the impact of UV radiation on finfish raised in hatcheries (e.g., Bell and Hoar, 1950; Dunbar, 1959); however, until recently, little documentation has appeared in the literature concerning UV stress upon pelagic fishes. Marinaro and Bernard (1966) concluded from their study that solar UV radiation may be a stressful agent for pelagic fish eggs which are often located near the surface of the water, especially those eggs spawned during the winter. For this study comparisons of survival were made among three groups: eggs exposed to full sunlight, eggs exposed to sunlight filtered by plexiglass or glass (nominally opaque to UV-B radiation), and eggs incubated in the dark. Similarly, Pommeranz (1974) found that solar UV radiation increased the mortality of eggs from the flatfish _Pleuronectes platessa_, but concluded that the tolerance limit of the eggs would not be exceeded during the normal spawning season (December through March). Pommeranz conducted his study during periods of relatively high irradiance (April and May) with eggs artificially induced and fertilized.

Hunter et al. (1979) exposed anchovy and mackerel eggs and yolk-sac larvae to UV-B radiation. Depending on the "shape" of the spectrum transmitted by each of two different filter-types for the fluorescent sunlamp sources, 50% of the anchovy survived a cumulative UV-B exposure ranging from 31.0 to 91.2 $kJ \cdot m^{-2}$. The two types of filters used were cellulose triacetate and polystyrene. For mackerel the comparable LD_{50}'s were 65.5 and 125 $kJ \cdot m^{-2}$. It was suggested by the authors that the differences in the dose-response relation between the two filter-types can be attributed to the transmission of shorter and more actinic radiation by the polystyrene filter. The relative, spectral response of the larvae and eggs appears to be consistent with an action spectrum similar to a generalized DNA action spectrum (Setlow, 1974). This was further corroborated by the result of a study reported in the same article by Hunter et al. (1979) which utilized a filter combination of polystyrene 666-U and cellulose triacetate with the FS40 sunlamps. In addition, the authors described histological and morphological effects of UV-B radiation in anchovy and mackerel larvae. They reported significant damage to both the brain and eye of the anchovy. Lesions occurred in the mackerel larvae, also; however, the severity was not as great as in the anchovy. Exposure of the larvae to UV-B radiation also resulted in the retardation of growth and in the disperal of melanosomes in the melanophores.

In a more recent report by Hunter et al. (1981) anchovy larvae were exposed to various levels of UV exposure from natural sunlight for 12 consecutive days. The larvae suffered significant mortality from February through October. Furthermore, exposure of the larvae to UV radiation inhibited their growth. However, the authors concluded that due to three factors, the seasonality of spawning for anchovy, the vertical distribution of the larvae, and the high mortality resulting from non-UV-radiation causes, anchovy populations may not be greatly affected by anthropogenic enhancement of solar UV-B radiation. Furthermore, Kaupp and Hunter (1981) noted that the solar irradiance level (320-500 nm) on a clear day in March (at 33°N latitude) was sufficient to insure maximal photorepair of potential UV damage for anchovy larvae at any depth in the ocean.

Zooplankton

Huntsman (1925) noted that exposure of lobster larvae to direct sunlight depressed the survival rate of the organisms. Three exposure levels for the larvae (being reared in glass jars) were utilized. The first was exposed to direct plus diffuse sunlight, the second to diffuse only, and the third had the sunlight excluded. It was found that larvae had decreased survival when reared in the "light" or "shade" when compared to those reared in the "dark". Similar results were obtained whether the experiment was performed behind window glass or exposed on the roof, indicating that it was probably not the UV-B portion of the solar spectrum which produced the impact. Comparable experiments with many other species of invertebrates were described in the same report. With only one exception the detrimental impact of sunlight was noted.

Klugh (1929, 1930) and Harvey (1930) utilized filters to determine which region of the solar spectrum stressed the marine invertebrates. In the first series of experiments described by Klugh (1929) four species of marine copepods, *Calanus finmarchicus*, *Tortanus discaudatus*, *Eurytemora herdmani*, and *Acartia clausi*; a caprillid, *Aeginella longicornis*; a hydroid, *Tubullaria crocea*, and the eggs of a fish, *Lophius piscatornis*, were exposed in quartz tubes behind glass filters. One filter-type transmitted visible, but not UV radiation; a second type transmitted both visible and UV radiation; the third basically transmitted only UV radiation. The results of the exposure of these animals to the sun behind the three types of filters (plus a control kept in the dark) led to the conclusion that it was the UV component of solar radiation that was stressful. Those animals normally found at the surface of the ocean during the day (*A. clausi* and the eggs of *Lophius*) were resistant to the UV stress; whereas the remainder of the experimental animals (which normally are found at depth or at the surface only in dim light) exhibited a dramatic sensitivity to solar UV radiation. In a subsequent report (Klugh, 1930) the author performed similar experiments with the eggs of the squid *Loligo pealii*. The eggs normally are enclosed in a gelatinous capsule and may be suspended near the surface of the ocean. The prediction of intermediate-to-low sensitivity of the developing eggs was borne out by the experiments.

A similar experiment was performed by Harvey (1930). Adult copepods, *Calanus finmarchicus*, were exposed through glass jars to either full sunlight or diffuse sunlight. Control animals were kept in the dark. There was a significant impact of the sunlight on the heartbeat of the animals, resulting in depressed heart rates. Subsequent studies which filtered the sunlight through red, green, or blue filters demonstrated that the shorter wavelengths were more effective in causing mortality in *C. finmarchicus* (Harvey, 1930).

More recently investigators have reported on the impact of simulated solar UV-B radiation upon the marine copepod *Acartia clausii* (Karanas et al., 1979) and upon shrimp larvae, crab larvae, and euphausids (Damkaer et al., 1980, 1981). Results of the report by Karanas et al. (1979) indicated that UV-B radiation increased mortality of the copepod populations and decreased fecundity of the survivors (Figure 6). It appeared as if the male adults were more sensitive than the females (Figure 7) and, in general, that younger larval stages were more sensitive than the older stages. The authors concluded that, currently, sufficient solar UV-B

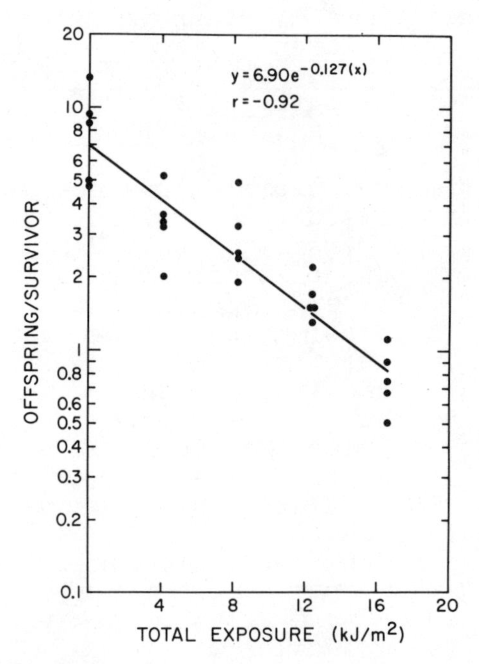

Figure 6. Relationship between number of offspring produced per survivor of irradiated C4-adult age group and fluence as fitted by a least-squares regression. Original data are shown. (From Karanas et al., 1979.)

radiation penetrates the upper region of the marine water column to inhibit the development of A. clausii. In a more recent paper Karanas et al. (1981) exposed larval stages of A. clausii to one of three different levels of UV-B exposure. The animals were then reared to sexual maturity, and adult virgin survivors were mated according to prior exposure. Larvae from each type of cross were separated, reared to maturity and counted, providing information on the survival capability of non-irradiated offspring from the survivors of the different mating types. The number of eggs and the number of living larvae produced were reduced by parental exposure to UV-B radiation (Figure 8). Multiple curvilinear regression analysis yielded the following relationship between UV-B exposure and fecundity for separate mating pairs of copepods:

$$[EGGS] = 61.04 \exp -(0.004D_m + 0.005D_f)$$

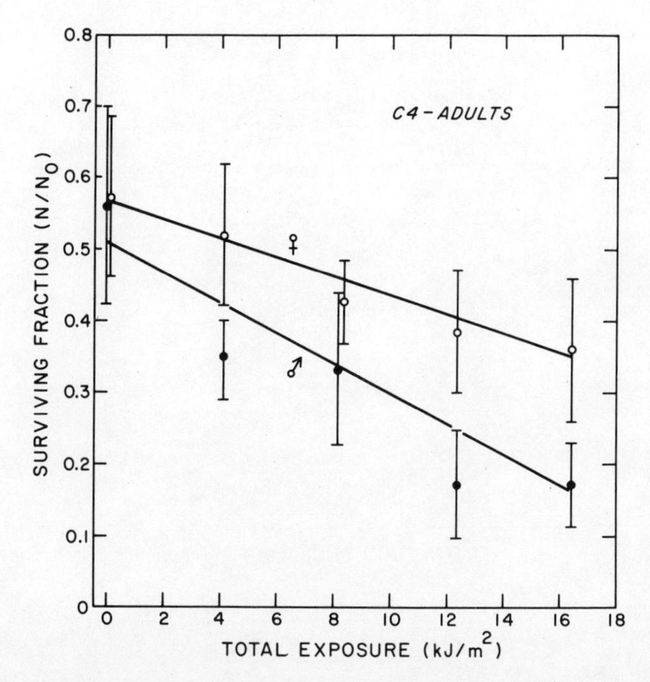

Figure 7. Least-squares regressions between male or female surviving fractions and fluences after exposure of C4-adult age group. Regression intercepts are not significantly different ($p>0.05$). Slopes of regressions (males $=-0.023$, females $=-0.014$) differ significantly ($p<0.05$). Mean and standard deviation of original data are shown. UV-B irradiance was 1.52 $W \cdot m^{-2}$. (From Karanas et al., 1979.)

and

$$[LARVAE] = 54.03 \exp -(0.030D_m + 0.028D_f)$$

where D_m represents the DNA-weighted level of exposure for the parental males and D_f is a comparable measurement for the parental females. It is of interest to note that neither the number of eggs produced nor the number of larvae developing from each mating pair was significantly affected by whether it was the male or the female parent which was irradiated. Also of interest is the fact that when both parents were irradiated at a specific level, about the same results were obtained as when a single parent was irradiated at a level equivalent to the sum of the levels of the irradiated pair (e.g., an exposure level of 25 EFF_{DNA} $J \cdot m^{-2}$

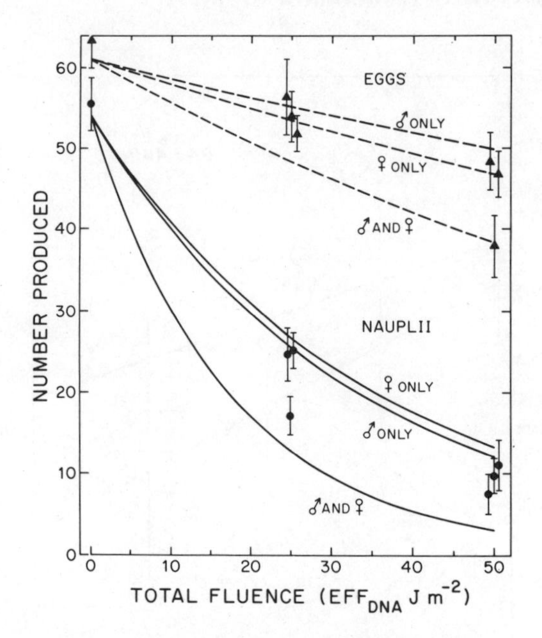

Figure 8. Multiple curvilinear regression analyses illustrating the production of eggs and living nauplii by mating pairs of Acartia clausii. The curves predict the regression responses for the following equations:

$$[EGGS] = 61.04 \exp -(0.004D_m + 0.005D_f)$$

and

$$[NAUPLII] = 54.03 \exp -(0.030D_m + 0.028D_f)$$

when only the male parent was irradiated ($D_f = 0$), only the female parent was irradiated ($D_m = 0$), or when both parents were irradiated equally ($D_m = D_f$). Data points represent sample mean (± st. error). N = 84 for each triad of curves. (From Karanas et al., 1981.)

for both parents produced about the same results as an exposure of 50 EFF_{DNA} $J \cdot m^{-2}$ for the male parent mated to a non-exposed female; an additive effect).

In a series of experiments by Damkaer et al. (1980, 1981) shrimp larvae, crab larvae, and euphausids were exposed to low levels of UV-B radiation under laboratory conditions. Beyond threshold levels, activity, developmental rates, and survival were depressed. Also, the total-dose response varied with dose-rate (i.e. reciprocity did not hold). The authors concluded that late in the season of surface occurrence for these species the threshold levels approximate levels of solar UV-B radiation when weighted by an analytic representation (Green et al., 1974) of an erythemal action spectrum; however, the probability that shrimp larvae would receive a lethal irradiance level is low for the first half of the season of surface occurrence, even with a 44% increase in damaging UV radiation.

Benthic animals

Clark and Kimeldorf (1970, 1971) observed that exposure of the sea anemone *Anthopleura xanthogrammica* to UV radiation in the 248-400 nm waveband can elicit retraction of the tentacles. The efficiency of radiation in the 320-400 nm region was very low; however, a sharp rise in efficiency occurred below 320 nm with a peak at 280 nm. Lees and Carter (1972) noted that the sea urchin *Lytechinus anamesus* also responded to UV exposure. The authors reported that the animal masked with or heaped shells on the aboral surface of the body in response to 254 nm and 360 nm radiation as well as direct and indirect sunlight. Unfortunately they did not determine whether it was the UV component of the sunlight which stimulated the covering response. In a report by Jokiel (1980) the UV component of sunlight was screened out and compared with the results following exposure to the full solar spectrum in an experiment concerning the UV sensitivity of coral-reef epifauna. The author found that "shade-loving" sponges, bryozoans, and tunicates were killed by one-to-two days exposure to solar UV radiation. A black sponge, *Chondrosia chucalla*, was the only common cryptic species found resistant to solar radiation. The author concluded that avoidance appears to be an important UV defense mechanism for many coral-reef organisms.

As long ago as 1938 the impact of UV-B radiation upon sea urchin eggs was reported (Giese, 1938; Hollaender, 1938). Hollaender (1938) studies the relative efficiency of UV radiation on the parthenogenic activation of eggs from the sea urchin *Arbacia punctulata*. The author found that 280 nm radiation was not effective when compared with 265, 240, 230, and 226 nm radiation. Of the wavelengths utilized by Giese (1938) (254, 265, 280, 302, 313, and 366 nm) the author noted that 280 nm radiation was most effective in retarding egg cleavage for the sea urchin *Strongylocentrotus purpuratus*, with 254, 265, and 203 nm radiation somewhat less effective, and 313 nm radiation even less effective. Even at the highest exposure levels of 366 nm radiation the eggs cleaved at nearly normal, control rates. Giese (1939) also studied the effect of UV radiation upon sperm from this species of sea urchin. It was noted that when irradiated sperm were used to fertilize unirradiated eggs, the cleavage in the eggs was retarded. The delay was proportional to total exposure and 265 nm radiation was more effective than 280 nm radiation. The relatively high efficiency of 265 nm radiation indicated that cleavage delay resulting from exposure of sperm to UV radiation was a nucleic-acid effect. Giese (1946) continued his determination of action spectra for sea urchin sperm and eggs, a study which corroborated the fact that retardation of egg cleavage by irradiated eggs was, most likely, a protein event, and retardation of egg cleavage by irradiated sperm was a nucleic-acid event. Wells and Giese (1950) found that the mechanism for potential UV-induced cleavage delay could undergo photorepair with wavelengths shorter than 430 nm being most effective. Giese (1964) has written an extensive review of these works, as well as studies of UV effects on other animals and animal cells.

Bacterioplankton

Eighty to ninety percent of the marine and estuarine pelagic bacteria are unattached cells (bacterioplankton). These cells play a significant role in the maintenance of energy flow within the ecosystem (Sieburth, 1976, 1979). ZoBell and McEwen (1935) exposed seawater nominally containing only bacterioplankton to midsummer sunlight at La Jolla, California (33°N latitude). The samples were contained in open petri dishes. The sunlight diminished the bacterial concentration in the dishes, even though previous observations on the vertical, diurnal, and seasonal distribution of bacteria in the sea failed to demonstrate any stress on the bacteria due to solar radiation.

In an article concerning the determination of marine primary production by the radiocarbon-uptake method Steemann Nielsen (1952) concluded that exposure of samples to solar UV and visible radiation inhibits bacterial respiration and reproduction. In a subsequent study by Vaccaro and Ryther (1954) the authors failed to note any bacteriocidal activity for solar radiation when the sample-bottles were located at a depth of 4 cm below the sea surface. However, the study by Vaccaro and Ryther utilized dissolved-oxygen content and colony counts on nutrient seawater agar as criteria for impact. Both of these criteria are significantly less sensitive to environmental stress than more recently established techniques.

Recently Thomson et al. (1980 b) noted that marine bacterial populations in laboratory microcosms, when exposed to UV-supplemented visible radiation, displayed an overall decrease in total numbers, an increase in the proportion of pigmented cells, a decrease in the number of cellulolytic microorganisms and an increase in respiration. The authors concluded that it was radiation in the UV-B waveband that was the stressful agent. Moehring (1980) followed the study by Thomson et al. with an investigation into the impact of UV-B radiation upon glucose heterotrophic potential of bacterioplankton within flow-through microcosms established in a fiberglass greenhouse. The fiberglass served as a UV cutoff filter transmitting solar radiation in the visible region. The microcosms were exposed to four levels of UV-B enhancement utilizing sunlamp/filter systems (1.92×10^2, 6.58×10^3, 7.02×10^3, and $7.78 \times 10^3 J \cdot m^{-2} \cdot d^{-1}$; 290-320 nm). Utilizing epifluorescent microscopy, electronic counting and radiocarbon uptake as tools, the author determined the velocity maximum for uptake ($V_{max} \cdot cell^{-1}$), turnover times (T_t), and assemblage affinity for glucose plus its natural concentration ($K_t + S_n$). Cell concentration, chlorophyll a, and temperature were also monitored for the microcosms. It was concluded that the heterotrophic activity of the bacterioplankton assemblage was altered by UV-B radiation; however the change was linked primarily to effects on other trophic levels. There was limited evidence of some direct effect. $V_{max} \cdot cell^{-1}$ and ($K_t + S_n$) were negatively correlated with UV-B exposure. Reduction in actual velocity of glucose utilization with accumulated UV-B dose was accompanied by an increase toward the potential velocity of the system.

SUMMARY

The impact of solar UV radiation upon marine organisms has been noted for many years. By comparing results of studies performed in full sunlight with those that excluded solar UV radiation it has been demonstrated that current levels of UV radiation can inhibit photosynthesis in phytoplankton and increase morbidity and mortality in zooplankton and in coral-reef epifauna. These results have been repeated in the laboratory, where radiation in the UV-B waveband has also been shown to depress phytoplankton growth rates, decrease community diversity, increase morbidity and mortality in finfish larvae, delay echinoid egg cleavage, depress copepod fecundity, and elicit faunal behavioral responses.

For all of the laboratory studies cited in this review, great caution must be exercised when relating the results to a natural body of water. To be considered are such factors as the depth of the attenuating water column, degree of vertical mixing, the ratio of photosynthetically active radiation to UV-B radiation, behavioral responses, seasonality of exposure, and the prehistory of the target organisms. Although near-surface organisms are impacted by UV-B radiation, the impact on the total, productive water column is still open to question.

REFERENCES

Bell, G.M., and W.S. Hoar. 1950. Some effects of ultraviolet radiation on sockeye salmon eggs and alevins. Can. J. Res. 28:35-43.

Caldwell, M.M. 1968. Solar ultraviolet radiation as an ecological factor for alpine plants. Ecol. Monogr. 38:243-268.

Calkins, J., and D. S. Nachtwey. 1975. UV effects on bacteria, algae, protozoa, and aquatic invertebrates. In: Impacts of Climatic Change on the Biosphere, Part I: Ultraviolet Radiation Effects [eds. D. S. Nachtwey et al.], 5-3 to 5-8. U.S. Dept. Transportation, DOT-TST-75-55, Washington, D.C.

Calkins, J., and T. Thordardottir. 1980. The ecological significance of solar UV radiation on aquatic organisms. Nature 283:563-566.

Clark, E.D., and D. J. Kimeldorf. 1970. Tentacle responses of the sea anemone Anthopleura xanthogrammica to ultraviolet and visible radiations. Nature 227:856-857.

Clark, E.D., and D. J. Kimeldorf. 1971. Behavioral reactions of the sea anemone, Anthopleura xanthogrammica, to ultraviolet and visible radiations. Radiat. Res. 45:166-175.

Damkaer, D. M., D. B. Dey, G.A. Heron, and E.F. Prentice. 1980. Effects of UV-B radiation on near-surface zooplankton of Puget Sound. Oecologia 44:149-158.

Damkaer, D.M., D. B. Dey, and G. A. Heron. 1981. Dose/dose-rate responses of shrimp larvae to UV-B

radiation. Oecologia. 48: 178-182
Dunbar, C.E. 1959. Sunburn in fingerling rainbow trout. Prog. Fish Cult. 21:74.
Gessner, F., and A. Diehl. 1951. Die Wirkung der natürlichen Ultraviolettstrahlung auf die Chlorophyllzerstörung von Planktonalgen. Arch. Mikrobiol. 15:439-454.
Giese, A.C. 1938. The effects of ultra-violet radiations of various wave-lengths upon cleavage of of sea urchin eggs. Biol. Bull 75:238-247.
Giese, A.C. 1939. Ultraviolet radiation and cell division. J. Cell. Comp. Physiol. 14:371-382.
Giese, A.C. 1946. Comparative sensitivity of sperm and eggs to ultraviolet radiations. Biol. Bull. 91: 81-87.
Giese, A. C. 1964. Studies on ultraviolet radiation action upon animal cells. In: Photophysiology, vol. 2 [ed. A. C. Giese], 203-245. Academic Press, New York.
Green, A.E.S., and J. H. Miller. 1975. Measures of biologically effective radiation in the 280-340 nm region. In: Impacts of Climatic Change on the Biosphere, Part I: Ultraviolet Radiation Effects [eds. D. S. Nachtwey et al.], 2-60 to 2-70. U.S. Dept. Transportation, DOT-TST-75-55, Washington, D.C.
Green, A.E.S., T. Sawada and E. P. Shettle.1974. The middle ultraviolet reaching the ground. Photochem. Photobiol. 19: 251-259.
Halldal, P. 1964. Ultraviolet action spectra of photosynthesis and photosynthetic inhibition in a green and a red alga. Physiol. Plant. 17:414-421.
Halldal, P. 1966. Light as a controlling factor. In: Marine Biology, vol. 2 [ed. C. H. Oppenheimer], 37-83. New York Academy of Sciences, New York.
Halldal, P. 1967. Ultraviolet action spectra in algology. A review. Photochem. Photobiol. 6: 445-460.
Halldal, P. 1968. Photosynthetic capacities and photosynthetic action spectra of endozoic algae of the massive coral Favia. Biol. Bull. 134:411-424.
Halldal, P. 1979. Effects of changing levels of ultraviolet radiation on phytoplankton. In: The Ozone Layer [ed. A. K. Biswas), 21-34. Pergamon Press, New York.
Halldal, P. and O. Taube. 1972. Ultraviolet action and photoreactivation in algae. In: Photophysiology, vol. 7 [ed. A. C. Giese) 163-188. Academic Press, New York.
Harris, G. P. 1978. Photosynthesis, productivity and growth: The physiological ecology of phytoplankton. Arch. Hydrobiol. Beih. 10:1-171.

Harvey, J. M. 1930. The action of light on Calanus finmarchicus (Gunner) as determined by its effect on the heart rate. Contrib. Can. Biol. 5: 85-92.

Haxo, F. T. 1970. Phytosynthetic action spectra of marine phytoplankton. Department of the Navy, Project No. NR105-443, Final Report for N00014-67-A-0109-0005.

Haxo, F. T., and L. R. Blinks. 1950. Photosynthetic action spectra of marine algae. J. Gen. Physiol. 33: 389-422.

Hollaender, A. 1938. Monochromatic ultra-violet radiation as an activating agent for the eggs of Arbacia punctulata. Biol. Bull. 75:248-265.

Hunter, J. R., J. H. Taylor, and H. G. Moser. 1979. Effect of ultraviolet irradiation on eggs and larvae of the northern anchovy, Engraulis mordax, and the Pacific mackerel, Scomber japonicus, during the embryonic stage. Photochem. Photobiol. 29: 325-338.

Hunter, J. R., S. E. Kaupp, and J. H. Taylor. 1981. Effects of solar and artificial ultraviolet-B radiation on larval northern anchovy, Engraulis mordax. Photochem. Photobiol. (In press).

Huntsman, A. G. 1925. Limiting factors for marine animals I. The lethal effect of sunlight. Contrib. Can. Biol. 2:83-88.

Jerlov, N. G. 1976. Irradiance. In: Marine Optics 127-150. Elsevier Scientific, Amsterdam.

Jitts, H. R., A. Morel and Y. Saijo. 1976. The relation of oceanic primary production to available photosynthetic irradiance. Aust. J. Mar. Freshwater Res. 27:441-454.

Jokiel, P. L. 1980. Solar ultraviolet radiation and coral reef epifauna. Science 207: 1069-1071.

Jones, L. W., and B. Kok. 1966. Photoinhibition of chloroplast reactions. 1. Kinetics and action spectra. Plant Physiol. 41:1037-1043.

Karanas, J. J., H. Van Dyke, and R. C. Worrest. 1979. Midultraviolet (UV-B) sensitivity of Acartia clausii (Copepoda). Limnol. Oceanogr. 24:1104-1116.

Karanas, J. J., R. C. Worrest, and H. Van Dyke. 1981. Impact of UV-B radiation (290-320 nm) on the fecundity of Acartia clausii (Copepoda). Mar. Biol. (In press).

Kaupp, S. E., and J. R. Hunter. 1981. Photorepair in larval anchovy, Engraulis mordax. Photochem. Photobiol. 33:253-256.

Klugh, A. B. 1929. The effect of the ultra-violet component of sunlight on certain marine organisms. Can. J. Res. 1:100-109.

Klugh, A. B. 1930. The effect of the ultra-violet component of the sun's radiation upon some aquatic organisms. Can. J. Res. 2: 312-317.

Lees, D. C., and G. A. Carter. 1972. The covering response to surge, sunlight, and ultraviolet light in Lytechinus anamesus (Echinoidea). Ecology 53: 1127-1133.

Lorenzen, C. J. 1979. Ultraviolet radiation and phytoplankton photosynthesis. Limnol. Oceanogr. 24: 1117-1120.

Marinaro, J. Y., and M. Bernard. 1966. Contribution a l'etude des oeufs et larves pélagiques de Poissons méditerranéens I. Note preliminarie sur l'influence léthale du rayonnement solaire sur les oeufs. Pelagos 6:49-55.

McLeod, G. C. 1958. Delayed light action spectra of several algae in visible and ultraviolet light. J. Gen. Physiol. 42:243-250.

McLeod, G. C. and J. Kanwisher. 1962. The quantum efficiency of photosynthesis in ultraviolet light. Physiol. Plant. 15:581-586.

Moehring, M. P. 1980. Influence of ultraviolet-B radiation on the heterotrophic activity of estuarine bacterioplankton. Ph.D. Thesis, Oregon State University.

Nachtwey, D. S., and R. D. Rundel. 1981. Ozone change: Biological effects. In: Stratospheric Ozone and Man [eds. F. A. Bower and R. B. Ward]. CRC Press, West Palm Beach, Florida. (In press).

Pommeranz, T. 1974. Resistance of plaice eggs to mechanical stress and light. In: The Early Life History of Fish [ed. J. H. S. Blaxter] 397-416. Springer-Verlag, New York.

Setlow, R. B. 1974. The wavelengths in sunlight effective in producing skin cancer: A theoretical analysis. Proc. Nat. Acad. Sci. USA 71:3363-3366.

Shibata, K. 1969. Pigments and a UV-absorbing substance in corals and a blue-green alga living in the Great Barrier Reef. Plant Cell Physiol. 10:325-335.

Sieburth, J. McN. 1976. Bacterial substrates and productivity in marine ecosystems. Annu. Rev. Ecol. Syst. 7:259-285.

Seiburth, J. McN. 1979. Sea Microbes. Oxford University Press, New York.

Sivalingham, P. M., T. Ikawa, Y. Yokohama and K. Nisizawa. 1974. Distribution of a 334 UV-absorbing-substance in algae, with special regard of its possible physiological roles. Bot. Mar. 17: 23-29.

Smith, R. C., and K. S. Baker. 1980a. Stratospheric ozone, middle ultraviolet radiation, and carbon-14 measurements of marine productivity. Science 208: 592-593.

Smith, R. C., and K. S. Baker. 1980b. Biologically effective dose transmitted by culture bottles in ^{14}C productivity experiments. Limnol. Oceanogr. 25:364-366.

Smith, R. C., K. S. Baker, O. Holm-Hansen and R. Olson. 1980. Photoinhibition of photosynthesis in natural waters. Photochem. Photobiol. 31:585-592.

Steemann Nielsen, E. 1952. The use of radioactive carbon (C^{14}) for measuring organic production in the sea. J. Cons. Cons. Int. Explor. Mer. 18:117-140.

Steemann Nielsen, E. 1964. On a complication in marine productivity work due to the influence of ultraviolet light. J. Cons. Cons. Int. Explor. Mer. 29:130-135.

Thomson, B. E., H. Van Dyke and R. C. Worrest. 1980b. Impact of UV-B radiation (290-320 nm) upon estuarine bacteria. Oecologia (Berl.) 47:56-60.

Thomson, B. E., R. C. Worrest and H. Van Dyke. 1980a. The growth response of an estuarine diatom (Melosira nummuloides [Dillw.] Ag.) to UV-B (290-320 nm) radiation. Estuaries 3:69-72.

Vacarro, R. F. and J. H. Ryther. 1954. The bacteriocidal effects of sunlight in relation to "light" and "dark" bottle photosynthesis experiments. J. Cons. Cons. Int. Explor. Mer. 20:18-24.

Van Dyke, H., and B. E. Thomson. 1975a. Response of a simulated estuarine community to UV irradiation. In: Impacts of Climatic Change on the Biosphere, Part I: Ultraviolet Radiation Effects [eds. D. S. Nachtwey et al.] 5-95 to 5-113. U. S. Dept. Transportation, DOT-TST-75-55, Washington, D.C.

Van Dyke, H., and B. E. Thomson. 1975b. Response of model estuarine ecosystems to UV-B radiation. In: Impacts of Climatic Change on the Biosphere, Part I: Ultraviolet Radiation Effects [eds. D. S. Nachtwey et al.] 5-9 to 5-10. U.S. Dept. Transportation, DOT-TST-75-55, Washington, D. C.

Wells, P. H., and A. C. Giese. 1950. Photoreactivation of ultraviolet light injury in gametes of the sea urchin Strongylocentrotus purpuratus. Biol. Bull. 99:163-172.

Wolniakowski, K. U. 1980. The physiological response of a marine phytoplankton species, Dunaliella tertiolecta, to mid-wavelength ultraviolet radiation. M.S. Thesis. Oregon State University.

Worrest, R. C., D. L. Brooker and H. Van Dyke. 1980. Results of a primary productivity study as affected by the type of glass in the culture bottles. Limnol. Oceanogr. 25:360-364.

Worrest, R. C., B. E. Thomson and H. Van Dyke. 1981b. Impact of UV-B radiation upon estuarine microcosms. Photochem. Photobiol. 33: 861-867.

Worrest, R. C., H. Van Dyke and B. E. Thomson. 1978. Impact of enhanced simulated solar ultraviolet radiation upon a marine community. Photochem. Photobiol. 27:471-478.

Worrest, R. C., K. U. Wolniakowski, J. D. Scott, D. L. Brooker, B. E. Thomson and H. Van Dyke. 1981a. Sensitivity of marine phytoplankton to UV-B radiation: Impact upon a model ecosystem. Photochem. Photobiol. 33:223-227.

Yoshida, T., and P. M. Sivalingham. 1970. Isolation and characterization of the 337 mμ UV-absorbing substance in red alga, Porphyra yezoensis Ueda. Plant Cell Physiol. 11:427-434.

ZoBell, C. E., and G. F. McEwen. 1935. The lethal action of sunlight upon bacteria in sea water. Biol. Bull 68:93-106.

ASSESSMENT OF EFFECTS OF UV RADIATION ON MARINE FISH LARVAE

John R. Hunter (1), Sandor E. Kaupp (2) and
John H. Taylor (2)
National Marine Fisheries Service
Southwest Fisheries Center
La Jolla, California 92038 (1) and
University of California San Diego,
Center for Human Information Processing
La Jolla, California 92093 (2)

ABSTRACT

Uncertainties in estimating the biologically effective UV dose for northern anchovy larvae, *Engraulis mordax*, are discussed including action spectrum used to weight UV dose, dose reciprocity, photorepair, relative sensitivity of life stages, and biological criteria for measuring UV effects. A preliminary estimate of UV induced losses to larval anchovy standing stock is presented. This calculation takes into account seasonal changes in larval abundance, UV radiation and average cloud cover, vertical distribution, and penetration of UV radiation into the habitat. The possible effects of ozone diminution on other populations of larval fish and the uncertainties in making an impact assessment on natural marine populations are discussed. Present evidence indicates that anchovy and other larval fish populations may be under some UV stress today but predicted levels of ozone diminution will probably not have a major effect on larval populations. Imprecisely defined habitat characteristics and the unknown effect of a small augmentation on high natural mortality rates are major barriers to accurate assessment of ozone decline on marine fish populations.

INTRODUCTION

Most of the world's important marine fish stocks, tunas, mackerels, flatfishes, pilchards, anchovies, cods and many others, produce pelagic eggs and larvae. Eggs of these fishes are small (about 1 mm diameter), translucent and neutrally buoyant. The translucent larvae hatch in an immature state. The epidermis is only a few cells thick (Bullock and Roberts 1975; O'Connell 1981), no scales and few mucus secreting cells occur in the integument (O'Connell 1981), and little pigmentation exists. Owing to the absence of protective integument or pigmentation, UV can penetrate deep into the body producing lesions in the eye, brain, and other tissues (Hunter et al. 1979). Consequently, most pelagic eggs and larval stages of fishes studied to date show a high sensitivity to UV radiation and often die if exposed in shallow containers to solar UV (Table 1). Although the UV sensitivity of these translucent pelagic larvae seems certain, it is not known if significant effects on larval populations would occur if ozone were substantially reduced.

The objective of this paper is to evaluate the uncertainties in making an assessment of UV effects on natural larval populations. We begin by considering the uncertainties underlying our estimate of the biologically effective UV dose for northern anchovy larvae, *Engraulis mordax*; we discuss action spectra used for weighting the dose, dose reciprocity, photorepair, relative sensitivity of life stages, and biological criteria for UV effects. We then present a preliminary estimate of UV effects on the larval anchovy population and discuss the uncertainties underlying the estimate. In a final section, we relate our findings on anchovy to other pelagic marine fish larvae.

In this paper we do not provide detailed descriptions of our methods because most of the biological data and methods used in this paper have been presented before (Hunter et al. 1979; Hunter et al. 1981; Kaupp and Hunter 1981). We also present some new and unpublished data, but the methodology, apparatus and radiometric techniques are the same as in the previous papers.

TABLE 1. UV-B sensitivity of fish eggs and larvae.

Species	Life stage	Days exposed	UV-B dose[1] ($Jm^{-2}_{bio.eff.}$)	Percent Survival	Reference
Engraulis mordax (northern anchovy)	eggs & yolk-sac larvae	4	1150	50.0	Hunter et al. 1979
Cynoscion nobilis (white sea bass)	eggs & yolk-sac larvae	4	1450[2] (est.)	50.0	Hunter et al. Unpublished
Scomber japonicus (Pacific mackerel)	eggs & yolk-sac larvae	4	2350	50.0	Hunter et al. 1979
Pleuronectes platessa (plaice)	eggs	17	390[3] (est.)	50.0	Pommeranz 1974
		2.5	100[3] (est.)	50.0	Pommeranz 1974
Engraulis mordax (northern anchovy)	eggs & yolk-sac larvae	12	605	50.0	Hunter et al. 1981
Sardina pilchardus (pilchard)	eggs	1	54[4] (est.)	52.0 (at hatch)	Marinaro & Bernard 1966
Diplodus annularis (porgy)	eggs	1	240[4] (est.)	94.8 (at hatch)	Marinaro & Bernard 1966
Trachurus symmetricus (jack mackerel)	eggs	1	260[4] (est.)	73.6 (at hatch)	Marinaro & Bernard 1966

1 The UV dose is weighted as indicated in the text.
2 The LD_{50} was estimated from data determined for UV doses less than 1000 $Jm^{-2}_{bio\ eff.}$
3 The dose was calculated from the date, latitude, and cloud cover presented in the study (Pommeranz 1974).
4 The doses were calculated for clear days for the days of exposure and latitude (Baker et al. 1980).

BIOLOGICALLY EFFECTIVE DOSE

The biologically effective UV dose we use to assess effects of UV radiation on anchovy in the sea is 605 $Jm^{-2}_{bio.eff.}$ (from Hunter et al. 1981, Fig. 1). This is the cumulative UV-B dose (weighted by a DNA action spectrum modified for anchovy mortality) that produced a 50% mortality of anchovy larvae after 12 daily exposures (egg stage to larval age 12 days). Data from experiments using only solar UV-B (290-320 nm) and from others using artificial UV-B (285-320 nm) were combined to calculate the dose response line using probit analysis (Finney 1952).

Major factors in evaluating the biological reliability and applicability of a biologically effective UV dose are: the action spectrum used to weight the dose-

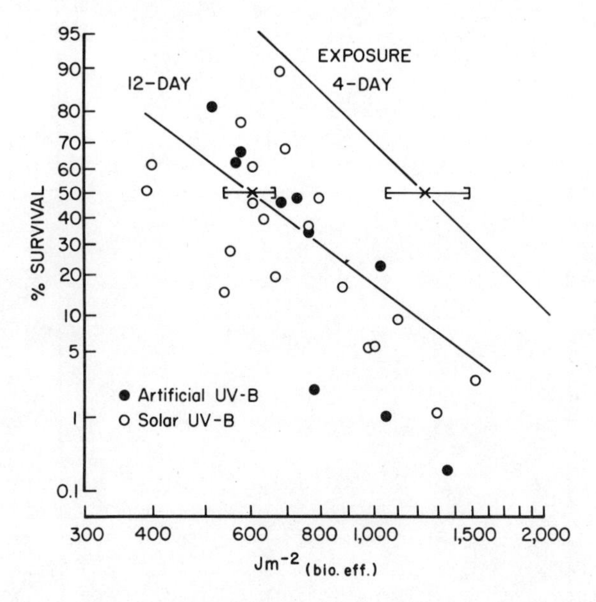

Figure 1. Relation between percent survival of northern anchovy larvae (probit scale) and cumulative bio. eff. weighted UV-B dose (log scale) for 12 daily exposures (lower line, and points) and for 4 daily exposures (upper line). Regression line for 12 day experiment based on combined data for artificial UV radiation (solid points) and solar radiation (open circles); each point is a mean for 10 containers; regression of survival probit (Y) for 12 days on log dose (X) is Y = 4.275X-6.89. Biologically effective UV dose for northern anchovy larvae is derived from 12 day experiment.

response data; dose reciprocity, photorepair mechanisms; the relative UV sensitivity of life stages; and the biological criteria used to measure effects. We consider our estimate of the biologically effective UV dose in terms of each of these factors. Statistical uncertainties are not considered because these are small relative to the biological uncertainties.

1. Spectral sensitivity to UV-B

Interpretation of all UV damage depends on the choice of an appropriate action spectrum for weighting biological effects. It is essential to weight biological effects because: ozone diminution results in an increase in radiant energy and change in spectrum in the most bio-active portion of the UV band; spectral irradiance in the sea varies with water type and depth; and significant differences exist between artificial spectra used in most laboratory work (filtered sunlamps) and the solar spectrum.

We used broad band spectroscopy (Nachtwey in Smith et al. 1980) to identify the appropriate action spectrum to weight dose-response data for anchovy larvae. This technique involves selection of an action spectrum and checking for consistency between measured biological effect and dose weighted by the action spectrum when biological effects are measured under different broad band UV spectra. This technique is valid as long as the law of reciprocity holds; that is, the biological effect is independent of dose rate over the range of doses under study (Smith et al. 1980), a point we consider in the next section.

Hunter et al. (1979) weighted their dose-response data for anchovy (4 daily exposures at high dose rates) by the DNA action spectrum of Setlow (1974) using the analytical fit of Green and Miller (1975) and brought along a common line dose-response data for three different artificial UV-B spectra. More recent data taken at lower dose rates (12 daily exposures) conducted under solar and artificial spectra did not converge when weighted by the Green and Miller equation (Hunter et al. 1981). The data for the 12-day experiment converged when weighted by an action spectrum resembling the DNA action spectrum but giving a higher weight to energy from 300-320 nm (Fig. 2).

This change in the spectrum from the DNA action spectrum greatly affected the weighting of the solar

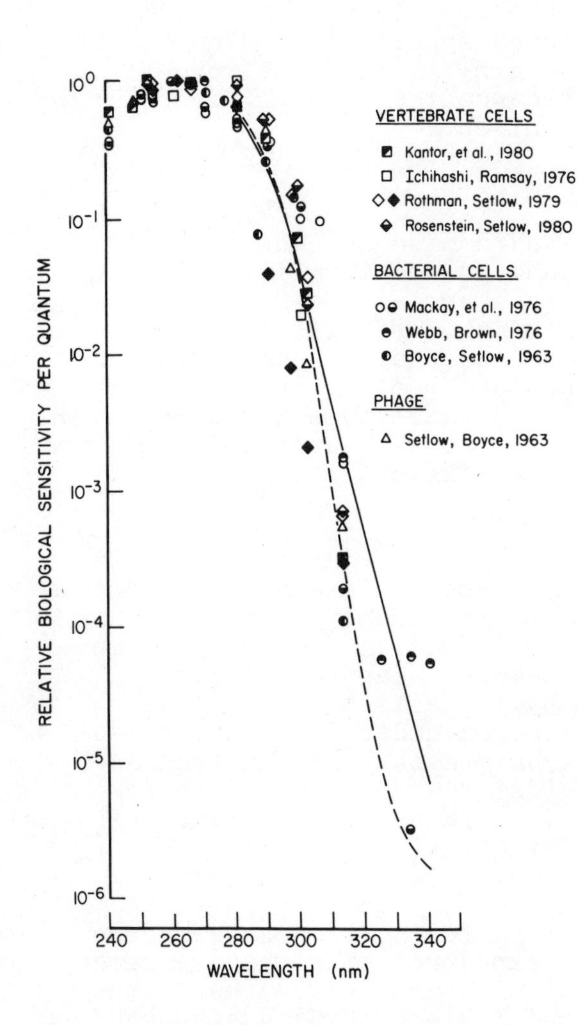

Figure 2. Comparison of action spectrum with data for the killing of vertebrate and bacterial cells, and inactivation of a phage; dotted line, Green and Miller (1975) fit to the DNA action spectrum of Setlow (1974); solid line, empirical action spectrum used to weight dose-response data for anchovy; and points from UV action spectra in literature normalized to the wavelength yielding the maximum response in the particular study. Points for vertebrate cells include: unscheduled DNA synthesis (Ichihashi and Ramsay, 1976; for 10 $J \cdot m^{-2}$ dose) (□), and inactivation (Kantor, et al., 1980; average values for normal and DNA excision-repair deficient strains) (◪) of human fibroblast; DNA pyrimidine dimer formation (◇) and killing (◆) of Chinese hamster cells. (Rothman and Setlow, 1979); and the killing of haploid frog cells (Rosenstein and Setlow, 1980; average of data with and without photoreactivating light) (⬘) For bacterial cells: Killing of Salmonella typhimurium (rec A) (○), and E. coli (rec A) (◒) (Mackay, et al., 1976; average of stationary phase and exponentially growing cells); stationary phase E. coli (B/r and B/r Hcr) (◓) (Webb and Brown, 1976); of E. coli (15 TAU) (◐) (Boyce and Setlow, 1963). For the inactivation of phage T_4B (△) (Boyce and Setlow, 1963).

data, but had a negligible effect on dose-response data taken under different artificial spectra. Thus, weighting by our modification of the DNA action spectrum brought data along two common lines, one for the high dose rates used in the earlier work, and one for low dose rates (12 daily exposures) which we use as the spectrum to weight all data. We will refer to this weighting as biologically effective (bio. eff.) in the rest of the paper. To convert our data (12 daily exposures) to unweighted Jm^{-2}, multiply solar data by 1900 and data from artificial sources (FS40 sunlamps and cellulose triacetate plastic filters) by 15.7.

The major assumption underlying our use of a DNA action spectrum modified for anchovy mortality is that the difference in survival between experiments using solar UV and those employing artificial UV was caused by a difference in spectral irradiance. On the other hand, higher solar mortality might be caused by dose rate effects that may have occurred under variable solar radiation but not under the uniform UV dose rates used in the laboratory. If this were the case, the preferable biological effective UV dose is one based on the solar data weighted by the DNA action spectrum. This change produces only minor changes in the predictions of effects in the sea (Appendix Table 4). Even the most extreme criteria for a biological dose (DNA weighting of laboratory dose-response data) alters our estimate of the relative effect of ozone decline on anchovy standing stock by less than a factor of 2. (The derivation of these calculations is explained later in the section "Preliminary Assessment of Effects of UV on Larval Anchovy Populations.") Use of the modified DNA action spectrum does not appear to be a major source of potential error.

2. Dose reciprocity

Unpublished data indicate dose rate effects occur in anchovy larvae, that is, dose reciprocity did not hold over a broad range of UV dose rates. In these experiments we used artificial UV sources and measured survival at age 12 days (procedures and radiometry are described in Hunter et al. 1979 and 1981). An example of these data is given in Table 2 and additional data are provided in Appendix Table 1. In the example, many fewer larvae survived at age 12 days when a cumulative dose of 530 Jm^{-2}bio. eff. was given at a high dose rate (41 $Jm^{-2}h^{-1}$bio. eff., two daily exposures) for 6 h (see Hunter et al. 1979 for the daily UV exposure

TABLE 2. Effect of UV-B dose rate on survival of anchovy larvae.

Cumulative UV-B dose Jm^{-2} bio. eff.	Percent survival[1] ± 2 X SE[2]	Dose rate[3] $Jm^{-2}h^{-1}$ bio.eff.	Number of daily UV exposures[4]	Larval ages when exposed to UV (days)
517	81.1 ± 23.8	6.6	12	0-12
530	3.5 ± 1.9	40.6	2	1-2
530	22.2 ± 6.7	40.6	2	4-5
559	63.4 ± 9.5	7.1	12	0-12
587	67.6 ± 19.0	7.5	12	0-12

[1] Normalized to survival in controls; survival measured at age 12 days; and N = 10 in each experiment.

[2] ± 2 X standard error of mean survival; N = 10 in each treatment.

[3] Summer fluence rates exceed 50 $Jm^{-2}h^{-1}$ bio. eff. at local apparent noon in La Jolla, Ca. (33°N lat.)

[4] Each daily UV exposure was 7.0 h (see Hunter et al. 1979 for daily pattern of UV irradiance).

scheme) than when comparable cumulative doses were given at much lower dose rates. At the same cumulative dose, from 2.5 to 17 times as many larvae can be expected to survive to age 12 days at a lower dose rate (calculated from dose-response line, Figure 1, where dose rate = 6.8 $Jm^{-2}h^{-1}{}_{bio.eff.}$ for 12 daily exposures with cumulative dose = 530 $Jm^{-2}{}_{bio.eff.}$) than at the higher dose rate (41 $Jm^{-2}{}_{bio.eff.}$).

The range of cumulative doses over which dose reciprocity holds is uncertain because differences in UV sensitivity among life stages makes interpretation difficult. Nevertheless, it appears that dose reciprocity may hold over the range of dose rates used to calculate the biologically effective dose for 12 days (6-20 $Jm^{-2}h^{-1}{}_{bio.eff.}$). This dose-response line was formed by varying daily dose rates among treatments, although expressed as a cumulative dose for 12 days. These dose rates occur naturally at our latitude and we use a daily dose, not cumulative dose, to estimate effects on natural populations. Dose rate effects may be of great importance in other studies where a cumulative dose based on very high dose rates is used to forecast events under natural conditions.

Important from the standpoint of UV assessment is that dose rate effects occurred at fluence rates that can occur on clear solar days. We have measured fluence rates in excess of 50 $Jm^{-2}h^{-1}{}_{bio.eff.}$ at local apparent noon, on clear summer days, at our latitude (33°N). We suspect that the differences in variability in survival between experiments using solar and those employing artificial sources of UV (Figure 1) could be caused by interactions between life stage sensitivity and variable solar fluence rates. The variable results obtained by Pommeranz (1974) in experiments on solar UV exposure of plaice eggs could also be interpreted in this way (Table 1).

3. Photorepair

Photorepair of UV-B induced lesions could be a significant factor in interpretation of biological effects of UV radiation in the sea. Photorepair has been demonstrated in larval anchovy at high dose rates (25 to 80 $Jm^{-2}h^{-1}{}_{bio.eff.}$) by shifting the period of UV exposure from the first 6 h to the last 6 h of a 12 h white light day, but the same procedure had no effect at low dose rates (6.4 to 10.1 $Jm^{-2}h^{-1}{}_{bio.eff.}$) Kaupp and Hunter 1981). At high dose rates, photorepair was indicated by survival and growth of anchovy larvae but at the low dose rates, repair presumably kept up with damage; consequently, shifting the period of photorepair fluence within the day had no effect.

Kaupp and Hunter (1981) also determined the relation between UV induced mortality and the intensity of photoreactive fluence (320-500 nm) (Figure 3). These experiments demonstrated that the level of photorepair fluence was sufficient to fully stimulate repair in all past laboratory work on anchovy. Kaupp and Hunter (1981) also found that even at the relatively high dose rate used (29 $Jm^{-2}h^{-1}{}_{bio.eff.}$), the photorepair fluence needed to fully stimulate repair was about 10% of that available from the sun during a clear equinoctial day at 33°N latitude. Just beneath the water surface, the ratio of photorepair fluence to UV-B fluence is about 100/1. This ratio will increase with depth in the sea (Smith and Baker, 1979). Thus photorepair will operate maximally for anchovy larvae at any depth, even under the added stress of enhanced UV-B from ozone depletion.

4. Life stage sensitivity

Unpublished data, collected using the same procedures and artificial UV sources outlined by Hunter et al. (1979; 1981) demonstrate that major differences exist in the UV sensitivity of anchovy at different life stages. Eggs, yolk-sac larvae (age 1-2 days), and larvae after the initiation of feeding (age 4-5 days) were given 2 daily exposures of UV-B and survival measured at age 12 days (Figure 4). Anchovy eggs were the least sensitive to UV radiation. No effect could be detected at the highest cumulative dosage used (328 $Jm^{-2}{}_{bio.eff.}$). Except for the lowest dosage tested (109 & 137 $Jm^{-2}{}_{bio.eff.}$), the survival of all other stages were statistically different from the controls ($P < .05$). Yolk-sac larvae were over twice as sensitive to UV as were 4-5 day-old larvae presumably because of a higher sensitivity during embryonic development. The high UV tolerance of egg stages indicates that the chorion may protect the developing embryo from UV radiation. This may be an adaptation to a near surface occurrence. Ahlstrom (1959) and Ahlstrom and Stevens (1976) observed that anchovy eggs are sometimes much closer to the sea surface than are anchovy larvae.

5. Biological criteria for effects

The criterion used to measure UV effects is a major determinant of the predicted biological effective

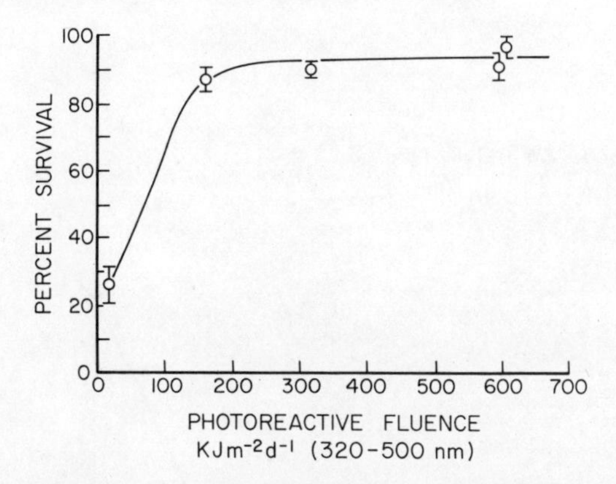

Figure 3. Survival of embryonic northern anchovy after 4 daily exposures to UV-B in relation to daily dosage of photoreactive fluence (320-500 nm in $KJm^{-2}d^{-1}$); UV-B fluence was constant at 189 $Jm^{-2}d^{-1}{}_{bio.eff.}$. Points are the mean survival of 15 containers (normalized to control) and bars are the 95% confidence intervals.

dose. In our earlier work (Hunter et al., 1979) we established a dose-response relation based on four daily UV exposures using the survival of larvae at age 4 days as the criterion for the biologically effective dose. Fifty percent survival occurred at about twice the cumulative dose, as when survival at age 12 days was the criterion (Figure 1). Our data on life stage sensitivity indicate that none of the larvae surviving the LD_{50} dose at age 4 days would survive to age 12 days. Apparently, the UV damage received during the embryonic period subsequently affected survival once larvae had to capture live prey instead of subsisting on yolk. By using a criterion of survival at age 12 days we may have again underestimated the effect of UV radiation on survival. Problems of this type will persist until UV-exposed larvae are reared to adulthood. Limits can be set on the total days of possible UV exposure, however. At age 20 days larvae begin daily vertical migrations to the surface at night and descend below significant penetration of UV in the day (Hunter and Sanchez 1976). In addition to survival, we have also examined effects of UV radiation on growth and histological evidence of damage, but have not used these criteria for an estimate of biologically effective dose.

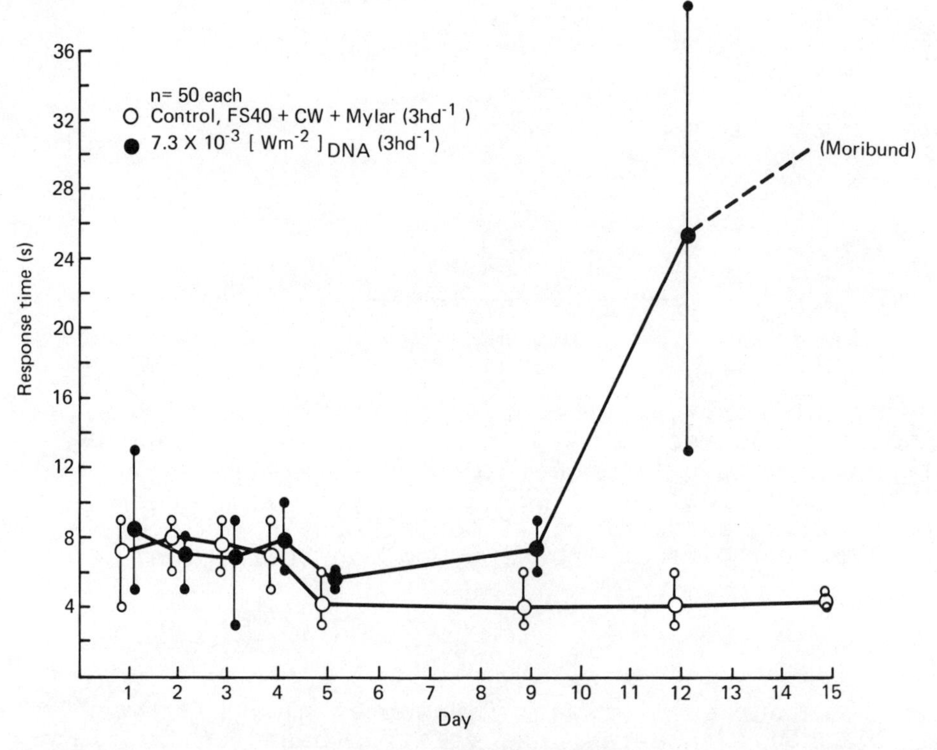

Figure 4. Sensitivity to UV-B radiation of various life stages of northern anchovy survival (probit scale) measured at age 12 days after 1-2 daily exposures in the egg, or yolk-sac (age 1-2 days), or first-feeding (age 4-5 days) stages. Each point is mean survival in 10 containers plotted at the cumulative UV-B dose (log scale); UV sources were FS40 sunlamps filtered by cellulose triacetate plastic sheet.

In all our work, the mean length of larvae was determined for each treatment at the end of every experiment. The length of UV-exposed larvae, expressed here as a fraction of the length of larvae in the controls, was used as a measure of growth. Growth of UV-treated larvae increased with the numbers of larvae that survived a UV treatment indicating a close linkage between survival and growth. This trend was consistent under all experimental conditions, including widely different dose rates, number of daily UV exposures, life stages, experiment duration and species (Figure 5). That anchovy and mackerel larvae exposed to UV were smaller than the controls at the end of the embryonic stage (age 4 days), indicates that growth was directly inhibited by UV radiation (Hunter et al. 1979). This may explain the overall consistency of the relation

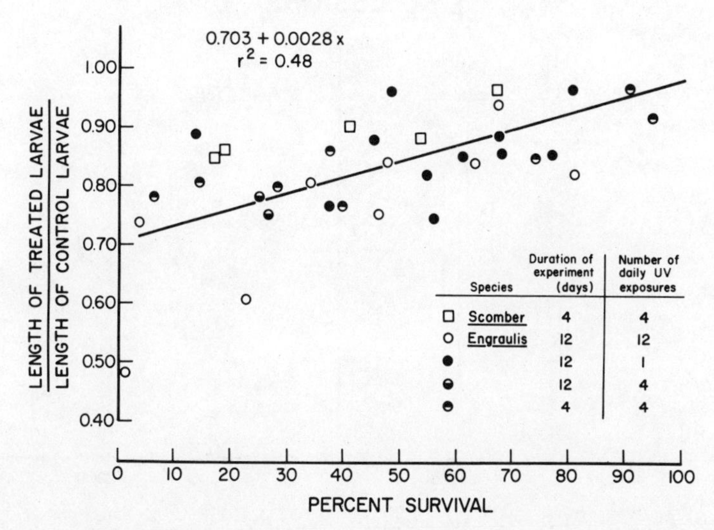

Figure 5. Growth of UV-B treated larval northern anchovy and Pacific mackerel as a function of the percentage of larvae surviving the same treatment. Growth is expressed nondimensionally, as the ratio of length of UV-treated larvae to that of the control (measured at the end of the experiment). Data are from a variety of experiments using artificial UV-B. Each point is mean value for 10 or 15 containers.

between growth and survival. However, reduction in growth in older larvae could also be caused by a failure to obtain sufficient food owing to UV damage to sensory or other organ systems.

Inhibition of growth was a sensitive assay of UV effects. Statistically significant differences in larval size between control and treatment groups were always detected even at the lowest UV doses (Hunter et al. 1979; 1981. In the sea, reduction in growth could increase mortality from predation because it increases the duration of the most vulnerable life stages.

High dose rates of UV-B radiation over a 4-day period induced lesions in the eye, brain and olfactory bulb in larval anchovy and mackerel (Hunter et al. 1979). The incidence of lesions increased with dosage (Figure 6). That UV-B penetrates deep into the body of larvae in these early embryonic stages is not surprising, as little pigmentation exists in either species and the epidermis is only a few cells thick. Larvae with

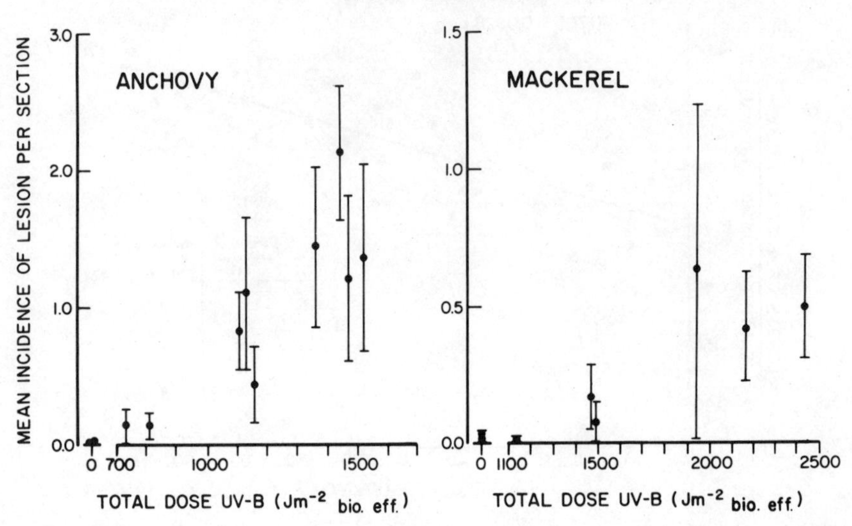

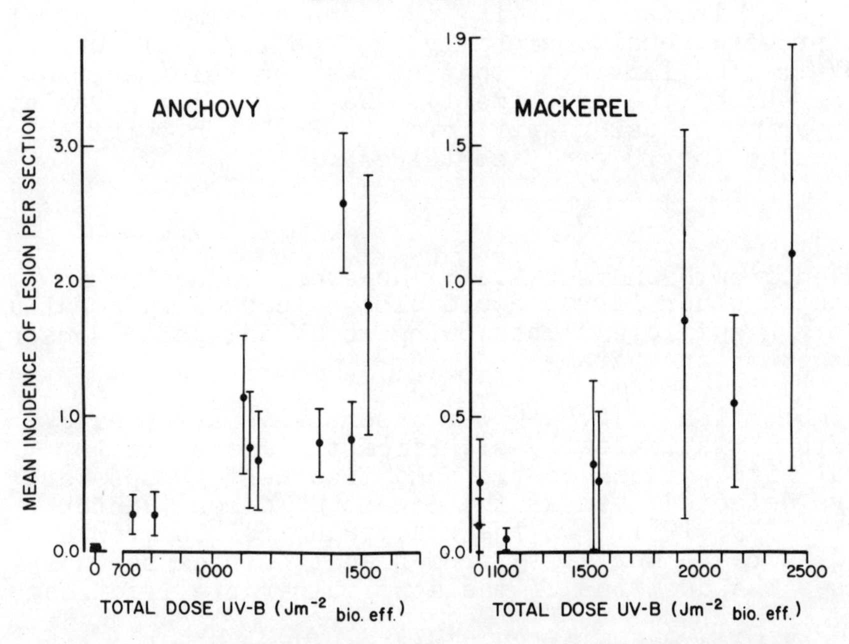

Figure 6. Mean incidence of lesions in the eye and brain of northern anchovy and Pacific mackerel that survived 4 daily doses of UV-B radiation (weighted by a empirical fit to DNA action spectrum; see Fig. 2) during the embryonic period. Incidence of lesions determined at age 4 days; points are means for treatments, and vertical bars are two standard errors.

lesions in the brain are probably incapable of feeding and die once all yolk is absorbed. At the high dose rate employed in these experiments, no larvae would survive to 12 days. Among the surviving larvae in our solar experiments (12 daily UV exposures), the incidence of lesions was much lower, and some UV-exposed larvae showed incipient signs of starvation. This work indicated that solar radiation can produce lesions in the brain and eye of larval anchovy. The incidence at age 12 days did not increase with dosage, presumably because of differential mortality rates in the UV treatments.

6. Conclusions

We have discussed the major factors affecting the biologically effective UV dose for anchovy larvae. Biological uncertainties produced by use of the modified DNA action spectrum, dose rate effects, and photorepair appear to be relatively minor. Solar testing for UV effects eliminates most of the objections in the use of an approximate action spectrum. Although dose reciprocity does not hold for all dose rates, it appears to hold for the ranges used in our dose-response line. In addition, the rates used are the same as those that occur naturally. Similarly, photorepair does not bias our results because sufficient photorepair fluence existed in all experiments, and no decrement of photorepair fluence relative to UV fluence is expected under natural conditions.

The greatest uncertainties are associated with differences in the life stage UV sensitivity of anchovy and the criterion used to measure effects. Our estimate of biologically effective dose may be conservative. High sensitivity of yolk-sac larvae and their sensitivity to higher dose rates over 2 days indicates that our biological effective dose, based on average sensitivity of all stages at low dose rates, may underestimate effects close to the sea surface. Similarly, a change in biological criteria for UV effects, for example, a decrement in growth, or measurement of the results of UV radiation over longer periods, would lower the effective dose and thereby elevate predicted losses to the anchovy larval population to present levels of UV fluence.

PRELIMINARY ASSESSMENT OF EFFECTS OF UV ON LARVAL ANCHOVY POPULATIONS

Realistic assessment of the effects of UV-B on the anchovy larvae population requires knowledge of seasonal changes in larval abundance, incident UV-B, spatial and temporal variation in the depth of penetration of UV-B in the habitat, vertical distribution of larvae in the sea, rates of vertical mixing of eggs and larvae, and natural rates of mortality. At present, many of these parameters can only be approximated or are unknown for anchovy and much less is known for most other larvae. In this section we present the preliminary assessment of possible effects of UV-B on larval anchovy populations of Hunter et al. 1981. These calculations will be used in subsequent sections to illustrate uncertainties in making an assessment of UV effects on larval populations.

Assessment of Effects

In this calculation the assumption is made that no vertical mixing of eggs and larvae occurs and that the proportion of larvae at any depth is the same as that determined by Ahlstrom (1959) (the equation used to fit Ahlstrom's data and all other documentation for these calculations are given as footnotes to Table 3). We use as the biological effective dose the daily equivalent of LD_{50} for 12 daily exposures, i.e. 605/12 = 50 $Jm^{-2}\ d^{-1}$ bio.eff..

It is assumed that all larvae die that exist at depths that receive a daily dose equal to or greater than 50 Jm^{-2} bio.eff. for 1 month, whereas those that receive lower doses survive. The UV-B daily dose at the water surface is estimated using the analytical estimates of Baker et al. (1980) adjusted for average cloud cover using the equation of Bener (1964). Average cloud cover data were 18-year monthly means (Renner 1979) for an area in the Southern California Bight (30° - 35°N, 115°-120°W) which is in the center of the anchovy spawning region (Ahlstrom 1967). The equations and the attenuation coefficient of Smith and Baker (1979) for moderately productive ocean water (0.5 mg Chla m^{-3}) are used to estimate the depth of penetration of UV-B in the sea at various ozone concentrations.

The calculations of Hunter et al (1981) indicate that at the present time about 13% of the annual production of larvae could be lost because of UV-B

TABLE 3. Proportion of northern anchovy annual larval standing stock affected by UV-B radiation at present (0.32 cm) and at 25% reduction (0.24 cm) in ozone concentration.

Month	.32 cm ozone (ambient)				.24 cm ozone (25% reduction)			
	Daily fluence UV-B (Jm^{-2} bio. eff.)[1]	Depth of LD_{50}[2] (m)	Larvae at or above LD_{50}[3]		Daily fluence UV-B (Jm^{-2} bio. eff.)[1]	Depth of LD_{50}[2] (m)	Larvae at or above LD_{50}[3]	
			Percent[4]	Percent weighted by relative larval abundance[5]			Percent[4]	Percent weighted by relative larval abundance[5]
J	25	ø	ø	ø	40	ø	ø	ø
F	52	0.11	3.76	0.62	82	1.34	13.32	2.17
M	94	1.74	15.09	2.53	149	2.95	19.65	3.29
A	150	3.03	19.90	3.96	238	4.21	23.48	4.67
M	158	3.17	20.37	2.70	250	4.35	23.85	3.16
J	167	3.32	20.85	1.65	265	4.50	24.28	1.92
J	162	3.24	20.59	1.16	257	4.42	24.06	1.36
A	147	2.97	19.72	0.37	233	4.16	23.32	0.44
S	111	2.20	16.96	0.30	176	3.40	21.09	0.37
O	65	0.72	9.73	0.11	103	1.95	15.98	0.18
N	30	ø	ø	ø	48	ø	ø	ø
D	20	ø	ø	ø	32	ø	ø	ø
Σx				13.40				17.56

[1]The sea surface irradiance is taken from Baker et al. (1980) coefficients for the model of Green et al. (1974) for 33°N and adjusted for 18-year average cloud cover (1961-1978) for the Southern California Bight (30-35°N, 115-120°W; Renner 1979) using the equation of Bener (1964) and then weighted by a modification of the DNA action spectrum.

[2]Depth in sea where the daily UV-B fluence is equivalent to the LD_{50}. The depth of UV-B penetration was calculated for moderately productive water (0.5 mg Chla m^{-3}) using the downwelling irradiance diffuse attenuation coefficients of Baker and Smith (pers. comm.); K_T = 0.3630 for 0.32 cm ozone and K_T = 0.3702 for 0.24 cm ozone.

[3]50 Jm^{-2}bio. eff.

[4]The monthly percent of larvae at or above the LD_{50} depth was calculated using the vertical distribution data for nothern anchovy larvae of Ahlstrom (1959) fitted by the equation, $y = 11.44x^{0.5}$, $r^2 = 0.99$, for the data from the surface to 28 meters, where y is the cumulative percent of larvae from the surface to depth x, in meters.

[5]10-year average larval abundance from Lasker and Smith (1977) was used to weight the percent of larvae affected by UV-B for each month; see Figure 7, upper graph.

mortality and this may increase to about 18% at a 25% reduction in ozone (Table 3 and Figure 7). The increase in the estimated UV-B mortality for the larval population is nearly a linear function of ozone concentration, with mortality increasing about 0.2% for each percent decrease in ozone (Figure 8). Thus ozone

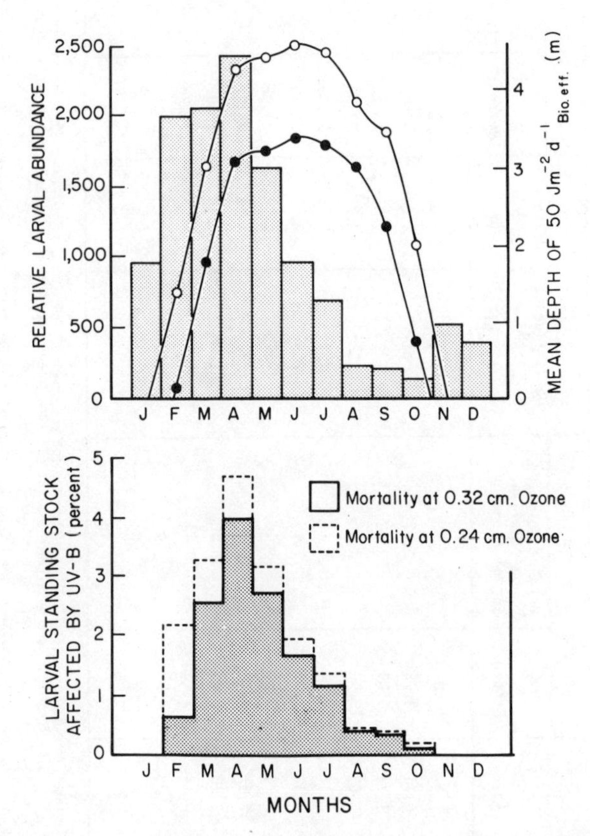

Figure 7. (upper) Relative abundance of larval anchovy per month (bars) and mean depth of UV-B penetration of 50 $Jm^{-2}d^{-1}{}_{bio.eff.}$ in the sea per month at present (solid circles) and at 25% reduction in ozone (open circles). 50 $Jm^{-2}d^{-1}{}_{bio.eff.}$ is the daily equivalent of the LD_{50} for anchovy larvae cumulated over 12 days. Data sources are given in footnotes to Table 3. (lower) Percent of annual larval anchovy standing stock affected by UV-B (50 $Jm^{-2}d^{-1}{}_{bio.eff.}$) per month at present ozone concentration (0.32 cm optical ozone depth, solid line) and at 25% reduction in ozone (0.24 cm optical ozone depth, dotted line) according to calculations given in Table 3.

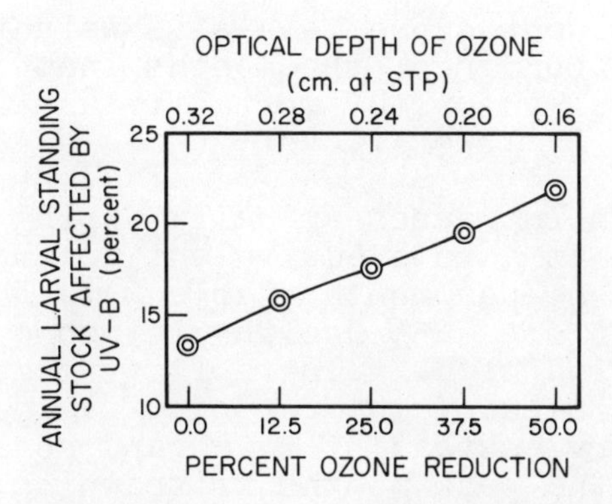

Figure 8. Estimated annual loss of larval anchovy standing stock (percent) as a function of optical depth of ozone. The method of calculation is illustrated in Table 3.

decline produced only a minor increase in predicted UV-B losses to the annual larval standing stock, but a loss of 13% at present ozone levels seems high. If such a UV-B mortality exists, it may have little effect on the population because natural rates of mortality from other causes are high; the rates decline from about 53% per day for eggs (Smith and Lasker 1978) to about 17% per day for 12-day-old anchovy larvae (Zweifel and Smith 1981). In addition, increases in larval mortality over the first few weeks of life cannot be directly translated into year class success. Formation of a year class is a process that continues throughout the larval and juvenile stages. In anchovy, this period may extend over 8-12 months.

Habitat Uncertainties

In this section we discuss how uncertainties in characterizing the habitat of larval anchovy affect our estimate of the effect of UV radiation on larval anchovy populations.

1) Seasonality of larval production. The portion of anchovy larvae in the water column affected by UV-B (portion of larvae at or above the depth where the daily UV-B dose = $50 Jm^{-2}_{bio.eff.}$) is highest in June (21% at 0.32 cm ozone; and 24% at 0.24 cm ozone) and declines with the seasonal changes in UV radiation (Figure 7, Table 3). Although the predicted rate of UV mortality is highest in June, the abundance of anchovy larvae (shaded area, Figure 7, upper) is relatively low (Lasker and Smith 1977). When the loss rate

per month at present ozone levels is weighted by the monthly larval abundance, the months that contribute most to the total annual UV mortality are March, April and May (Figure 7, lower).

UV radiation increases annually during the anchovy spawning season at a rate equivalent to about a 25% reduction in ozone per month (Table 3). Thus, the principal effect of a 25% decline in ozone is to advance the period of UV vulnerability 1 month earlier in the spawning season; the period of vulnerability would begin in February instead of March and remain at elevated levels through the peak months of spawning. For large larval losses to occur under reduced ozone conditions requires that losses be high under present ozone conditions because of the sharp seasonal increase in UV that occurs throughout the spawning season. This greatly reduces the relative impact of ozone decline in our calculation.

Since our estimate of relative abundance of larvae is based on long-term averages, the major uncertainty in this parameter is the inter-annual variation in larval production over the peak months of spawning. Larval abundance between the first and second half of the four months of maximum spawning varies by about a factor of two, with higher production occurring with about equal frequency in either of the two periods. (Stauffer, G., unpublished data, S.W.F.C., La Jolla, Ca 92038). We do not consider larval abundance a major uncertainty in our calculations for anchovy, but it is a major uncertainty in species where the seasonality of annual larval production is poorly defined.

2) Vertical distributions. One of the principal reasons that the relative effect of increased UV radiation from a 25% ozone decline is minor is that anchovy larvae are abundant at other depths as well as in the upper few meters. A 25% reduction in ozone in moderately productive waters (0.5 mg Chla m^{-3}) is equivalent to about a 1 m increase in the depth of penetration of the biologically effective dose for anchovy larvae. An increase in depth of penetration of 1 m does not have a major effect on a population distributed over about 60 meters. The form of the relation between larval abundance and depth in the sea is a critical determinant of the effect of UV on the population. Unfortunately, this parameter is not measured with the accuracy required for UV assessment work.

Vertical distribution studies usually include a near surface net tow and additional tows taken at depth increments of about 10 m or greater. Spacing of depth intervals of 1-3 m are not used because ship movements and swell height introduce errors of that magnitude. A 25% reduction in ozone would increase the depth of penetration of the biologically effective dose by 1 m over a 1-3 m range. Thus the resolution required for UV assessment is finer than the depth range over which a net tow integrates. This produces an uncertainty for anchovy of perhaps 2X because the cumulative percentage of larvae at a depth could range from the central value of 12% (2.5m) to 0 at the surface and up to 34% at the next (deeper) depth interval (9.5 m).

Other uncertainties of possibly much greater magnitude include size or stage specific differences in vertical distribution and diel and seasonal changes in vertical distribution. Ahlstrom (1959) points out that the vertical distribution of anchovy and other larvae in the mixed layer may change with the depth of the mixed layer (Figure 9). In southern California, the depth of the mixed layer is highly variable, but generally it is deeper in the winter than in the summer. We have no data on size specific differences in depth distribution for larvae less than 10 mm but believe such differences may occur.

3) Vertical mixing. A major uncertainty is the extent of passive or active interchange of larvae between the UV impacted surface layers and greater depths. In our calculations we assumed that larvae present in the upper few meters remained there for 12 days. Significant interchange of eggs and larvae between the surface and greater depths probably occurs. Smith (1973) and Hewitt (1980) have measured the horizontal dispersion of sardine and anchovy eggs and yolk-sac larvae, but rates of vertical dispersion are unknown. Larvae at the onset of feeding (4.0 mm standard length) become active swimmers and interchange of larvae from the surface to greater depths is probably a function of active movements. They swim intermittently throughout the day at speeds ranging from 0.5-1.0 body lengths/sec (Hunter 1977). Hence, 4 mm larva could swim a distance of 80 m or more in 1 day. Although the swimming patterns of larvae are serpentine, redundant, and probably largely determined by food distribution (Hunter and Thomas 1974), a larvae could easily swim from the surface to depths beyond significant penetration of UV in a day or vice versa. Random vertical movements in the water

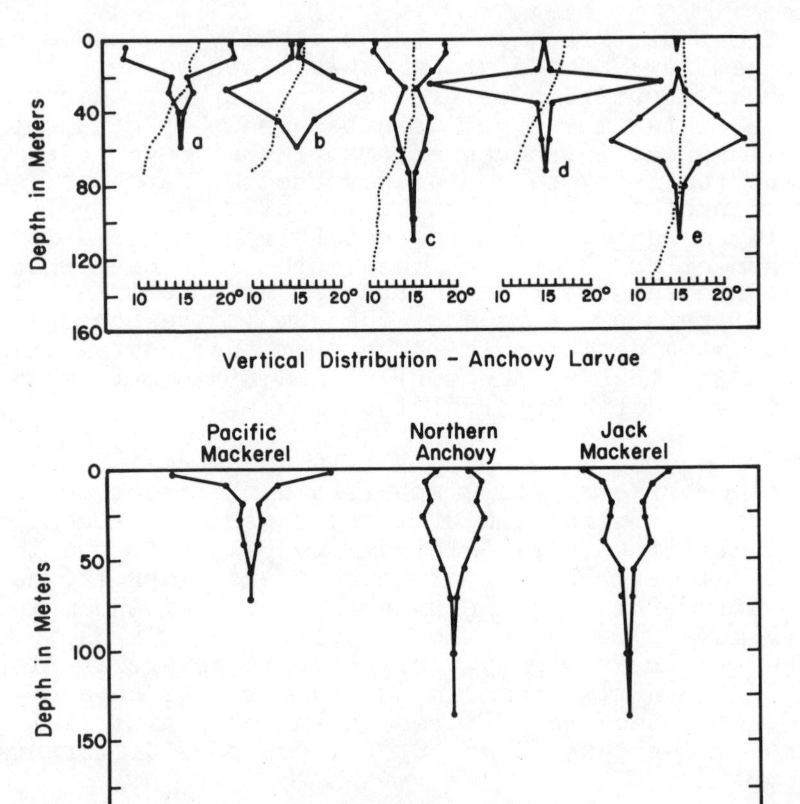

Figure 9. Vertical distribution of various larval fishes from Ahlstrom (1959). (Upper): vertical distribution in anchovy in five separate vertical series with superimposed temperature profiles. (Lower): average vertical distribution of northern anchovy, Pacific mackerel and jack mackerel.

column could greatly reduce the total cumulative dose received by individuals. For these reasons, our calculations may overestimate the potential losses of the standing stock at present and reduced ozone concentrations.

4) Penetration of UV into the habitat. In general, coastal waters of California are free from the influence of major river systems and from the effects of shallow silt bottom habitats. Consequently, surface Chl-a concentration is the best index of penetration of UV radiation in the habitat (Smith and Baker, 1979).

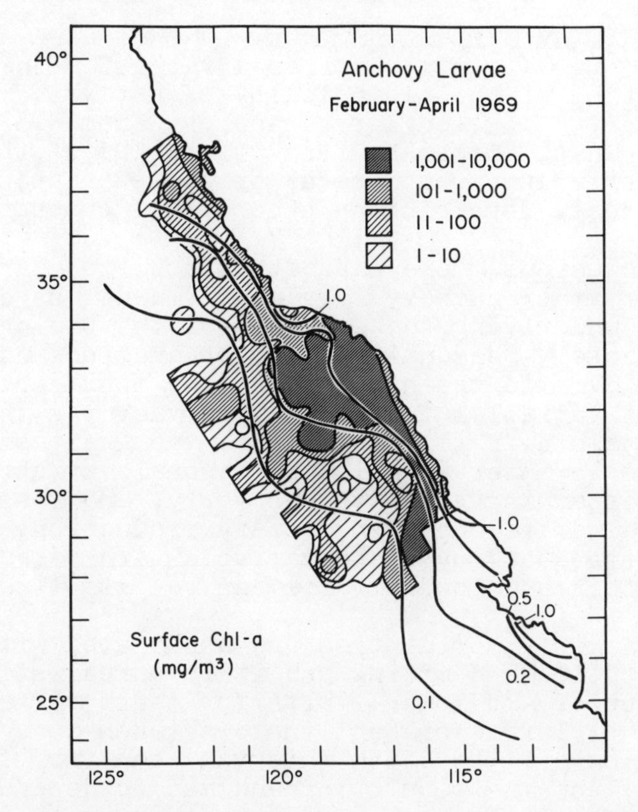

Figure 10. Abundance of anchovy larvae (number/10 m^2 surface area) in relation to surface chlorophyll concentration (mg Chla m^{-3}) in February-April 1969. Anchovy larval abundance is from unpublished larval survey data, Southwest Fisheries Center, La Jolla, Ca. and chlorophyll from Owen and Sanchez (1974).

Measurement of surface Chla concentrations and larval anchovy abundance in the peak months of spawning (February-April) indicate that most anchovy larvae occur at surface Chl-a concentrations between 0.2 and 1.0 mg Chla m^{-3}, although some are found at concentrations as low as 0.1 mg Chla m^{-3} (Figure 10). We used 0.5 mg Chla m^{-3} because this is close to the annual mean for the Southern California Bight and there is a measured diffuse attenuation coefficient (Smith and Baker 1979). This introduces a potential error because the mean Chl concentration for the habitat weighted by larval abundance may be different. We expect that 0.5 mg would be close to the weighted value (Figure 10).

The range of Chla concentration for the habitat (0.1 to 1.0 mg Chla m^{-3}) implies a variation in depth of penetration of the biologically effective UV dose (50 $Jm^{-2}h^{-1}$ $_{bio.eff.}$) or larval anchovy from 4.5 m to 1.1 m in March. Thus, in March, from 24% to 12% of the larvae in the water column would be at or above the biologically effective dose, depending on chlorophyll concentrations.

5) Conclusions. The parameters used to characterize the UV habitat of anchovy larvae are much less certain than those underlying the estimate of the biologically effective dose. Seasonality of reproduction and penetration of UV into the habitat are the most certain of the habitat variables for anchovy larvae. As they are based on means over time and space, the greatest uncertainty lies in inter-annual and seasonal variability; e.g., late spawning years versus early, and seasonal and inter-annual differences in primary production. Uncertainties introduced by these variables are perhaps 2X and definite upper and lower limits can be established.

Much larger uncertainties are associated with vertical distribution and mixing and it is unrealistic to establish upper and lower limits for these variables. The number of larvae present in tows taken at a 2 m depth (representing a 0-5 m depth interval) can vary from 0.25% to 27% of the population and may be dependent on the depth of the mixed layer (Figure 9, upper; Ahlstrom, 1959). Thus seasonal changes in vertical distribution, or possible biases in calculating the average vertical distribution, could either increase, or reduce to nearly zero, estimated UV effects. Similarly, vertical mixing could reduce estimated UV effects to zero at present or reduced ozone concentrations. We conclude that the parameters of vertical distribution and mixing could account for the full range of possible UV effects on the population independent of all other variables. In other species, the other habitat parameters could introduce similar uncertainties because of the lack of accurate information.

POSSIBLE EFFECTS OF OZONE DEPLETION ON OTHER FISH LARVAE

It is beyond the resources of any program to study the UV vulnerability of every habitat and species of larval fish in the coastal waters of California or any other major zone of the world's oceans. Thus, it is important to relate the information on anchovy to other species in order to evaluate potential risk of ozone decline to larval fishes in general. In this section we discuss the potential risk of UV damage to other abundant larval fishes in California coastal waters.

Most abundant fish larvae

Annual ichthyoplankton surveys conducted over the last 25 years document the abundance and distribution of larval fishes in California waters. These surveys extend from central Baja California (Mexico) to San Francisco Bay, California and from 240 nautical miles at sea to within a few nautical miles of the coast (total area surveyed is about 215,000 sq. n. mi). The surveys sample larvae of nearly all major California pelagic fish populations as well as many others. The species composition and rank order of abundance of species remains relatively constant over the years (Ahlstrom 1965).

All of the 15 most abundant species taken in these surveys occur at depths where penetration of UV radiation is minimal, although many also occur in the upper 5 m (the region we judge to be of significance for UV effects in waters of moderate Chla concentration (Table 4). In these 15 species, the percentage of the population that is taken in a tow at 2 m depth (representing 0-5 m depth range) ranges from 0% for several species of lanternfish and 1 species of deep sea smelt to 34% for Pacific mackerel (Ahlstrom 1959).

The vertical distribution and the season of larval abundance are key factors in estimating the impact of ozone decline on the population. To evaluate how differences in these parameters affect vulnerability to ozone decline, we have estimated effects of UV for two additional species, Pacific mackerel and jack mackerel. Both species are among the 15 most abundant and are of commercial importance. Their seasonal abundance and vertical distribution differ from anchovy. Pacific mackerel are summer spawners and have a more shallow vertical distribution than anchovy (Kramer 1960; Ahlstrom 1959); jack mackerel are predominately spring spawners and have a deeper vertical distribution than anchovy (Farris 1961; Ahlstrom 1959) (Figure 9).

For Pacific mackerel, we have dose-response data for only a 4-day exposure to UV-B (Hunter et al. 1979) and we have used the ratio of anchovy LD_{50} for 4-day exposure to 12-day exposure to adjust the 4-day dose of mackerel to one for 12 days. Since embryonic Pacific mackerel had $\frac{1}{2}$ the sensitivity of embryonic northern anchovy, the LD_{50} dose for 12 days would be twice that for anchovy larvae, 102 $Jm^{-2}d^{-1}{}_{bio.eff.}$ Pacific

TABLE 4. Fifteen most abundant fish larvae taken in the California Current Region in 1955[1] and their occurrence in the upper 5 m[2], listed in order of abundance.

Species and common name	Rank abundance	Total number of larvae taken (x 100)	Percent of total larvae	Percent[2] of species vertical distribution in upper 5m
Engraulis mordax (northern anchovy)	1	140	39.0	11.8
Merluccius productus (Pacific hake)	2	60	16.7	0.7
Sebastes sp. (rockfishes)	3	29	8.2	3.3
Citharichthys sp. (sanddabs)	4	20	5.7	0.6
Leuroglossus stilbius (California smooth tongue)	5	15	4.2	0.2
Sardinops caerulea (Pacific sardine)	6	14	3.9	17.2
Trachurus symmetricus (jack mackerel)	7	13	3.7	11.0
Triphoturus mexicanus (lanternfish)	8	13	3.7	0.7
Vinciguerria lucetia lanternfish)	9	13	3.5	1.5
Stenobrachius leucopsarus (lanternfish)	10	7	2.1	0.0
Diogenichthys laternatus (lanternfish)	11	5	1.3	0.0
Bathylagus wesethi (snubnosed black smelt)	12	3	1.9	0.0
Lampanyctus ritteri (lanternfish)	13	2	0.6	0.0
Scomber japonicus (Pacific mackerel)	14	2	0.5	34.4
Protomyctophina sp. (lanternfish)	15	2	0.5	0.0
All 15 species combined	-	338	94.5	-

[1]From Ahlstrom, 1965.
[2]From data of Ahlstrom (1959) recalculated using standard techniques of Smith and Richardson (1977).

mackerel larvae are found more inshore than the northern anchovy larvae, and 1.0 mg Chla m^{-3} is more representative of these more productive waters. Hence, the UV energy does not penetrate as deeply into the water column. Our calculations indicate that the relative effect of a 25% ozone reduction (ratio of percent effected at 0.24 cm ozone/percent effected at 0.32 cm ozone) for Pacific mackerel larvae is similar to that for anchovy larvae. This ratio was 1.38 for Pacific mackerel and 1.31 for anchovy (Table 5, Appendix Table 2). The calculation also indicates that the higher UV tolerance of Pacific mackerel larvae to anchovy larvae (2X) is commensurate with the shallow depth distribution and summer spawning pattern of the mackerel. Thus, UV tolerance of larvae may be predictable from vertical distribution and seasonal spawning pattern (temporal and spatial).

To illustrate this point, the UV sensitivity of jack mackerel larvae was predicted assuming their UV sensitivity, relative to habitat characteristics, was similar to northern anchovy and Pacific mackerel larvae. Larval jack mackerel vertical distribution (Figure 9; Appendix Table 3) is deeper than that of the larval anchovy (0.95X or Pacific mackerel (0.33X). They have a higher mean seasonal UV-B exposure relative to anchovy larvae (1.25X) but lower than that of the Pacific mackerel (0.90X). Jack mackerel larvae develop in clearer ocean water (about 0.1 mg Chla m-3) than either the anchovy or Pacific mackerel larvae, hence, UV penetrates deeper into the water column than for either the larval anchovy (2.08X) or the larval Pacific mackerel (4.19X). Taking these factors into account, jack mackerel larvae may have a UV LD_{50} of about 125 $Jm^{-2}d^{-1}{}_{bio.eff.}$, that is, relative to larval anchovy, (0.93) (1.25) (2.08) (50 $Jm^{-2}d^{-1}{}_{bio.eff.}$) = 123 $Jm^{-2}d^{-1}{}_{bio.eff.}$, and relative to Pacific mackerel larvae, (0.33) (0.90) (4.19) (102 $Jm^{-2}d^{-1}{}_{bio.eff.}$) = 127 $Jm^{-2}d^{-1}{}_{bio.eff.}$. This calculated estimate of UV sensitivity translates into a relative effect of 25% ozone reduction (ratio of the percent effected at 0.24 cm ozone/percent effected at 0.32 cm ozone) of 1.95 (Table 5; Appendix Table 3).

The increase in loss to the larval standing stock for a major reduction in ozone (25%) is less than a factor of 2 for larval anchovy, Pacific mackerel, and that hypothesized for jack mackerel. This suggests to us that predicted levels of ozone reduction will have only minor influence on these larval populations. This conclusion may apply to most of the abundant larvae taken in coastal waters of California because these three species represent the range in types of vulnerable

TABLE 5. Proportion of annual larval standing stock affected by UV-B radiation at present and 25% reduced ozone concentration (0.32 and 0.24 cm optical depth of ozone, respectively)

Species	LD_{50} for 12 daily exposures (Jm^{-2} bio.eff.)	Months of Peak spawning	Percent affected 0.32 cm ozone	Percent affected 0.24 cm ozone	% affected by 0.32 cm ozone / % affected by 0.24 cm ozone
Engraulis mordax [1] (northern anchovy)	50	F,M,A,M	13.4	17.6	1.31
Scomber japonicus [2] (Pacific mackerel)	130	A,M,J,J,A	17.8	24.5	1.38
Trachurus symmetricus [3] (jack mackerel)	125	M,A,M,J	7.8	15.2	1.95

[1] documentation for calculation in Table 3.

[2] documentation for calculations in Appendix Table 2.

[3] documentation for calculations in Appendix Table 3.

vertical distributions and spawning seasons.

Neustonic larvae

Clearly larval fishes that occur only in the upper few meters of the sea are potentially the most vulnerable to ozone depletion. Surface (neuston) tows give an indication of the species present near the surface, but must be compared to tows taken at the same locality at greater depths to identify the exclusively neustonic larvae. Anchovy larvae are a prime example. They are the most abundant fish larvae in neuston tows taken along the coast of California (Table 6), but this species is also abundant at other depths. Larvae identified as exclusively neustonic by the data of Ahlstrom and Stevens (1976) include those of saury, flyingfish, sablefish, and a few species of lanternfishes. Flyingfish, saury, and sablefish larvae are heavily pigmented. Moser (MS, SWFC, La Jolla, Ca.) points out that heavy melanistic pigmentation is found universally among neustonic fish larvae and that neustonic larvae are larger and more mature at hatching than are other pelagic larvae. All of these factors probably substantially reduce the potential for UV damage in neustonic forms.

UV sensitivity data are not available for marine organisms identified as being exclusively neustonic. However, the cuticle of the marine insect, *Halobates*, which exists on the surface of the open sea, is over a hundred times as UV absorbent as that of other related forms from freshwater and mangrove habitats (Cheng et al. 1978). It seems likely that neustonic forms are tolerant of UV radiation, nevertheless they should be studied because the entire population experiences significant UV radiation and a change in ozone could have important consequences.

DISCUSSION

Our estimates of the effects of ozone diminution on larval populations are obviously simplistic, contain many uncertainties and assumptions and should be regarded as an *index* of possible losses to UV radiation. The estimated loss to the annual standing stock is an index of effects and should not be confused with actual mortality rates. The estimated loss to anchovy standing stock seems small relative to the very high rates of natural mortality. More importantly, losses of anchovy increased by less than a factor of 2 even under a severe ozone reduction (25%). We conclude that anchovy,

TABLE 6. Five most abundant fish larvae taken in a neuston net[1] (upper 0.5 m) compared to a net towed obliquely from 200 m depth to the surface (Ahlstrom & Stevens 1976).

Species	Neuston net: Occurences	Neuston net: Number of specimens	Oblique tow: Occurrences	Oblique tow: Number of specimens	Occurrence in neuston as percent of total occurrence
COASTAL STATIONS[2]					
Engraulis mordax (northern anchovy)	25	3,041	36	10,671	29
Cololabis saira (Pacific saury)	49	1,095	5	8	91
Oxyporhamphus micropterus (small wing flyingfish)	8	785	3	6	73
Anoplopoma fimbria (sablefish)	7	468	0	0	100
Tarletonbeania crenulanis (lanternfish)	12	296	31	130	28
All collections combined	105	7,031	105	19,823	
OFFSHORE STATIONS[3]					
Cololabis saira (Pacific saury)	25	347	1	1	96
Cololabis adocetus (saury)	6	158	0	0	100
Lampadena urophaos (lanternfish)	7	144	16	63	30
Taaingichthys minimus (lanternfish)	14	142	3	3	82
Exocoetidae several sp. (flyingfish)	12	64	0	0	100
All collections	43	1,027	43	3,899	

[1]Neuston tows were not quantified, hence quantitative comparisons to oblique cannot be made.

[2]From Puget Sound to southern Baja California (Mexico) from shore to 240 nautical miles offshore.

[3]From Puget Sound to southern Baja California (Mexico) at distances greater than 240 nautical miles from the coast.

mackerel, and most other fish larvae living in the upper mixed layer may be under some UV stress today, but projected levels of ozone decline (16%) probably will not have a catastrophic effect.

The sensitivity of our biologically effective dose for anchovy may be insufficient because we used a major effect (50% mortality) and a relatively short period to assess the effects of UV exposure (12 days). On the other hand, increasing the sensitivity of our criteria for effects would elevate projected UV losses at present and depleted ozone levels, and not greatly affect our estimate of the relative impact of ozone diminution on the larval population. Our calculation of the impact of UV radiation is a much better index of relative effects than of the absolute proportion of the standing stock that could be lost. Most of the possible changes in parameters or assumptions affect the absolute larval losses but have little effect on the relative change caused by increased UV radiation. Exceptions to this generalization include possible interactions between ozone reduction and other environmental variables. For example, a late anchovy spawning season, coupled with less chlorophyll production, or a shallow mixed layer that caused a compaction of the vertical distribution of larvae, might increase UV mortality in a given year, but the reverse situation could also occur. It is conceivable that through such interactions ozone reduction might amplify the natural variation in larval mortality over the years. It is uncertain if such events could ever be detected or if they would affect the size of the incoming year classes.

We believe that the largest uncertainties in assessment of UV impact on all species are the habitat parameters; vertical distribution, vertical mixing, penetration of UV into the habitat, and seasonal occurrence. The impact of ozone reduction, even in the most UV-sensitive species, could be negligible or zero depending on the characteristics of the habitat. Habitat data are better for anchovy larvae than for other marine larvae, but the uncertainties are still large. Vertical mixing alone could reduce to zero the impact of UV radiation on anchovy larval populations.

The key habitat parameter for assessment of UV effects is vertical distribution. To be greatly affected by ozone decline requires that the UV sensitive stages of a species must occur primarily at depths of 5 m or less depending on the attenuation of UV in the habitat. Most larval fishes, indeed most marine organisms, do not have vertical distributions of this type. The vertical distributions of many larvae, including anchovy, jack mackerel and Pacific mackerel, penetrate into deeper strata, though they also occur at the surface. Species that are restricted in their distribution to the upper 5 m may be the most vulnerable to UV radiation. These neustonic species comprise a limited group of species specialized for near-surface existence (flyingfish, saury and others). The UV tolerance of exclusively neustonic larvae have not been studied. This remains the weakest element in our general conclusion that the impact of ozone decline on larval fishes is probably small.

We hope that our research may affect the directions of future UV impact studies on the marine environment. On the basis of our experience, we make the following recommendations. Experiments using solar-UV radiation should be included in studies of UV tolerance or UV action spectrum for the species should be determined. UV dose rates and exposure periods should be kept within the bounds of anticipated or existing solar UV patterns. Cumulative dose should not be used to project effects from experiments employing high dose rates. Most importantly, the critical habitat parameters must be known or measurement of these parameters must be included as part of the work.

ACKNOWLEDGEMENTS

This work was supported in part by a contract from the United States Environmental Protection Agency (EPA).

REFERENCES

Ahlstrom, E. H. 1959. Vertical distribution of pelagic fish eggs and larvae off California and Baja California. Fish. Bull. U.S. 60: 107-146.

Ahlstrom, E. H. 1965. Kinds and abundance of fishes in the California Current Region based on egg and larval surveys. Calif. Coop. Oceanic Fish. Invest. Rept. 10: 31-52.

Ahlstrom, E. H. 1967. Co-occurrence of sardine and anchovy larvae in the California Current Region off California and Baja California. Calif. Coop. Oceanic Fish. Invest. Rept. 11: 117-135.

Ahlstrom, E. H. and E. Stevens. 1976. Report of neuston (surface) collections made on an extended CalCOFI cruise during May 1972. Calif. Coop. Oceanic Fish. Invest. Rept. 18: 167-180.

Baker, K.S., R. C. Smith and A.E.S. Green. 1980. Middle ultraviolet radiation reaching the ocean surface. Photochem. Photobiol. 32: 367-374.

Bener, P. 1964. Investigations of the influence of clouds on UV sky radiation. Tech. Note No. 3 U.S. Air Force.

Boyce, R., and R. Setlow. 1963. The action spectra for ultraviolet-light inactivation of systems containing 5-bromouracil-substituted deoxyribonucleic acid I. *Escherichia coli* 15 T A U. Biochim. Biophys. Acta 68: 446-454.

Bullock, A. M. and R. J. Roberts. 1975. The dermatology marine teleost fish. I. The normal integument. Oceanog. Mar. Biol. Ann. Rev. 13: 383-411.

Cheng, L., M. Douek and D. A. I. Goring. 1978. UV absorption by gerrid cuticles. Limnol. Oceanogr. 23: 554-556.

Farris, D. A. 1961. Abundance and distribution of eggs and larvae and survival of larvae of jack mackerel (*Trachurus symetricus*). Fish. Bull, U.S. 61: 247-279.

Finney, D. 1952. Probit Analysis. A Statistical Treatment of the Sigmoid Response Curve

Green, A.E.S., T. Sawada and E. Shettle. 1974. The middle ultraviolet reaching the ground. Photochem. Photobiol. 19: 251-259.

Green, A.E.S. and J. Miller. 1975. Measures of biologically effective radiation in 280-320 nm region. In CIAP Monogr. 5, Part I, Chap. 2, 60-69. U.S. Dept. Transp., Climatic Impact Assessment Program. Wash.D.C.

Hewitt, R. 1980. The value of pattern in the distribution of young fish. ICES Symp. on Early Life History of Fish, Woods Hole, Mass., April 1979, Rapp. P.-v. Reun. Cons. int. Explor. Mer. 178: (In press).

Hunter, J. R. 1977. Behavior and survival of northern anchovy, *Engraulis mordax*, larvae. Calif. Coop. Oceanic Fish. Invest. Rept. 19: 138-146.

Hunter, J. R., S. E. Kaupp and J. H. Taylor. 1981. Effects of solar and artificial UV-B radiation on larval northern anchovy, *Engraulis mordax*. Photochem. Photobiol. (In press).

Hunter, J. R. and C. Sanchez. 1976. Diel changes in swim bladder inflation of the northern anchovy, *Engraulis mordax*. Fish. Bull. U.S. 74: 847-855.

Hunter, J. R., J. H. Taylor and H. G. Moser. 1979. The effect of ultraviolet irradiation on eggs and larvae of the northern anchovy, *Engraulis mordax*, and the Pacific mackerel, *Scomber japonicus*, during the embryonic stage. Photochem. Photobiol. 29: 325-328.

Hunter, J. R. and G. L. Thomas. 1974. Effect of prey distribution and density on the searching and feeding behavior of larval anchovy, Engraulis mordax Girard. In Early Life History of Fish [ed. J.H.S. Blaxter]. 559-574. Springer-Verlag, New York.

Ichihashi, M. and C. A. Ramsay.1976. The action spectrum and dose response studies of unscheduled DNA synthesis in normal human fibroblasts. Photochem. Photobiol. 23: 103-106.

Kantor, G. J., J.C. Sutherland and R. B. Setlow. 1980. Action spectra for killing non-dividing normal human and Xeroderma pigmentosum cells. Photochem. Photobiol. 31: 459-464.

Kaupp, S. E.. and J. R. Hunter. 1981. Photorepair in the larval anchovy, Engraulis mordax. Photochem. Photobiol. 33: 253-256.

Kramer, D. 1960. Development of eggs and larvae of Pacific mackerel and distribution and abundance of larvae 1952-56. Fish. Bull U.S. 60: 393-438.

Lasker, R. and P. E. Smith. 1977. Estimation of the effects of environmental variation on the eggs and larvae of the northern anchovy. Calif. Coop. Oceanic Fish. Invest. Rept. 19: 128-137.

Mackay, D., A. Eisenstark, R. B. Webb, and M. S. Brown. 1976. Action spectra for lethality in recombinationless strains of Salmonella typhimurium and Escheri-coli. Photochem. Photobiol. 24: 337-343.

Marinaro, J. and M. Bernard. 1966. Contribution a l'Etude des Oeufs et Larves Pélagiques de Poissons mediterranéens. I. Note preliminaire sur l'influence léthale du rayonnement solaire sur les oeufs. Pelagos 6: 49-55.

O'Connell, C. P. 1981. Development of organ systems in the northern anchovy, Engraulis mordax, and other teleosts. Amer. Zool.

Owen, R. W. and C. K. Sanchez. 1974. Phytoplankton pigment and production measurements in the California Current Region, 1962-1972. NMFS Data Rept. 91.

Pommeranz, T. 1974. Resistance of plaice eggs to mechanical stress and light. In The Early Life History of Fish [ed. J.H.S. Blaxter], 397-416. Springer-Verlag, New York.

Renner, J.A. (1979, unpublished). In Synoptic Weather Data System. Dept. of Commerce, Natl. Mar. Fish. Serv., Southwest Fish. Center, La Jolla, Ca.

Rosenstein, B. S., and R. B. Setlow. 1980. Photoreactivation of ICR 2A frog cells after exposure to monochromatic ultraviolet radiation in the 252-313 nm range. Photochem. Photobiol. 32: 361-366.

Rothman, R. H. and R. B. Setlow. 1979. An action spectrum for cell killing and pyrimidine dimer formation in Chinese hamster V-79 cells. Photochem. Photobiol. 29: 57-61.

Setlow, R. B. 1974. The wavelengths in sunlight effective in producing skin cancer: a theoretical analysis. Proc. Nat. Acad. Sci. U.S.A. 71: 3363-3366.

Setlow, R. and R. Boyce. 1963. The action spectra for ultraviolet-light inactivation of systems containing 5-bromouracil-substituted deoxyribonucleic acid II. Bacteriophage T_4. Biochem. Biophys. Acta 68:455-461.

Smith, P. E. 1973. The mortality and disperal of sardine eggs and larvae. Rapp. P.-v. Reun. Cons. perm. Int. Explor. Mer. 164: 282-292.

Smith, P. E. and S. L. Richardson. 1977. Standard techniques for pelagic fish egg and larval surveys. FAO Fisheries Tech. Paper, No. 175.

Smith, P. E. and R. Lasker. 1978. Position of larval fish in an ecosystem. Rapp. P.-v. Reun Cons. Int. Explor. Mer. 173: 77-84.

Smith, R. C. and K. S. Baker. 1979. Penetration of UV-B and biologically effective dose-rates in natural waters. Photochem. Photobiol. 29: 311-323.

Smith, R. C., K. S. Baker, O. Holm-Hansen, and R. Olson. 1980. Photoinhibition of photosynthesis in natural waters. Photochem. Photobiol. 31: 585-592.

Stouffer, G., unpublished data. S.W.F.C., La Jolla, CA. 92038.

Webb, R. B. and M. S. Brown. 1976. Sensitivity of strains of Escherichia coli differing in repair capability to far UV, near UV and visible radiations. Photochem. Photobiol. 24: 425-432.

Zweifel, J. R. and P. E. Smith. 1981. Estimates of abundance and mortality of larval anchovies (1951-1975): methods and results. ICES Symp. on Early Life History of Fish. Woods Hole, Mass., April 1979. Rapp. P.-v. Reun. Cons. Int. Explor. Mer. 178: (In press).

Appendix Table 1

Data used to estimate UV-B dose rate effect and UV sensitivity of specific life stage in larval northern anchovy.

Experiment date	Cumulative dose (Jm^{-2} bio.eff.)	Expected survival[1] (percent)	Percent survival normalized to control[2]		Daily rate Jm^{-2} bio.eff.	Number of daily UV exposures	Larval age when exposed to UV (days)[4]
			Mean	2 X SE[3]			
5.15.80	142	99.6	98.6	8.8	10.9	2	1 & 2
	142	99.6	103.0	8.7	10.9	2	4 & 5
7.8.80	179	94.6	88.2	8.5	13.7	2	1 & 2
	179	94.6	85.2	9.6	13.7	2	4 & 5
7.22.80	293	75.5	52.8	15.3	22.5	2	1 & 2
	293	75.5	85.8	12.7	22.5	2	4 & 5
7.12.79	322	87.9	10.6	16.8	24.7	2	1 & 2
	322	87.9	84.8	14.5	24.7	2	4 & 5
5.1.80	322	87.9	51.4	6.8	24.7	2	1 & 2
	322	87.9	86.6	11.8	24.7	2	4 & 5
6.26.80	530	59.4	3.5	1.9	40.6	2	1 & 2
	530	59.4	22.2	6.7	40.6	2	4 & 5
6.28.79	644	45.2	26.7	12.6	24.7	4	eggs 0,1 & 2
	644	45.2	26.2	17.6	24.7	4	3,4,5 & 6

[1]Estimated from dose response line for 12 daily exposures to UV-B (Hunter et al. 1981).

[2]Survival measured at age 12 days.

[3]2X standard error of the mean where N = 10 in each treatment in an experiment.

[4]Dose rate applied for 6 h each day; UV radiation increased and decreased from 6 h rate in two 15-min. steps of .5X the 6 h dose rate and .7X 6 h dose rate.

[5]Ages 0-3 days is the embryonic period (egg and yolk-sac stages); feeding begins at age 4 days.

Appendix Table 2

Proportion of Pacific mackerel annual larval standing stock affected by UV-B radiation at present (0.32 cm) and at 25% reduction (0.24 cm) in ozone concentration

Months	.32 cm ozone (ambient)				.24 cm ozone (25 % reduction)			
	Daily fluence UV-B (Jm^{-2} bio eff.)[1]	Depth of LD_{50}[2] (m)	Larvae at or above LD_{50}[3]		Daily fluence UV-B (Jm^{-2} bio eff.)[1]	Depth of LD_{50}[2] (m)	Larvae at or above LD_{50}[3]	
			Percent[4]	Percent weighted by relative larval abundance[5]			Percent[4]	Percent weighted by relative larval abundance[5]
J	25	Ø	Ø	Ø	40	Ø	Ø	Ø
F	52	Ø	Ø	Ø	82	Ø	Ø	Ø
M	94	Ø	Ø	Ø	149	0.58	22.95	0.99
A	150	0.60	23.25	3.22	238	1.30	30.48	4.22
M	158	0.68	24.31	3.34	250	1.37	31.09	4.28
J	167	0.77	25.35	4.07	265	1.46	31.79	5.11
J	162	0.72	24.79	2.33	257	1.41	31.43	2.95
A	147	0.57	22.81	4.31	233	1.26	30.21	5.71
S	111	0.13	13.61	0.48	176	0.84	26.09	0.93
O	65	Ø	Ø	Ø	103	0.01	6.31	0.28
N	30	Ø	Ø	Ø	48	Ø	Ø	Ø
D	20	Ø	Ø	Ø	32	Ø	Ø	Ø
Σx				17.75				24.47

[1] The sea surface irradiance is taken from Baker et al. (1980) coefficients for the model of Green et al. (1974) for 33°N, and adjusted for 18-year average cloud cover (1961-1978) for the Southern California Bight (30-35°N, 114-120°W; Renner 1979) using the equation of Bener (1964) and then weighted by an empirical fit to the DNA action spectrum.

[2] Depth in sea where the daily UV-B fluence is equivalent to the LD_{50}. The depth of UV-B penetration was calculated for productive water (1.0 mg Chla m^{-3}) using the downwelling irradiance diffuse attenuation coefficients of Baker and Smith (pers. comm.); KT = 0.6404 for 0.32 cm ozone and K_T = 0.6531 for 0.24 cm ozone.

[3] 102 Jm^{-2} DNA eff. This daily dose was calculated by assuming the sensitivity of Pacific mackerel was the same relative to Northern anchovy as was determined for 4 daily exposures to UV-B (Hunter et al. 1979).

[4] The monthly percent of larvae at or above the LD_{50} depth was calculated using the vertical distribution data for Pacific mackerel larvae of Ahlstrom (1959) fitted by the equation y = 27.81 x 0.353, r^2 = 0.98, for the data from the surface to 28 meters, where y is the cumulative percent of larvae from the surface to depth x, in meters.

[5] 5-year average larval abundance from Kramer (1960) was used to weight the percent of larvae affected by UV-B for each month.

Appendix Table 3

Proportion of jack mackerel annual larval standing stock affected by UV-B radiation at present (0.32 cm) and at 25% reduction (0.24 cm) in ozone concentration.

Month	.32 cm ozone (ambient)				.24 cm ozone (25% reduction)			
	Daily fluence UV-B (Jm^{-2} bio eff.)[1]	Depth of LD_{50}[2] (m)	Larvae at or above LD_{50}[3]: Percent[4]	Larvae at or above LD_{50}[3]: Percent weighted by relative larval abundance[5]	Daily fluence UV-B (Jm^{-2} bio eff.)[1]	Depth of LD_{50}[2] (m)	Larvae at or above LD_{50}[3]: Percent[4]	Larvae at or above LD_{50}[3]: Percent weighted by relative larval abundance[5]
J	25	Ø	Ø	Ø	40	Ø	Ø	Ø
F	52	Ø	Ø	Ø	82	Ø	Ø	Ø
M	94	Ø	Ø	Ø	149	1.22	8.90	1.57
A	150	1.24	8.97	2.03	238	4.35	16.83	3.81
M	158	1.58	10.13	2.60	250	4.67	17.45	4.49
J	167	1.95	11.23	1.86	265	5.06	18.17	3.01
J	162	1.75	10.64	0.98	257	4.86	17.80	1.64
A	147	1.11	8.48	0.31	233	4.21	16.55	0.60
S	111	Ø	Ø	Ø	176	2.33	12.31	0.04
O	65	Ø	Ø	Ø	103	Ø	Ø	Ø
N	30	Ø	Ø	Ø	48	Ø	Ø	Ø
D	20	Ø	Ø	Ø	32	Ø	Ø	Ø
Σx				7.78				15.16

[1] The sea surface irradiance is taken from Baker et al. (1980) coefficients for the model of Green et al. (1974) for 33°N and adjusted for 18-year average cloud cover (1961-1978) for the Southern California Bight (30-35°N, 114-120°W; Renner 1979) using the equation of Bener (1964) and then weighted by an empirical fit to the DNA action spectrum.

[2] Depth in sea where the daily UV-B fluence is equivalent to the LD_{50}. The depth of UV-B penetration was calculated for low productive water (0.1 mg Chla m^{-3}) using the downwelling irradiance diffuse attenuation coefficients of Baker and Smith (pers. comm.); $K_T = 0.153$ for 0.32 cm ozone and $K_T = 0.150$ for 0.24 cm ozone.

[3] 125 Jm^{-2} bio. eff. The daily dose was used assuming jack mackerel have the same sensitivity as Northern anchovy larvae and Pacific mackerel (see text for explanation).

[4] The monthly percent of larvae at or above the LD_{50} depth was calculated using the vertical distribution data for jack mackerel larvae of Ahlstrom (1959) fitted by the equation $y = 11.44\ x\ 0.5$, $r^2 = 0.99$, for the data from the surface to 28 meters, where y is the cumulative percent of larvae from the surface to depth x, in meters.

[5] 4-year average larval abundance from Farris (1961) was used to weight the percent of larvae affected by UV-B for each month.

Appendix Table 4

Effect of action spectrum selection and UV sources used to estimate UV-B mortality in northern anchovy larval standing stock.

UV source	Action spectrum used to weight data	LD_{50} dose[1] $Jm^{-2}d^{-1}$ bio.eff.	Percent losses to annual larval standing stock		0.24 cm ozone losses
			0.32 cm ozone	0.24 cm ozone	0.32 cm ozone losses
Artificial[2]	Green & Miller, DNA	50	6.59	12.11	1.84
Artificial	Hunter et al., modified DNA	50	9.48	14.68	1.55
Solar	Green & Miller DNA	30	11.69	17.72	1.52
Solar & artificial	Hunter et al., modified DNA	50	13.40	17.56	1.31

[1] cumulative 12-day dose/12 days.

[2] Westinghouse FS-40 sunlamps® filtered by CTA plastic.

THE EFFECT OF UV-B IRRADIATION ON THE INTEGUMENT OF THE MARINE FLATFISH *PLEURONECTES PLATESSA* L.

Alistair M. Bullock

Dunstaffnage Marine Research Laboratory
P. O. Box 3
Oban, Argyll, Scotland

INTRODUCTION

Although it is universally recognised by dermatologists that UV-B is the component of the spectrum responsible for inducing cellular destruction in mammalian skin (e.g. Magnus 1976) the effect of UV-B on the fish integument has received little attention to date. This is probably due in part to the misconception common among many marine biologists that ultraviolet light does not penetrate seawater significantly. While this may be true of coastal waters high in particulate matter the work of Steeman Nielsen (1964), Jerlov (1968), Calkins (1975) and Smith and Baker (1978) indicates that UV-B can readily penetrate clear oceanic water to biologically significant depths.

In relation to morphological changes in the skin following UV-B irradiation the most relevant work appears to be that of Bell and Hoar (1950) who demonstrated severe degeneration of both epidermal and dermal components in the skin of the sockeye salmon *Oncorhynchus nerka*. Dunbar (1959) reported that rainbow trout fingerlings when moved to outdoor tanks began to lose equilibrium, darkened and developed necrotic areas of the dorsum following three days exposure to bright sunlight. Fingerlings exposed to artificial ultraviolet developed morphologically similar lesions to those of the 'sunburned' groups. Crowell and McCay (1930) and Allison (1960) report similar results for brook trout and lake trout fingerlings respectively.

Of these investigators only Bell and Hoar (1950) were able to quantify the UV dose to which the fish were subjected using the Finsen unit as their standard measurement and from the results postulated that as the intensities used in their study were comparable to those found in nature the effects observed would be of significance in the survival of hatchery fry when moved outdoors.

As part of an ongoing program of aquaculture research at this laboratory an investigation into the effect of UV-B radiation on the skin of the marine flatfish (Pleuronectes platessa L.) was initiated. This species was selected because of its benthic nature thus making it particularly useful in radiation experiments, the cumulative dose being determined more accurately due to the fishes natural instinct to remain upon the stratum.

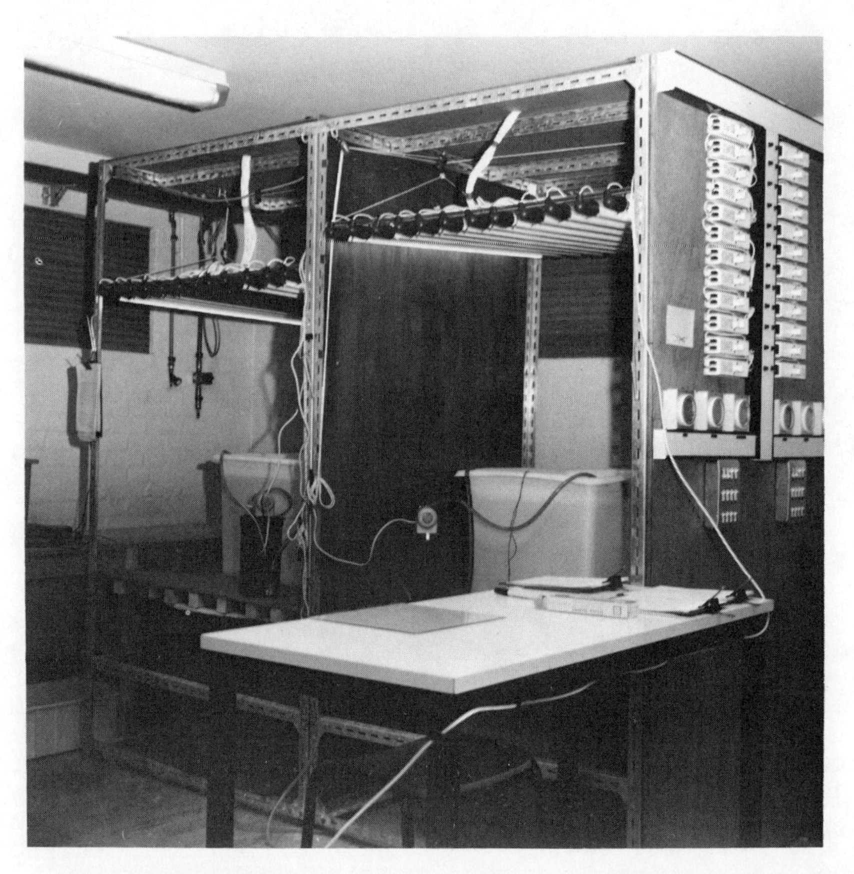

Figure 1. Irradiation array.

MATERIALS AND METHODS

1. Irradiation array

The apparatus (Fig. 1) consisted of an aluminium framework providing support for two banks of twelve fluorescent tubes mounted at 7 cm intervals on plywood sheeting. Maximum reflectivity of the tubes was attained by sheets of metalized PVC fixed behind the tubes. The length of irradiation was controlled by linking the tubes in series to automatic time switches thus making it possible, by incorporating groups of daylight fluorescent tubes (Thorn 40W Daylight-Coolwhite), to simulate an approximate diurnal rhythm. Intensity of irradiation was controlled by raising or lowering the tube bank via a pulley system.

2. Experimental design

The irradiation chamber comprised 12 circular plastic containers fixed in a 4 x 3 format to a perspex base. Holes were drilled in the walls of each to allow an adequate interchange of aerated water between containers. Twelve 1^+ year group plaice (*Pleuronectes platessa* L.) approximately 8 cm in length were used in each experiment, one fish being placed in each container at least 12 h prior to irradiation. To prevent fish swimming free the chamber was covered with wide mesh netting. All experiments were carried out at 10°C at a water depth of 40 cm in 120 l white polythene tanks using a recirculating seawater filtration system. Plots of UV irradiance taken prior to the experiments showed no loss of intensity at any position on the tank base.

3. Histology

Fish were fixed in 10% buffered formalin and processed for histology. Sections of skin were cut at 6 µm and stained with Haematoxylin and Eosin or Mallorys Trichrome Method.

4. UV-B measurement

During the initial stages of the present project an attempt was made to measure the UV-B emission from the fluorescent tubes (Phillips TL 40W/12) using a Kipp and Zonen solarimeter thermopile coupled to a Hewlett-Packard digital voltmeter. This method however proved unsatisfactory. There was for instance some doubt as to the sensitivity of the detector within the UV-B range of spectral activity; it also proved impractical to waterproof the device for underwater measurements. Eventually this technique was abandoned in favor of a portable UV digital radiometer manufactured by Macam Photometrics (Macam Photometrics Ltd., Livingstone, Scotland.) This gives a digital display through five full scale decades reading from 0-19.99 x 1 $\mu W\ cm^{-2}$ through 0-19.99 x 10^4 $\mu W\ cm^{-2}$ with a sensitivity of 0.1 $\mu W\ cm^{-2}$. The detector head uses a Ga As (P) solid state photodiode, waterproofed and cosine corrected for underwater measurements. The peak wavelength response is at 310 nm with a bandwidth of 34 nm. Calibration is carried out at regular intervals at 313 nm using a mercury vapor discharge lamp as the source. The radiation is fed into a monochromator to isolate the spectral lines and the instrument is calibrated using NBS standards within $\pm$ 1%. The temperature coefficient of the detector is $\pm$ 0.3% per ^{o}C at 313 nm.

RESULTS

In order fully to understand the cellular events taking place following irradiation it is first necessary briefly to describe the morphology of the normal plaice epidermis (Fig. 2). It consists of a layer of 6-10 cells in thickness incorporating three major cell types, the most common being the Malpighian cell. Within the outer zone large, often clear cells can be seen, these are the goblet or mucous cells which are capable of copious mucus secretion and provide the fish with a protective mucoid layer. Acidophilic granule cells (AGC's)usually situated within the basal cell layer are common in this particular species but their specific function is at present unclear.

Following a cumulative dose of 63 mJ cm^{-2} given at an intensity of 30 $\mu W\ cm^{-2}$ the fish were killed at intervals ranging from 30 min to 72 h post-irradiation (p.i.). At 30 min p.i. a significant response was observed when compared with the normal non-irradiated control skin. The most prominent feature was the engorged appearance of the AGC's, the majority having moved into the supra-basal cell layer (Fig. 3). By 1 h p.i. a number of AGC's had ruptured, releasing their contents into the surrounding epidermis, the resulting cell (Fig. 4) appearing in section as a clear vacuole not to be confused with the overlying, but much larger mucous cells.

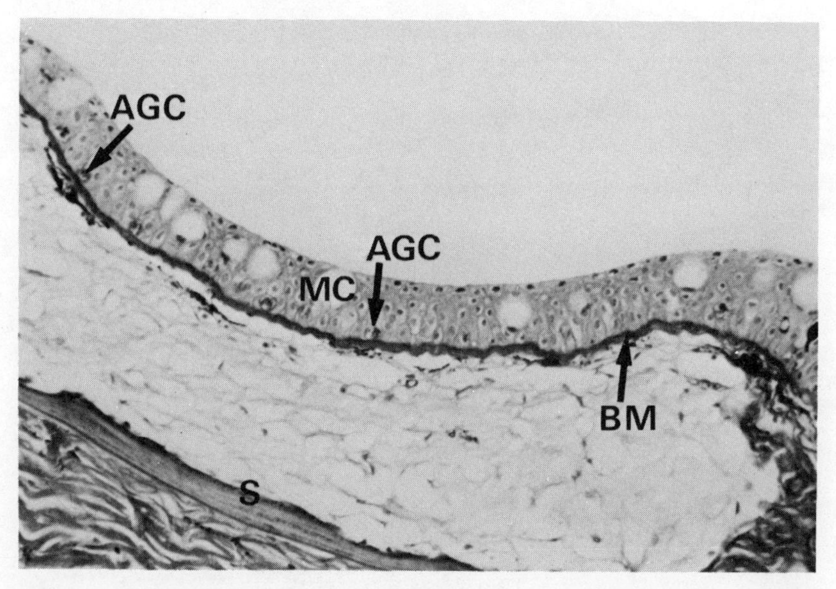

Figure 2. Normal plaice epidermis. Note the position of the acidophilic granule cells (AGC) abutting the basement membrane (BM). Mucous cells (MC) Scale (S) Mallorys Trichrome Stain (MTS) x 320.

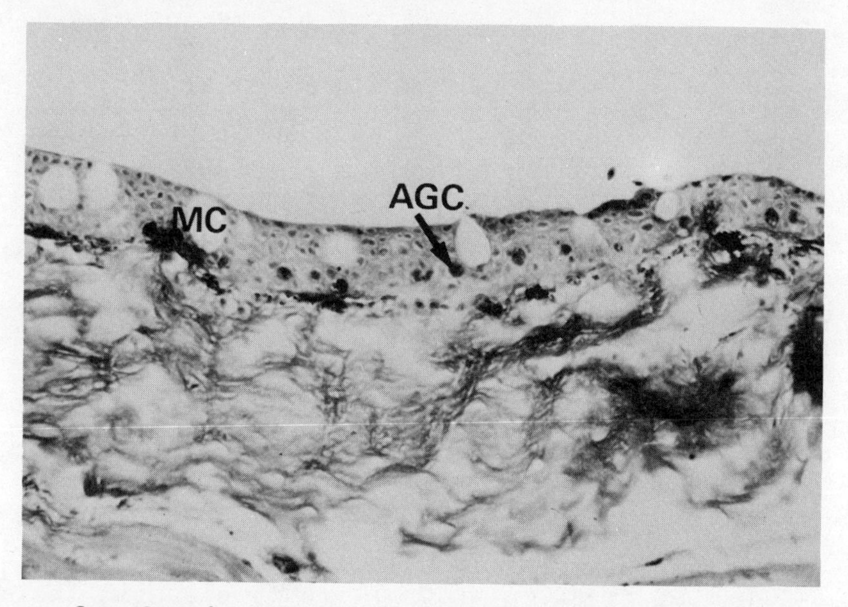

Figure 3. 30 min post-irradiation. The AGC's are packed with granules and hence more densely staining and have moved into the suprabasal zone. Mucous cell (MC) MTS x 320.

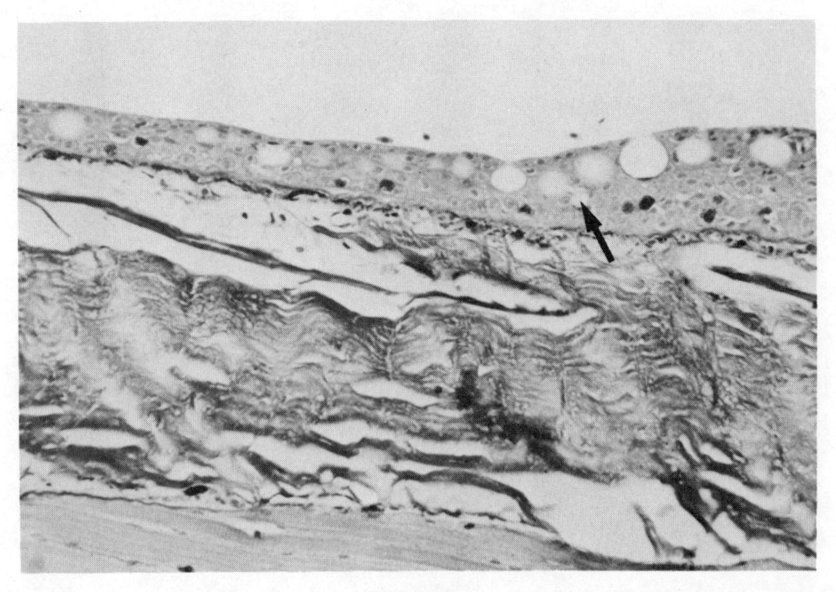

Figure 4. 1 h post-irradiation. Although similar to 30 minutes p.i. a number of AGC's have ruptured releasing their contents into the epidermis. Vacuolated AGC arrowed. MTS x 320.

By 6 h p.i. (Fig. 5) the AGC's often appeared vacuolated although maintaining, in most instances, a granular appearance and many were seen within the upper layer of the epidermis. There was also evident at this stage a concomitant swelling of the mucous cells accompanied by a marked irregularity of the epidermal surface. This was attributed to the engorged mucous cells forcing the upper epidermal layer into a rugose pattern. At 8 h p.i. haematoxylin and eosin staining revealed the presence in moderate numbers of the piscine equivalent of the mammalian 'sunburn' cell.

Cell necrosis within the upper half of the epidermis was well advanced by 24 h p.i. with epidermal separation within the mid-zone particularly prominent (Fig. 6). By 32 h p.i. the upper layer had sloughed leaving the lower zone of cells undergoing necrosis. It was particularly evident at this stage that juvenile mucous cells whose genesis would be within the basal/supra-basal layer were being stimulated into activity (Fig. 7), possibly as a 'secondary defence mechanism' against irradiation trauma. By 48 h p.i. epidermal destruction was virtually complete (Fig. 8) leaving only a

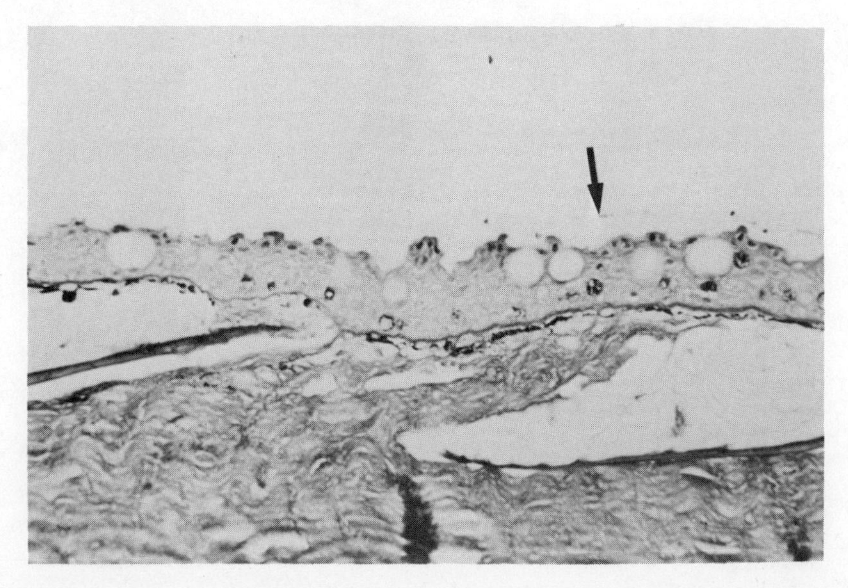

Figure 5. 6 h post-irradiation. The epidermal surface has become very irregular and the mucous cells are producing copious amounts of mucus some of which can be seen on the surface (arrowed). The AGC's are less granular, the occasional cell apparent within the upper layer. MTS x 320.

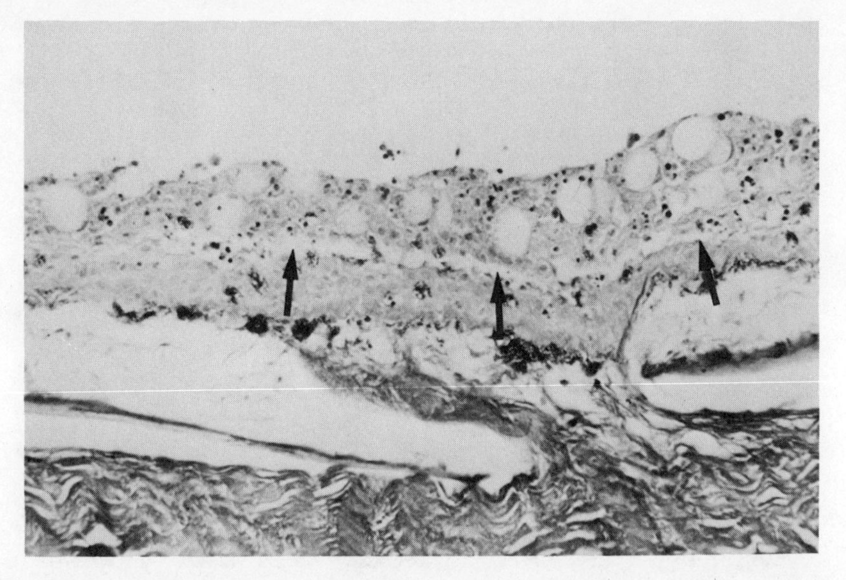

Figure 6. 28 h post-irradiation. Cellular necrosis is particularly evident within the upper epidermal layer. Epidermal separation within the mid-zone is prominent (arrowed) MTS x 320.

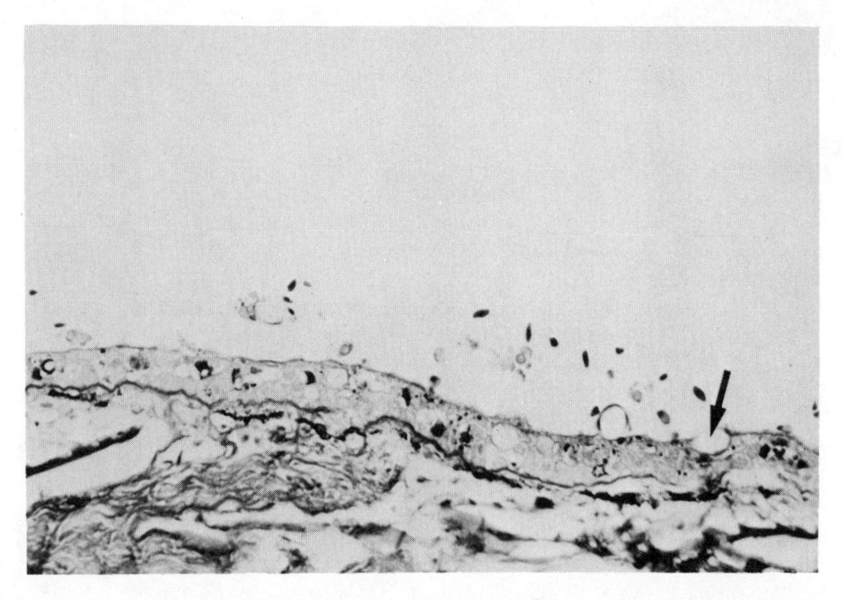

Figure 7. 32 h post-irradiation. The upper epidermis has now sloughed leaving a layer of cells undergoing necrosis. An immature mucous cell (arrowed) can be seen on the surface. MTS x 320.

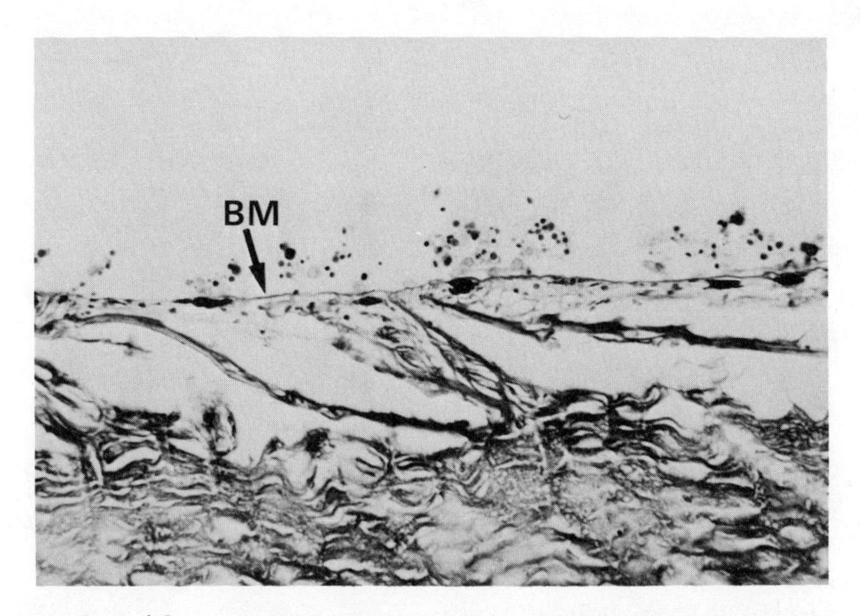

Figure 8. 48 h post-irradiation. Epidermal destruction is now virtually complete leaving the basement membrane (BM) intact. MTS x 320.

relatively few necrotic Malpighian cells sloughing from the basement membrane.

DISCUSSION

The results reported here are limited to one cumulative dose of UV-B irradiation only, viz. 63 mJ cm^{-2}. This level was selected in order to determine the cellular response in fish skin when equated to the mammalian MED on virginally white skin, reckoned at this latitude (55°) to be between 72 and 144 mJ cm^{-2} at noon in mid-summer (Johnson pers. comm.).

Suspected skin trauma due to ultraviolet radiation is not unknown by fish pathologists (Roberts 1978) and if one extrapolates from these results it becomes apparent that given clear water conditions and maximum penetration of solar UV-B the implication in terms of skin trauma to farmed fishes is profound. If such a dose can induce total destruction of the epidermis within such a short period then it follows that lower doses must be capable of at least precipitating cellular necrosis within the upper epidermal layer thus depriving the fish of its protective outer layer and subsequently exposing the fish to a variety of infectious agents should water conditions permit.

Water depths of 40 cm are not uncommon in aquaculture systems and at least one clinical case of a "light" problem affecting the dorsal and pectoral fins of salmonid fry held in outdoor tanks has been reported in Scotland this year (Roberts pers. comm.) following a long period of uninterrupted bright, clear weather.

REFERENCES

Allison, L.N. 1960. "Sunburning" fingerling lake trout with ultraviolet light and the effect of a Niacin-fortified diet. Prog. Fish Cult. 22 (3) 114-116.
Bell, M.G. and W. S. Hoar. 1950. Some effects of ultraviolet radiation on sockeye salmon eggs and alevins. Can. J. Research Sect. D. 28 (1) 35-43.
Calkins, J. 1975. Measurements of the penetration of solar UV-B into various natural waters. CIAP Monograph 5, Part 1 Chapter 2 (Appendix E) 267-298.
Crowell, M.F. and C. M. McCay. 1930. The lethal dose of ultraviolet light for brook trout (Salvelinus

fontinalis). Science 72 No. 1875 582-583.
Dunbar, C.E. 1959. Sunburn in fingerling rainbow trout. Prog. Fish Cult. 21 (2) 74.
Jerlov, N.G. 1968. In Optical Oceanography Elsevier Oceanography Series 5. Elsevier Publishing Co. London.
Magnus, I.A. 1976. In Dermatological Photobiology Blackwell Scientific Publications, London.
Roberts, R. J. 1978. In Fish Pathology Bailliere Tindal, London.
Smith, R. C. and K. S. Baker. 1978. Penetration of UV-B and biologically effective dose rates in natural waters. Photochem. Photobiol. 29 311-323.
Steemann Nielsen, E. 1964. On a complication in marine productivity work due to the influence of ultra-violet light. J. du Conseil. 29 No 2. 130-135.

ASSESSMENT OF THE INFLUENCE OF ENHANCED UV-B ON MARINE PRIMARY PRODUCTIVITY

Raymond C. Smith and Karen S. Baker

Visibility Laboratory
Scripps Institution of Oceanography
University of California, San Diego
La Jolla, California 92037

ABSTRACT

A simple model, utilizing recent data, is used to estimate the potential impact of increased UV-B on aquatic primary productivity. The model is then utilized to emphasize major uncertainties in our present predictive capabilities. These (generally non-linear) uncertainties, which are much larger than the current linear conclusions drawn from our simple model, include: (1) the identification and quantification of the relevant photoprocesses and their biological weighting functions as well as the limitation of ^{14}C techniques as currently and routinely used; (2) the possibility of community structure change, due to increased UV-B stress, and the consequent potential for tropodynamic changes in the food chain; (3) the influence of water column mixing on the determination of biologically effective doses and photosynthetically available radiant energy incident on and available to phytoplankton.

INTRODUCTION

It is currently estimated that certain continued human activities may reduce the long-term global average thickness of the stratospheric ozone layer by 16.5% (NAS, 1979). It is generally accepted that this reduction in the ozone layer will lead to an increase in solar ultraviolet radiation reaching the earth's surface (Green et al 1980, 1974). This increased UV-B (280 - 320 nm) radiation, when weighted by the different relative effectiveness of its constituent wavelengths for producing damage, has been called damaging ultraviolet (DUV) radiation.

UV-B and hence DUV, can penetrate to ecologically significant depths in natural waters (Jerlov, 1950; Lenoble, 1956; Smith and Baker, 1979). Thus, there is concern that reduced stratospheric ozone, and a consequent increase in UV-B at the earth's surface, may have an adverse effect on aquatic ecosystems (NAS, 1979; CIAP, 1975), including phytoplankton productivity.

There is strong evidence that UV-B adversely affects phytoplankton productivity, at least as it is currently measured, and this evidence will be reviewed below. Conclusions that can be drawn from this evidence, with respect to possible adverse ecological consequences of increased DUV on natural phytoplankton populations, are limited by major uncertainties. These uncertainties include identifying and quantifying the relevant photoprocesses active when UV-B is incident on phytoplankton; measuring the response of natural phytoplankton populations to DUV; and accurately estimating effective biological dose rates for plankton populations that are naturally mixed within the water column.

We introduce a simple model, which utilizes recent experimental data, for estimating the potential impact of increased UV-B on aquatic primary productivity. This model summarizes our current state of knowledge and serves to highlight the major uncertainties in our present predictive capabilities. Our intent is to focus on the key limitations of the model, drawing particular attention to the potential non-linear aspects of assessing possible adverse effects of UV-B on aquatic ecosystems.

THEORETICAL BACKGROUND

An ultimate objective is to quantitatively evaluate the spectral effect of UV-B on the primary productivity of natural phytoplankton populations. Such a description requires a measure of the biologically effective irradiance incident on the phytoplankton. This in turn requires a knowledge of the penetration of UV-B in natural waters, and the calculation of biologically effective dose-rates as a function of depth. The essentials of these calculations have been discussed in detail by Smith and Baker (1979) and by Smith et al. (1980) and are reviewed here for completeness.

(a) Downward global flux

The downward global flux, $E_d(0+,\theta,\lambda)$, just above the ocean surface (0+) is given as the sum of the direct, $E_{sun}(0^+,\theta,\lambda)$, and the diffuse, $E_{diff}(0^+,\theta,\lambda)$, components,

$$E_d(0^+,\theta,\lambda) = E_{sun}(0^+,\theta,\lambda) + E_{diff}(0^+,\theta,\lambda), \quad (1)$$

where θ is the sun zenith angle and λ the wavelength. The total downward irradiance just below the surface (0^-) is calculated by

$$E_d(0^-,\theta,\lambda) = t(\theta) \cdot E_{sun}(0^+,\theta,\lambda) + t_d \cdot E_{diff}(0^+,\theta,\lambda) \quad (2)$$

where $t(\theta)$ is the transmittance of the air-sea interface as calculated using Fresnell's equations and t_d = 0.94 is the transmittance of the air-sea interface for a uniform radiance distribution (Preisendorfer, 1976; Austin , 1974).

The downward spectral irradiance at any depth z underwater, $E_d(z,\theta,\lambda)$, can be calculated from the spectral irradiance just beneath the ocean surface, $E_d(0^-,\theta,\lambda)$, by

$$E_d(z,\theta,\lambda) = E_d(0^-,\theta,\lambda) \exp -K_T(\lambda)\cdot Z \quad (3)$$

where $K_T(\lambda)$ is the total diffuse attenuation coefficient for irradiance.

Semi-empirical analytic formulae have been developed by Green (Green et al. 1974; Shettle and Green,1974; referred to as GSS; Green, Cross and Smith, 1980; referred to as GCS) for calculating the global UV-B (280-340 nm) reaching the earth's surface, $E_d(0+,\theta,\lambda)$, as a function of wavelength, solar angle, aerosol thickness, surface albedo and, in particular, ozone thickness ω_{oz}.

Both the GSS and GCS semi-empirical models have been fitted to our measured spectral, $E_d(0^+,\theta,\lambda)$, and total irradiance data (Baker, Smith and Green, 1980; referred to as BSG; Smith and Baker, 1980). These data are distinguished from previous observational data, having been obtained for a marine atmosphere and at equatorial latitudes which have small sun zenith angles. This data permitted development of a model for the UV-B radiation reaching the ocean surface.

Use of the BSG model allows our actual spectral irradiance measurements to be reliably extrapolated to any sun angle giving the data all the computational advantages of an analytic formula. The fit, for solar zenith angles from 0° to 60°, is within the estimated experimental accuracy of our data ($\approx$ 20%). The fit becomes poorer for large sun zenith angles (low sun elevations). However, at these angles, the contribution of solar irradiance to the biological dose becomes small and the error due to this fit is insignificant.

(b) Relative biological efficiency

It is now generally recognized that the relative biological efficiency for the effect of UV-B on phytoplankton photosynthesis plays a key role in assessing the potential impact of increased UV-B on aquatic photosynthesis (NAS, 1979). The biologically effective irradiance must be based upon a weighting function, $\epsilon(\lambda)$ that takes account of the wavelength dependency of biological action. Ideally a biological weighting function or action spectrum, $\epsilon(\lambda)$, should be determined for the organism one is studying. However, a weighting function can be generalized and treated as an approximation for all organisms (Nachtwey, 1975; Setlow, 1974). While there is a great variety of physiological and morphological phenomena attributable to UV-B radiation, the principal chromatophores for this diversity of plant responses may be similar. Giese (1964), discussing the response of animal cells to UV, and Caldwell (1971), discussing adverse plant responses to UV-B, suggest that action spectra for these various responses commonly conform to the absorption spectra of either proteins or nucleic acids. Jones and Kok (1966), studying the photoinhibition of chloroplast reactions, suggest that UV-B radiation has a multiplicity of deleterious effects in photosynthesis. However, their data suggest that the modes of action of inhibiting light have similar spectral shapes and they give an action spectrum for photoinhibition in chloroplasts.

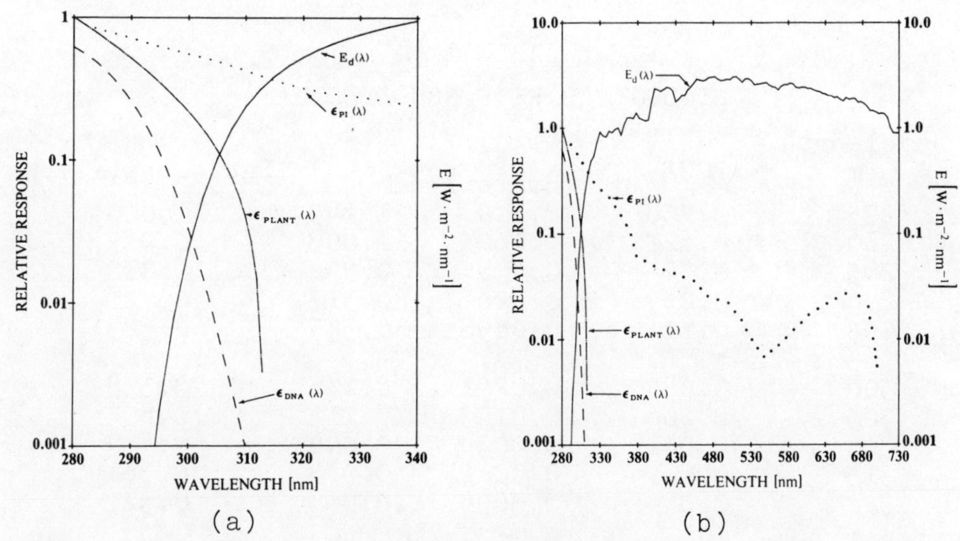

Figure 1. Relative biological efficiencies $\epsilon(\lambda)$ (left hand ordinate): for Setlow's (1974) average action spectrum for biological effects involving DNA(---); for Caldwell's (1971) generalized action spectrum for plants (- . -); and for Jones and Kok's (1966) action spectrum for photoinhibition of chloroplasts (....). The latter data have been adapted from Jones and Kok's data by renormalizing their data to 1.0 at 275 nm. The right hand ordinate (corresponding to the solid curve) gives the noon spectral irradiance for wavelength comparison of $\epsilon(\lambda)$ and $E_d(0^+,90,\lambda)$. (a) 280-340 nm, (b) 280-730 nm.

In our previous discussions, we have chosen to work with four possible action spectra: (1) Setlow's (1974) DNA action spectrum; (2) Hunter et al. 1980 modification of Setlow's DNA action spectrum for damage to the eggs and larvae of northern anchovy; (3) Caldwell's (1971) generalized action spectrum for adverse UV-B effects on plants; and (4) an action spectrum for chloroplast inhibition adopted from Jones and Kok (1966). These four biological response spectra, as normalized for our purposes, are plotted in Fig. 1 and listed in Table 1. It should be clearly noted that, to date, the appropriate weighting function (or functions) for the effect of UV-B on natural phytoplankton populations is uncertain.

TABLE 1
BIOLOGICAL RESPONSE SPECTRA

Wavelength [nm]	DNA	DNA (Hunter)	Plant	Photoinhibition
275	0.758	0.819	1.343	1.000
280	0.621	0.662	1.003	0.917
285	0.445	0.451	0.732	0.812
290	0.259	0.240	0.517	0.721
295	0.111	0.0898	0.349	0.623
300	0.033	0.0291	0.218	0.558
305	0.006	0.0103	0.117	0.492
310	0.001	0.0034	0.040	0.427
315	0.0	0.0012	0.0	0.361
320	0.0	0.0004	0.0	0.321
325	0.0	0.0	0.0	0.294
330	0.0	0.0	0.0	0.279
335	0.0	0.0	0.0	0.259
340	0.0	0.0	0.0	0.236
345	0.0	0.0	0.0	0.206
350	0.0	0.0	0.0	0.196
360	0.0	0.0	0.0	0.138
370	0.0	0.0	0.0	0.105
380	0.0	0.0	0.0	0.066
390	0.0	0.0	0.0	0.057
400	0.0	0.0	0.0	0.052
410	0.0	0.0	0.0	0.047
420	0.0	0.0	0.0	0.043
430	0.0	0.0	0.0	0.042
440	0.0	0.0	0.0	0.041
450	0.0	0.0	0.0	0.037
460	0.0	0.0	0.0	0.034
470	0.0	0.0	0.0	0.026
480	0.0	0.0	0.0	0.025
490	0.0	0.0	0.0	0.024

TABLE 1

BIOLOGICAL RESPONSE SPECTRA (CONT)

Wavelength [nm]	DNA	DNA (Hunter)	Plant	Photoinhibition
500	0.0	0.0	0.0	0.022
510	0.0	0.0	0.0	0.019
520	0.0	0.0	0.0	0.014
530	0.0	0.0	0.0	0.010
540	0.0	0.0	0.0	0.008
550	0.0	0.0	0.0	0.006
560	0.0	0.0	0.0	0.008
570	0.0	0.0	0.0	0.009
580	0.0	0.0	0.0	0.011
590	0.0	0.0	0.0	0.013
600	0.0	0.0	0.0	0.015
610	0.0	0.0	0.0	0.018
620	0.0	0.0	0.0	0.019
630	0.0	0.0	0.0	0.022
640	0.0	0.0	0.0	0.023
650	0.0	0.0	0.0	0.025
660	0.0	0.0	0.0	0.028
670	0.0	0.0	0.0	0.030
680	0.0	0.0	0.0	0.030
690	0.0	0.0	0.0	0.019
700	0.0	0.0	0.0	0.005

(c) Biologically effective dose (no mixing)

The biologically effective dose is given by

$$E_B(\theta)\left[W\cdot m^{-2}\right]_{\epsilon(\lambda)} = \int E(\theta,\lambda)\left[W\cdot m^{-2} nm^{-1}\right] \cdot \epsilon(\lambda)\cdot d\lambda[nm] \quad (4)$$

where $\epsilon(\lambda)$ is the relative biological efficiency for the biological effect under study. We use the notation $[W\cdot m^{-2}]_{\epsilon(\lambda)}$ to indicate a physically measured (or measurable) absolute irradiance weighted by a relative function. It is important to realize that the choice of a normalizing wavelength for $\epsilon(\lambda)$ is arbitrary. Use of different normalizing wavelengths leads to widely varying values for biologically effective irradiance thus prohibiting significant comparison. However, if $\epsilon(\lambda)$ and its wavelength normalization are specified (e.g. as in Table 1), then quantitative comparisons can be made between biologically effective dose-rates. Thus the resultant units $[W\cdot m^{-2}]_{\epsilon(\lambda)}$, while arbitrary, are precisely defined and are useful for a comparison with other dose-rates in the same units obtained for different situations, e.g., different depths, different ozone thicknesses, or by different investigators.

We have chosen to renormalize the data of Jones and Kok so that the data are contiguous throughout the spectrum and so they are easily comparable to the Setlow and Caldwell biological efficiencies. We chose to set $\epsilon_{PI}(275) = 1.0$ since there was a small maximum at this wavelength and because, for ecological purposes, wavelengths less than 275 nm appear to be insignificant at the earth's surface.

The total daily biologically effective dose can be calculated by integrating $E_B(z,\theta)$ for all sun angles, after converting sun angles to time, over the course of a day, viz.,

$$E_{TB}\left[J\cdot m^{-2} day^{-1}\right]_{\epsilon(\lambda)} = \int E_B(\theta)\left[J\cdot s^{-1}\cdot m^{-2}\right]_{\epsilon(\lambda)} \cdot dt[s]. \quad (5)$$

This biological effective dose is for a fixed depth in the water column. Thus, for marine organisms, use of

Eq. 5 implies that mixing rates are slow compared to the time an organism receives DUV radiation. This "no-mixing" assumption is usually a "worst case" situation; i.e. in an unmixed water column near surface marine organisms receive a maximum biological dose. It is shown that "complete mixing" can confer considerable protection to a marine organism from DUV radiation.

(d) Biological effective dose (complete mixing)

Kullenburg (this volume) has discussed the role of vertical mixing in relation to DUV radiation and presented examples to show that mixing times vary over a wide range in the sea. Complete mixing, where mixing rates are rapid compared to the time an organism receives DUV radiation, is the opposite environment situation from no-mixing. Thus, consideration of these two extreme situations will indicate the limits of DUV radiation incident on a marine organism.

Morowitz (1950) and Zepp and Cline (1977) have calculated the average biological dose for complete mixing; i.e., for an organism that spends an equal amount of time at each depth in the mixed layer. Smith and Baker (1979) have shown that for some biological weighting functions there exists an effective attenuation coefficient, K_ϵ . Combining these results the total average biological dose within the mixed layer can be shown to be

$$E_B(\text{complete mixing}) = \frac{E_B(0-)}{K_\epsilon \cdot Z_m} (1 - \exp -K_\epsilon \cdot Z_m) \tag{6}$$

where $E_B(0-)$ is the biologically effective dose (for a weighting function $\epsilon(\lambda)$) just beneath the water surface, K_ϵ the effective diffuse attenuation coefficient for E_B, and Z_m is the depth of the mixed layer. Using the same notation the biologically effective dose at depth Z under conditions of no mixing is

$$E_B(Z;\ \text{no mixing}) = E_B(0-)\ \exp(-K_\epsilon \cdot Z) \tag{7}$$

It is of some interest to compare the ratio of biological dose for the limiting conditions of complete mixing to no mixing ,

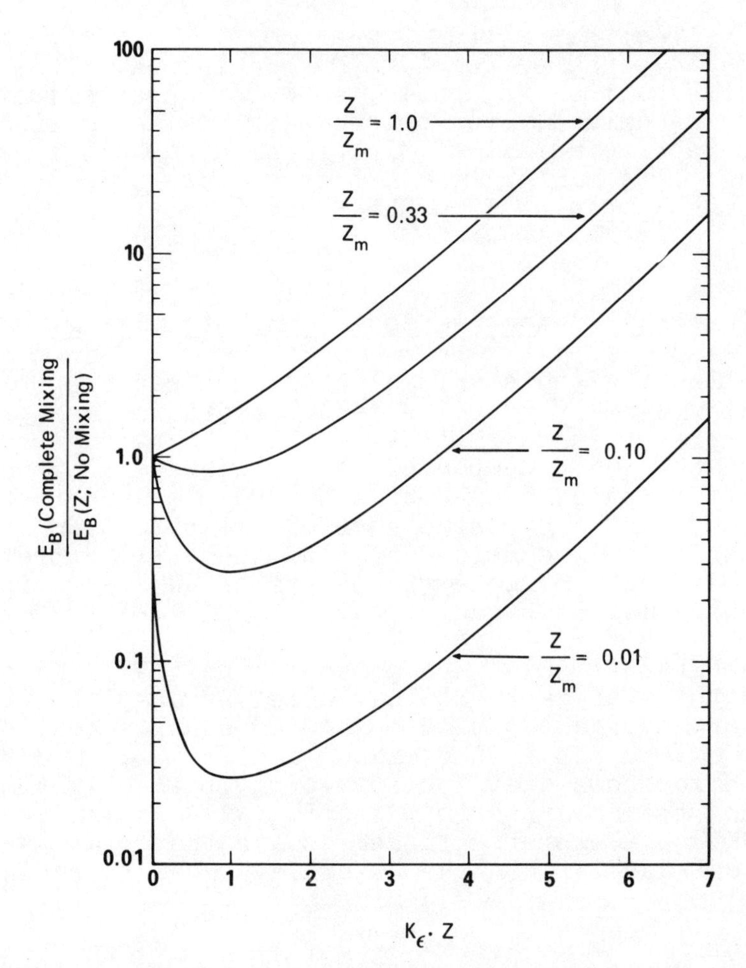

Figure 2. Ratio, R, of biologically effective dose, E_B, assuming complete mixing to E_B assuming no mixing in the water column vs. depth of the organism receiving the dose. This depth is expressed in terms of the effective attenuation length for the biologically effective dose under consideration. The ratio of the depth of the organism ($K_\epsilon \cdot Z$) to the mixed layer depth $K_\epsilon \cdot Zm$) is shown as a parameter varying from Z/Zm = 1.0 to 0.01. When R>1 an organism will experience greater DUV in a completely mixed water column than in a stable water column. Conversely, when R<1 an organism will receive some "protection" from water column mixing.

$$R = \frac{E_B \text{ (complete mixing)}}{E_B(Z; \text{ no mixing})} = \frac{1}{\exp(-K_\epsilon \cdot Z)} \cdot \frac{1 - \exp(-K_\epsilon . Zm)}{K_\epsilon \cdot Zm} \tag{8}$$

Figure 2 shows a plot of Eq. 8 with the ratio of the depth of the organisms to mixed layer depth (Z/Z_m) as a parameter. Consider, for example, an organism in ocean water with a mixed layer depth (Z_m) of 50 meters. The curves indicate that the biological dose at 5 meters would be as much as three times less, and at 0.5 meters more than 30 times less, for complete mixing compared to no mixing. Thus, near surface (few attenuation lengths) organisms can receive considerable "protection" from mixing ($R<1$). Conversely, with increasing depth (or as the depth of interest approaches the depth of the mixed layer) mixing exposes otherwise deep organisms to increased DUV ($R>1$). Smith and Baker (1979) have shown that K_{DNA}^{-1} is approximately six meters in the clearest ocean waters and about 2.5m in moderately productive waters containing average quantities of dissolved organic material. Mixed layer depths range from a few to a few hundred meters. Thus, in general, $K_\epsilon \cdot Z/K_\epsilon . Zm \leq \frac{1}{10}$ and we can expect mixing to reduce biologically effective doses in the top attenuation lengths of the ocean.

Equations analogous to Eqs. 6-8 can be written by replacing E_B with the photosynthetically available radiant energy (PAR) an organism is exposed to,

$$\text{PAR(complete mixing)} = \frac{\text{PAR}(0-)}{K_{PAR} . Zm} \left[1\text{-}\exp(K_{PAR} \cdot Z_m)\right] \tag{9}$$

$$\text{PAR}(Z; \text{ no mixing}) = \text{PAR}(0^-) \exp(-K_{PAR} . Z) \tag{10}$$

where K_{PAR} is defined by Eq. 10. In general, $K_{PAR} < K_\epsilon$ and $K_{PAR} \cdot Z/K_{PAR} . Zm \leq 1$, so while mixing may reduce the average PAR a phytoplankton "sees" it is as likely to increase PAR, especially at depths below a few attenuation lengths.

A further consideration is the depth to the euphotic layer ($\approx 4.61\ K_{PAR}^{-1}$) relative to the depth of the mixed layer (Z_m). Indeed, the above equations are a modified form of Sverdrup's (1953) critical depth model where it has been postulated that the critical depth must be greater than the depth of mixing for net positive production to occur. Conversely, if the depth of the mixed layer is greater than the critical depth then turbulence sweeps the algal cells below the euphotic zone, column photosynthesis becomes less than column respiration, and there can be no crop increase.

The above discussion indicates that the rate and depth of mixing is an important consideration, not only for determining the average radiant energy available for photosynthesis but also, for estimating DUV radiation. It can be seen that the potential costs and/or benefits to phytoplankton are functions of the relevant photoprocess and their weighting functions ($\epsilon(\lambda)$), as determined by the relevant diffuse attenuation coefficients ($K_{\epsilon}(\lambda)$). Thus, the natural variations in ocean mixing can alter estimates of productivity models by at least plus or minus an order of magnitude.

(e) Diffuse attenuation coefficient for irradiance

We have previously (Smith and Baker, 1979) presented preliminary results of underwater spectral irradiance measurements and the determination of the diffuse attenuation coefficient for irradiance, $K_T(\lambda)$, in spectral regions from 300 to 400 nm. We have also presented a classification of natural waters (Smith and Baker, 1978), based on data from a wide range of ocean water types, for determining $K_T(\lambda)$ (350 to 700 nm) in terms of chlorophyll concentrations. Most recently (Baker and Smith, in press) we have developed a descriptive and predictive classification of natural waters, whose dissolved and suspended material is primarily of biogeneous origin, for determining $K_T(\lambda)$ (300 to 750 nm) in terms of chlorophyll concentration and dissolved organic material (DOM).

This work will not be reviewed here. The essential point, however, is that once $K_T(\lambda)$ is determined, either by direct measurement or from our classification model, the spectral irradiance at any depth can be determined (Eq. 3) from a knowledge of the surface irradiance $E_d(0^-, \theta, \lambda)$. Then, if the relative biological efficiency $\epsilon(\lambda)$ is known or can be selected for comparative purposes, the biologically effective dose at any depth can be determined from Eqs. 4 and 5 or 6.

UV-B and simulated in-situ primary productivity

The amount of UV-B reaching the ocean's surface has long been suspected as a factor influencing primary

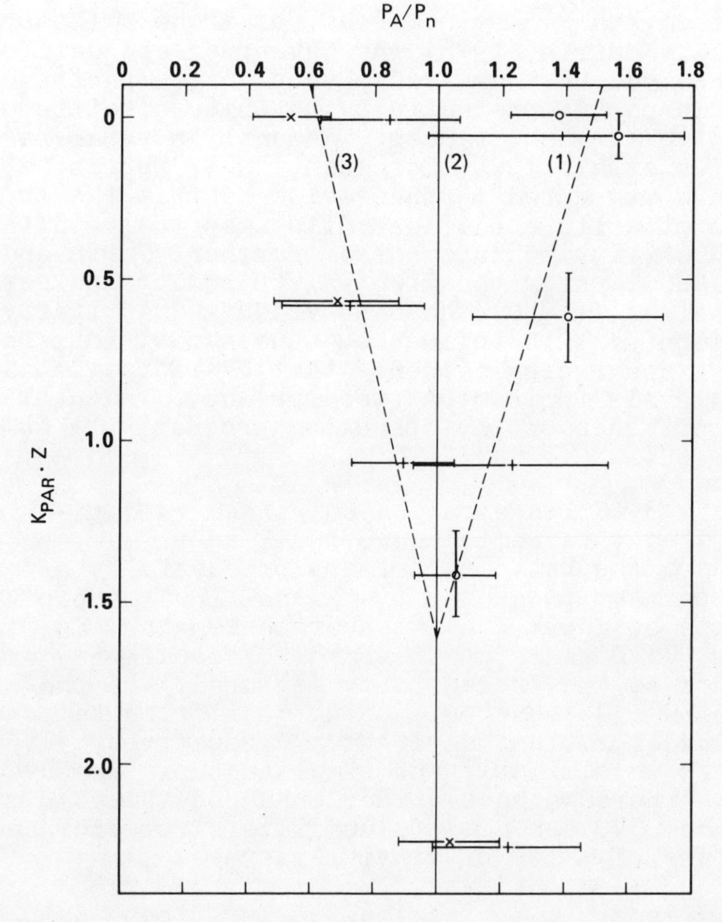

Figure 3. Ratio of natural to altered productivity as a function of depth (in attenuation lengths K·Z). Open symbols refer to productivity altered by exclusion of MUV using mylar ([]) or vinyl (O). Crosses and pluses refer to productivity altered by enhancement of MUV using FS40 lamps (x) or FS40 lamps filtered with CTA (+). Each point represents the average of several experiments.

productivity. Harris (1978) reviewed the literature with respect to photoinhibition of phytoplankton photosynthesis and references numerous authors who have suspected a role of UV in photoinhibition from depth profile studies in which the photoinhibition effect appeared to be related more to the spectral characteristics, rather than absolute intensity, of incident irradiance.

In addition, there are numerous researchers (see references in Harris, 1978) who have produced data showing that increased UV increases phytoplankton photoinhibition by comparing productivity in culture bottles having different UV transmittances. Recently several researchers (Jitts et al. 1976; Lorenzen, 1979; Worrest et al. 1980) have presented further evidence that the transmittance of culture bottles influences the results of ^{14}C productivity measurements. Further, Smith and Baker (1980), using the spectral transmittance data of Worrest et al. (1980), have provided a quantitative description of this influence. The sum of this research is convincing evidence that UV radiation, at the present levels incident at the surface of natural waters, has an influence on phytoplankton productivity as currently measured.

Figure 3 (Smith et al. 1980) presents data which demonstrates this influence of UV-B on primary productivity. Here the ratio of primary productivity altered (Pa) to primary productivity natural (Pn) is plotted as a function of depth (in attenuation lengths, $K_{PAR} \cdot Z$) for different UV-B radiation regimes. Productivity natural, Pn, refers to the productivity measured, via the simulated in-situ ^{14}C technique, using quartz incubation bottles under ambient daylight. Productivity altered, Pa, refers to similarly and simultaneously measured productivity where we have either enhanced the UV-B radiation using FS40 sunlamps (plus filters) or excluded UV-B by use of mylar or vinyl filters.

In agreement with previous workers these data indicate that when UV-B is enhanced then $Pa/Pn < 1$, whereas when UV-B is excluded then $Pa/Pn > 1$. The fact that excluding natural UV-B increases productivity indicates that present levels of UV-B at the ocean's surface are acting as an inhibitor of productivity in static bottle ^{14}C measurements. The distinguishing feature of our productivity work was the inclusion of contemporaneous spectral irradiance data, including the UV-B region of the spectrum. These optical data allowed the quantitative evaluation of biological dose, and hence some inference with respect to $\epsilon(\lambda)$, to be made.

The data and details with respect to inferring a relative biological weighting function for UV-B damage to phytoplankton productivity have been described elsewhere (Hunter et. al. (1979; Smith et al. 1980; Smith and Baker, 1980). Here we briefly review this procedure and emphasize the assumptions and uncertainties involved. We do not know the biological photoprocess (or combination of photoprocesses), with its effective biological efficiency $\epsilon(\lambda)$, which is responsible for the UV-B effect on photosynthesis as shown in Fig. 3. However, we can assume a plausable biological efficiency for some postulated photoprocesses, $\epsilon(\lambda)$, calculate the biological effective dose for this photoprocess, and compare the measured biological effects with the calculated dose rates to see if the results are consistent with the original postulate.

This procedure is schematically diagramed in Fig. 4. The lower left of this diagram shows a plot of relative response versus wavelength where, for illustrative purposes, we have shown three possible weighting functions, ϵ_i, and three different irradiance regimes, $E(\lambda)$. The biological effective dose, for a particular choice of ϵ_i and $E(\lambda)$, is the area under the product of these curves (see Eq. 4). As a consequence, for any given radiation regime, the biological effective dose, E_B, will change depending on the choice of $\epsilon(\lambda)$.

We assume that the measured biological effect (in this example reduced productivity) will be some reasonably behaved function of biological dose (lower right of

(1) Assume plausible biological efficiency for postulated photoprocess - *i.e.* $\epsilon(\lambda)$.

(2) Calculate biological effective dose for this photoprocess.

(3) Compare measured effect with (vs) calculated dose rate.

(4) Check consistency of results.

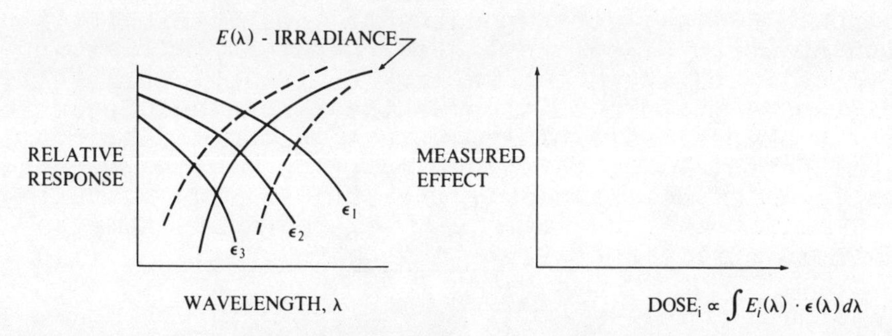

Figure 4. Schematic diagram illustrating technique of broad-band spectroscopy used to infer spectral character of effective biological efficiency, $\epsilon(\lambda)$. See text for details.

schematic diagram). After carrying out the first three steps outlined in Fig. 4, we check for "consistency" of results by plotting the measured effect versus the postulated dose.

This technique, of postulating an $\hat{\epsilon}(\lambda)$ and checking for consistency between the measured biological effect and the dose calculated using $\epsilon(\lambda)$, could be termed broad-band spectroscopy (Nachtwey, private communication). The technique is appropriate provided the law of reciprocity (i.e., the photoeffect of a given total radiation dose is independent of the dose rate) holds. We have assumed that reciprocity holds for the range of doses considered in our experiments. It is probable that this assumption is incorrect.

It should be emphasized that this technique does not prove one has selected the correct weighting function to describe the photoprocess under investigation. Rather it eliminates those weighting functions that are inconsistent with the results and gives a general indication of the relative spectral weighting consistent with the biological effect. Use of additional radiation regimes, with different wavelength cutoffs, would allow a more selective determination of a consistent weighting function.

From our productivity work (Smith et al. 1980; Smith and Baker, 1980) it was shown that weighting functions heavily weighted toward the UV-B region were inconsistent with our data. As a working hypothesis we assumed that the changes in photosynthesis observed for our different UV-B irradiance regimes could be described as photoinhibition "due to" a biological weighting function, $\epsilon_{PI}(\lambda)$, corresponding to the photoinhibition action spectrum of chloroplasts presented by Jones and Kok (1966), and shown in Fig. 1. Following Harris (1978) we expressed photoinhibition as the percentage decrease from the highest rate of ^{14}C uptake per unit chlorophyll at some depth, P_{max}, to values of productivity per unit chlorophyll at depths nearer the surface, P, i.e.,

$$\text{Photoinhibition} = PI = \frac{P_{max} - P}{P_{max}} \tag{11}$$

Figures 5 show plots of photoinhibition vs. total daily biological dose (Smith et al. 1980). The circles are for the natural radiation regime, the crosses for the enhanced radiation regime. As a comparative

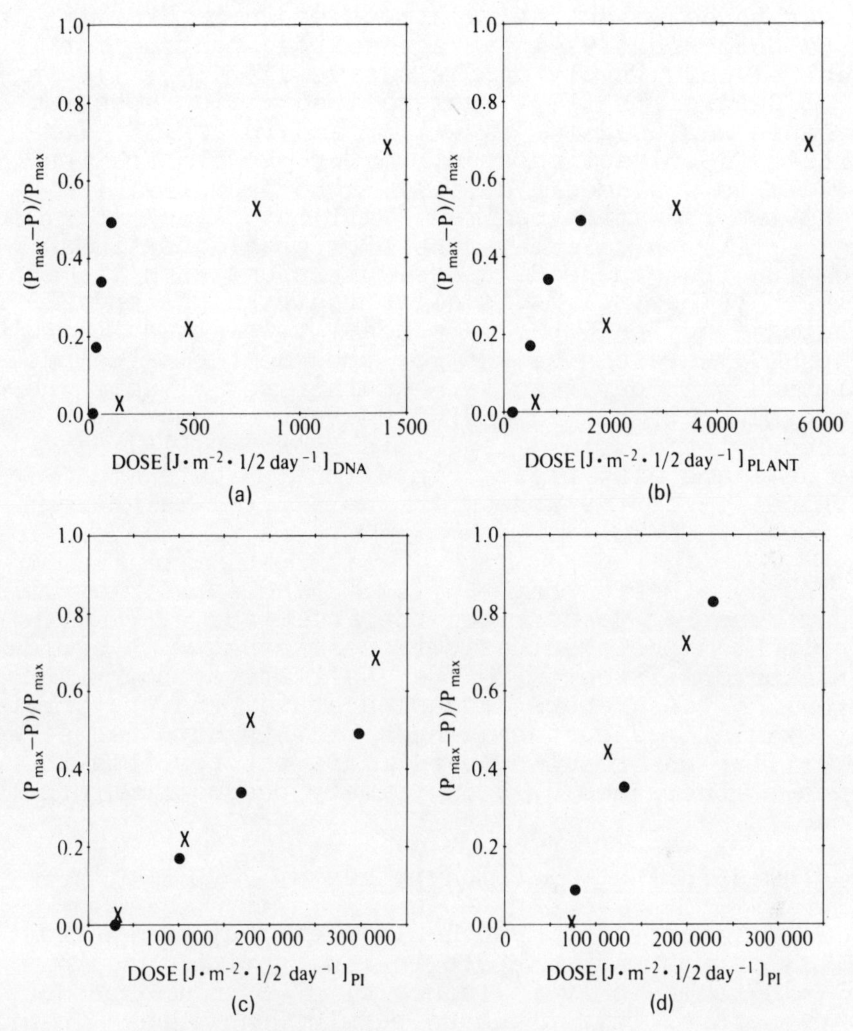

Figure 5. Photoinhibition (Eq. 11) versus biological dose weighted by a biological efficiency $\epsilon(\lambda)$. The circles denote natural radiation regimes, the crosses represent altered radiation regimes. The first three figures (5a,5b,5c) the crosses represent an enhanced MUV radiation regime whereas in the last figure (5d) the crosses represent a regime where the MUV radiation has been excluded. The weighting efficiencies used for the four figures are as follows: 5a) DNA, 5b) Plant, 5c) Photoinhibition, 5d) Photoinhibition (after Smith et al. 1980).

demonstration we have plotted the total daily biological dose for three weighting functions shown in Fig. 1. In Fig. 5a Setlow's (1974) average action spectrum for biological effects involving DNA was used ($\epsilon_{DNA}(\lambda)$); in Fig. 5b Caldwell's (1974) generalized action spectrum for plants was used ($\epsilon_{plant}(\lambda)$); and in Fig. 5c Jones and Kok's (1966) action spectrum for photoinhibition of chloroplasts was used ($\epsilon_{PI}(\lambda)$). The fact that the data for the two radiation regimes "fall into line" (Fig. 5c) using $\epsilon_{PI}(\lambda)$ to calculate the biological dose is an indication that these data are consistent with the above working hypothesis. It is clear that the DNA and PLANT weighting functions, which are heavily weighted toward shorter MUV wavelengths, do not produce biological doses consistent with our results. Another set of data where the crosses represent a radiation regime where UV-B was excluded using a mylar filter, is shown in Fig. 5d. These data are also brought into line by determining a biological effective dose with the $\epsilon_{PI}(\lambda)$ weighting function.

The hypothesis that $\epsilon_{PI}(\lambda)$ is the appropriate biological function to describe the influence of UV-B on phytoplankton photosynthesis is consistent with our data. It is also consistent with the qualitative observations of previous researchers discussed above. Before critically examining the consequences of this hypothesis, we will utilize our results to estimate the possible influence of increased UV-B on primary productivity in the oceans.

Model based on the $\epsilon_{PI}(\lambda)$ hypothesis

If we accept the hypothesis that photoinhibition, with its corresponding biological weighting function $\epsilon_{PI}(\lambda)$, is the only influence of UV-B on the productivity of natural phytoplankton populations then a quantitative estimate can be made with respect to the potential impact of increased UV-B levels. We emphasize again that our objective in presenting this simple model is to highlight the major uncertainties in our present predictive capabilities. As will be discussed in the following section, the model does not account for possibly significant non-linear biological effects.

The impact of UV-B on natural phytoplankton depends not only on $\epsilon_{PI}(\lambda)$ but also on the penetration of UV-B into natural waters. The penetration of UV-B into water depends upon the attenuation by water and by the dissolved and suspended material in the water, including the phytoplankton. Thus UV-B, or alternatively the daily biological effective dose E_{TB} (Eq. 5), will penetrate to different depths in different water types (Smith and Baker, 1979; Baker and Smith, 1981 in press). We have already anticipated this by plotting our results in Fig. 3 in terms of attenuation lengths.

Here we consider the attenuation of photosynthetically available radiation (PAR), i.e. Eq. 10,

$$K_{PAR} = \frac{-1}{Z} \ln \left[\frac{PAR(Z)}{PAR(0^-)}\right] \tag{12}$$

The relationship between K_{PAR} and the attenuation of photoinhibiting radiant energy, K_{PI}, is discussed in detail elsewhere (Baker and Smith, in preparation). It is sufficient here to recognize that, by characterizing depth in terms of a unitless parameter $(K \cdot Z)$, the effect of different water types, to a first order approximation, become normalized (Smith, 1981). Thus, as seen in Table 2, influence of reduced or enhanced UV-B is limited to the upper 1.5 attenuation lengths, for either highly productive or oligotrophic waters.

An analogous normalization has been made by many researchers (e.g. Harris, 1978; Morel, 1978) discussing primary productivity. Here the chlorophyll concentration is taken as a measure of phytoplankton biomass and is used to define productivity per unit biomass; i.e.,

$$P_B\left[\frac{mgC \cdot day^{-1}}{mgChl}\right] = \frac{\text{concentration carbon fixed per day}}{\text{chlorophyll concentration}} \tag{13}$$

Morel (1978, Fig. 3) has shown that P_B varies by less than a factor of ten while biomass and production range over more than three orders of magnitude.

Figure 6 presents an idealized plot of P_B versus $K_{PAR} \cdot Z$. The solid curve (2) is representative of a typical productivity versus depth curve under existing natural conditions. The light line ($\phi = 1/8 = 0.125$) represents the minimum quantum requirement (Rabinowitch and Godvindjee, 1969; Morel, 1978), that is the maximum quantum yield. The parallel curve represents a times ten lower quantum yield. The data of Morel (1978) and others show that most "real world" data fall within these limits. Our idealized curve (2) is mid-way (on a logarithmic scale) between these limits, and it will

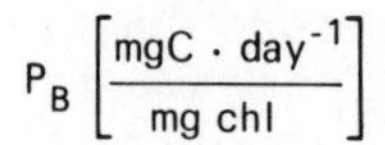

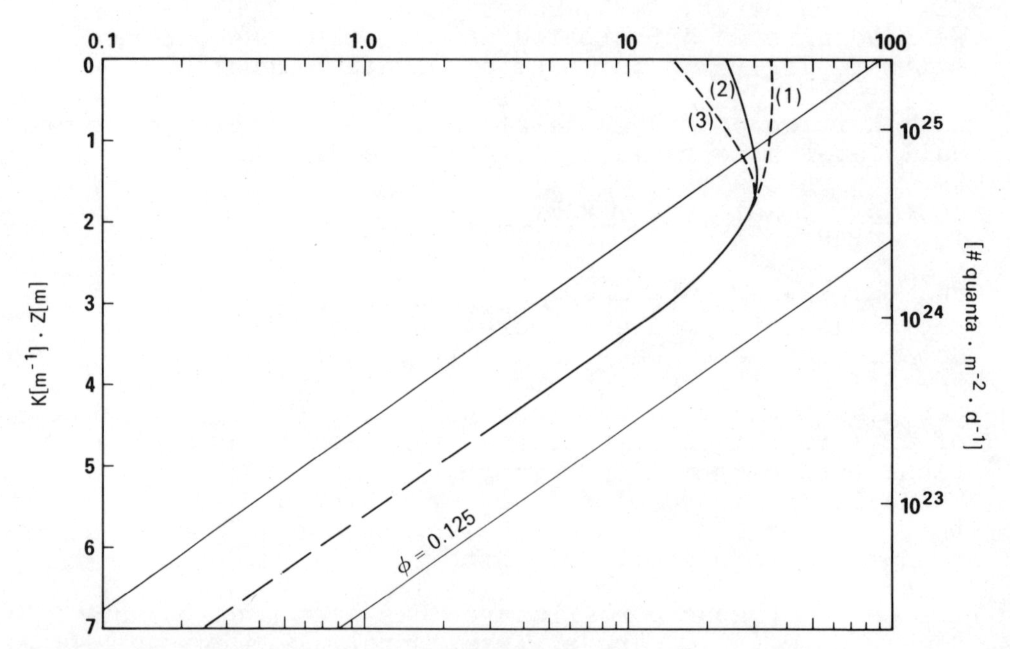

Figure 6. Idealized plot of productivity per unit biomass, P_B, versus depth in attenuation lengths, $K_{PAR} \cdot Z$. Curves 1, 2, and 3 where radiation regimes correspond to similarly labeled curves Fig. 2 and represent: (1) UV-B excluded; (2) UV-B at its normal (ambient) level; (3) UV-B enhanced. The light line (ϕ = 0.125) represents the minimum quantum requirement for photosynthesis. The parallel light line represents a ten times lower quantum yield. See text for details.

be shown that the conclusions based upon the following calculations are independent of the particular choice of where the curve is placed.

The dashed (1) and dotted curves (3) have been derived from our results shown in Fig. 3. The Pa curve for "enhanced" UV, for example, has been calculated by multiplying the ratio of Pa/Pn times the P_B values for the "normal" curve. The P_B for "excluded" UV-B has been similarly calculated using the Pa/Pn ratio for reduced UV-B from Fig. 3. Values for curves (1), (2), and (3) are given in Table 2. Because we have normalized both productivity and depth these curves are representative of the full range of ocean water types, from clear low productivity waters to more turbid highly productive waters.

Equation 12 shows the relationship between attenuation lengths and the surface to depth ratio of photosynthetically available radiant energy. The euphotic zone, those depths above which the net phytoplankton photosynthesis is positive, is frequently taken as the depth at which there is 1% of the surface PAR. This 1% level corresponds to $K_{PAR} \cdot Z = 4.61$ (since $\exp -4.61 = 0.01$). Thus, an integral of the P_B curves in Fig. 6 to a depth of 4.6 attenuation lengths is a measure of the average productivity per unit biomass within the euphotic zone.

Table 3 shows the tabulation of these integrals for the three curves given in Fig. 6. Also presented are the percentage increase (excluded UV-B) or decrease (enhanced UV-B) in P_B from the present natural levels at each depth. For example, near the surface ($K_{PAR} \cdot Z = 0.1$) enhanced UV-B would cause a decrease in productivity per unit biomass of 35.4%. The integrated decrease in productivity per unit biomass would be 25.4% to a depth of one attenuation length. The estimated reduction in P_B for the whole euphotic zone would be 9.0%. Analogously if no UV-B were incident at the earth's surface, P_B would increase by 11.6%.

It should be noted that these percentage comparisons between normal P_B and the P_B to be expected with enhanced or excluded UV-B remain unchanged if the curves are shifted toward the higher ($\phi = .125$) or lower ($\phi = 0.0125$) limits. Thus these conclusions with respect to the effect of UV-B on P_B are independent of our choice of the "normal" curve.

While we have not previously presented the above $\epsilon_{PI}(\lambda)$ model, we have published (Smith and Baker, 1980) conclusions based on accepting the hypothesis that $\epsilon_{PI}(\lambda)$ adequately describes the spectral weighting of photoinhibition on phytoplankton photosynthesis, as measured by the static bottle ^{14}C technique. In summary the model predicts that a 25% reduction in ozone thickness (the enhancement simulated for the data in Fig. 3) will cause a 9% reduction in primary productivity, regardless of absolute productivity level. In other words the total amplification factor (discussed below) for ϵ_{PI} is approximately one third ($A \approx 1/3$).

TABLE 2

Ratio of natural to altered productivity as a function of depth corresponding to curves 1, 2 and 3 of Eq. 3.

$(K\,Z)_i$	(1) EXCLUDED UV / NORMAL UV	(2) ENHANCED UV / NORMAL	(3) ENHANCED UV / EXCLUDED UV
0.1	1.45	.645	.445
0.3	1.39	.695	.500
0.5	1.33	.740	.557
0.7	1.27	.790	.622
0.9	1.21	.840	.694
1.1	1.15	.885	.770
1.3	1.09	.930	.853
1.5	1.03	.980	.952
1.7	1.00	1.000	1.000
1.9	1.00	1.000	1.000
2.1	1.00	1.000	1.000

TABLE 3

Productivity per unit biomass [mgC · day^{-1})mgChl)] corresponding to curves 1, 2 and 3 of Fig. 6 and percentage increase (a decrease) for excluded (and enhanced) UV.

	P_B(EXCLUDED UV)		P_B(NORMAL UV)	P_B(ENHANCED UV)	
$(K \cdot Z)_i$	P_B	Σ%	P_B	P_B	Σ%
0.1	35.0	44.9	24.2	15.6	- 35.4
0.3	35.0	41.9	25.2	17.5	- 32.9
0.5	35.0	38.8	26.3	19.5	- 30.5
0.7	34.9	35.7	27.5	21.7	- 27.9
0.9	34.5	32.5	28.5	23.9	- 25.4
1.1	34.0	29.3	29.6	26.2	- 22.8
1.3	33.0	26.1	30.3	28.1	- 20.3
1.5	31.8	22.8	30.9	30.3	- 17.8
1.7	30.0	20.1	30.0	30.0	- 15.6
1.9	28.0	18.1	28.0	28.0	- 14.1
2.1	26.0	16.6	26.0	26.0	- 12.9
2.3	23.1	15.4	23.1	23.1	- 12.0
2.5	20.3	14.5	20.3	20.3	- 11.3
2.7	17.8	13.8	17.8	17.8	- 10.7
2.9	15.2	13.3	15.2	15.2	- 10.3
3.1	13.2	12.8	13.2	13.2	- 10.0
3.3	10.3	12.5	10.3	10.3	- 9.75
3.5	8.4	12.3	8.4	8.4	- 9.55
3.7	6.8	12.0	6.8	6.8	- 9.40
3.9	5.4	11.9	5.4	5.4	- 9.28
4.1	4.6	11.8	4.6	4.6	- 9.18
4.3	3.8	11.7	3.8	3.8	- 9.10
4.5	3.1	11.6	3.1	3.1	- 9.03

(a) Limitations of the $\epsilon_{PI}(\lambda)$ Model

The above model quantitatively predicts, in agreement with Lorenzen's (1979) qualitative estimate, that a reduction in stratospheric ozone will have an impact less than the variance in the accuracy of current techniques used to measure phytoplankton productivity. We now turn to a critical analysis of the prediction based on the above model.

A key consideration, when discussing the potential impact of ozone reduction, is the spectral nature of the relevant biological photoeffects; i.e. the characteristics of $\epsilon(\lambda)$. A reduction in the ozone thickness will increase irradiance only in the 290-320 nm region. As a consequence, only photoprocesses with weighting functions primarily in the UV-B region are likely to be significantly influenced by an ozone change. This is shown in Fig. 7 where we have plotted the relative biological efficiencies, $\epsilon_{PI}(\lambda)$ and $\epsilon_{DNA}(\lambda)$, and the downward spectral irradiance, $E_d(\lambda)$, for two stratospheric ozone concentrations against wavelength. The effective biological dose (Eq. 4) is the product of the appropriate biological efficiency $\epsilon(\lambda)$ and the spectral irradiance $E_d(\lambda)$ summed over each infinitesimal wavelength interval. The change in biological dose is appreciable for ϵ_{DNA} (which is primarily in the UV-B region) and is relatively small for ϵ_{PI} (which has a significant response in the visible region of the spectrum).

This comparison between weighting functions can be made quantitative by consideration of an amplification factor (McDonald, 1971), A,

$$A = \frac{\Delta P/P}{\Delta \omega/\omega} \tag{14}$$

where $\Delta P/P$ is the percentage change in biological effect and $\Delta\omega/\omega$ is the percentage change in ozone thickness. More recently (Green et al. 1976; Rundel and Nachtwey, 1978) this amplification factor has been subdivided into two components. First the radiation amplification factor, R,

$$R = \frac{\Delta E_\epsilon/E_\epsilon}{\Delta \omega/\omega} \tag{15}$$

where $\Delta E_\epsilon/E_\epsilon$ is the percentage change in biologically effective dose ($\epsilon(\lambda)$ weighted). Second, the biological amplification factor, B,

$$B = \frac{\Delta P/P}{\Delta E_{\epsilon} / E_{\epsilon}} \tag{16}$$

Smith and Baker (1980) have shown that $R_{PI} \approx 1/10$ whereas R_{DNA} varies from 1.8 to 4.0 as the ozone thickness is reduced from 0.32 to 0.20 atm.cm. Thus, the radiation amplification factor is 20 to 40 times larger for ϵ_{DNA} than ϵ_{PI}. It can be seen that the amplification factor is a non-linear function of $\epsilon(\lambda)$. Such non-linearities are the crux of our present uncertainties.

The strong dependence of the amplification factor $\epsilon(\lambda)$, and hence presumably on the associated photoprocess, focuses attention on the importance of properly

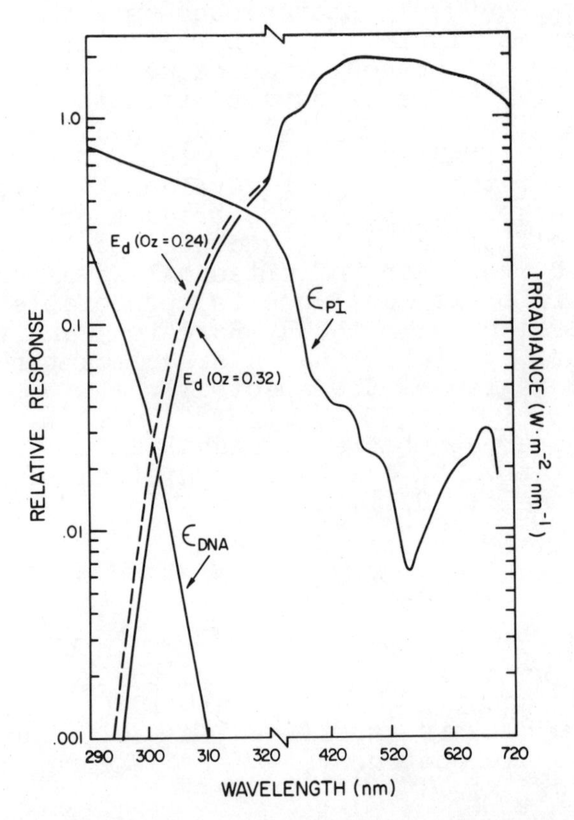

Figure 7. Relative biological efficiencies, $\epsilon_{PI}(\lambda)$ and $\epsilon_{DNA}(\lambda)$, and downward spectral irradiance, $E_d(\lambda)$, for two stratospheric ozone concentrations plotted against wavelength. Note that the wavelength scale is expanded below 320 nm in order to show greater detail in the UV-B region of the spectrum (after Smith and Baker, 1980).

identifying and quantitatively determining the "correct" $\epsilon(\lambda)$ for increased UV-B effects. If photoprocesses other than photoinhibition can potentially affect natural phytoplankton populations, then the relevant $\epsilon(\lambda)$ for these alternative photoprocesses must be identified in order to estimate possible detrimental effects of increased UV-B. For example, UV-B absorption and damage to DNA may not become evident for incubation times that are short compared to times of cell division (which is typical for routine ^{14}C measurements). Thus, the present ^{14}C technique is not adequate to assess photoprocesses which may become evident with time frames longer than the limited incubation period of this method.

As a consequence the uncertainties in the model are small compared to our uncertainties associated with the choice of a "correct" $\epsilon(\lambda)$, to describe the effects of increased UV-B on phytoplankton productivity. Changes in the shapes of the curves given in Fig. 6, or the experimental results given in Fig. 3, would only change the predictions from the model by some relatively small percentage. On the other hand, possible longer time scale photoeffects, with weighting functions in the UV-B region, may have amplification factors 20 to 40 times larger than that found for $\epsilon_{PI}(\lambda)$.

(b) Impact at higher trophic levels

If biological systems were inherently linear, a 9% reduction in primary production would be translated through the food chain with a 9% reduction at each level. Ryther's (1969) description of food chains is illustrative of possible non-linear trophic relationships in assessing the productivity of the ocean. Table 4 (adapted from Ryther, 1969) presents an estimate of world fish production based on the flow of material through the food chain. The overall estimate is based on separating the oceans into three provinces and: (1) an estimate of the amount of carbon fixed annually in each oceanographic province; (2) an estimate of the efficiency of transfer of material up through the food chain; (3) an estimate of the number of trophic levels between phytoplankton and commercial fish in the respective oceanographic provinces. This description by Ryther has been criticized by Alverson et al. (1970) who claim the overall estimate may be in error by plus or minus two orders of magnitude.

For the purpose of estimating the possible impact of increased UV-B on world productivity the key point is not Ryther's estimate per se, but rather how the number of trophic levels and the trophic efficiency influence the overall fish production. A change in the number of trophic levels or a factor of two change in trophic efficiency can cause an order of magnitude change in the output resource.

Worrest et. al. (in press) has suggested that enhanced UV-B may alter the community structure of phytoplankton populations. If a shift in community structure caused a change in tropodynamics, the resultant impact could easily be more significant than the estimated decrease in phytoplankton production due to increased photoinhibition.

Table 4

PROVINCE	PERCENTAGE OF OCEAN	AREA [10^6 km^2]	MEAN PRODUCTIVITY [gm C / m^2 · year]	APPROXIMATE RANGE OF CHLOROPHYLL CONCENTRATION [mg · m^{-3}]	TOTAL PRODUCTIVITY [10^9 tons C / year]	(%)	TROPHIC LEVELS	TROPHIC EFFICIENCY (%)	"FISH" CARBON [10^6 tons / year]	(%)	FISH PRODUCTION 10^6 tons/year [Fresh weight]	(%)
OPEN OCEAN	90	326	50	0.01 - 0.1	16.3	(82)	5	10	16	(0.6)	1.6	(0.6)
COASTAL ZONE	9.83	36	100	0.1 - 1.0	3.6	(18)	3	15	12	(43)	120	(43)
UPWELLING AREAS	0.17	0.6	300	1.0 - 10	0.18	(0.5)	1½	20	16	(57)	160	(57)
TOTAL		362.6			20	(100)			28.2	(100)	282	(100)

Adapted from J.H. Ryther 1969 Photosynthesis and fish production in the sea. Science. **166**: 72-76.

SUMMARY

A simple quantitative model, consistent with current fixed bottle ^{14}C productivity data, is used to calculate a decrease in phytoplankton productivity due to the projected increase in UV-B. The simple $\epsilon_{PI}(\lambda)$ model provides an estimated 5% decrease in primary production for a 16% decrease in stratospheric ozone.

These results are then used to focus attention on the possible non-linear uncertainties overlooked in the simple model: (1) the role of $\epsilon(\lambda)$ in predictive models and the limitations of ^{14}C techniques as currently and routinely used; (2) the possibility of community structure change, due to increased UV-B stress, and the consequent potential for tropodynamic changes in the food chain; (3) the influence of mixing on estimates of biologically effective doses and PAR. Any, or all, of these uncertainties are potentially orders of magnitude more important than "linear conclusions" drawn from our simple model.

We conclude that future research should focus on these, generally non-linear, potential impacts if a realistic assessment of enhanced UV-B on phytoplankton production is to be made.

ACKNOWLEDGEMENTS

The work was supported by the Stratospheric Impact Research and Assessment Program of the U. S. Environmental Protection Agency, EPA Grant No. R806-4-89-02, and is a contribution to the research encouraged by the IAPSO Working Group on Optical Oceanography.

REFERENCES

Alverson, D.L., A.R. Longhurst and J.A. Gullard. 1970. How much food from the sea?. Science. 168: 503-505.

Austin, R.W. 1974. The remote sensing of spectral radiance from below the ocean surface. Chp. 14 in Opt. Aspects of Oceanography. N.J. Jerlov and E. Steemann Nielson [Eds.] Academic Press, London and New York.

Baker, K.S., R.C. Smith and A.E.S. Green. 1980. Middle ultraviolet radiation reaching the ocean surface. Photochem. and Photobiol. 32: 367-374.

Caldwell, M.M. 1971. Solar ultraviolet radiation and the growth and development of higher plants in Photophysiology [Ed. by Giese]. 6:131-177. Academic Press, New York.

CIAP. 1 975. Impacts of climate changes in the biosphere. Monograph 5, Part 1. Ultraviolet radiation effects. Dept. of Transportation. PB-247725.

Giese, A.C. 1964. Studies on ultraviolet radiation action on animal cells in Photophysiology [Ed. A.C. Giese] 7:203-241.

Green, A.E.S., T. Sawada and E.P. Shettle. 1974. The middle ultraviolet reaching the ground. Photochem. and Photobiol. 19:251-259.

Green, A.E.S., K.R. Cross and L.A. Smith. 1980. Improved analytic characterization of ultraviolet skylight. Photochem. and Photobiol. 31:59-65.

Harris, G.P. 1978. Photosynthesis, productivity and growth: The physiological ecology of phytoplankton. Arch. Hydrobiol. Beih. 10:1-171.

Hunter, J.H., J.H. Taylor and H.G. Moser. 1979. The effect of ultraviolet irradiation on eggs and larvae of the northern anchovy, Engraulis mordax, and the pacific mackerel, Scomber japonicus, during the embryonic stage. Photochem. and Photobiol. 29: 325-338.

Jerlov, N.G. 1950. Ultraviolet radiation in the sea. Nature 166:111.

Jitts, H.R., A. Morel and Y. Saijo. 1976. The relation of oceanic primary production to available photosynthetic irradiance. Aust. J. Mar. Freshwater Res. 27:441-454.

Jones, L.W. and B. Kok. 1966. Photoinhibition of chloroplast reactions. 1. Kinetics and action spectra. Plant Physiol. 41:1037-1043.

Lenoble, J. 1956. Etude de la pénétration de l'ultraviolet dans la mer. Ann. Geophs. 12:16.

Lorenzen, C.J. 1979. UV radiation and phytoplankton photosynthesis. Limnol. and Oceanogr. 24:1117-1120.

McDonald, J.E. 2 March 1971. Statements submitted before the House Subcommittee on Transportation Appropriations.

Morel, A. 1978. Available, usable, and stored radiant energy in relation to marine photosynthesis. Deep-sea Research 25:673-688.

Morowitz, H.J. 1950. Absorption effects in volume irradiation of microorganisms. Science 111:229.

National Academy of Sciences. 1979. Protection against depletion of stratospheric ozone by chlorofluorocarbons. National Research Council, Washington, D.C.

Nachtwey, D.S. 1975. Linking photobiological studies at 254 nm with UV-B, CIAP Monograph 5, Impacts of climatic change on the biosphere, Part 1, Ultraviolet radiation effects [Eds: D.S. Nachtwey, M.M. Caldwell and R.H. Biggs] DOLT-TST-75-55, U.S. Dept. of Transportation, Washington, D.C.

Preisendorfer, R.W. 1976. Hydrologic Optics. U.S. Dept. of Commerce.

Rabinowitch, E. and Godvindjee. 1969.

Rundel, R.D. and D. S. Nachtwey. 1978. Skin cancer and ultraviolet radiation. Photochem. and Photobiol. 28:345-356.

Ryther, J.R. 1969. Photosynthesis and fish production in the sea. Science 166:72-76.

Setlow, R.B. 1974. The wavelengths in sunlight effective in producing skin cancer: a theoretical analysis. Proceedings of National Academy of Sciences, U.S.A. 71(9):3363-3366.

Shettle, E.P. and A.E.S. Green. 1974. Multiple scattering calculation of the middle ultraviolet reaching the ground. Appl. Opt. 13:1567-1581.

Smith, R. C. 1981. Remote sensing and depth distribution of ocean chlorophyll. J. Mar. Ecology (In press).

Smith, R.C. and K.S. Baker. 1978. Optical classifications of natural waters. Limnol. and Oceanogr. 23:260-267.

Smith, R.C. and K.S. Baker. 1979. Penetration of UV-B and biologically effective dose-rates in natural waters. Photochem. and Photobiol. 29:311-323.

Smith, R.C. and K.S. Baker. 1980. Stratospheric ozone, middle ultraviolet radiation and ^{14}C measurements of marine productivity, Science 208: 592-593.

Smith, R.C., K.S. Baker, O. Holm-Hansen and R. Olson. 1980. Photoinhibition of photosynthesis in natural waters. Photochem. and Photobiol. 31:585-592.

Sverdrup, H.U. 1953. On conditions for the vernal blooming of phytoplankton. J. Cons. Explor. Mer. 18: 287-295.

Worrest, R.C., D.L. Brooker and H. Van Dyke, 1980. Results of a primary productivity study as affected by the type of glass in the culture bottle, Limnol. and Oceanogr. 25:360-364.

Worrest, R.C., K.U. Wolniakowski, J.D. Scott, D.L. Brooker, B.E. Thomson and H. Van Dyke (in press). Sensitivity of marine phytoplankton to UV-B radiation: Impact upon a model ecosystem. Photochem. and Photobiol.

Zepp, R.G. and D.M. Cline. 1977. Rates of direct photolysis in aquatic environments. Environ. Sci. Technol. 11:359.

MODELING LIGHT LOSS VERSUS UV-B INCREASE FOR ORGANISMS WHICH CONTROL THEIR VERTICAL POSITION IN THE WATER COLUMN

John Calkins

Department of Radiation Medicine (1) and
School of Biological Sciences (2)
University of Kentucky, Lexington, KY 40536

Quantitation of solar UV effects on aquatic systems is most often based on models which assume the exposed organisms are located in the water column in a manner independent of the incident solar UV (Zaneveld, 1975, Lorenzen, 1979, Hunter et al., this volume, and Smith and Baker, this volume). Planktonic organisms are in fact seldom static in the water; plants and animals move up and down in the water in complex ways which often show high correlations with the incident solar radiation (Barcelo and Calkins 1980 a & b; Barcelo and Calkins, 1979; Barcelo, 1980). Since solar UV has been demonstrated to be highly injurious, and is not a new agent in the environment, it is possible that a significant portion of the plankton control their position in such a way that they avoid overexposure to damaging solar UV radiation. Positioning control through swimming ability, bouyancy changes and utilization of turbulence or water currents is well known.

If it is assumed that aquatic organisms control their vertical position in the water column to maintain a given "safe" level of UV-B exposure then, from measurements such as ours (Calkins and Thordardottir, this volume), the effect of increased UV-B upon the availability of visible light can be readily computed. Figure 1 illustrates diagrammatically the reduction of visible light which would accompany the repositioning of an organism in the water column to a new level maintaining a constant UV-B dosage, assuming (for convenience in illustration) that the incident UV-B were doubled.

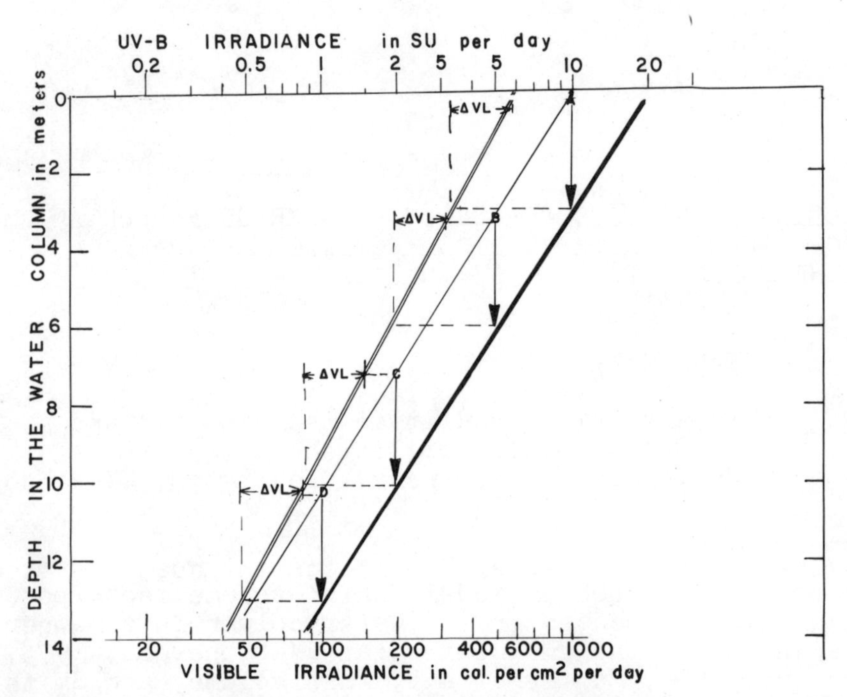

Figure 1. A diagrammatic representation of visible light loss for organisms sensing and responding to UV-B upon an increase in UV-B irradiance at the water surface. Attenuation of visible (double line) and UV-B (light single line) with depth in the water column approximates conditions at a productive Icelandic station. Species of different UV-B tolerance are assumed to position themselves at levels where they would receive 10, 5, 2 and 1 SU per day (positions A, B, C, and D); upon an increase in UV-B incident at the water surface (heavy line) all species would be forced to move down the same distance (arrows) to reestablish the tolerable level of UV-B and would, because of attenuation of visible light by the water column, also receive reduced visible light. The loss of light (Δ VL) will be the same fraction of the available light regardless of the depth to which the species has adjusted. It should be especially noted that this model implies that <u>all</u> organisms in the euphotic zone will be equally affected rather than the common model which suggests that UV-B effects would be confined to the near-surface zone.

It should be especially noted that with semilogarithomic attenuation of both green and UV-B as we usually observe, the percentage loss in visible light available to the organism will be exactly the same regardless of its initial position. It is often assumed that the organisms deep in the water column are "shielded" from solar UV-B effects and only the near-surface organisms would be affected. The extreme sensitivity of photosynthesis to UV-B exposure which Lorenzen (1979) demonstrated suggests that organisms quite deep in the euphotic zone could be affected by UV-B. Table I shows the change in visible (green) light which would accompany an adjustment of position upon a 10% increase in UV-B dosage. In the case of the marine stations of low productivity the loss of visible irradiance would be about 3%; however, for the stations showing a relatively high photosynthetic productivity the loss of visible irradiance would exceed 5%; similar computations for other areas which have been surveyed are listed. While the calculated reductions of visible light seem rather small, if they reduced the efficiency of a significant portion of primary photosynthesis and also the feeding

TABLE I

The change in visible light available for photosynthesis following compensation for a 10% increase in UV-B.

Location	Station	Loss of visible (%)
Iceland[1]	25	3
Iceland[1]	50	5.2
Chesapeake Bay[2]	D	2.3
San Diego[3]	C	4.1
Puerto Rico[2]	C	4.3

1. From Calkins and Thoradottir, this volume.

2. From Calkins 1975.

3. From Calkins, this volume.

by zooplankton, then the ultimate productivity of many ocean areas, especially arctic and antarctic where light for photosynthesis is minimal, would be depressed probably by at least the amount of visible light which would be lost. Considering the present high utilization and frequent over utilization of marine resources, a 3% reduction of marine productivity would be a serious loss.

ACKNOWLEDGEMENTS

The work on which this report was based was supported in part by the Office of Water Research and Technology, U. S. Department of Interior under the provisions of Public Law 88-379.

REFERENCES

Barcelo, J. A. 1980. Photomovement, pigmentation and UV-B sensitivity in Planaria. Photochem. Photobiol. 32: 107-109.

Barcelo, J. A. and J. Calkins. 1979. Positioning of aquatic microorganisms in response to visible light and simulated solar UV-B irradiation. Photochem. Photobiol. 29: 75-83.

Barcelo, J. A. and J. Calkins. 1980a. The relative importance of various environmental factors on the vertical distribution of the aquatic protozoan Colipis spiralis. Photochem. Photobiol. 31: 67-73.

Barcelo, J. A. and J. Calkins. 1980b. The kinetics of avoidance of simulated solar UV radiation by two arthropods. Biophys J. 32: 921-930.

Calkins, J. 1975. Measurements of penetration of solar UV-B into various natural waters. CIAP Monogr. 5 Part 1, Chapter 2, Appendix E, 2-267 to 2-296.

Lorenzen, C. J. 1979. UV radiation and phytoplankton photosynthesis. Limnol. and Oceanogr. 24: 1117-1120.

Zaneveld, J. R. V. 1975. Penetration of ultraviolet radiations into natural waters. CIAP Monogr. 5, Part 1, Chapter 2.4: 2-108 to 2-166.

EFFECTS OF UV-B RADIATION ON *THALASSIOSIRA PSEUDONANA*: A PRELIMINARY STUDY

M. L. Geiger (1), D. R. Norris (1), J. H. Blatt (2) and R. D. Petrilla (1)

Department of Oceanography and Ocean Engineering (1) and Department of Physics and Space Sciences (2) Florida Institute of Technology, Florida Institute of Technology
Melbourne, Florida 32901

ABSTRACT

Effects of differing levels of ultraviolet-B radiation on cell concentrations, maximum specific growth rates, and pigment content of the diatom *Thalassiosira pseudonana* were investigated. Increased UV-B compared to the base UV-B level (intended to simulate the natural level of UV-B present in the environment) tended to show detrimental effects in all parameters. When compared to our normal culture conditions (UV-B excluded), even the base level of UV-B radiation showed detrimental effects in all parameters except chlorophyll *a* concentrations, the only tested parameter which increased with the addition of UV-B radiation.

METHODS

Aliquots of stock cultures of *Thalassiosira pseudonana* in the first day of logarithmic growth, were pipetted into 15 sterile culture tubes each containing 10 ml of medium. These inoculated tubes were divided into five groups of three tubes each to be irradiated at different wavelengths and intensities as described below. The culture medium was an enriched sea water type, designated MML-1.

An irradiance monitor, designed by J. Blatt, was constructed to sense UV and visible radiation (Blatt et al. in preparation). The detector is equipped with two photodiodes, an R-403 solar blind phototube with a spectral response of 185-320 nm and an R-420 UV visible tube with a spectral response of 185-850 nm (Hammamatsu TV Co., Ltd.). Output from the detector was recorded in relative units (millivolts).

Two irradiation cabinets were utilized, one for stock cultures and non-UV controls and a second for ultraviolet irradiation. Both were approximately 4 ft. x 2 ft. x 2 ft. and contained a series of fluorescent lamps with the lamp ballasts mounted on the outside. The stock culture chamber contained four cool white (F40CW) fluorescent tubes which were on a 15:9 hour light: dark cycle (note Figure 1). A fan with accompanying air ports was installed in each chamber to remove excess heat, temperature was maintained at about 25°C.

The ultraviolet irradiation chamber contained three cool white tubes and two Westinghouse (FS40) fluorescent sunlamps. While the cool white lamps were on a 15:9 hour cycle, the sunlamps were on only 6 hours near the middle of the visible light cycle. All UV lamps had a preliminary burning period of 100 hours in order to stabilize their output (Sisson and Caldwell, 1975).

Two types of filters were used: Mylar type 92S (Dupont Co.) with a cutoff at 312 nm and a double thickness (0.13 nm) of Kodacel type TA401 (Kodak Co.) which has a cutoff at 287 nm. The Mylar filtered sample tubes were considered to represent exposure to UV-A while the Kodacel filtered samples were considered to be exposed to both UV-B and UV-A. The Mylar filtered replicates served as a control at each UV-B treatment level in order to monitor any effects due to longer wavelength UV radiation. UV transmission (from 260 to 400 nm) of the filters (also including one wall of a culture tube) was measured on a Beckman DB-GT grating spectrophotometer.

In order to determine the potential effect of a selected level of increased UV-B radiation a base level of UV-B was established. Combining the known spectral response of our detector and the spectral irradiance produced by the sun and the sunlamps it was possible to estimate the general relationship of our simulator to natural conditions. Using filtered sunlamp irradiance as specified by Westinghouse Corp. 1977, we could simulate within our chamber conditions approximating

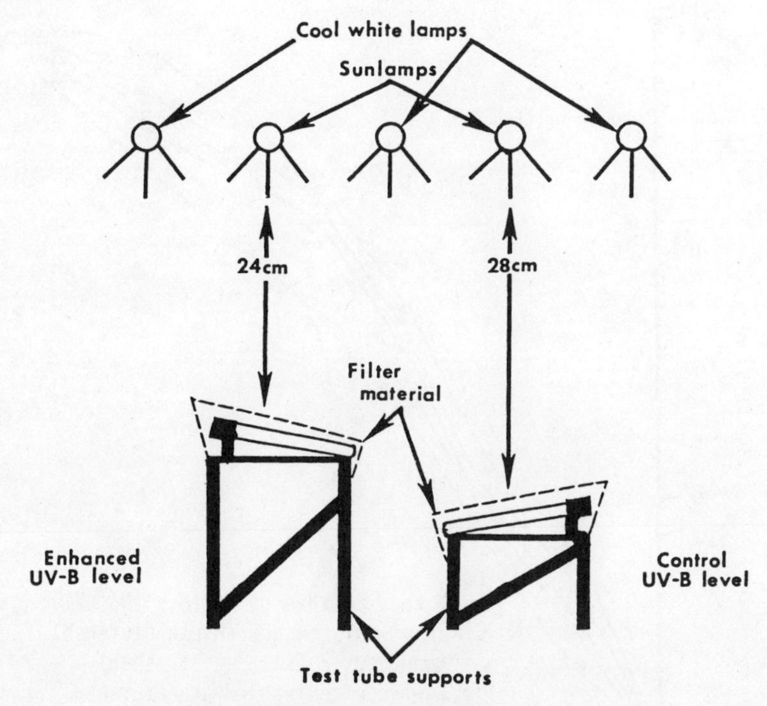

Figure 1. Lighting and filter configuration for UV-B irradiation.

solar irradiance in water ranging in depth from 6 m to less than 1 m for Jerlov's (1976) water types I to 5. At this base level of UV-B radiation there were placed two sets of three replicate culture tubes, one filtered by Mylar and the other by a double thickness of Kodacel.

Along with responses at the base level, a "moderately" increased UV-B level was investigated using another double set of three replicates, filtered as the base set but placed closer to the sunlamps. An additional set of replicate tubes was placed in the stock culture chamber, at a level of visible illumination equivalent to the UV exposed cultures to serve as a secondary control to determine effects of zero UV-B radiation. This set of replicates was also filtered by one layer of Mylar to maintain equivalent conditions. Kodacel filters were changed for each run to compensate for solarization caused by the UV radiation.

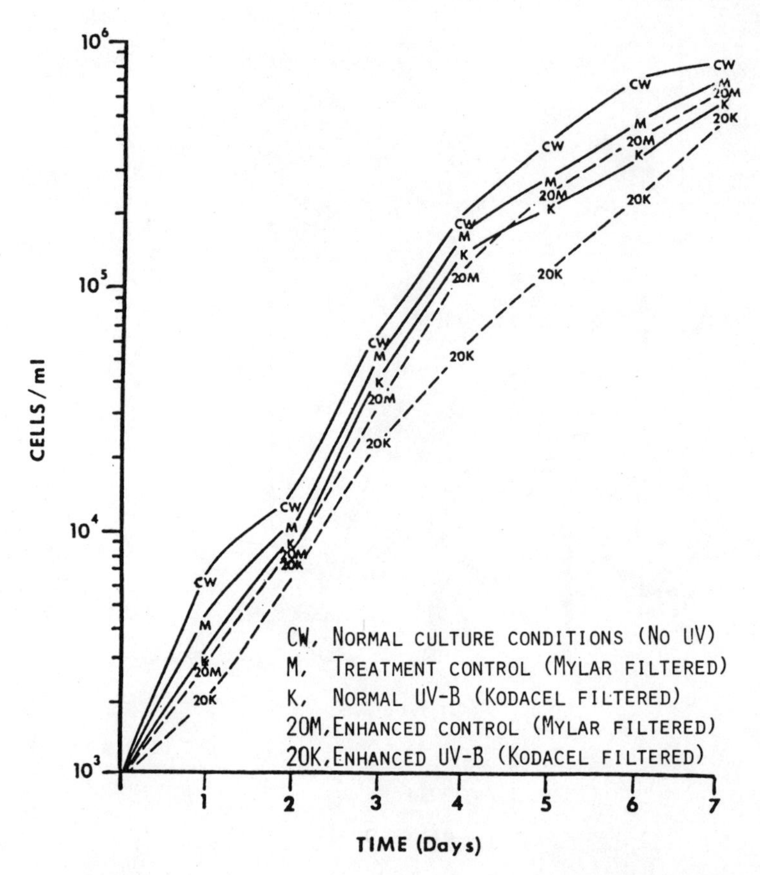

Figure 2. Growth curves for Trial 1.

Each experimental tube began with a concentration of about 10^3 cells · ml^{-1}. Two hemacytometer counts per day were made on all replicates for the six days following the end of the daily UV irradiation period. The maximum specific growth rate was determined as suggested in Standard Methods (1975).

Chlorophyll _a_ and phaeophytin _a_ analyses (Standard Methods, 1975) were performed on all treatment replicates after the last cell counts. The instrument used was a Turner Model 111 Fluorometer equipped with high sensitivity door, and filters for light emission (CS2-64) and excitation (McCarthy Scientific Co. No. 100-6300 substituted for CS5-60). All analyses were done on the 10X window orifice. The fluorometer was calibrated with chlorophyll solutions of known concentrations (analyzed spectrophotometrically).

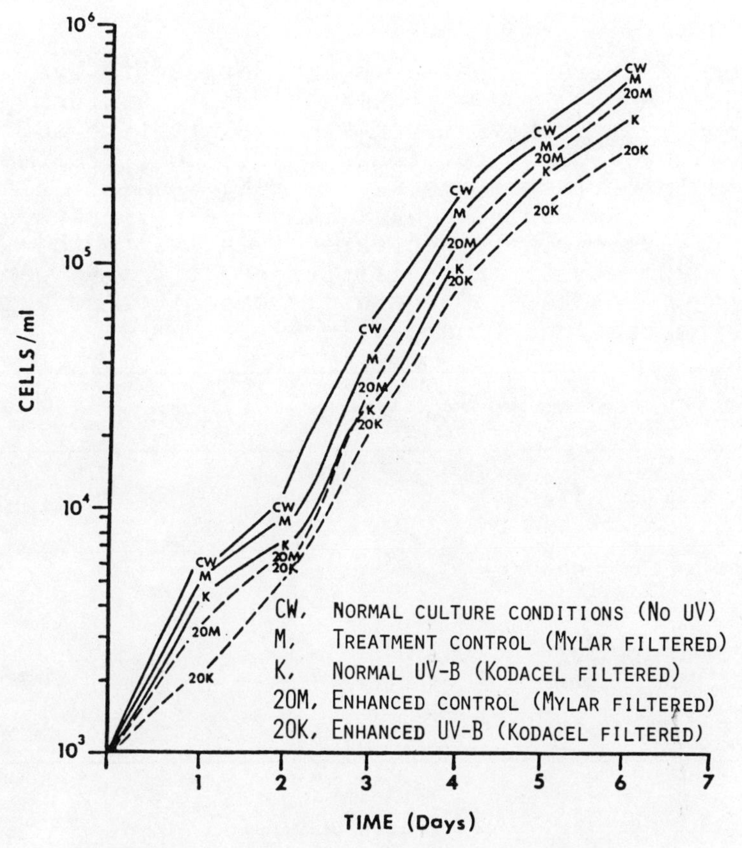

Figure 3. Growth curves for Trial 4.

RESULTS

We report here the results of two runs comparing no UV, base level UV-B, and enhanced UV-B level. Figures 2 and 3 illustrate our growth curves. One trial showed differences between cell counts at both base and enhanced UV-B levels relative to controls, however, neither was significant at the $P < 0.05$ level. Significant differences between cell counts did occur in the second run (Trial 4), at the $P < 0.01$ for the base level and at $P < 0.05$ for the enhanced UV-B level. However, there was no significant difference at the $P < 0.05$ level between the base and enhanced UV treatment in both trials. The apparent effect of increased UV-B was 6.9% (23.6 - 16.7) decrease in cell concentrations for Trial 1 and 18.3% (52.2 - 33.9) decrease for Trial 4 (Table 1).

TABLE 1

Comparison of Filter Treatments and Representative Effects for Trials 1 and 4. CW, Normal culture conditions plus Mylar filter; M, Treatment control (Mylar filtered, no UV-B); K, Kodacel filtered (Base UV-B); 20M, Treatment control (Mylar filtered, no UV-B); 20K, Kodacel filtered (enhanced UV-B). Positive (+) values were increases and negative (-) values were decreases relative to the controls. *No significant differences at the 95 percent probability level between the differences (or effect of UV-B) at the base and enhanced level.

PARAMETER	COMPARISON	PERCENT DIFFERENCE TRIAL I	PERCENT DIFFERENCE TRIAL 4	THIS PERCENT REPRESENTS THE EFFECT OF:	DIFFERENCE TRIAL I	DIFFERENCE TRIAL 4	THIS PERCENT REPRESENTS THE EFFECT OF:
CELL CONCENTRATION	M VS. K	-16.7	-33.9	UV-B AT BASE LEVEL	- 6.9*	-18.3*	ENHANCED UV-B
	20M VS. 20K	-23.6	-52.2	UV-B AT ENHANCED LEVEL			
	CW VS. M	-16.6	-14.7	UV-A AT BASE LEVEL	-13.9	-28.9	BASE UV-B COMPARED TO NO UV-B
	CW VS. K	-30.5	-43.6	UV-A + B AT BASE LEVEL			
	CW VS. 20M	-23.0	-16.4	UV-A AT ENHANCED LEVEL	-18.2	-43.5	ENHANCED UV-B COMPARED TO NO UV-B
	CW VS. 20K	-41.2	-59.9	UV-A + B AT ENHANCED LEVEL			
MAXIMUM SPECIFIC GROWTH RATE	M VS. K	- 2.4	-13.1	UV-B AT BASE LEVEL	- 8.4*	+ 5.5*	ENHANCED UV-B
	20M VS. 20K	-10.8	- 7.6	UV-B AT ENHANCED LEVEL			
	CW VS. M	- 5.6	- 7.2	UV-A AT BASE LEVEL	- 2.2	-12.1	BASE UV-B COMPARED TO NO UV-B
	CW VS. K	- 7.8	-19.3	UV-A + B AT BASE LEVEL			
	CW VS. 20M	-22.3	-13.2	UV-A AT ENHANCED LEVEL	- 8.4	- 6.7	ENHANCED UV-B COMPARED TO NO UV-B
	CW VS. 20K	-30.7	-19.9	UV-A + B AT ENHANCED LEVEL			
CHLOROPHYLL *a* CONCENTRATION PER CELL	M VS. K	+33.3	+25.0	UV-B AT BASE LEVEL	-27.1*	-33.6*	ENHANCED UV-B
	20M VS. 20K	+ 6.2	- 8.6	UV-B AT ENHANCED LEVEL			
	CW VS. M	- 7.7	+ 7.7	UV-A AT BASE LEVEL	+30.8	+26.9	BASE UV-B COMPARED TO NO UV-B
	CW VS. K	+23.1	+34.6	UV-A + B AT BASE LEVEL			
	CW VS. 20M	+23.1	+34.6	UV-A AT ENHANCED LEVEL	+ 7.7	-11.5	ENHANCED UV-B COMPARED TO NO UV-B
	CW VS. 20K	+30.8	+23.1	UV-A + B AT ENHANCED LEVEL			

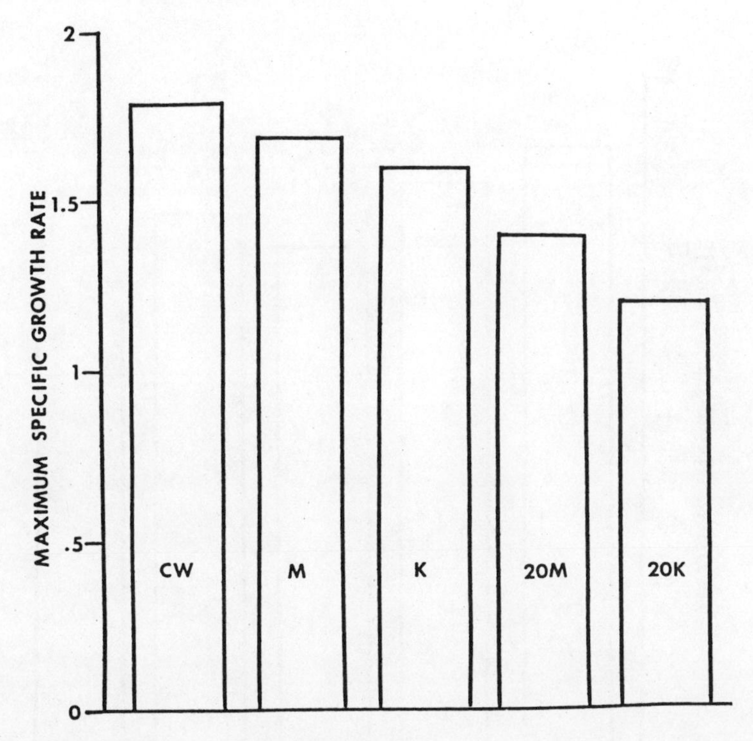

Figure 4. Maximum specific growth rates for Trial 1. CW, normal culture conditions (no UV); M, treatment control (Mylar filtered); K, base UV-B (Kodacel filtered); 20 M enhanced control (Mylar filtered); 20 K, enhanced UV-B (Kodacel filtered).

The maximum specific growth rates are shown in Figs. 4 and 5. The apparent effect of increased UV-B was 8.4% (10.8 - 2.4) decrease for Trial 1 but a relative 5.5% (13.1 - 7.6) increase in growth rate for Trial 4 (Table 1). There was no significant difference ($P < 0.05$) between filter treatments at either UV-B level.

Figures 6 and 7 present the mean chlorophyll a concentrations per cell. The apparent effect of enhanced UV-B compared to the normal level for Trial 1 was a 27.1% (33.3 - 6.2) decrease in chlorophyll a content; for Trial 4 this relative value was 33.6% (25 + 8.6) decrease (Table 1). There was no significant difference ($P < 0.05$) between filter treatments at either UV-B level. The chlorophyll a to phaeophytin a ratio in both trials were all ≥ 1.7 indicating little

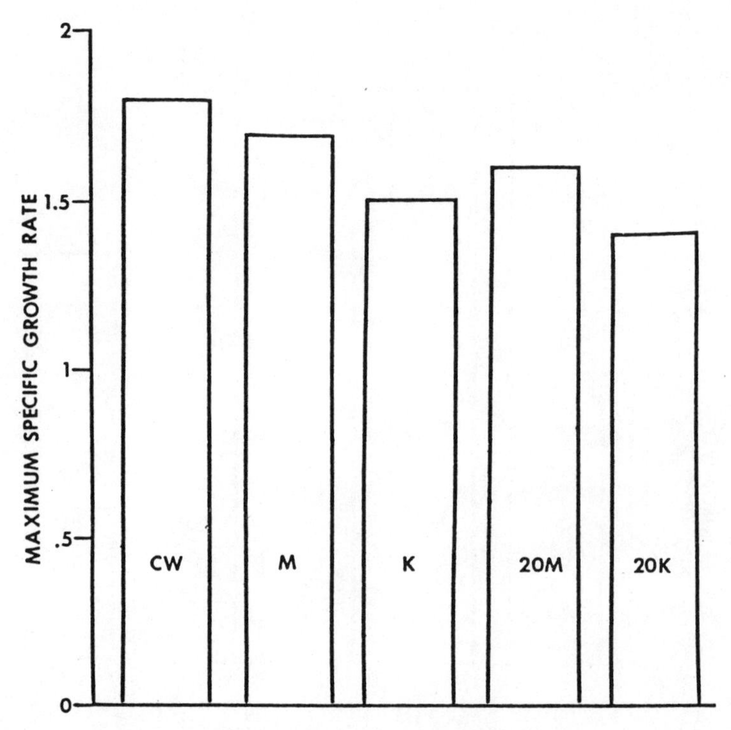

Figure 5. Maximum specific growth rates for Trial 4. Legend as in Fig. 4.

or no degradation of chlorophyll _a_ due to enhanced UV-B.

DISCUSSION

The effects of UV plus filter treatments compared to no UV conditions (Table 1) indicate that UV-A radiation is detrimental to cell division and maximum specific growth rate. Clayton (1977) reports that the delay or prevention of cell division is a sensitive response of cells to light. In general, it is believed that the response to UV radiation is due to the effects on proteins and DNA (cf. Halldal and Taube, 1972). There are several repair mechanisms present to reverse DNA damage by UV radiation including excision repair in the absence of light and photoreactivation, an enzymatic process requiring light of wavelengths between 300 and 500 nm. Neither repair process is completely effective and can therefore leave residual damage. Another mechanism, which might diminish the

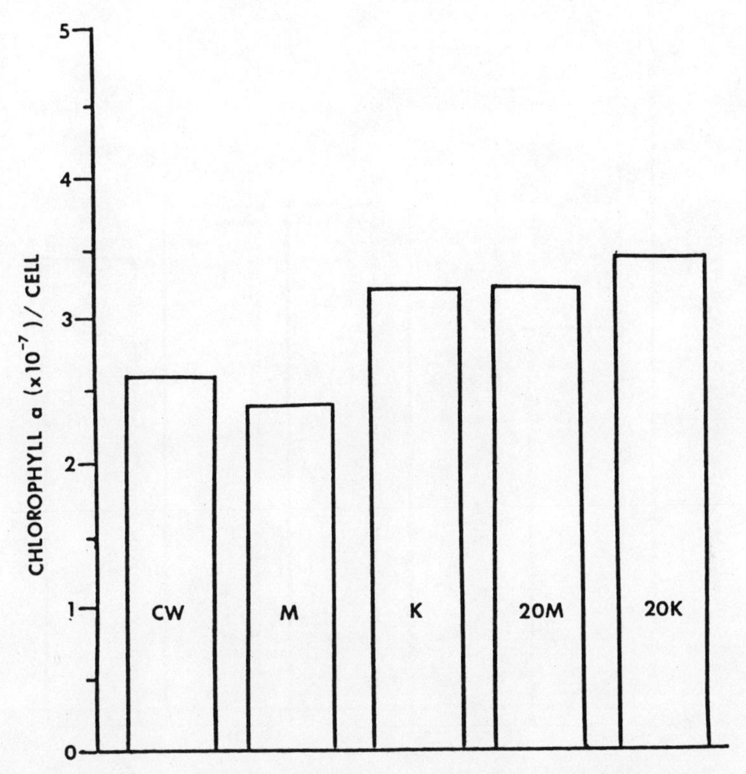

Figure 6. Mean chlorophyll a concentrations (μg x 10^{-7}· cell^{-1}) for Trial 1. Legend as in Fig. 4.

effects of UV-B in this experiment, is photoprotection, a possible interaction effect involving the UV-A and UV-B radiation. Photoprotection is a phenomenon in which irradiation by UV-A decreases the sensitivity of certain cells to UV-B radiation (Clayton, 1977). Though cell concentrations did decrease in this study, the results are not more definitive perhaps due to the fact that the amount of UV-B exposure of the organisms did not produce enough damage to be statistically significant, considering the high variability inherent in our system.

We believe that cell concentration is a better measure of the gross effects of UV-B on *T. pseudonana* than the maximum specific growth rate. The maximum specific growth rate seems to be a better measure of the rate limiting nutrients present (Standard Methods, 1975; Parsons and Takahaski, 1973). Cell division,

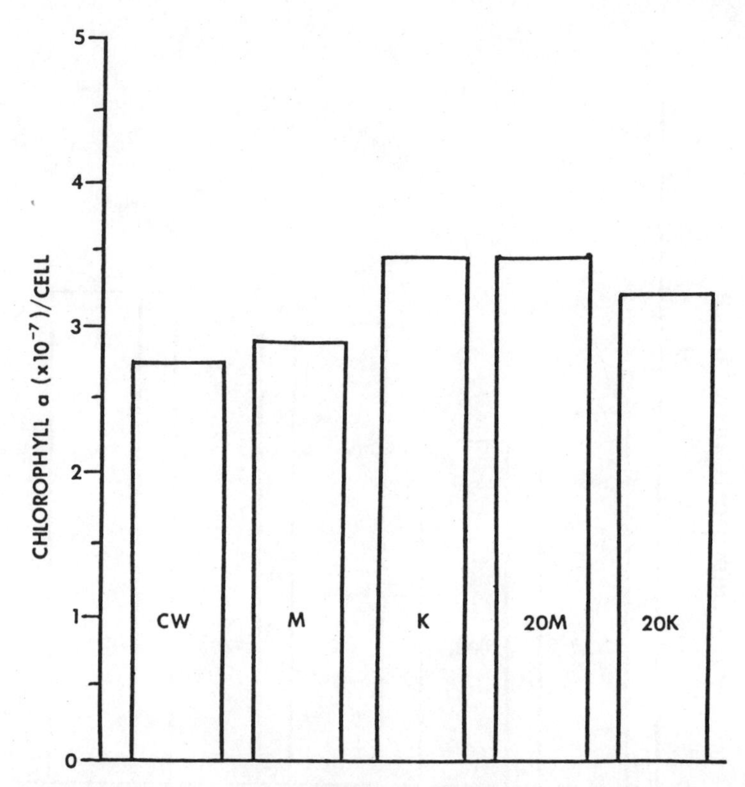

Figure 7. Mean chlorophyll a concentrations ($\mu g \times 10^{-7} \cdot cell^{-1}$) for Trial 4. Legend as in Fig. 4.

although normally accompanied by growth, is not invariably proportional to it (Eppley and Strickland, 1968). Cell concentration would seem to act as an integrator that represents the sum of responses over a period of time.

Our base UV irradiance level was comparable to the lower value of Worrest et al. (this volume) who found *Thalassiosira pseudonana* to be the most sensitive of seven phytoplankton species. Though direct comparisons cannot be made, our results do not disagree with those reported by Worrest et al. (this volume). Calkins and Thordardottir (1980) report that *Thalassiosira gravida* and *T. polychorda* were relatively resistant (of six phytoplankton species) to UV-B exposure.

There appears to be a mechanism for regulating the cell chlorophyll content in many algal cultures and natural populations. It is believed that this chlorophyll regulatory mechanism is a function of light intensity and pigment synthesis. Yentsch and Lee (1966) suggest that the regulatory mechanism was related to visible wavelengths, but it may be applicable to UV wavelengths as well. Whether the effects of UV-B resulting in chlorophyll reduction is mediated by a direct destruction of chlorophyll already present or in some way blocks the synthesis of new chlorophyll is not known.

The response of *T. pseudonana* to enhanced UV-B in culture is expected to be the minimum response of the organism in a natural situation (provided avoidance strategies are not also important). Caldwell and Nachtwey (1975) describe strategies for coping with solar UV, such as biological repair mechanisms, noted above. Other approaches are avoidance of UV radiation by physical screening or by behavioral responses. Physical screening (absorption of UV radiation by superficial pigments) offers a possible explanation for the increase of chlorophyll *a* in the cells as a result of the addition of UV-B to standard culture conditions. Chlorophyll is rather insensitive to UV radiation (Halldal and Taube, 1972). Chlorophyll *a* concentration increased through 4-5 weeks at the lower levels of UV-B radiation in the studies by Worrest et al. (1978) and Worrest et al. (this volume).

Behavioral responses include the movement of the plankton deeper in the water column, or an alteration in the activity period. If phytoplankters cannot detect increased UV-B *per se* and avoid exposure by responding to day length or visible radiation intensity, then any increase in UV-B without a corresponding increase in other environmental cues could result in detrimental effects to them. Such effects will occur in a significant portion of the euphotic zone (Lorenzen, 1979; Thomson et al. 1980).

REFERENCES

Caldwell, M. M. and D. S. Nachtwey. 1975. Expected changes in solar UV radiation. *In*: Grobecker, A. J. [ed.]. Impacts of Climatic Change on the Biosphere. CIAP Monogr. 5 (DOT-TST-75-55), Part 1: Ultraviolet Effects. Department of Transportation, Washington, D. C. 1.4-1.10.

Calkins, J. and T. Thordardottir. 1980. The ecological significance of solar UV radiation on aquatic organisms. *Nature* *283*: 563-566.

Clayton, R. K. 1977. Light and living matter. Vol. 2 Kreiger Publishing Company, New York.

Eppley, R. W. and J.D.H. Strickland. 1968. Kinetics of marine phytoplankton growth. In: Droop, M. R. and E. J. F. Wood [eds.]. Advances in Microbiology of the sea. Academic Press, New York. 23-62.

Halldal, Per and Örn Taube. 1972. Ultraviolet action and photoreactivation in algae. In: A. C. Giese [ed.], Photophysiology, Vol. 7. Academic Press, New York. 163-188.

Jerlov, N. G. 1976. Mar. Opt. Elsevier Scientific Publishing Company.

Lorenzen, C. J. 1979. Ultraviolet radiation and phytoplankton photosynthesis. Limnol. Oceanogr. 24: 1117-1120.

Parsons, T. and M. Takahaski. 1973. Biological oceanographic processes. Pergamon Press, New York.

Sisson, W. B. and M. M. Caldwell. 1975. Lamp/filter systems for simulation of solar UV irradiance under reduced atmospheric ozone. Photochem. Photobiol. 21: 453-456.

Standard methods for the examination of water and waste water (Fourteenth edition). 1975. American Public Health Association, Washington, D. C.

Thomson, B. E., R. C. Worrest, and H. VanDyke. 1980. The growth response of an estuarine diatom (Melosira nummuloides [Dillw.] Ag.) to UV-B (280-320 nm) radiation. Estuaries 3 (1): 69-72.

Westinghouse Corp. 1977. Product Information. Atlanta, Georgia.

Worrest, R. C., H. VanDyke, and B. E. Thomson. 1978. Impact of enhanced simulated solar ultraviolet radiation upon a marine community. Photochem. Photobiol. 27: 471-478.

Yentsch, C. S. and R. W. Lee. 1966. A study of photosynthetic light reactions and a new interpretation of sun and shade phytoplankton. J. of Mar. Res. 24: 319-337.

THE EFFECTS OF ULTRAVIOLET IRRADIATION ON PHOTOSYNTHESIS BY *RUPPIA MARITIMA L.* (WIDGEON GRASS)

Gary N. Wells (1) and D. S. Nachtwey (2)

Department of Biological Sciences
Florida Institute of Technology
Melbourne, Florida 32901 (1) and
NASA Johnson Space Center, Houston, Texas
77058 (2)

ABSTRACT

Net photosynthesis by Widgeon grass, *Ruppia maritima* L. is significantly reduced following exposure to simulated solar UV-B radiation at a constant dose rate of 2.0 sunburn units hr^{-1}. A 36% inhibition of photosynthesis is observed when plants are exposed to ambient levels of UV-B (approximately 25 sunburn units day^{-1} on a clear day in the vicinity of the Johnson Space Center). Photosynthetic inhibition increased linearly to about 20-25% with increasing exposure of UV-B to 15 sunburn units (approximately 60% of the surface irradiance). Above 15 sunburn units its inhibition increases more slowly to 40-45% at 50 sunburn units. This study indicates that net photosynthesis by Widgeon grass is very sensitive to UV-B irradiation, that ambient levels of UV-B have the potential to significantly reduce net photosynthesis and, there does not appear to be a threshold level of UV-B exposure below which no significant reduction in photosynthesis is observed.

INTRODUCTION

The amount of ozone in the stratosphere is a major determinant of the amount of ultraviolet radiation penetrating to the earth's surface. A reduction of the ozone layer will necessarily result in an increase in the solar ultraviolet radiation in the 290-320 nm waveband (UV-B). Projections by the Panel on Stratospheric Chemistry and Transport et al. (NAS, 1979) suggested that an ozone layer reduction will result in a significant increase in solar UV-B radiation. Ozone reduction leading to increased incidence of solar UV-B radiation has the potential of becoming a major environmental variable with respect to the biotic component. To assess the potential impact on the biosphere of an ozone decrease/UV-B radiation increase, a basic understanding of the sensitivity of both terrestrial and aquatic organisms to UV-B is required. With very few exceptions, studies to date do not allow an accurate assessment of the impact of increased UV-B radiation on any specific organism in nature.

With respect to terrestrial plants, a major biotic component, Halldal (1964), Van and Garrard (1975) and Sisson and Caldwell (1977) Brandle et al. (1977) report a significant reduction in net photosynthesis by plants subjected to high doses of UV-B irradiation. Similar studies have not been performed with aquatic agiosperms.

The objectives of this study, therefore, are to determine the effect that solar UV-B radiation at present surface irradiances, has on photosynthesis in *Ruppia maritima*, a marine angiosperm, and to assess the possible impact an increase in UV-B irradiance will have on this grass.

MATERIALS AND METHODS

Plants

Widgeon grass, *Ruppia maritima* was collected from shallow tidal pools (10 to 20 cm deep) adjacent to a lagoon at the eastern end of Galveston Island. Leaf length was often longer than the depth of the water resulting in the leaf floating horizontally on the surface. These plants were returned to the laboratory and maintained for one week under constant illumination in 20 gallon tanks containing Instant Ocean (29°/oo salinity).

Simulated solar UV-B

UV-B radiation was supplied by six FS-40 fluorescent sunlamps (Westinghouse Corporation) filtered by a cellulose triacetate film, Kodacel TA-401 (Eastman Kodak Co.). UV-B intensities were measured by a Robertson-Berger sunburn ultraviolet meter with a lamp to sensor distance of 29.2 cm. The FS-40 sunlamps provided UV-B radiation at a dose rate of 2 sunburn units (SU)$\cdot$hr^{-1} after the lamps were preburned for 72 hours. The spectral irradiance of this lamp system is shown in Figure 1.

Irradiation procedures

Leaf sections (2 cm) were stripped of epiphytes, rinsed and placed in plastic petri dishes (100 x 20 mm) containing approximately 30 ml filtered seawater and covered with a strip of Kodacel to filter out UV-C. Control plants were similarly prepared but were covered with Mylar to filter out both UV-B and UV-C (Fig. 1). In all experiments, exposure time determined the total UV-B dose received. No visible radiation was available during UV-B exposures.

Carbon fixation

After exposure to UV-B, leaf sections were submerged in 5.0 ml filtered seawater and equilibrated in the light at an intensity of 600 μE m^{-2}sec^{-1} for 8 minutes at 26°C. Following equilibration, the seawater was discarded, fresh seawater (5.0 ml) was added and 15 μCi of [^{14}C] bicarbonate in 15 μl (specific activity, 46 mCi/mmole) was introduced to the incubation medium. Leaf sections were allowed to fix radioactive bicarbonate for 15 minutes and the reaction terminated with 100% methanol.

Extraction of radioactive compounds

Following ^{14}[CO_2] fixation, the leaf sections were homogenized in 1.0 ml of 100% methanol and the homogenate clarified in a clinical centrifuge. The pellet was extracted 3 times with 100% methanol and the alcohol extracts pooled. Equal volumes of ether and water were added to the alcohol extract to partition the chlorophyll. The chlorophyll layer was quantitatively removed, diluted to 10 ml in a volumetric flask and total chlorophyll determined according to the method of Strain and Svec (1966). The alcohol insoluble pellet was washed 3 times (once overnight) with distilled water and the water extracts pooled with the water/methanol fraction. The aqueous fraction was acidified with formic acid and a 0.1 ml aliquot assayed for radioactivity in a Packard tri-carb scintillation counter. The rate of photosynthesis was determined according to Goldman et al. (1971).

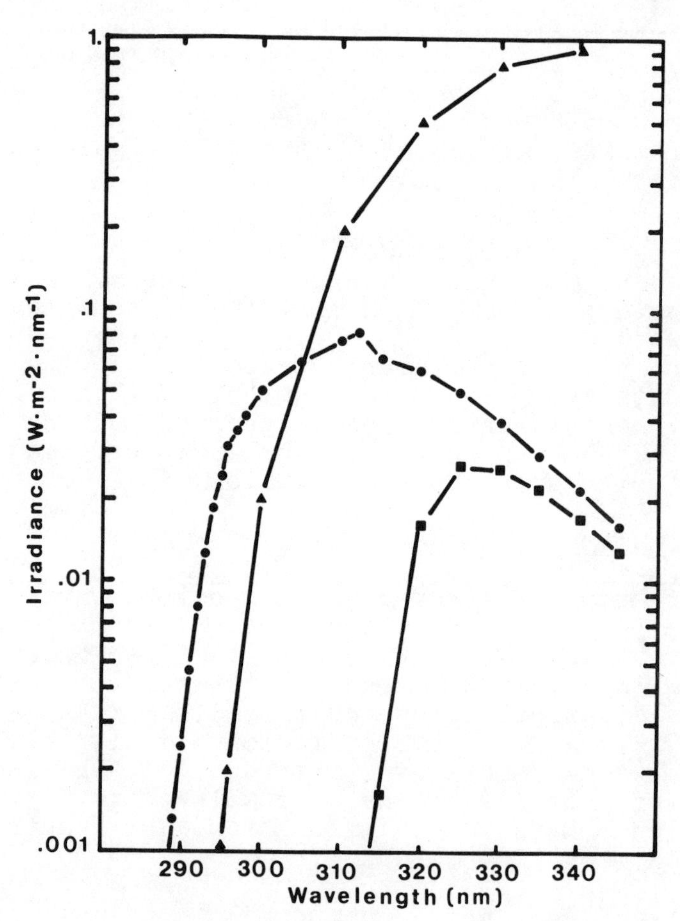

Figure 1. Spectral irradiance of six Westinghouse FS-40 sunlamps. UV-B treatment was achieved by filtering the lamps with one layer of Kodacel TA-401, 5 mil, (●——●). Irradiation control was achieved by filtering the lamps with one layer of Mylar Type A, 10 mil, (■——■). Relative global radiation (sun and sky), 20° Zenith Angle (▲——▲).

RESULTS AND DISCUSSION

The surface irradiance levels of solar UV-B received for July 14, 1977 are shown in Figure 2. These exposure rates, as a function of time, are typical for clear summer days in the Johnson Space Center area. The dose rate increase exponentially until a maximum of about 4 sunburn units hr^{-1} is received by midday. The differential dose rates received

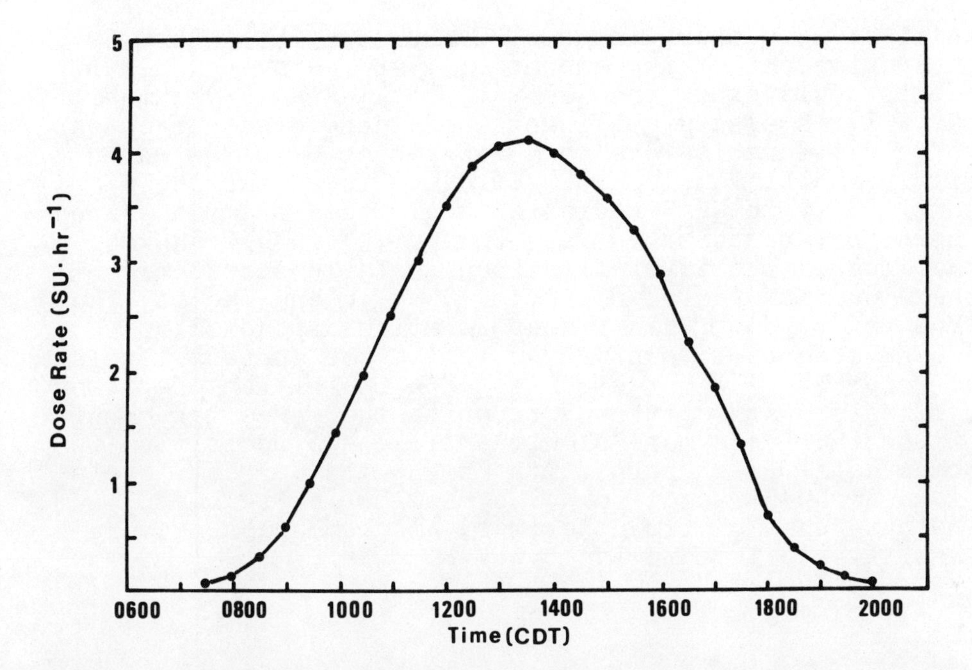

Figure 2. Surface irradiance of solar UV-B. The amount of UV-B was determined by a Robertson-Berger UV meter for July 14, 1977. Total dose received, 25.5 sunburn units.

prior to and following the midday peak are a function of the solar zenith angle. By integrating the area under the curve in Figure 1, it was determined that the total UV-B exposure for this date was 25.5 SU. The dose rate as well as the total surface irradiance levels will vary from day to day and from season to season as a function of cloud cover, air particulates, solar zenith angles and ozone layer thickness. The data in Figure 2 indicates that the earth's biotic component is currently experiencing biologically active levels of UV-B radiation.

To determine the effect current levels of solar UV-B radiation have on photosynthesis and the possible impact an increase in UV-B radiation may have, seagrass leaves were exposed at a constant dose rate of 2.0 $SU{\cdot}hr^{-1}$ to increasing amounts of total UV-B radiation. The results of these experiments (Figure 3) show that current levels of UV-B radiation have an alarmingly high inhibitory effect on Widgeon grass photosynthesis. A 36% inhibition of photosynthesis was observed in plants exposed to a total of 25 sunburn units of UV-B radiation. This is the equivalent of the amount of UV-B radiation the plants would experience in a typical clear-sky summer day. It is also significant that there does not appear to be a tolerable level of UV-B radiation below which little inhibition occurs. Evidence for a threshold sensitivity to UV-B radiation would be a sigmoid shaped inhibitory response curve. In fact, Figure 2 shows that there is a initially linear relationship between the amount of photosynthesis reduction and the level of UV-B radiation. This linear relationship holds up to a UV-B dose of 15 sunburn units (60% of current levels of solar UV-B) which elicits a 25% inhibition of photosynthesis. Above 15 sunburn units the percentage inhibition increases more slowly until a 43% inhibition is observed at 50 sunburn units (twice the current levels of solar UV-B radiation. The results in Figure 3 indicate that photosynthesis by Widgeon grass is very sensitive to irradiation by UV-B.

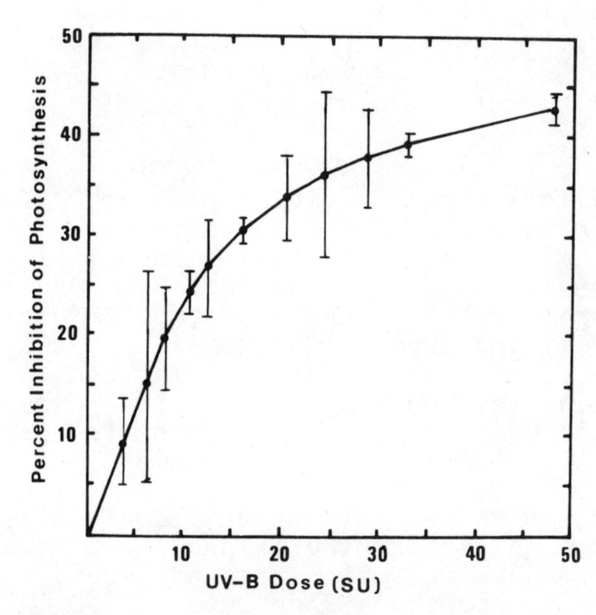

Figure 3. Effect of UV-B irradiation on photosynthesis by Ruppia maritima. The average photosynthetic rate for the Mylar control was 657.0 μg C·mg Chl^{-1} · hr^{-1}. average dose rate was 2.05 sunburn units · hr^{-1}.

It is not clear, however, whether photosynthetic inhibition by simulated UV-B in the absence of visible light is an accurate assessment of the actual effects of ambient solar UV-B. Evidence by Van and Garrard (1975) suggests that it is not. They show that plants belonging to the C^4 pathway which displayed no apparent reduction in net photosynthesis when subjected to high levels of UV-B in the greenhouse exhibited significant reduction in growth chamber trials. The differential responses of the plants in greenhouse and growth chamber experiments was attributed to a lower visible light intensity in the growth chamber experiments. While these results may suggest a photoinduced protective mechanism operating in these plants, it is more likely the greater capacity for photosynthesis by C^4 plants increases their tolerance to high ambient levels of solar UV-B.

Another likely source of error comes from the fact that a spectral difference exists between solar UV radiation and that simulated by the FS-40 sunlamp (Figure 1). Although the irradiation system allows for simulation of the total irradiance available it does not simulate either the spectral difference between solar and FS-40 irradiation or underwater spectral changes (Zaneveld, 1975). This results in the Robertson-Berger meter overestimating the amount of biologically effective UV-B (Smith and Calkins, 1976) under natural conditions.

Caldwell (1971) has suggested protective mechanisms which may be employed by terrestrial plants. These include absorption of UV-B radiation by cuticular and epidermal waxes, cell wall materials and flavonoids, photoreactivation and nastic movement. These extracellular and cellular components may serve as UV-B screens, reducing the UV-B radiation to only a fraction of the incident radiation. The increased sensitivity that Widgeon grass shows toward UV-B radiation may reflect the fact that seagrasses, in contrast to terrestrial plants, lack a waxy cuticle and have a photosynthetically active epidermis.

Experiments like those depicted in Figure 3, therefore, may not be an accurate assessment of what is happening in a natural condition but they may give some insight into the effectiveness of a UV-B screening mechanism and consequently allow for an evaluation of the sensitivity of the plant to UV-B exposure.

REFERENCES

Brandle, J. R., W. F. Campbell, W. B. Sisson and M. M. Caldwell. 1977. Net photosynthesis, electron transport capacity, and ultrastructure of Pisum sativum L. exposed to ultraviolet-B radiation. Plant Physiol. 60: 165-169.

Caldwell, M. M. 1971. Solar ultraviolet radiation and the growth and development of higher plants. In A.C. Giese [ed.], Photophysiology, Vol. 6, Academic Press, N. Y. 131-177.

Goldman, C. R., E. S. Metsen, R. A. Vollenweider and R. G. Wetzel. 1971. Measurements (in situ) on isolated samples of natural communities. In R. A. Vollenweider [ed.] IBP Handbook No. 12 Primary Production in the Aquatic Environment. Blockwell, Oxford. 70-73.

Halldal, P. 1964. Ultraviolet action spectra of photosynthesis and photosynthetic inhibition in a green and red alga. Physiol. Plant. 17: 414-421.

National Academy of Sciences, Panel on Stratospheric Chemistry and Transport et al. 1979. In Stratospheric Ozone Depletion by Halocarbons: Chemistry and Transport, NAS, Washington, D. C. p. 5.

Sisson, W. B. and M. M. Caldwell. 1977. Atmospheric Ozone Depletion: reduction of photosynthesis and growth of a sensitive higher plant exposed to enhanced UV-B radiation. J. Exp. Bot. 28: 691-705.

Smith, R. C. and J. Calkins. 1976. The use of the Robertson meter to measure the penetration of solar middle-ultraviolet radiation (UV-B) into natural waters. Limnol. Oceanogr. 21: 746-749.

Strain, H. and W. A. Svec. 1966. Extraction, separation, estimation, and isolation of the chlorophylls. In L. P. Vernon and G. R. Sneely [eds.] The Chlorophylls. Academic Press, New York. 21-66.

Van, T. K. and C. A. Garrard. 1975. Effects of UV-B radiation on the net photosynthesis and the rate of partial photosynthetic reactions of some crop plants. In Imports of Climatic Changes on the Biosphere. Climatic Import Assessment Program Monogr. 5: 125-145.

Zaneveld, J. R. V. 1975. Penetration of ultraviolet radiation into natural waters. In CIAP Monograph 5. Part 1, 2.4: 108-157.

EFFECTS OF UV RADIATION ON PHOTOSYNTHESIS OF NATURAL POPULATIONS OF PHYTOPLANKTON

C. W. Modert (1), D. R. Norris (1), J. H. Blatt (2) and R. D. Petrilla (1)

Department of Oceanography and Ocean Engineering (1) Department of Physics and Space Sciences (2) Florida Institute of Technology Melbourne, Florida 32901

ABSTRACT

The effect of ultraviolet radiation on the gross primary productivity of natural populations of phytoplankton was studied via the oxygen light and dark bottle method. Levels of UV less than present day amounts reaching sea level had variable effects on the photosynthetic rate. It appeared that phytoplankton are able to utilize UV-B and/or UV-A in a beneficial way up to a threshold level. UV-B radiation less than the incident ambient level caused a decrease in productivity in phytoplankton taken from the upper 0.1 m.

INTRODUCTION

The effect of ultraviolet radiation on natural phytoplankton populations has received little attention (cf. Lorenzen, 1979; Calkins and Thordardottir, 1980; and Smith and Baker, 1980). We have attempted to determine the effects of relative increases of ultraviolet radiation, mainly UV-B, on the rate of photosynthesis in natural populations of lagoonal phytoplankton. This study was conducted during the summer (June-August, 1977) at the Florida Institute of Technology Anchorage located on the Indian River approximately 5 km from the main campus (28° 05'N, 87° 37'W). The Indian River is a shallow lagoon on the east coast of Florida separated from the Atlantic Ocean by a barrier island. Further details of this study are given in Modert (1977).

MATERIALS AND METHODS

An irradiance monitor designed by J. Blatt, was constructed to monitor bands in the UV and visible ranges (Blatt et al. in preparation).

Water samples were collected from the upper 10 cm of the lagoon. Surface water was used to obtain the phytoplankton which presumably had been exposed to the highest intensities of solar ultraviolet radiation. These samples were filtered through a 202 μm mesh to minimize the number of grazing zooplankters. Additional water samples were collected for salinity analysis and determination of phytoplankton density and species composition; the plankton samples were preserved with formalin. Water temperature was recorded at the time of each collection. A one liter preserved phytoplankton sample was allowed to settle at least 72 hours and then concentrated (Dodson and Thomas, 1964) to 300 ml. Counts and identifications were made using a Sedgewick Rafter cell and a Whipple field reticule in a Leitz HM-Lux microscope. Chains of cells were recorded separately but as "1" each.

The rate of photosynthesis was determined by the dissolved oxygen method. Dissolved oxygen was determined by the azide modification of the Winkler titration and converted to gross productivity units (mg C$\cdot$ $m^{-3}$$\cdot$ hr^{-1}) using the equations of Strickland and Parsons (1972). Fourteen 250 ml Erlenmeyer flasks served as the incubation vessels. Four flasks were Vycor, the remaining 10 were Pyrex. Vycor is a quartz glass which has a higher transmittance of UV and visible radiation than standard laboratory glassware. We initially believed, as indicated in literature from the manufacturer, that the Vycor flasks would transmit about 90% of the radiation in the UV-B to visible ranges. It was later determined, after measuring the transmittance directly (Beckman Model DB-GT Grating Spectrophotometer), that the Vycor flasks transmit somewhat less in the UV-B range (Fig. 1). Two Pyrex flasks were used as initial bottles, four as dark, and the remaining four along with the four Vycor flasks served as light bottles. Two dark and two each of the Pyrex and Vycor bottles were placed in a control

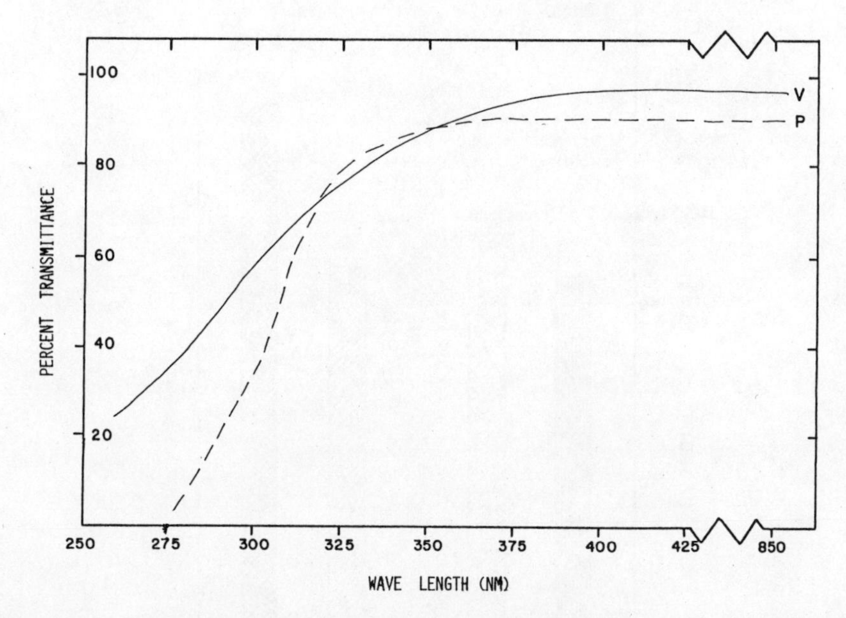

Figure 1. Transmittance curves for the Vycor (V) and Pyrex (P) flasks.

incubator while a like set was placed in a UV enhanced incubator. Lagoonal water from 1 m depth was continuously pumped through the incubators. The control was radiated by solar energy only while the UV enhanced incubator was radiated by an unfiltered Westinghouse FS40 (UV) fluorescent sunlamp in addition to natural sunlight. The UV lamp output had been stabilized by burning 100 hours prior to incubation. Incubation varied from three to six hours around solar noon, depending on cloud conditions.

Radiation intensities in the ultraviolet and visible bands were recorded hourly during the incubation period. For a set of six runs (days), the FS40 sunlamp was adjusted to provide moderate increase in UV-B (+UV-A) radiation within the UV enhanced Vycor flasks relative to that within the control Vycor flasks. A similar adjustment was made to provide a large increase in the UV radiation for a set of five runs.

Salinity values during the course of the study remained quite stable, ranging from 29.5 to 32.0 o/oo. Similarly, temperatures at the collection site had a small variation (28.0-30.0°C) over the three month period. Water temperatures in both incubators gradually

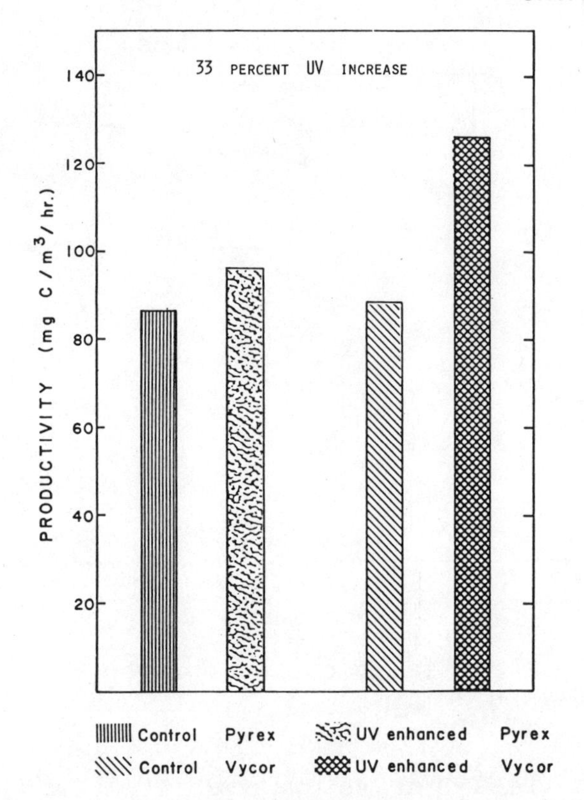

Figure 2. Gross primary productivity for 14 June 1977. Effect on moderately increasing UV-B (+UV-A) over the control level.

rose during each incubation period throughout the study, the increases ranged from 2.5° to 4.0°C. However, since the temperatures in the control and the UV enhanced incubators were always the same, temperature increases have been discounted as causing the observed differences in the gross primary productivity between the two incubators.

RESULTS

The phytoplankton counts ranged from 135×10^4 cells $\cdot$ l^{-1} on 9 June to 26×10^4 cells $\cdot$ l^{-1} on 4 July. The dominant species was usually *Skeletonema costatum* or *Nitzschia closterium*; *Gonyaulax monilata* was dominant on one sampling day.

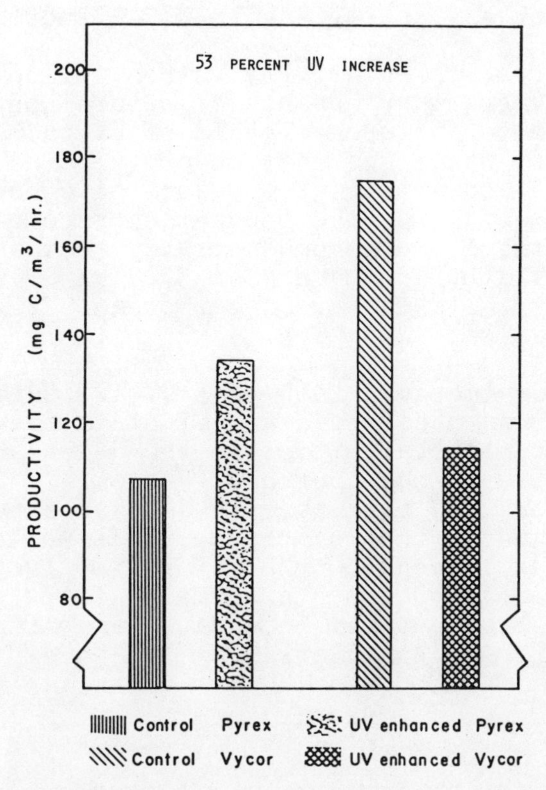

Figure 3. Gross primary productivity for 3 August 1977. Effect of a large increase of UV-B (+UV-A) over the control level.

Typical results of the effect of a moderate increase of UV-B (+UV-A) over the control level are shown in Figure 2. This sample was run on 14 June, the incubation period was six hours, skies were clear to partly cloudy. The production rate of the control Vycor was 88 mg $C \cdot m^{-3} \cdot hr^{-1}$, while that of the UV enhanced Vycor was 126 mg $C \cdot m^{-3} \cdot hr^{-1}$, giving an increase of 43%. The control Pyrex had a production rate of 86 mg $C \cdot m^{-3} \cdot hr^{-1}$, while the UV enhanced was 97 mg $C \cdot m^{-3} \cdot hr^{-1}$, resulting in an increase of 13%.

Figure 3 illustrates results of the large increase of UV-B (+UV-A) over the control level. The sample was run on 3 August, 1977, incubation was four and one-half hours under clear to partly cloudy skies. The production rate was 175 mg $C\ m^{-3} \cdot hr^{-1}$ and 114 mg $C \cdot m^{-3}\ hr^{-1}$ for the control and UV enhanced Vycor respectively, showing a decrease of 35%. The rate was 108 mg $C \cdot m^{-3} \cdot hr^{-1}$ and 134 mg $C \cdot m^{-3} \cdot hr^{-1}$ for the control and UV enhanced

Pyrex respectively, resulting in a 24% increase.

Gross primary productivity values reflect considerable variation in the response of natural populations of phytoplankton to different levels of UV radiation. In this study, it was found that moderately increasing the UV-B (+ UV-A) above that of the control consistently had a beneficial effect on the production rate of the phytoplankton (Table 1). The productivity in the UV enhanced Vycor increased by an average of 37.8% ± 29.9 over the control Vycor while the average increase for the UV enhanced Pyrex was 22% ± 8.5.

The larger increase in UV-B (+UV-A) radiation consistently had a detrimental effect in the UV enhanced Vycor but a beneficial effect in the UV enhanced Pyrex flasks relative to the controls (Table 1). In the UV enhanced Vycor, the productivity decreased by an average 23.4% 12.6 below that of the controls while the UV enhanced Pyrex increased by 21.4% ± 20.6 over the controls Figure 4 presents a summary of the results using Vycor flasks at the "moderate" and "large" increased levels of UV radiation.

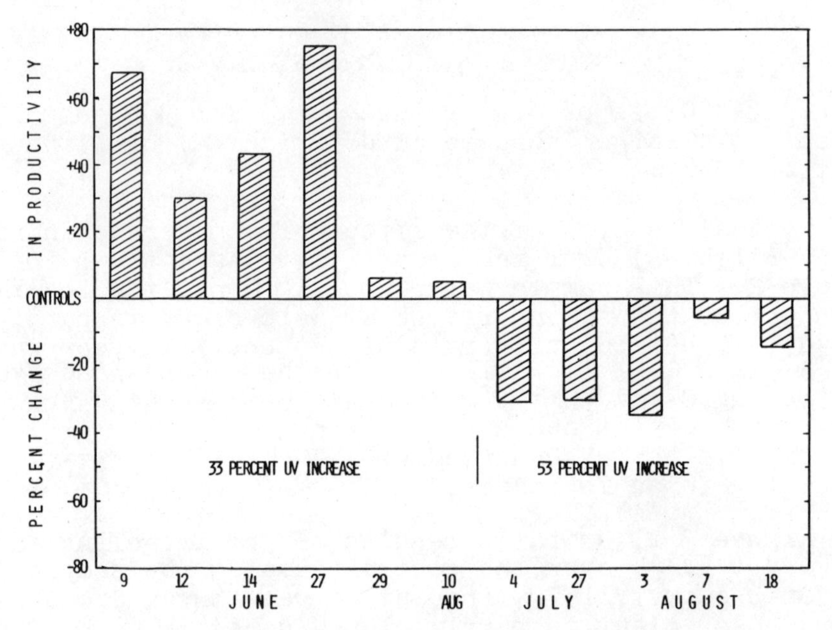

Figure 4. Summary of results using Vycor flasks. Positive values (found after moderate increase in UV-B) were increases and negative values (found after large increase in UV-B) decreases relative to controls.

TABLE 1

Summary of results, determined by finding the percent difference in productivity between the control and UV enhanced flasks. Plus (+) values were increases and negative (-) values were decreases relative to the controls.

DATE	VYCOR	PYREX
	Moderate UV-B increase	
9 June	+68	+283
12 June	+30	+22
14 June	+43	+13
27 June	+75	+34
29 June	+6	+26
10 August	+5	+15
	Large UV-B increase	
4 July	-31	+55
27 July	-31	+7
3 August	-35	+24
7 August	-6	+3
18 August	-14	+18

DISCUSSION

The results of this study demonstrate that ultraviolet radiation affects the production rate of natural phytoplankton populations. We believe that these results are primarily due to the effects of UV-B radiation. The irradiance from the sunlamp in the UV-A and visible regions is small compared to that from the sun at sea level. Irradiance was increased less than 10% and less than 2% in the UV-A and visible respectively (Modert, 1977). Nevertheless, the UV enhanced Pyrex consistently showed increases in productivity with both moderate and large UV-B (+UV-A) increases relative to the controls. The production rate in the UV enhanced Vycor consistently increased with a moderate increase in UV-B (+UV-A), but decreased with a large increase relative to control flasks.

These results suggest that the threshold for detrimental levels of UV radiation was bracketed in the case of the Vycor series. And, since we believe that even the higher enhanced level of UV radiation was less than surface ambient, our results suggest that today's natural levels are sufficient to reduce the photosynthetic rate in the near surface phytoplankton. That is, today's incident UV radiation level is above the threshold level for maximum photosynthesis. This observation is supported by the recent findings of Lorenzen (1979) and Smith et al. (1980).

The increase in productivity for the moderate UV enhanced Vycor series and the consistent increases in productivity for both the moderate and large UV enhanced Pyrex series indicate that the phytoplankters utilized UV-B and/or UV-A in a beneficial way. This suggests that UV quanta are being absorbed and somehow resulting in increased oxygen production. Worrest et al. (this volume, Fig. 1) have indicated that relative radiocarbon uptake increased at low UV-B fluence levels for some phytoplankton species.

A mechanism which may play a role in the survival of phytoplankton *in situ* is Langmuir circulation. The possible importance of this phenomenon to phytoplankton is presented by Smayda (1970). Langmuir vortices bring phytoplankton cells toward the surface where they are exposed to higher UV light intensities while at the same time they carry cells to depths of reduced UV radiation. While the two conditions (rising and sinking) are conflicting in this regard, the survival via sinking may be critical for a population at a given "moment" in time. Langmuir circulation may also explain the variability in response of natural phytoplankton populations mentioned above and by other authors (Lorenzen, 1975; Smith et al., 1980), since populations taken from low light intensities are more sensitive to UV than are surface populations (Steemann Nielsen, 1964).

REFERENCES

Calkins, John and Thorunn Thordardottir. 1980. The ecological significance of solar UV radiation on aquatic organisms. *Nature* *283*: 563-566.

Dodson, Anne and W. H. Thomas. 1964. Concentrating plankton in a gentle fashion. *Limnol. Oceanogr.* *9* (3): 455-456.

Lorenzen, C. J. 1975. Phytoplankton responses to UV radiation and ecological implications of elevated UV irradiance. *In* Impacts of Climatic Change on the Biosphere. CIAP Monogr. 5 (DOT-TST-75-55), Part L; Ultraviolet Effects. Department of Transportation, Washington, D. C. 5-83 to 5-91.

Lorenzen, C. J. 1979. Ultraviolet radiation and phytoplankton photosynthesis. Limnol. Oceanogr. 24 (6): 1117-1120.

Modert, C. W. 1977. The effects of ultraviolet radiation on the production rates of natural populations of estuarine phytoplankton. M. S. Thesis, Florida Institute of Technology.

Smayda, T. J. 1970. The suspension and sinking of phytoplankton in the sea. In H. Barnes [ed.] Oceanogr. Mar. Biol. Ann. Rev., Vol. 8: Allen and Unwin Ltd. 353-414.

Smith, R. C. and K. S. Baker. 1980. Stratospheric ozone, middle ultraviolet radiation, and carbon-14 measurements of marine productivity. Science 208: 592-593.

Smith, R. C., K. S. Baker, O. Holm-Hansen, and R. Olson. 1980. Photoinhibition of photosynthesis in natural waters. Photochem. Photobiol. 31: 585-592.

Steemann Nielsen, E. 1964. On a complication in marine productivity work due to the influence of ultraviolet light. J. Cons., Cons. Int. Explor. Mer. 29: 130-135.

Strickland, J.D.H., and T. R. Parsons. 1972. A practical handbook of seawater analysis. Fisheries Research Board of Canada, Bull. 165. Ottawa (2nd ed.).

THE EFFECT OF HIGH INTENSITY U.V. RADIATION ON BENTHIC MARINE ALGAE

M. Polne and A. Gibor

Biological Sciences and Marine Science Institute
University of California,Santa Barbara, Ca 93106

INTRODUCTION

We are using fluorescence microscopy in the determination of damage caused to algal cells by various chemical and physical treatments. Damage is manifest as a change in the fluorescence spectrum of the cells. Fluorescence microscopy exposes cells to short-blue (400-460 nm) and UV-A (315-400 nm) light. Shorter wavelengths in the UV-B range are eliminated by the glass microscope lenses and filters. We observed that exposure of algal tissue to the exciting beam of the fluorescence microscope also causes damage to cells which can be seen as a change in the fluorescence spectrum over time during exposure. We have attempted to determine the degree of damage to the algae as a result of this exposure and to describe the progress of the damage to the cells.

MATERIAL AND METHODS:

Plants of *Porphyra perforata*, *Ulva taneiata*, *Enteromorpha intestinalis*, *Gigartina exasporata*, *Callophyllis violacea*, *Halymenia hollenbergii*, and *Macrocystis pyrifera*, were collected at campus-point (University of California, Santa Barbara). The plants were brought to the laboratory in sea water within 20 minutes of collection and observed immediately under a fluorescence microscope with a temperature controlled microscope stage maintained at 16 °C. Other samples of tissue were held in the laboratory in low light for several days before being exposed to the exciting light of the fluorescence microscope.

A Reichert microscope equipped with a HBO 200 W Osram arc lamp was used in the epiillumination mode. Power levels at the tissue surface were 0.1-0.9 Watt M^{-2}. Parallel experiments were conducted by irradiating algal tissues under a germicidal lamp (GE 30 watt, 253.7 nm peak emission), at 25 cm distance from the source (0.02 Watts M^{-2}). The pattern of damage under germicidal irradiation was identical to that observed by irradiation with the exciting beam of the fluorescence microscope.

Changes in the fluorescence of the algal pigments were recorded in two ways: 1. By photomicrography on Ektachrome 160 (Tungsten) film. 2. By the use of a microspectrofluorometer (Nanometric, Inc., model 127-0143, Nanospec/10S). Either the total fluorescence spectrum or the kinetics of change of a single peak in the spectrum were monitored.

The viability of the cells was determined using the following criteria: 1. Recovery of normal fluorescence and remaining normal for the following two weeks. 2. By uptake of, and concentration of neutral-red dye or the exclusion of Evans-blue dye.

RESULTS AND DISCUSSION:

The following results were obtained by color photomicrography and visual observations. All irradiations at 370 nm and shorter were found to be damaging to the experimental plants within the first 10 minutes of irradiation. Fresh blades collected directly under the noon sun at the beach were the least sensitive, and did not show signs of damage until after 10 minutes of exposure. Subtidal algae such as *Gigartina*, *Callophylis* and *Halymenia* were generally more sensitive while low light adapted plants were most sensitive, changing their fluorescence level after 1-3 minutes of irradiation.

It is interesting to note that red algae in general

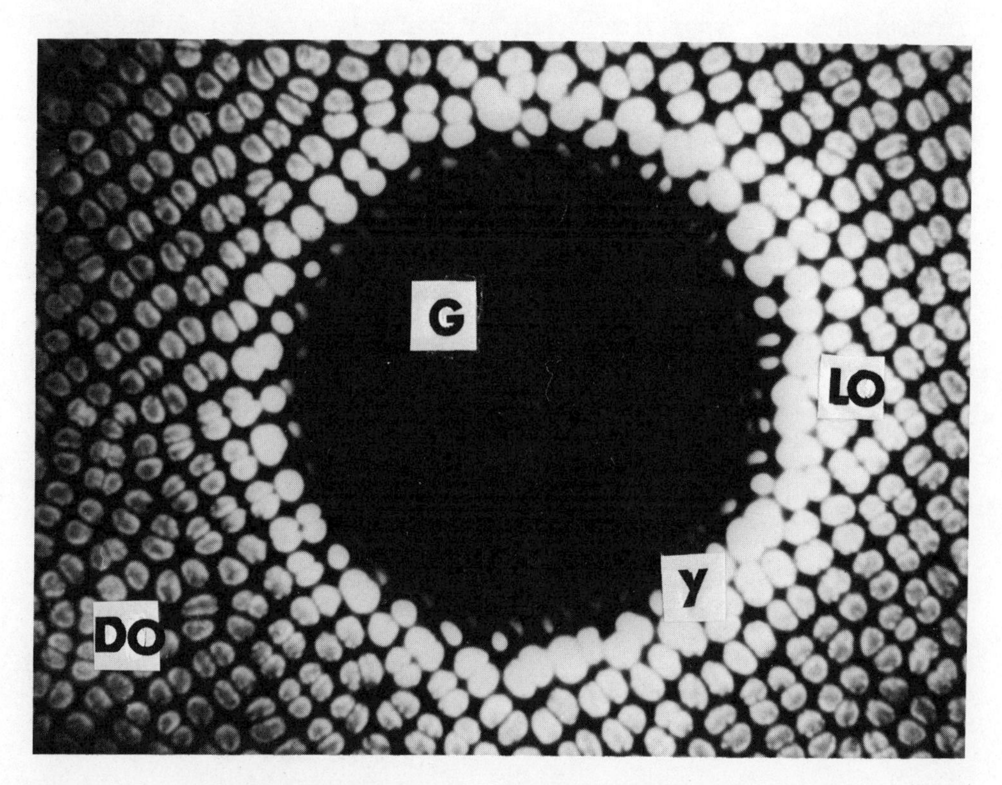

Fig. 1. Porphyra tissue in which a spot was irradiated through a 40X objective lens, for 20 minutes, then photographed through the 10X objective. Center cells are most damaged and fluoresce green (G). Edge of "burned spot" is composed of yellow (Y) and light-orange (LO) cells. Healthy cells in the background fluoresce dark-orange (DO).

are less sensitive to this radiation than green or brown algae; which might be attributed to the protective action of the secondary pigments which they contain (Abeliovich and Shilo 1972). Changes in the fluorescence spectra occurred in all the irradiated algae. We present here as an example the changes which take place in Porphyra cells during irradiation.

The first noticeable change of fluorescence was seen as a slight brightening of the dark-orange tone of fluorescence. This change was reversible and returned to normal after an overnight incubation in the laboratory. Exposure of 20-30 minutes to the exciting beam resulted in a spot composed of shrunken cells with green fluorescence. In Porphyra the spot of damaged, shrunken, green fluorescing cells (Fig. 1 -G-) was surrounded by concentric rings of less damaged cells forming a gradient from most to least damaged. There were green (G), yellow (Y), light-orange (LO), and dark-orange (DO), fluorescing cells corresponding to the less intensely irradiated areas of the tissue (Fig. 1).

Cells with green fluorescence did not recover. These cells lost their fluorescence and by the next day were completely bleached. Yellow and bright-orange cells also shrank and eventually died. The outer ring, of dark-orange, slightly damaged cells recovered.

The following results were obtained by measuring the fluorescence of single cells with the microspectrofluorometer. A comparison between the fluorescence spectra for healthy and damaged cells of Porphyra is shown in Fig. 2. Non-irradiated cells (stars) had a peak of fluorescence at 720 nm, corresponding to antenna chlorophyll, (Ley et al. 1976). After 20 minutes of exposure to UV (circles) the 720 nm peak dropped considerably and a new peak appeared at 578 nm, apparently Phycoerythrin fluorescence became dominant. After 37 minutes of irradiation (triangles) fluorescence was green,

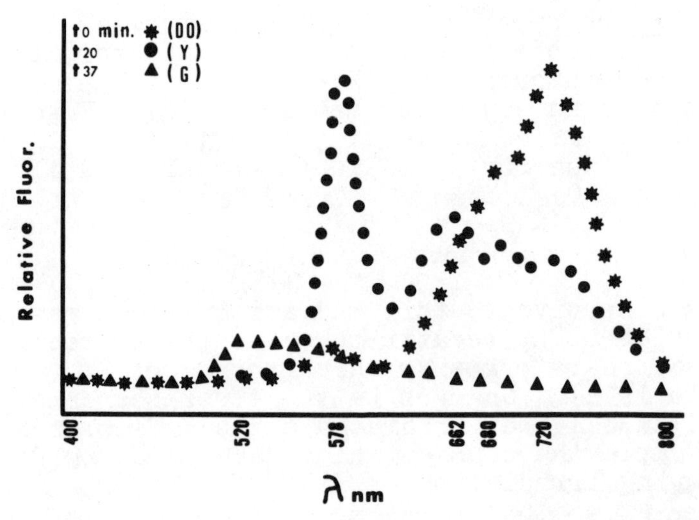

Figure 2. Fluorescence spectrum of a healthy Porphyra cell (stars) compared with the effect of 20 minutes (circles) and 37 minutes (triangles) of irradiation. The 720 nm peak gave way to the 578 nm peak, then both subsided.

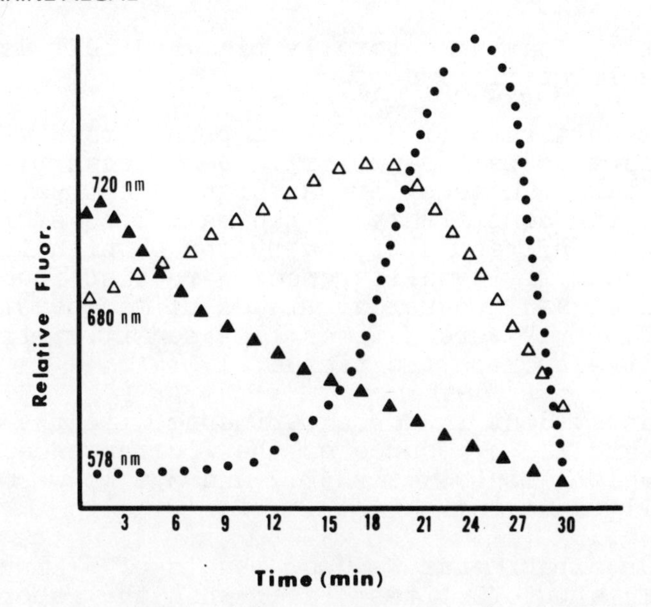

Figure 3. Kinetics of fluorescence change during irradiation of Porphyra cells. 720 nm is the first to peak and to decline. 680 nm starts increasing immediately. The fluorescence of 578 nm is the last to develop and the shortest to exist.

cells were shrunken and a single low peak fluorescence at 520 nm remained.

The pigments seemed to degrade sequentially as described in Fig. 3. First there was a slight increase in fluorescence at 720 nm, resulting in a brighter red-orange color. This was followed by the rapid decay of the 720 nm peak and a simultaneous increase in fluorescence at 680 nm gradually achieving the initial level of the 720 nm peak in about 10-20 minutes. As the 680 nm peak declined a peak at 578 nm achieved its maximum at about 24 minutes. It probably represents fluorescence from the red algal pigment Phycoerythrin. When this peak was at full strength the cells were fluorescing bright yellow. The 578 nm peak subsequently decreased as the cytoplasm shrank away from the cell walls. The gradually decreasing 680 nm peak could still be detected even in the most shrunken cells which had lost all orange and yellow fluorescence. This red fluorescent disappeared within 24 hours after irradiation, leaving a totally bleached cell as seen through a light microscope.

The sensitivity of different plants to UV irradiation appears to be non-uniform. Deep ocean plants are at least twice as sensitive as high intertidal dwellers. Moreover, the sensitivity of the same species is different when preconditioned by low or high light adaptation. Sensitivity differences were also found between sexual and non-sexual blades of Macrocystis, where sporophyls were 3 times less sensitive than non-reproductive sporophytic tissue.

As was evident in our experiments with the germicidal lamp and the UV source of the fluorescence microscope, wavelengths between 253.7 and 370 nm were all potentially damaging.

In our experiments we used levels of UV irradiation which were about 10 times higher than the reported present levels in nature. Our wet plants were severely damaged in less than 10 minutes of exposure to these levels of radiation. However exposed air dried low tide plants were resistant to the radiation. A significant increase in UV irradiation of wet plants is liable to cause irreversible damage to these cells.

ACKNOWLEDGEMENTS

This work was supported by the NOAA office of Sea-Grant, Department of Commerce, under grant number NOAA 448668-22557.

REFERENCES

Abeliovich, A. and M. Shilo. 1972. Photooxidative death in blue-green algae, J. Bact., Vol. III (3), 682-689.

Bener, P. 1972. Approximate values intensity of natural ultraviolet radiation for different amount of atmospheric ozone. European Research Office, US Army, London. Contract number DAJA 37-68-C-1017, 4-59.

Ley, A.C. and W. L. Butler. 1976. The efficiency of energy transfer from Photosystem II to Photosystem I in Porphyridium cruentum. P.N.A.S.

Stair, R. 1969. Measurements of natural ultraviolet radiation, historical and general introduction. In: The Biologic Effect of Ultraviolet Radiation. [F. Urbach ed.] Pergamon Press, Oxford.

THE EFFECTS OF UV-B IRRADIATION ON HIGHER PLANTS

M. Tevini, W. Iwanzik and U. Thoma

Botanical Institute II
University of Karlsruhe
F.R.G.

INTRODUCTION

Recently it has been established that chlorofluoromethanes and other gases cause a reduction of the ozone layer. Measurements by several groups predicted a reduction of between 7.5% (NAS 1976) and more recently 16% (NAS 1979). As a consequence, a displacement of the solar spectrum to shorter wavelengths and increased intensity in the UV-B waveband (280-320 nm) are to be expected. Additionally, the intensity of the UV-B reaching the earth depends on several other parameters, e.g., position of sun, season, geographical location, height above sea level, etc., so that calculations are only valid for one particular place.

Bener made calculations of the spectral distribution of the global ultraviolet radiation based on measurements in the Swiss Alps at varying wavelengths, solar altitudes and with 6 different values for the ozone concentration (Bener 1968, 1972).

More recently, Green (1976) and Green et al. (1980) have recalculated Bener's procedure and thus provided an improved analytical characterization of the daylight based on more precise radiative transfer calculations of Braslau and Dave (1973) and Dave and Halpern (1975). Using Green's Model, Caldwell developed a computer program with input parameters (weighting functions, altitude, albedo, solar zenith angle, ozone thickness, aerosol absorption and aerosol scattering) which can be specified by the user all over the world (Caldwell, personal communication. We gratefully acknowledge that these programs were made available to us by Prof. M. Caldwell, Utah State University). With this program the diffuse, direct and global fluxes of ultraviolet radiation (290-360 nm) striking our geographical place and also the biological effectiveness of this radiation as weighted by the "DNA" and "plant" action spectra were calculated. We used the Caldwell program for the calculation of ozone reduction rates simulated in our lighting device and in the growth chamber (Table 1).

The biological effect of UV-B radiation on higher and lower plants is well reviewed by Caldwell (1971) and by Klein (1978). Until now, reductions in total weight and leaf area have regularly been observed under UV-B stress designed to simulate more than 25% ozone reduction (Krizek 1975; Sisson and Caldwell 1976, 1977; Van and Garrard 1976; Van et al. 1978; Basiouny et al. 1978; Tevini et al. 1979; Esser 1979; Teramura 1980; Teramura et al. 1980).

Moreover, the further vegetative growth of the plant before flowering is also highly dependent on the light in the visible range (PAR = photosynthetically active radiation) and the light-dark rhythm. Thus reduced fresh weight and inhibition of flower formation were found in Lycopersicum, Impatiens and Phaseolus when fluorescent - white light tubes of relative low light intensity combined with additional UV-B irradiation were used (Klein et al. 1965). One working group reports on the major effects on growth parameters (Teramura 1980), on photosynthesis, respiration and transpiration of soybean seedlings grown under four UV-B radiation flux densities and four PAR levels (Teramura et al. 1980). Different seasonal effects which might also come from different daylight-PAR were observed by Semeniuk and Stewart (1979) in the leaf morphology of Poinsettia.

The transmission of the epidermis may be an important factor to consider in the effect of UV-B on cellular and metabolic processes. Investigations into about 25 species with different epidermal properties showed a general light transmission of about 10%, whereby 95% to 99% of the UV irradiation was absorbed by more than half of the species tested (Robberecht and Caldwell 1978).

TABLE 1. Biologically weighted irradiance and daily dose of global radiation for several ozone layer densities as well as for the lighting device. All calculations were made according to Setlow (1974) for DNA and to Caldwell (1971) for generalized plant response (plant). In the last column, the daily sum of photosynthetic active radiation (PAR) is shown.

	Weighted irradiance (BUV) mW/m²		Daily effective dose J/m²		O_3-reduction %	PAR, daily sum J/m²
	DNA	plant	DNA	plant		
Global flux						
- Karlsruhe[a]	1.8	20.9	-	-	-	9.1×10^6
- Karlsruhe[b]	1.2	22.0	20.4	382.1	-	-
- Ozone 0.32 (cm · atm)[c]	4.0	63.0	81.9	1370.0	0	-
0.27	5.8	83.4	120.8	1847.2	15.6	-
0.24	7.5	99.9	156.8	2237.7	25.0	-
0.21	9.9	121.0	208.5	2743.1	34.4	-
0.18	13.5	148.5	285.5	3412.3	43.7	-
0.15	18.8	185.5	404.7	4323.8	53.1	-
0.12	27.2	236.5	597.5	5609.6	62.5	-
Lighting device						
- Control	5.9	44.8	509.8	3368.4	-	3.2×10^5
- Schott 3 mm	8.9	68.4	763.8	5909.8	30.1	3.2×10^5
- Schott 2 mm	13.0	96.4	1118.9	8326.4	41.1	3.0×10^5
- without filter[d]					DNA	
- high intensity	28.5	197.4	2459.8	17055.4	63.4	3.7×10^5
- medium intensity	17.9	124.4	1550.0	10749.9	51.6	3.7×10^5
- low intensity	9.4	65.5	815.6	5655.7	32.6	3.7×10^5
Growth chamber						
- Control	1.4	11.9	55.1	325.5	-	3.5×10^6
- UV-B enhanced	12.5	107.3	487.6	4369.0	40.3	3.5×10^6

[a] own measurements in Karlsruhe (49°N, 22.9.1980, 11. a.m.)
[b] theoretically calculated according to Green's model for Karlsruhe 49°, 22.9.1980)
[c] theoretically calculated according to Green's model for several ozone layer densities (Albedo 0.25; aerosol 1; solar zenith angle 25.5; altitude 115 m, 173 Julian date)
[d] small amounts of UV-C were not included in the calculation of biological effectiveness.

Assuming that the leaf surface might also be altered by UV-B irradiation we carried out stereoscan studies on this subject.

The purpose of the present study was to evaluate the major effects of UV-B on growth, composition and function of higher plants by simulating several O_3-reduction rates between 30% and 63%. We therefore compared four plant species (barley, bean, radish and Indian corn) under several UV-B intensities in respect to the protein-, lipid- and pigment content. UV-B effects on the biosynthetic function of the plants were studied in the photosynthetic system during greening of etiolated barley seedlings.

MATERIALS AND METHODS

Irradiation conditions

We used two different light sources in our experiments a) a self-built lighting device equipped with UV-B tubes combined with white light tubes. In the first part of the experiments the plants were irradiated in a climatically stable room without any filter (Chapter A), in the second part we used UV-stable Schott filters (Chapter B), b) a Xenon lamp combined with the same UV-B tubes and the same Schott filters, all installed in a program-regulated growth chamber. The measured irradiance of the lighting device and of the growth chamber was weighted in the UV-B waveband of 290-320 nm according to Setlow (1974) for DNA and to Caldwell (1971) for general plant response in order to receive the biological effective irradiance (Table 1).

The calculations of the global fluxes and of the expected values for biological effective irradiance for different ozone thickness were based on Green's model using computer programs with the following variable parameters: Latitude 49°, date 173 (Julian date) and 266 (=22.9.80), altitude 0.115 km, albedo 0.25 aerosol scaling coefficient 1. The calculated BUV values of the lighting device and the growth chamber were compared to expected BUV values according to Green's Model. As a basis for estimation of the simulated ozone reduction rates the DNA-weighting function was used.

Lighting device

The lighting device (100 x 70 cm) was equipped with 3 UV-B emitting tubes (Philips TL 40/12) combined with 4 white light tubes (Philips TL 40/29). The UV tubes could be regulated by a dimmer to give different UV-B intensities without causing spectral displacement. A large proportion of the UV-B range was filtered by a glass plate in the control compartment. The spectra of the lighting device without and with glass filter (control) are shown in Figure 1. All spectra were measured by a spectroradiometer type EG and G 550, in the UV region with 2 nm intervals in the PAR-region with 10 nm steps (Fig. 2, 3, 4). The distance between the lamps and the plants was 50 cm, the experimental chamber was kept at a constant temperature of 20°C ± 1°C. The plants were irradiated continuously at the following irradiation levels (see also Table 1).

low intensity (l) = 0.60 W/m^2 (280-320 nm)
≅ BUV 9.44 mW/m^2 ≅ - 32.6% O_3

medium intensity (m) = 1.16 W/m^2 (280-320 nm
≅ BUV 17.94 mW/m^2 ≅ -51.6% O_3

high intensity (h) = 2.30 W/m^2 (280-320 nm
≅ BUV 28.47 mW/m^2 ≅ -63.4% O_3

In the second part of our experiments we used the same lighting device but filtered the light source with Schott filters of two and three mm thickness. The spectra of these combinations are shown in Figure 2. The use of filter systems was necessary because the UV-B tubes emit rather small but biologically very effective amounts of UV-C, which were not included in the estimation of the BUV-radiation. Therefore, the ozone reduction rates for experiments with unfiltered radiation could be higher than calculated.

Growth chamber

The production model of the growth chamber (Weiss type 600 E/ +5 + 80 Ju-Pa) was modified for the purpose of our research. In addition to the Xenon lamp (Osram XQO 6000 W) 6 Osram-sun Ultra-Vitalux (300 W) and 2 UV-B tubes (Philips TL 40/12) were installed (Fig. 3). The experimental chamber was divided into 2 compartments by panes of glass. The control compartment was covered either by a) window glass with a low UV-B penetrability, or b) solarium glass, which at approximately 290 nm is more penetrable than window glass and corresponds

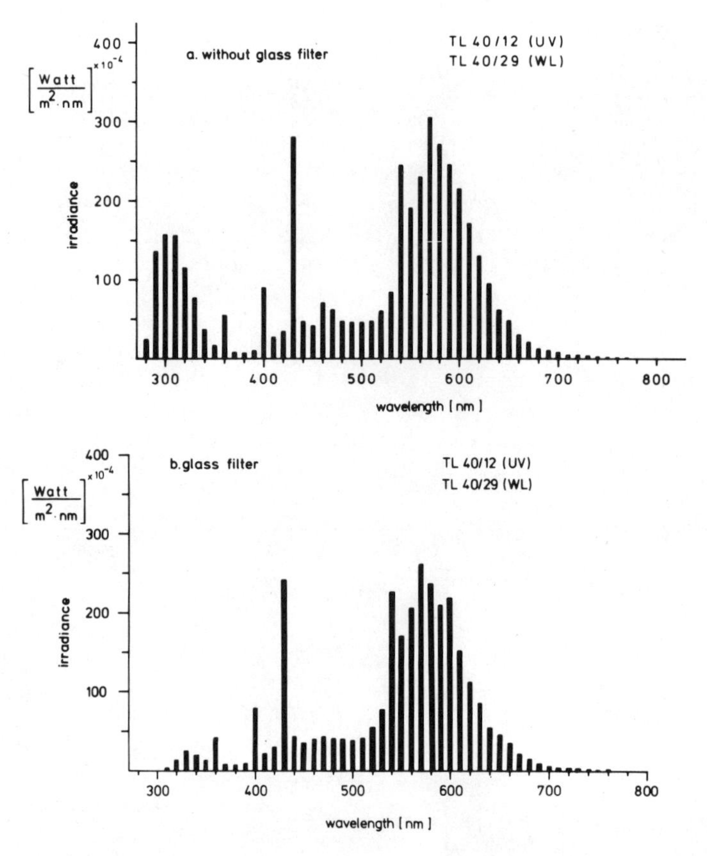

Figure 1. Complete spectra of the lighting device a) without filter b) with glass filter (control)

roughly to natural light conditions (Fig. 4). Above the UV-B compartment a Schott filter was inserted (WG 305 2 mm). The plants were grown in alternating light and dark (dark period 11 hours, light period 13 hours of which 7 hours had a maximum illuminance of 40,000 Lux and 3 hours of gradually increasing and decreasing light intensity. The daily sum of photosynthetically active radiation (PAR) was 3.46×10^6 J/m² in the UV-B enhanced and control part of the growth chamber.

The values for biological effective radiation were 12.5 mW/m² ≅ 40.5% O_3 - reduction in the UV-B enhanced part and 1.35 mW/m² in the control part (Table 1). The temperatures were set at: dark phase : 10 C at a relative humidity of 88%, light phase : 19°C at a relative

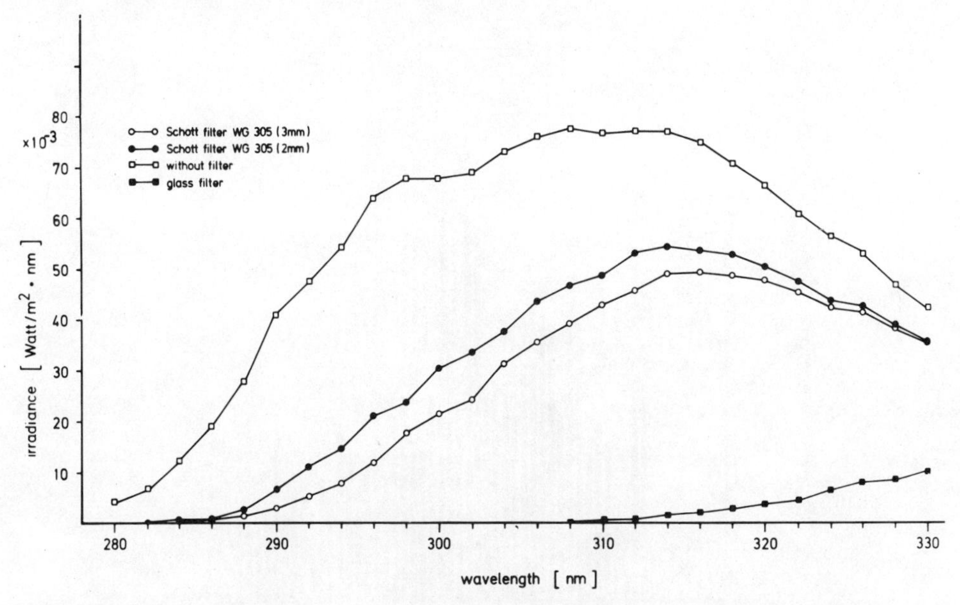

Figure 2. UV-B spectra of the lighting device with and without filters.

humidity of 86%. The plants were grown in soil and were exposed to these conditions for 7 days (Hordeum vulgare), 10 days (Zea mays), 17 days (Phaseolus vulgaris) or 16 days (Raphanus sativus) respectively.

Quantitative analyses

10 - 30 plants were cut 1 cm above the soil and the resulting plant material used to determine the growth parameters, fresh weight, dry weight and leaf area.

The soluble proteins were determined with the folincio-calteus reagent according to Lowry (1951) and according to the Kjeldahl method, which gave comparable results, showing that no interference with flavonoids had taken place.

For total lipid determination the plants were homogenized in an ethanol-chloroform mixture (2:1; v/v), and the total lipid content was determined gravimetrically.

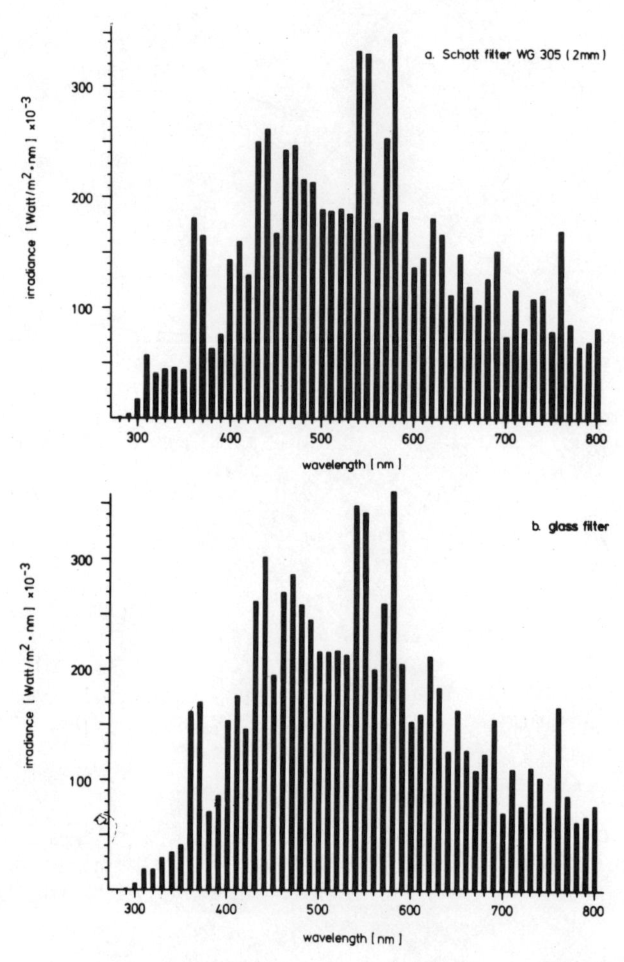

Figure 3. Complete spectra in the growth chamber a) Schott filter WG 305 (2 mm) b) Solarium glass.

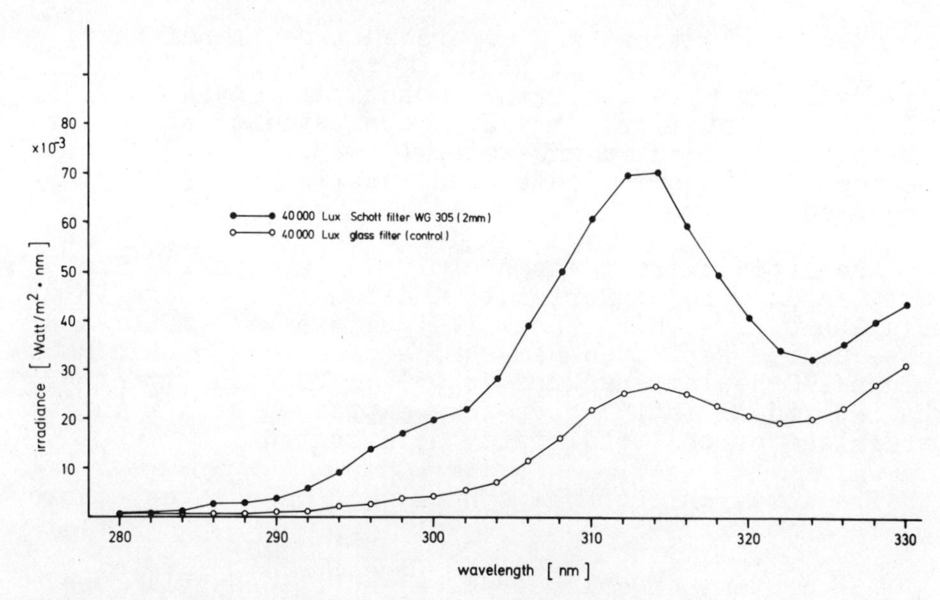

Figure 4. UV-B spectra in the growth chamber with Schott filter WG 305 (2 mm) and solarium glass (control) at 40,000 Lux.

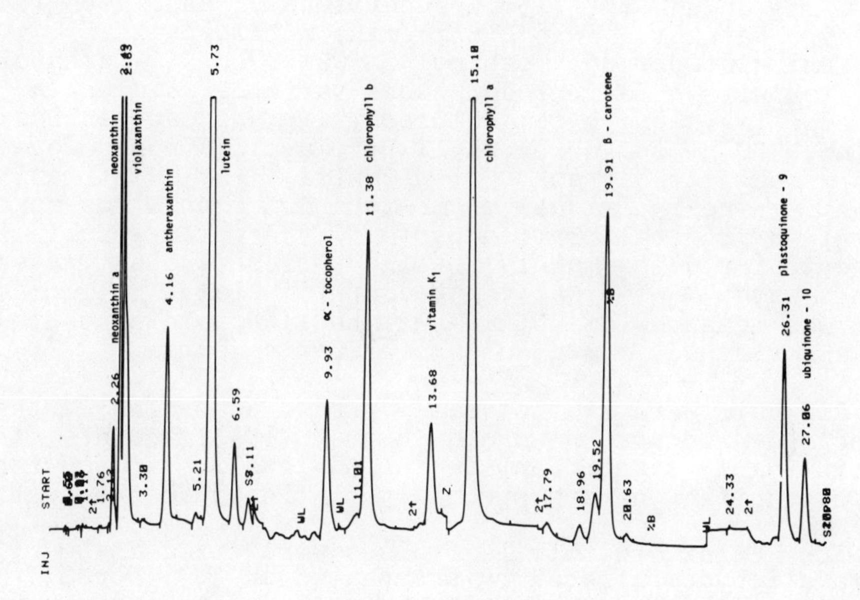

Figure 5. High pressure liquid chromatographic separation of pigments and plastidquinones of green barley leaves

The lipid extract was separated two-dimensionally on a prepared silica gel plate (Merck 5721) and the glycerolipids were determined quantitatively. The solvent for the first dimension consisted of chloroform, methanol, acetic acid and water (85:25:15:3; v/v), for the second dimension acetone and acetic acid (80:25; v/v) were used.

The lipid extract was hydrolised in alkali. The fatty acids were converted to their methyl esters in a solution of BF_3 in methanol (Metcalfe et al. 1966) and subsequently separated gas-chromatografically on a DEGS column (GC Hewlett Packard 7620 A). The quantities of the individual fatty acids were expressed as a relative percentage of the total fatty acid content.

The total chlorophyll content was calculated according to Arnon (1949).

The content of chlorophyll a and chlorophyll b as well as the total carotenoid content was estimated according to Ziegler and Egle (1965). Separation and determination of the pigments and of plastidquinones was done by high pressure liquid chromatography. (HPLC model: Hewlett Packard 1084 B) on a RP-8 (10μ 25 cm, ∅ 4.6 mm) column (Fig. 5). Conditions of chromatography: Rate of flow: 2.5 ml/min; mobile phase: gradient elution: 85% methanol in water on 95% methanol in water within 20 min; detector: variable wavelength detector (wavelength for chlorophyll a and carotenoids 430 nm, 470 nm for chlorophyll b; plastidquinones were determined at their absorption maxima. The quantitative amount of the individual pigments and quinones was ascertained from calibration curves. The water soluble pigments including mainly the flavonoids were extracted in hot water. After adding 3 drops of $Alcl_3$, the solution was measured at 405 nm and the flavonoid content was ascertained from a Rutin calibration curve.

All data were statistically proved according to the U-test of Wilcoxon, Mann and Whitney (1945) and are statistically significant on the 5% level. Values with lower significance are specially mentioned in the text.

For SEM examination leaf discs were fixed in 5% glutaraldehyde buffered by phosphate (pH 7.3) for 2 hours in darkness, dehydrated in a graded series of acetone and were then dried in a critical point drier using liquid CO_2. The discs were coated with gold in a vacuum evaporator, and examined in a Cambridge Stereoscan Microscope operating at 20 k.v.

RESULTS AND DISCUSSION

A. Irradiation in the lighting device without filters

In the first part of the investigation, different crop plants were exposed to low UV-B intensity (BUV 9.44 ≅ 32.6 O_3-reduction) to ascertain the sensitivity of the DNA of plants to UV-B radiation. Therefore, soon after germination, seedlings were exposed for 5 days (radish), 7 days (barley), 8 days (bean) or 10 days (maize) to continuous irradiation. The growth parameters -- fresh weight, dry weight and leaf area - as well as the chlorophyll, carotenoid, flavonoid protein and lipid content were quantitatively determined. In addition, the characteristic membrane glycerolipids were measured quantitatively and the fatty acid spectrum recorded.

Leaf parameters. All four plant species investigated reacted to increased UV-B exposure. Bean seedlings were especially sensitive, their fresh weight as well as their dry weight being approximately halved in comparison to the same number of control seedlings. The reduction of fresh weight and leaf area was almost as pronounced in radish seedlings. Both species are dicotyledons whose broad transversally situated leaves were especially exposed to UV-B irradiation.

The monocotyledonous plants, barley and especially maize, showed a much smaller reduction in leaf area and fresh weight. It is interesting to note that the decrease in the leaf area and the fresh weight was parallel, while barley, maize and radish seedlings exhibited a higher dry weight than the control plants. This fact contradicts the theory of a general inhibition of the synthesis pathways by UV-B. Rather it infers primary damage to the epidermal structure as a consequence of which a controlled gas and water exchange is not maintained. A deformed epidermis can be demonstrated in UV-B irradiated leaves by SEM (Fig. 6a-c). The epidermal cells from the upper and lower side of the leaves show a smooth surface which might be the consequence of a water deficiency induced by damaging UV-B effects on the guard cells.

Proteins, lipids and pigments. In all four plant species exposed to higher UV-B-density-fluxes a rise in

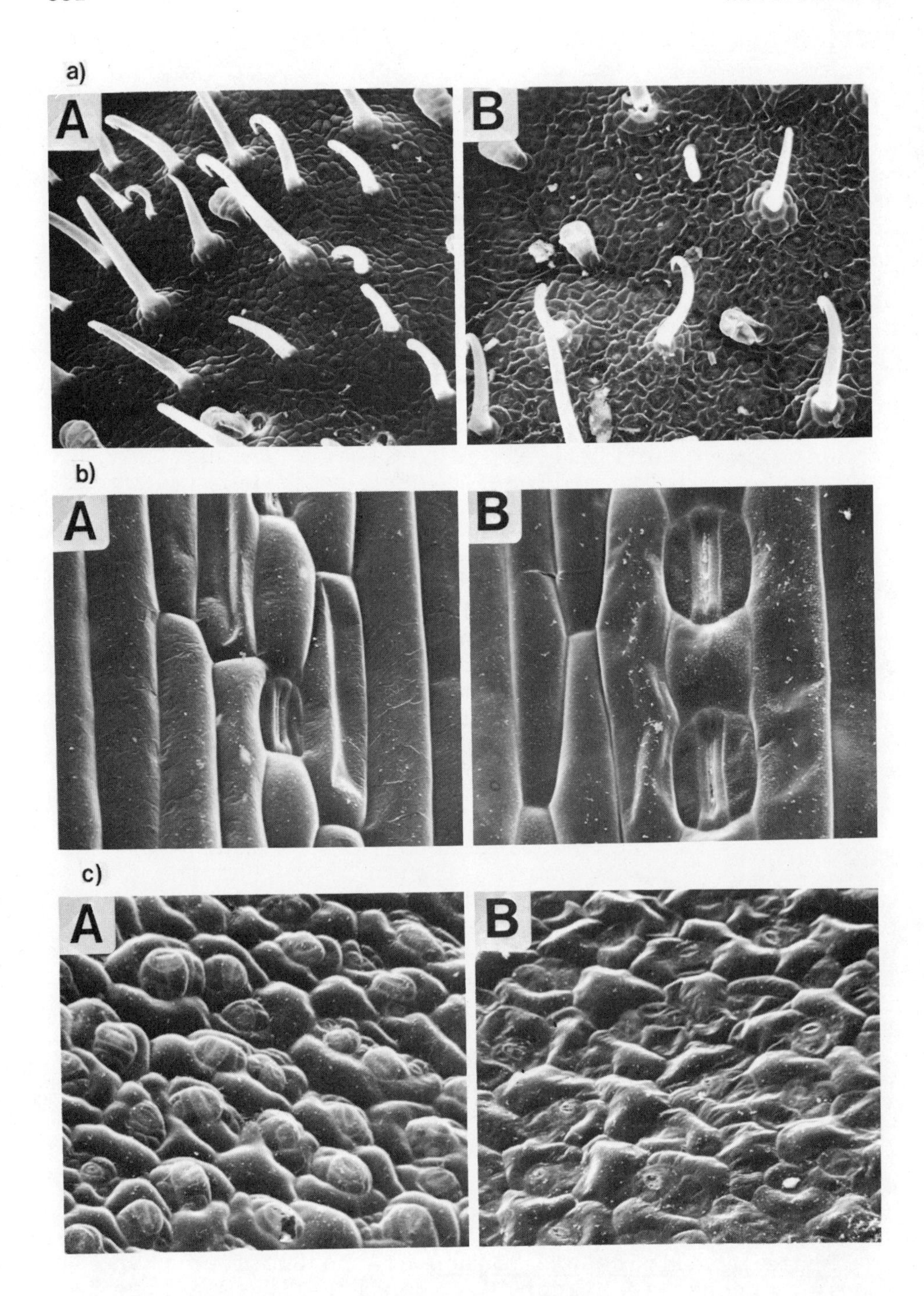
a)
A
B
b)
A
B
c)
A
B

the total protein was found compared to the control plants. The higher protein content only partly explains the increase in the dry weight of the plants. The extent to which the total biosynthesis of proteins is stimulated is presently unknown. It is possible that only the synthesis of the aromatic amino acids is enhanced. Their precursors also enter into the flavonoid biosynthesis. The determination of these amino acids is the subject of further investigation.

The concentration of the flavonoids rose similarly to that of the proteins and lipids in barley and radish seedlings; the bean is an exception (Fig. 7). Synthesis of these pigments, which are localised in the epidermis and mesophyll, is generally stimulated by blue light and ultraviolet radiation (Wellmann 1971).

A higher concentration of flavonoids is also always found in alpine plants which are exposed to increased ultraviolet irradiation (Caldwell 1968). It has been shown, therefore, that plants may protect themselves from increased UV irradiation by accumulating substances to absorb the radiation. In greenhouses under normal window glass, plants which have not adapted to solar UV-B radiation produce considerably less flavonoids and are therefore unprotected when suddenly exposed to UV irradiation present out-of-doors (Bogenrieder and Klein, 1978).

The pigments responsible for photosynthesis (chlorophylls and carotenoids) behaved contrary to the secondary plant substances. They diminished to a larger extent than the fresh weight or the leaf area. In this respect the maize seedlings showed the least sensitivity, and barley seedlings were clearly more sensitive than radish seedlings. The decline was especially drastic in the bean. The UV-B irradiated plants con-

←

Figure 6. SEM pictures of the epidermal surfaces of some leaves
a) Upper surface of bean leaves, A control, B UV-B irradiated all(230 x)
b) Leaf surface of barley, A control (700x), B UV-B irradiated (1200 x)
c) Upper surface of radish leaves, A control, B UV-B irradiated (all 550 x)

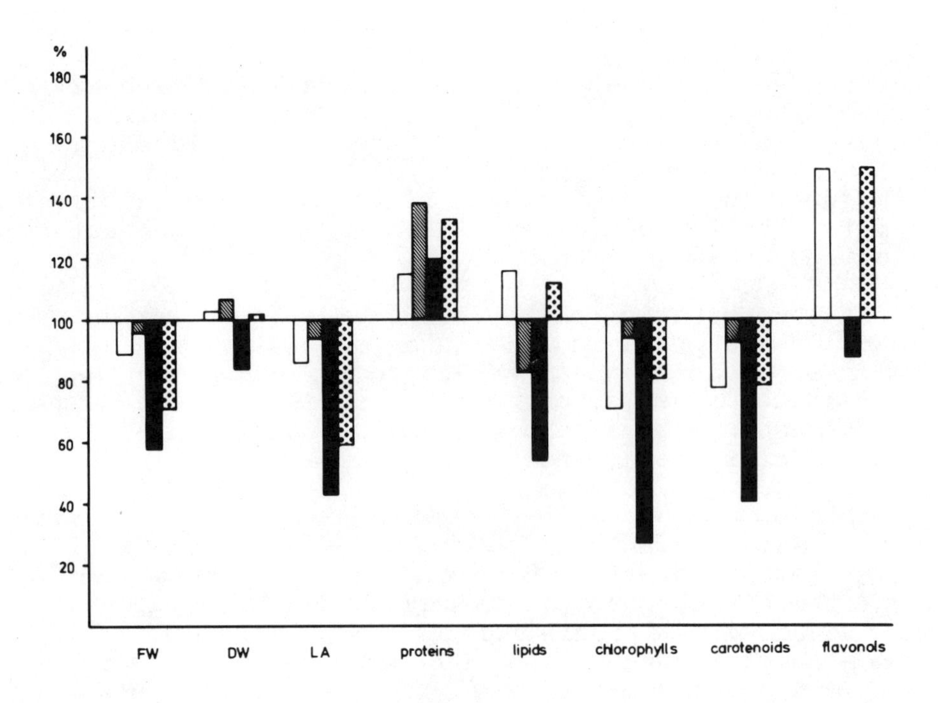

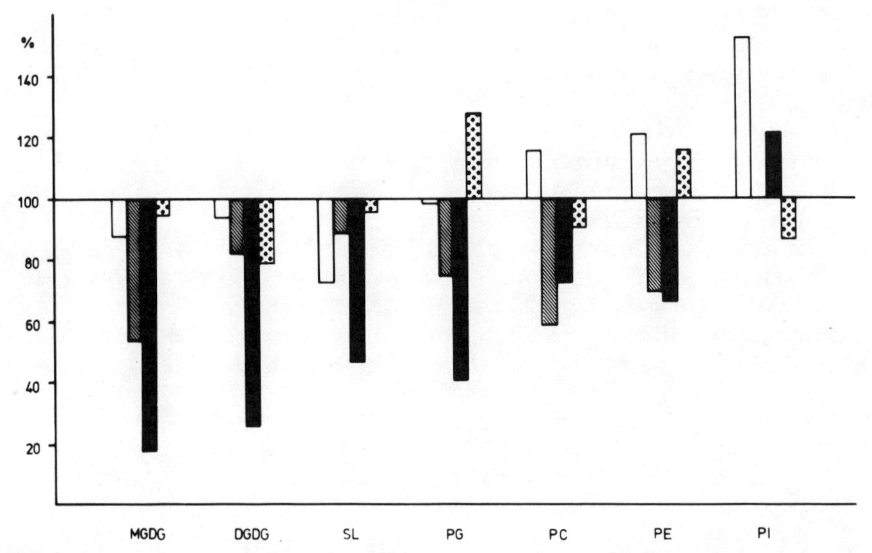

Figure 7. Influence of increased UV-B irradiation on the growth parameters and the content of some leaf compounds in barley, corn, bean and radish seedlings expressed as percentage of the controls (=100%) based on 100 plants.

tained only a quarter of the chlorophylls in oomparison to the control plants.

In general, the damaging influence affected the chlorophylls more than the carotenoids, although the latter represent a light-protection factor for the chlorophylls.

Due to the high sensitivity of the chlorophylls and carotenoids we assumed that the damaging effects of the UV-B irradiation could stem from its influence on the chloroplasts and therefore on photosynthesis. The effects on isolated chloroplasts will, therefore, be reported in later experimental data (Iwanzik and Tevini 1980, in preparation).

Glycerolipids. The glycerolipids analyzed here are diglycerides, which contain a sugar (glycolipid) or a phosphate residue and a further component (phospholipid) in an ester bond. These lipids represent important components of the membrane and fulfill mostly structural, but also functional purposes. Thus, the glycolipids MGDG and DGDG as well as the phospholipid PG are main components of the chloroplast. PE and the greater part of PC are situated, on the other hand, in the mitochondria and in plasma membranes. From the change in concentration of the individual lipids as a result of irradiation by UV-B radiation a differentiated influence on the various membrane systems could be deduced.

No clear, uniform picture was obtained from the 4 investigated plant species (Fig. 7). The concentration of the galactolipids MGDG and DGDG as well as the sulpholipid was diminished throughout. It is especially notable that the maize seedlings showed a reduction of almost 50% in their MGDG content, although the influence of the UV-B irradiation on the chlorophyll concentration was only very slight. Structural and functional lipids seem to be affected in a different way, at least in the case of maize. The chlorophyll and galactolipid formation were both inhibited to the same degree in bean seedlings.

Except in radish the MGDG to DGDG ratio was shifted in favor of DGDG. That could indicate the start of senescence. In yellowing tissue such a DGDG accumulation has been frequently observed (Tevini 1976).

The concentration of the phospholipids varied greatly. On the whole, the harmful effect of UV-B on them seems much less obvious. In barley and radish there was even a rise in their concentration (increase of PC, PE, PI, about 20-50%) in comparison to the control plants (Fig. 7). Whether this is linked to a rise in the number of mitochondria must still be investigated.

Fatty acids. The fatty acid spectrum of the total leaf extract showed a diminished linolenic acid concentration in all those plants exposed to high UV-B irradiation (Table 2). This decrease was particularly high in the very sensitive bean (14%), while in barley seedlings no significant decrease could be ascertained. The percentage of linoleic acid (bean, radish) and of lauric acid (bean, maize) increased at the cost of the linolenic acid.

Irradiation of barley with different irradiation intensities. In this part of the experiments the dependence of the UV-B effect on the intensity of the UV-B irradiation was ascertained for barley. Seedlings were exposed to 3 (high, medium and low, for BUV_{DNA} see Table 1) intensities.

The external appearance of the plants was very strongly affected by the irradiation intensity. Increasing damage can be very clearly correlated with rising UV-B irradiance. After exposure to the highest intensity (BUV 28.47 mW/m^2 simulating 63.40% O_3- reduction) the plants exhibited a stunted, crooked and irregular growth. Considerable signs of leaf bronzing appeared. The plants exposed to medium intensities (17.94 mW/m^2 simulating 51.5% O_3-reduction) were less effected, grew straighter and reached more than half the height of the control plants. The extent of the scorching was, however, only slightly less. After exposure to the lowest intensity (9.44 mW/m^2 simulating 32.6% O_3-reduction) only few differences in the appearance could be recognized between the UV-B irradiated plants and that of the control plants. They were slightly smaller and hardly showed any signs of scorching. Comparison of the growth parameters showed clearly that the fresh weight, dry weight and leaf area depend on the irradiation intensity (Fig. 8). The reduction in the fresh weight at high intensity was 60%, at medium intensity 27% and in low intensity only 11%. Similar values were obtained for the leaf area.

TABLE 2

Percentage change in the amount of certain fatty acids after UV-B irradiation in leaves of bean, barley, radish and corn. a = not significant

fatty acid	bean	barley	radish	corn
linolenic acid	-13.9	-2.0 a	-5.6	-6.9
linoleic acid	+4.7	-1.1 a	+4.7	+1.7 a
palmitic acid	+3.8	-2.0 a	-4.0	--
myristic acid	-1.5 a	+2.0 a	--	--
lauric acid	+10.0	--	--	+4.0

The reduction in the dry weight was much less severe. After exposure to high intensity it was lowered by 31.6%; irradiation of a low intensity results in a slight rise compared to the control plants. The negative effects of UV-B on the pigment content of the plants were especially pronounced (Fig. 8). As the intensity increased, the chlorophyll and carotenoid content was drastically reduced. Higher intensities lead to a reduction in the chlorophylls of about 70% and in the carotenoids of about 65% and even after irradiation with a low intensity the plants only contained 71 and 78% respectively compared to the control plants. Chlorophylls were reduced more than carotenoids. This makes a purely photooxidative destruction unlikely, because carotenoids acting as protective pigments for chlorophylls should be more susceptible to destruction. It is not clear whether biosynthetic capacity of chlorophylls and carotenoids decreases with rising intensity or whether the accumulation of the pigments into the membrane after completion of the synthesis is merely inhibited. This must be investigated.

No clear correlation can be drawn between the UV-B intensity and the increase or decrease in the content of proteins and lipids. A rise in the protein content could be seen at low and medium intensity, while the total lipid content only increased at the low intensity (Fig. 8).

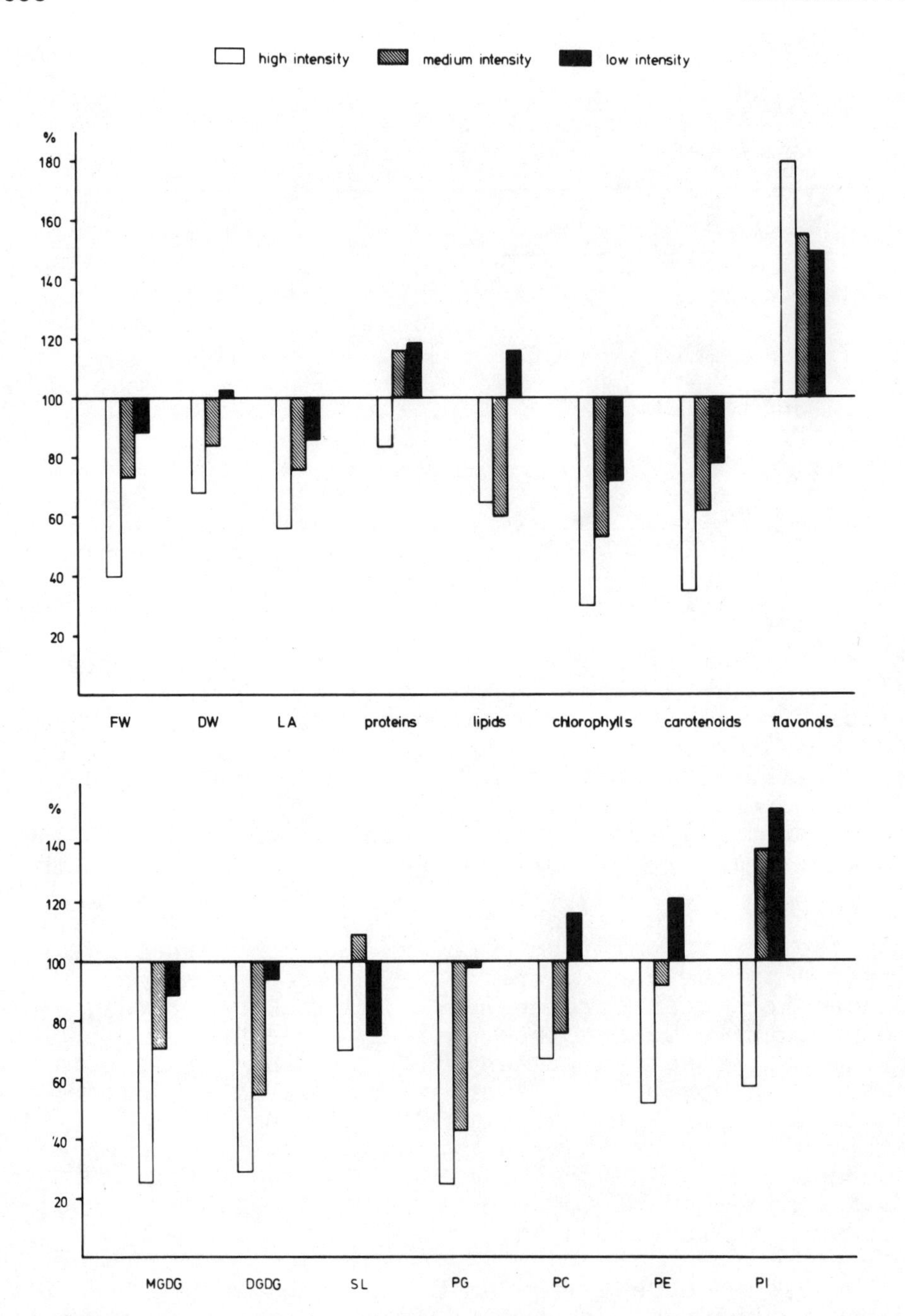

Figure 8. Influence of the different UV-B irradiation intensities on the growth parameters and the content of substances contained in barley seedlings based on 100 plants expressed as percentage of the controls (=100%).

Similar to the pigments, an especially steep drop in the content of the glycerolipids MGDG, DGDG and PG, which are chiefly localised in the chloroplasts, was found. It seems, therefore, natural to assume that the chloroplasts are an important target for the damaging effects of UV-B irradiation (Fig. 8).

Low intensity irradiation promoted the formation of the other phospholipids (PC is increased by 16%, PE by 21% and PI even by 52%, while high intensities, on the other hand, inhibited this accumulation (decrease of almost 50%) (Fig. 8).

It seems that retardation or inhibition of biosynthetic pathways by high UV-B irradiances is not a general rule. It is not surprising that the flavonoid content increased as the irradiation intensity rose. Plants irradiated with highest UV-B intensity contained twice as much flavonoids as control plants (Fig. 8).

B. Irradiation in the lighting device with filters

a) Effect on growth and composition of barley seedlings. When using the lighting device without filters it was impossible to prevent small amounts of UV-C reaching the plants. Therefore, we used different UV-B transmitting filters which cut off at 285 nm (Fig. 2). The biological effective irradiance (BUV_{DNA}) and the simulated O_3-reduction rates were:

glass filter (sol. glass)	5.91 mW/m²	
Schott filter (WG 305 3mm)	8.94 mW/m² ≅	30.1% O_3 reduction
Schott filter (WG 305 2 mm)	12.95 mW/m² ≅	41.1% O_3 reduction
without filter	28.47 mW/m² ≅	63.4% O_3 reduction

In these experiments we used solarium glass as a control because it has a better transmission in the UV-B region than window glass. Thus we attained nearly natural light conditions. The extent of UV-B damage appeared less pronounced than in the earlier experiments with window glass as a control.

High decreases in growth parameters and pigments could only be seen when the seedlings were irradiated without a filter. Fresh weight and the chlorophylls were lowered by about 40% compared to the control plants (glass filter). The chlorophyll a content dropped much more than the chlorophyll b content. Dry weight, leaf area and the carotenoids were decreased to about 80% (Fig. 9, 10). The protein content was only slightly influenced (Fig. 11), whereas the flavonoids were drastically increased to nearly 200% (Fig. 11).

The damage to the plants was much less when small amounts of UV-C, emitted by the UV tubes, were absorbed by filters (Schott WG 305 3 mm; 2 mm). Under these conditions the decrease in the growth parameters and pigments reached only 10 - 20% and there was hardly any difference between the two filters used (simulating 30 to 41% O_3-reduction respectively). Leaf area and dry weight were slightly increased when the Schott 2 mm filter was used, although these plants showed bronzing effects, which could not be seen with the 3 mm filter.

b) Development of the photosynthetic system. During the greening process of etiolated leaves a rapid formation of photosynthetic pigments occurs and the biosynthetic steps involved might also be inhibited by UV-B radiation. Therefore, etiolated, 7-day-old barley seedlings were irradiated in the lighting device with and without filters. Every hour for the first 8 hours and also after 24 hours the pigments were separated and quantitatively analysed by means of high pressure liquid chromatography.

In order to preserve clarity, only pigments which were most affected after 2, 5, 8 and 24 h irradiation are illustrated in Fig. 12. Up to the 8th hour the rise in content at all 4 irradiation conditions was almost similar, but lower at enhanced UV-B levels.

Higher amounts of chlorophyll a and b as well as more β-carotenes were synthesized in barley seedlings which were protected from UV-B radiation (glass filter) and least in completely unprotected ones (without filters). A distinct dependence on the intensity at moderate irradiation strengths only became apparent after 24 hours. The antheraxanthin content, conversely, fell more and more steeply under the influence of weak UV-B irradiation as the irradiation duration increased. This was especially notable after 24 hours without UV-B irradiation.

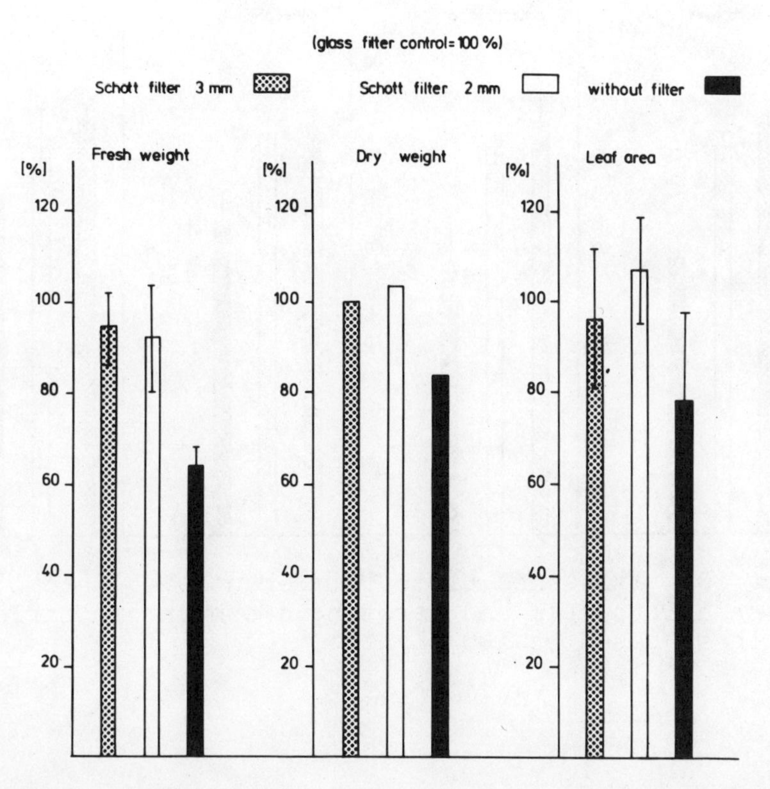

Figure 9. Influence of the different UV-B intensities on the growth parameters (fresh weight, dry weight, leaf area) in barley seedlings, based on 100 plants.

Due to the kinetics of the pigment formation it is not clear whether enzyme biosynthesis or the incorporation into the membrane is inhibited. The theory of intervention at the enzyme level is supported by the behavior of antheraxanthin, whose drop in concentration in light of high intensity and especially after short exposure was strongly inhibited. The accumulation of violaxanthin, which together with antheraxanthin and zeaxanthin are interconvertible in the so-called

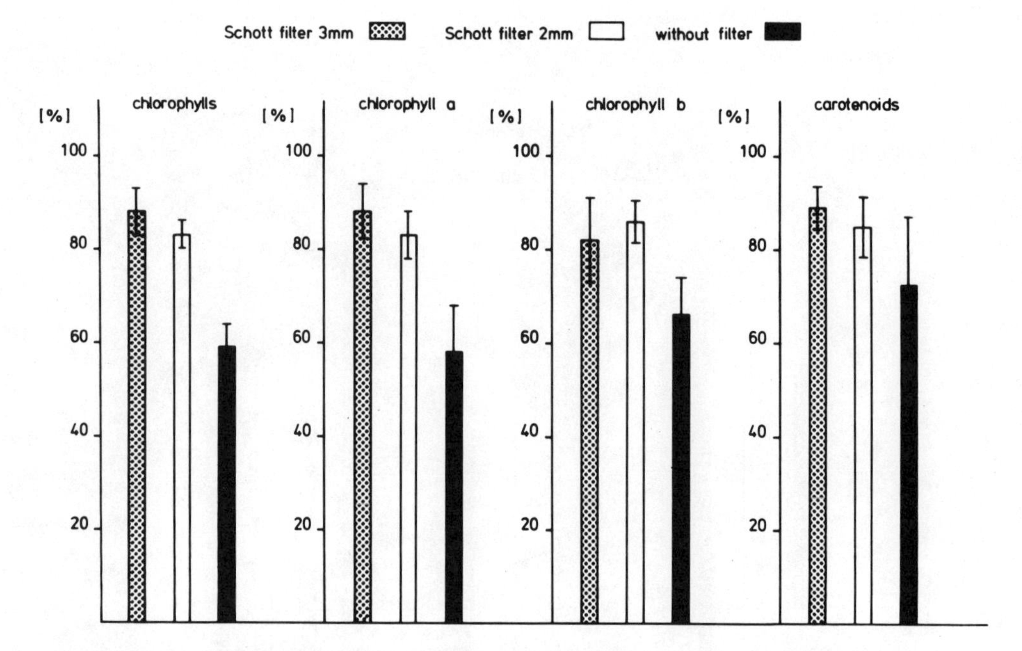

Figure 10. Influence of the different UV-B intensities on chlorophyll and carotenoid content in barley seedlings, based on 100 plants.

xanthophyll cycle (Siefermann-Harms 1977), proceeded accordingly in the opposite direction. The highest concentration of violaxanthin was not found in the plants protected by the glass filter after 24 hours, however, but in those exposed to really high intensity irradiation of 12.9 mW/m^2 (Table 3). No explanation for this can be put forward at present.

C. Irradiation in the growth chamber

a) Influence on the growth parameters and concentration of substances contained in the plants. The conditions in the growth chamber were set to reproduce normal daily rhythms of light and dark as well as temperature. The irradiation level, set at

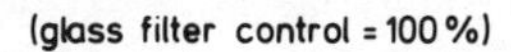

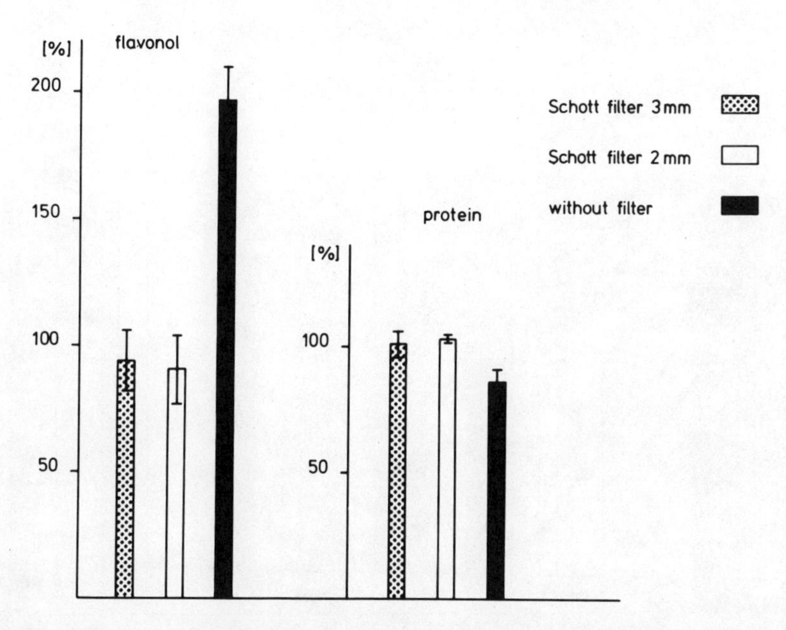

Figure 11. Influence of the different UV-B intensities on flavonol and protein content in barley seedlings, based on 100 plants.

TABLE 3

Content of antheraxanthin and violoxanthin in etiolated (0 h) and greening (24 h) barley leaves. Values in μg/1g FW

pigment	0 h	24 h			
		glass filter	Schott filter (3mm)	Schott filter (2mm)	without filter
antheraxanthin	26.5	5.7	9.3	10.3	13.6
violaxanthin	24.5	40.7	46.9	48.6	42.7

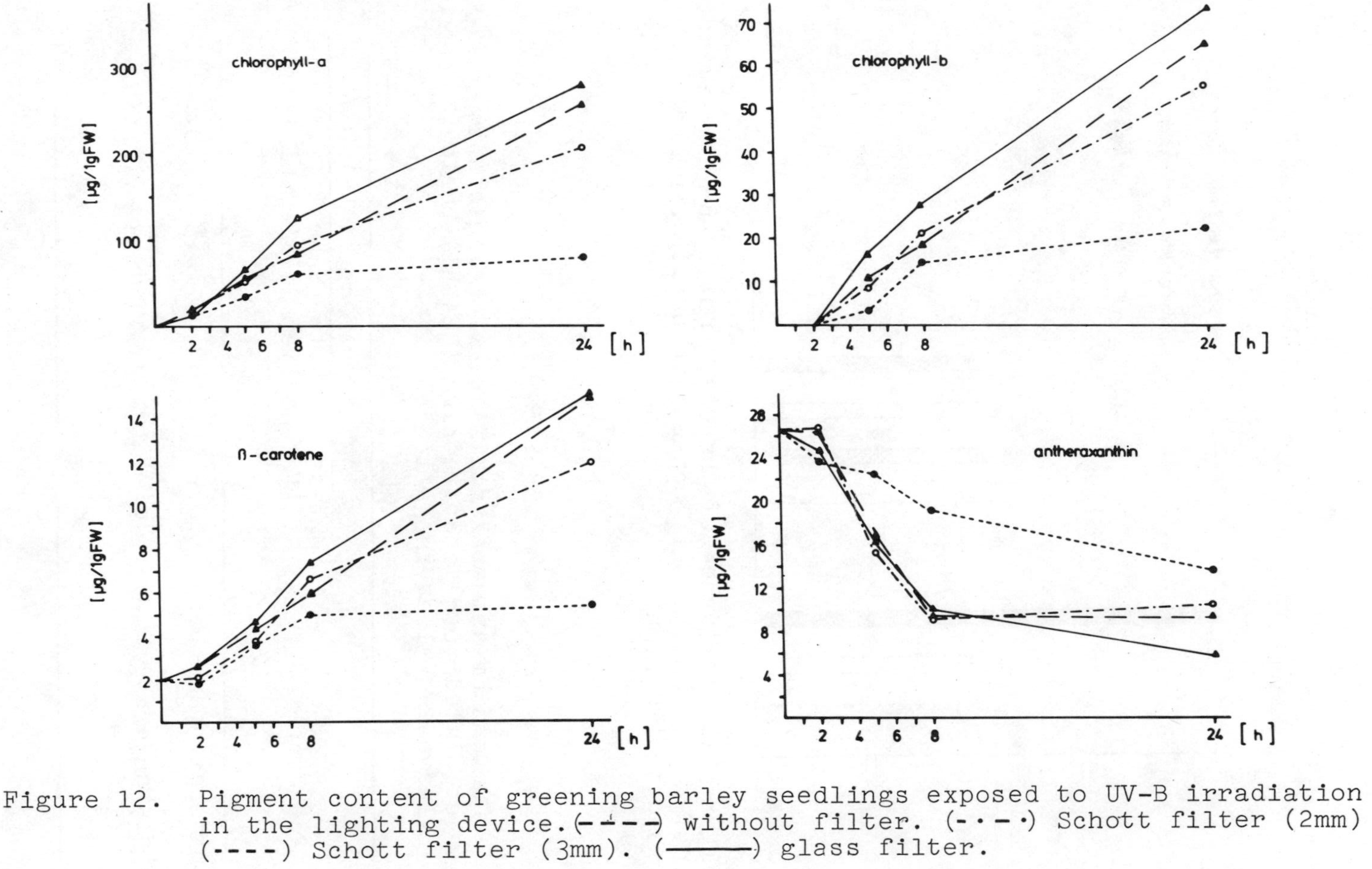

Figure 12. Pigment content of greening barley seedlings exposed to UV-B irradiation in the lighting device. (-•-•-) without filter. (-·-·) Schott filter (2mm) (----) Schott filter (3mm). (——) glass filter.

1.05 W/m² (≅BUV = 12.5 mW/m²) corresponds to a reduction of the ozone layer by about 40% (Table 1). Under these conditions, we obtained results for the four plant species which were largely similar to those in the lighting device. It was surprising, however, that the order of the plants with regard to their sensitivity to UV-B was not the same as in the lighting device. Radish seedlings were the most strongly affected under the conditions in the growth chamber. Their content in chlorophylls and carotenoids was reduced by about 30%. In addition, the protein content and the leaf area fell by about 25% compared to the control plants. (Fig. 13) The behavior of the barley seedlings was comparable to that in the lighting device with the exception of the diminished protein and lipid content in the growth chamber. Corn seedlings showed nearly the same resistance to the UV-B stress, although the fresh weight and the leaf area were reduced by about 25%. Bean seedlings were less influenced in the growth chamber than in the lighting device. Although the fresh weight and the dry weight were hardly changed under enhanced UV-B conditions, the leaf area, the chlorophylls and the carotenoids were diminished up to 30% when compared to the control plants. This behavior seems hardly explicable at present. The damaging or stimulating effects of UV-B irradiation were not so pronounced in the growth chamber as in the lighting device. This mitigation of effects could be due to the small amounts of UV-C radiation which could not be excluded from the lighting device without filters and which because of their high biological effectiveness, amplified the UV-B effects even more. Apart from that, the plants in the lighting device were exposed continuously whereas those in the growth chamber had a change to recover in the dark. However, the exceptionally high proportion of visible light probably plays an important role in repair mechanisms. The range between 310 and 520 nm in the spectrum has been shown to be especially effective in this regard (Jagger and Stafford 1962). This range of the irradiation spectrum was strongly represented in the growth chamber and only slightly in the lighting device (Figs. 1 and 4).

b) Pigment content during the greening process in barley seedlings. The rise in the content of chlorophylls and β-carotene in the course of the greening process is illustrated in Figure 14. The concentrations of these pigments were approximately 25% lower in UV-B-treated plants than in the control plants after

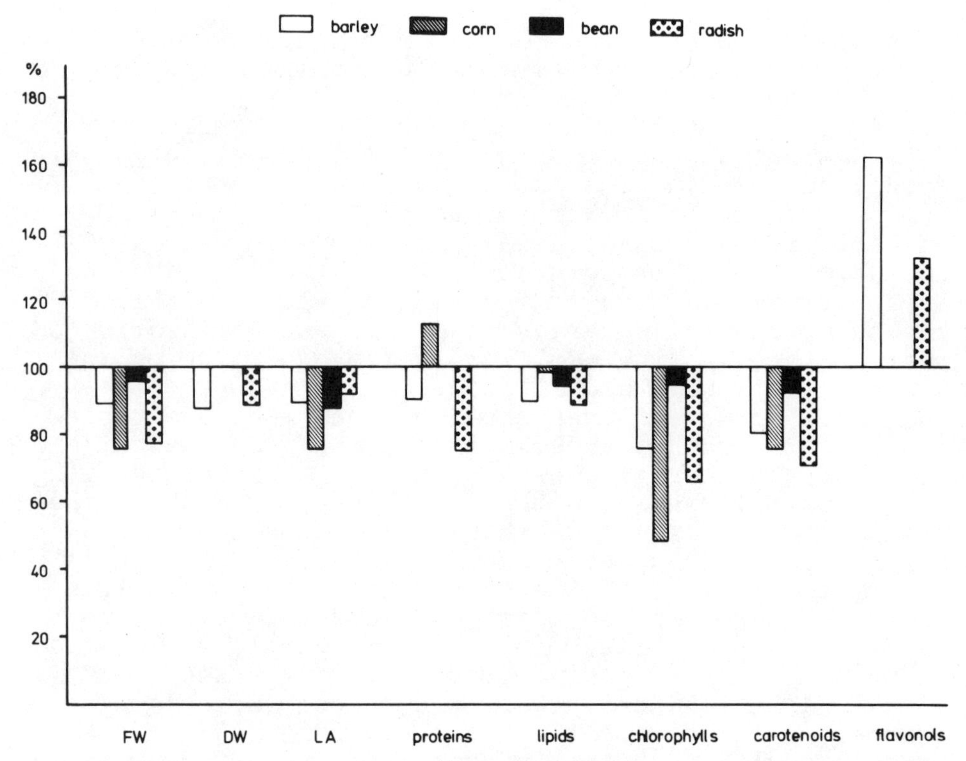

Figure 13. The influence of increased UV-B irradiation on the growth parameters and the concentration of substances contained in barley, maize, bean and radish seedlings based on 100 plants (control = 100%) grown in the growth chamber.

24 h. The effect was, therefore, much less pronounced than in the lighting device. The xanthophylls hardly differed in their content (after UV-B exposure) compared to that of the control plants.

IV Summarized discussion

UV-B irradiation, depending on its intensity, causes lasting damage or change in the growth, composition and function of plants. The plant species investigated, in this case bean, barley, corn and radish plants, are very susceptible to the effects of continuous UV-B irradiation, corn on the other hand, proved to be, at least apparently, resistant. In a natural light-dark cycle in the growth chamber, the damage is largely

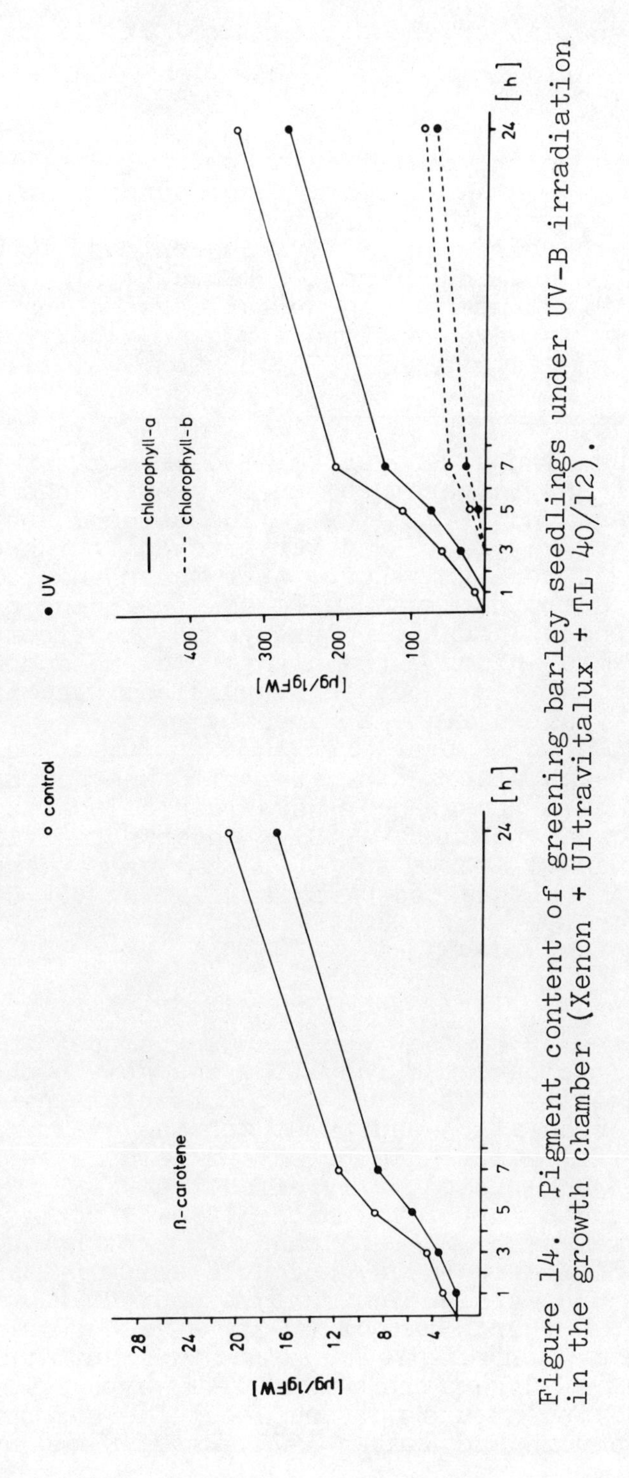

Figure 14. Pigment content of greening barley seedlings under UV-B irradiation in the growth chamber (Xenon + Ultravitalux + TL 40/12).

mitigated despite an almost identical irradiation intensity.

With respect to the protein content, a stimulation under low UV-B intensity is in accordance with results of field experiments (Esser 1980) and results from alpine plants (Pirschle 1941). In contrast to this, however, a reduction of the protein content was found under high UV-B intensities in barley seedlings. Similar results were reported for some other plants (Andersen and Kasperbauer 1973, Basiouny et al. 1978).

When the plants were irradiated in a natural light-dark rhythm and temperature range (growth chamber), the otherwise seriously damaging influence on the appearance of the sensitive plants was very much diminished. This appears to support the theory that the plants are more capable of adapting in the darkness or in the course of the light phase - possibly through the development of protective mechanisms. The higher flavonoid content measured could mean a better adaptation. The high proportion of visible light apparently gives the plant a better chance to recover from the harmful effects. In soybeans it was also shown, that with increasing levels of PAR the UV-B irradiances were less effective in reducing the growth (Teramura 1980). Under normal field conditions grave damage to plants can only be seen when they grow under enhanced UV-B irradiation for several weeks (Esser 1979 and 1980).

A comparison of the results with respect to the ozone reduction and the resulting changes in the irradiation intensity is quite difficult because of the different irradiation equipment and the evaluation of the plant efficiency. We based our calculation of the increase of the biologically effective UV-B irradiation after the reduction of the ozone layer thickness on Green's model using DNA and plant "weighting", factors according to Setlow (1974) and Caldwell (1971). Both weighting functions are commonly accepted in research literature. Bener's values (1968, 1972) are about 20% too low in comparison to those of other authors (Dave and Halpern 1976, Green et al. 1974), especially because of different assumptions about the aerosol content (Schulze and Kasten 1975, Braslau and Dave 1973).

SUMMARY

The purpose of the present study was to evaluate the major effects of enhanced UV-B irradiation on growth, composition and function of higher plants. We therefore irradiated plants with different UV-B intensities either continuously under a lighting device with and without filters or in a natural light dark rhythm in a growth chamber. The calculation of the biological effectiveness of the light sources and intensities used was based on DNA and plant weighting functions according to Setlow (1974) and Caldwell (1971). For the estimation of the corresponding ozone reduction rates we used Green's model, which simulates the amount of global irradiation dependent on the thickness of the ozone layer. The resulting O_3-reduction rates for our light sources range from 30% to 63%.

A. Lighting device without filters

The four plants cultivated in the lighting device, bean, barley, maize and radish, showed different responses to increased UV-B exposure (33% ozone reduction). Bean seedlings suffered the most harmful effects followed by radish and barley, whereas maize was only slightly influenced.

Deficits were to be found in all plant species, especially in the fresh weight (up to 45%), the leaf area (up to 60%), total lipid content of maize and bean seedlings (up to 50%), in lipids (chlorophylls up to 75%, carotenoids up to 60%). In addition, the glycerolipids, which are situated primarily in the chloroplasts (MGDG, DGDG, PG), were diminished by 80%. Bean seedlings suffered the most severe quantitative losses. The dry weight was somewhat lowered only in the case of bean seedlings, in all the other cases, it even increased slightly.

All the irradiated plants showed a rise in their protein content of up to 40% compared to the control plants. Barley and radish seedlings demonstrated a slightly higher total lipid content. The content of flavonoids increased in barley and radish seedlings by about 50% through UV-B irradiation.

The analysis of fatty acids showed a general decrease in the proportion of linolenic acid while the shorter and more saturated fatty acids increased.

The harmful effects were more extensive in barley seedlings with increasing irradiation intensity. High irradiances (≈ 63% or 52% ozone reduction respectively) sharply curtailed the growth of the seedlings with a reduction of fresh weight of 60% or 25% respectively, and of the leaf area of 45% or 67% respectively. Furthermore, scorching appeared regularly in the form of bronze leaf discoloration.

The fresh weights fell to the same degree as the chlorophyll and carotenoids. The protein content showed no significant dependence on the UV-B intensities. In contrast the flavonoid content of barley seedlings was raised parallel to increasing UV-B irradiance and reached 200% of the control, when using highest UV-B intensity. The amount of glycerolipds was reduced at high UV-B intensity. Low UV-B intensity, however, raised the content of the phospholipids PC, PE and PI up to 40%.

B. Lighting device with filters

Barley seedlings showed severe damage only after irradiation without filter protection. When UV-B-transmitting Schott filters (3 mm, 2 mm) were used growth parameters and pigments were only reduced by 10% to 20%.

During the development of the photosynthetic system in etiolated barley plants, high irradiation intensities prevented the accumulation of chlorophylls and β-carotene. After 24 hours of UV-B exposure without filter only 30% of the concentration in the control plants was attained. Lowered irradiation levels (41% - 30% O_3-reduction) did not cause such a high reduction in the above-mentioned pigment concentration (10 and 25% respectively). Antheraxanthin concentrations dropped off more slowly at high irradiation intensities.

C. Growth chamber

The harmful effects were not as extensive under the conditions in the growth chamber (alternating light and dark, low irradiation intensity) as in the lighting device.

The pigments of all plants were most severely reduced. The protein content rose only in corn; in barley and radish seedlings, in contrast, it was lowered by about 10 or 25% respectively. The flavonoid content rose in barley and radish plants to high amounts.

During the greening process of etiolated barley seedlings, a retarded formation of chlorophyll and β-carotenes was observed.

VI Abbreviations

MGDG = Monogalactosyldiglyceride

DGDG = Digalactosyldiglyceride

SL = Sulfoquinovosyldiglyceride

PG = Phosphatidylglycerol

PC = Phosphatidylcholine

PE = Phosphatidylethanolamine

PI = Phosphatidylinositol

LA = Leaf area

FW = Fresh weight

DW = Dry weight

BUV = Biologically effective UV-B radiation 290-315 nm)

ACKNOWLEDGEMENTS

We are very grateful to Martyn Caldwell for making his computer program for the calculation of ozone reduction rates available to us. We gratefully acknowledge the help of Heidi Hartmann, Melitta Heldt, Ernst Heene for technical assistance and of Liz Mole for translation and Ursula Widdecke for preparation of the manuscript.

REFERENCES

Andersen, R. and M. J. Kasperbauer. 1973. Chemical composition of tobacco leaves altered by near-ultraviolet and intensity of visible light. Plant Physiol. 51: 723-726.

Arnon, D. I. 1949. Copper enzymes in isolated chloroplasts, polyphenol-oxidase in Beta vulgaris. Plant Physiol. 24: 1-15.

Basiouny, F. M., T. K. Van, and R. H. Biggs. 1978. Some morphological and biochemical characteristics of C_3 and C_4 plants irradiated with UV-B. Physiol. Plant 42: 29.

Bener, P. 1968. Spectral intensity of natural ultraviolet radiation and its dependence on various parameters. In: The Biologic Effects of Ultraviolet Radiation, F. Urbach [ed.] Pergamon Press Oxford. 351-358.

Bener, P. 1972. Approximate values intensity of natural ultraviolet radiation for different amounts of atmospheric ozone. European Research Office, US Army, London. Contrakt Number DAJA 37-68-C-1017. 4-59.

Bogenrieder, A. and R. Klein. 1978. Die Abhängigkeit der UV-empfindlichkeit von der lichtqualität bei der Aufzucht (Lactuca sativa L.) Angewandte Botanik 52: 283-293.

Brandle, J. R., W. F. Campbell, W. B. Sisson, and M. M. Caldwell. 1977. Net photosynthesis, electron transport capacity and ultrastructure of Pisum sativum L. exposed to ultraviolet-B radiation. Plant Physiol. 60: 165-168.

Braslau, N. and J. V. Dave. 1973. Effect of aerosols on the transfer of solar energy through realistic model atmospheres. Part III: Ground level fluxes in the biologically active bands, 0.2850 - 0.3700 microns. IBM Research Report RC4308, IBM, Thomas J. Watson Research Center, Yorktown Heights, New York.

Caldwell, M. M. 1968. Solar ultraviolet radiation as an ecological factor for alpine plants. Ecological Monographs, Durham N. C., 38: 243-268.

Caldwell, M. M. 1971. Solar UV irradiation and the growth and development of higher plants. Photophysiology VI, [Giese, ed.] 131-177.

Dave, J. V. and P. Halpern. 1977. Effect of changes in ozone amount on the ultraviolet radiation received at sea level of a model atmosphere. In: Radiation in the Atmosphere, H.-J. Bolle [ed.] Science Press. 611.

Esser, G. 1979. Einfluß einer nach Schadstoffimission vermehrten Einstrahling von UV-B Licht auf den Ertrag von Kulturpflanzen (FKW 22), 1. Versuchsjahr, Bericht Batelle-Institute e.V. Frankfurt, BF-R-63. 575-1.

Esser, G. 1980. Einfluß einer nach Schadstoffimission vermehrten Einstrahlung von UV-B-Licht auf Kulturpflanzen, 2. Versuchsjahr. Bericht Batelle Institut e.V. Frankfurt, BF-R-63.984-I

Fox, F. M. and M. M. Caldwell. 1978. Competitive interaction in plant populations exposed to supplementary ultraviolet-β radiation. Oecologia (Berl.) 36: 173-190.
Green, A.E.S., K. R. Cross, and L. A. Smith. 1980. Improved analytical characterization of ultraviolet skylight. Photochem. Photobiol. 31: 59-65.
Green, A.E.S., T. Sawada, and E. P. Shettle. 1974. The middle ultraviolet reaching the ground. Photochem. Photobiol. 19: 251-259.
Jagger, J. and R. S. Stafford. 1962. Biological and physical ranges of photoprotection from ultraviolet damage in micro-organisms. Photochem. Photobiol. 1: 245-257.
Klein, R. M., P. C. Edsall, and A. C. Gentile. 1965. Effects of near ultraviolet and green radiations on plant growth . Plant Physiol. 40: 903-906.
Klein, R. M. 1978. Plant and near ultraviolet radiation. Bot. Rev. 44: 1-127.
Krizek, D. T. 1975. Influence of ultraviolet radiation on germination and early seedling growth. Physiol. Plant 34: 182-186.
Lindoo, S. J. and M. M. Caldwell. 1978. UV-B radiation induced inhibition of leaf expansion and promotion of anthocyanin production. Plant Physiol. 61: 278.
Lowry, O. H., N. T. Rosebrough, A. L. Farr and R. J. Randall. 1951. Protein measurement with the Folin Phenol Reagent. J. Biol. Chem. 193: 265-275.
Metcalfe, L. D., A. A. Schmitz, and J. R. Pelka. 1966. Rapid preparation of fatty acid methyl esters for gaschromatography analysis. Analyt. Chem. 38: 514-515.
NASA Reference Publication 1010. Aug. 1977. Chlorofluoromethanes and the Stratosphere. R. Hudson [ed.] NASA Inform. Office.
NAS, Committee on Impacts on Stratospheric Change et al. 1979. In: Protection against Depletion of Stratospheric ozone by chlorofluorocarbons. p. 62, National Academy of Sciences, Washington, D. C.
Pirschle, K. 1941. Weiter Beobachtungen über den Einfluß von langwelliger und mittelwelliger UV-Strahlung auf höhere Pflanzen, besonders polyploide und hochalpine Formen (Stellaria, Epilobium, Arenaria, Silene). Biol. Zentralblatt 61: 452-473.
Robberecht, R. and M. M. Caldwell. 1978. Leaf epidermal transmittance of ultraviolet radiation and its implications for plant sensitivity to ultraviolet radiation induced injury. Oecologia 32: 277-287.

Schulze, R. and F. Kasten. 1975. Der Einfluß der Ozonschicht der Atmosphäre auf die biologisch wirksame Ultraviolettstrahlung an der Erdoberfläche. Strahlentherapie 150: 219-226.
Semeniuk, P. and R. N. Stewart. 1979. Seasonal effect of UV-B-Radiation on poinsettia cultivars. J. Amer. Soc. Hort. Sci. 104: 246-248.
Siffermann-Harms, D. 1977. The xanthophyll cycle in higher plants. In: Lipids and Lipid Polymers in Higher Plants, M. Tevini and H. K. Lichtenthaler [eds.] Springer New York-Heidelberg. 218-229.
Sisson, W. B. and M. M. Caldwell. 1976. Photosynthesis, dark respiration, and growth of Rumex patientia L. exposed to ultraviolet irradiance (288 to 315) nanometers simulating a reduced atmospheric ozone column. Plant Physiol. 58: 563-568.
Sisson, W. B. and M. M. Caldwell. 1977. Atmospheric ozone depletion: reduction of photosynthesis and growth of a sensitive higher plant exposed to enhanced UV-B radiation. J. Exp. Bot. 28: 691-705.
Teramura, A. H. 1980. Effects of ultraviolet-B irradiances on soybean. I. Importance of photosynthetically active radiation in evaluating ultraviolet-B irradiance effects on soybean and wheat growth. Physiol. Plant 48: 333-339.
Teramura, A. H. 1980. Effects of ultraviolet-B irradiances on soybean. II. Interaction between ultraviolet-B and photosynthetically active radiation on net photosynthesis, dark respiration, and transpiration. Plant Physiol. 65: 483-488.
Teramura, A. H., S. V. Kossuth, and R. H. Biggs. 1978. Effects of UV-B-enhancement under contrasting PAR growth regimes on NCE, dark respiration, and growth in soybeans. Plant Physiol. 61: 74.
Tevini, M. 1976. Veränderungen der Glyko- und Phospholipidgehalte während der Blattvergilbung. Planta (Berl.) 128: 167-171.
Tevini, M., W. Iwanzik and D. Steinmoller. 1979. Die Wirkung von UV-B auf den Lipid-metabolismus von Nutzpflanzen. Kurzfassung. Arbeitstagung Pflanzliche Lipide, 5.-6.10, Köln, Botanisches Institut.
Tevini, M. and W. Iwanzik. 1980. The effects of UV-B-irradiation on Plants. Abstract 2nd Congress F.E.S.P.P. 27th July to 1st August. Santiago de Compostella, Spain.
Van, T. K. and L. A. Garrard. 1976. Effect of UV-B radiation on net photosynthesis of some C_3 and C_4 plants. Soil and Crop Science Society of Florida Proceedings 35. 1-3.

Vu, C. V., L. H. Allen, and L. A. Garrard. 1978. Effects of supplemental ultraviolet radiation (UV-B) on growth of some agronomic crop plants. Soil and Crop Science Society of Florida, Proceedings, Vol. 38, 59-63.

Wellmann, E. 1971. Phytochrome- Mediated Falvone Gylcoside synthesis in cell suspension cultures of Petroselinum Hortense after preirradiation with ultraviolet light. _Planta_ _101_: 283-286.

Ziegler, R. and K. Egle. 1965. Zur quantitativen Analyse der Chloroplastenpigmente. I. Kritische Überprüfung der spektralphotometrischen Chlorophyll-bestimmung. _Beitr_. _Biol_. _Pflanzen_ _41_: 11-63.

PRELIMINARY RESULTS REGARDING THE SPECTRAL EFFICIENCY OF UV ON THE DEPRESSION OF PHOTOSYNTHESIS IN HIGHER PLANTS

Arno Bogenrieder and Richard Klein

University of Freiburg, Biological Institute II
Schanzlestr. 1, D 7800 Freiburg, FRG

Compared to marine plants, terrestrial higher plants are even more exposed to UV-B radiation. This holds particularly true for plants growing at higher altitudes under conditions of enhanced UV intensities.

As already pointed out by numerous authors, UV irradiation lowers the rate of photosynthesis or, when applied too intensively, causes other damage to the irradiated plants (Bell and Merinova 1961; Halldal 1964; Jones and Kok 1966; Sisson and Caldwell 1976, 1977; Bogenrieder and Klein 1977, 1978). Despite the numerous suppositions about the spectral efficiency of UV-B radiation -- e.g. in the works compiled by Caldwell (1971) -- little is known about the actual mechanics of UV-B inhibition of photosynthesis in higher plants.

The methodical problems of such investigations are quite obvious:

- Net-photosynthesis of whole plants should be measured with high accuracy.
- A source of intense visible light should be available.
- A UV-lamp emitting intense UV radiation in a range of 200 to 400 nm is required.
- Selective filtering UV radiation is needed.
- The number of quanta of the effective UV radiation should be determined.

Nearly one year ago we started to build up an arrangement, which allows the irradiation of plants with monochromatic UV-B light during photosynthesis measurements. Net photosynthesis of whole plants (in hydroculture) is determined by measuring gas exchange. Our test plants are Rumex alpinus L. and Lactuca sativa L., both of which have been tested in earlier experiments (Bogenrieder and Klein 1977, 1978). These plants are enclosed into two fully climatised cuvettes of a gas exchange apparatus. One of the cuvettes is made of ordinary UV-absorbing Plexiglass (lower transmission limit at about 380 nm) the other of Quartz, which permits UV transmission. One test cuvette is irradiated with visible light only (Osram HQIL 400 W, wavelengths shorter than 380 nm filtered out) whereas the Quartz cuvette is subjected to additional UV treatment of selected wavelengths. The UV radiation is emitted by a Xenon arc (Osram XBO 450 W) and filtered by UV interference filters. UV intensities are measured as relative units by means of a Zeiss spectrophotometer with a photomultiplier and can therefore only be compared approximately. Later determinations will make it possible to transform these relative values into absolute energy values.

If plants of Rumex or Lactuca (exposed to UV for six hours and observed continuously during irradiation) show any decrease in net photosynthesis compared with UV-free controls, we consider this to be a UV effect.

Although we have made good progress in the course of our experimental work, we cannot yet offer a more detailed action spectrum. From what we have so far found out, though, the following conclusions can be drawn:

1) At wavelengths longer than 309 nm our test plants did not show any effect -- that is, within the range of intensities supplied by our Xenon arc. However, the highest intensities we could apply are presumably lower than those effective under outdoor conditions.

2) Radiation of shorter wavelengths results in a reduction of net photosynthesis. Among the wavelengths tested the 260 nm radiation is the most effective (see Figure 1).

3) The increase of irradiation intensities is accompanied by a decrease of the rates of net

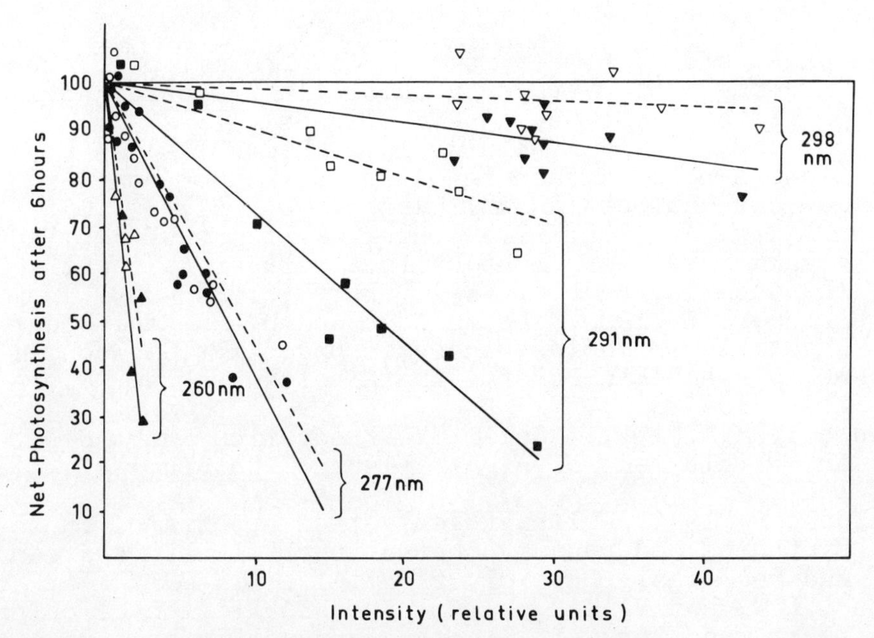

Figure 1. Net photosynthesis after 6 hours of irradiation with UV compared with non-irradiated controls.
Dashed lines: Rumex alpinus L.
Continuous lines: Lactuca sativa L.

photosynthesis. For both plants -- Rumex alpinus and Lactuca sativa -- preliminary calculations suggest a linear relation.

4) Regarding the sensitivity, there is a remarkable difference between the two plant species tested. When exposed to the same UV intensities as Lactuca, Rumex shows a stronger decrease in net photosynthesis.

This is especially true of radiation between 280 and 300 nm, that is, UV-B radiation. In this spectral range Rumex seems to be more sensitive, as was seen in earlier investigations (Bogenrieder and Klein 1977, 1978, 1980).

ACKNOWLEDGEMENTS

This research is financially supported by the Deutsche Forschungsgemeinschaft.

REFERENCES

Bell, L.N. and G. L. Merinova. 1961. The effect of dose wavelength of UV rays on Chlorella photosynthesis. Biofisika 6. No. 2, 159-164 (transl.).

Bogenrieder, A. and R. Klein. 1977. Die Rolle des UV-Lichtes beim sogenannten Auspflanzschock von Gewächschaussetzlingen. Angew. Bot. 51: 99-107.

Bogenrieder, A. 1978. Die Abhängigkeit der UV-Empfindlichkeit von der Lichtqualität bei der Aufzucht (Lactuca sativa L.). Angew. Bot. 52: 283-293.

Bogenrieder, A. 1980. Beeinflußt der UV-Ausschluß das Ergebnis von Photosynthesemessungen mit Gaswechselmeßanlagen? Ein Beitrag zum Küvettenproblem. Flora 169: 510-523.

Caldwell, M.M. 1971. Solar UV irradiation and the growth and development of higher plants. Photophysiol. 6: 131-177.

Halldal, P. 1964. Ultraviolet action spectra of photosynthesis and photosynthetic inhibition in a green and red alga. Physiol. Plant. 17: 414-421.

Jones, L.W. and B. Kok. 1966. Photoinhibition of chloroplast reactions. I. Kinetics and action spectra. Plant Physiol. 41: 1037-1043.

Sisson, W. B. and M. M. Caldwell. 1976. Photosynthesis, dark respiration and growth of Rumex patientia L. exposed to ultraviolet irradiance (288 to 315 nanometers) simulating a reduced atmospheric ozone column. Plant Physiol. 58: 563-568.

Sisson, W. B. 1977. Atmospheric ozone depletion: reduction of photosynthesis and growth of a sensitive higher plant exposed to enhanced UV-B radiation. J. Exp. Bot. 28: 104, 691-705.

POSSIBLE ERRORS IN PHOTOSYNTHETIC MEASUREMENTS ARISING FROM THE USE OF UV-ABSORBING CUVETTES: SOME EXAMPLES IN HIGHER PLANTS

Arno Bogenrieder and Richard Klein

University of Freiburg, Biological Institute II, Schänzlestr. 1 D 7800
Freiburg FRG

ABSTRACT

Young plants of Picea abies L., Achillea millefolium L., Lactuca sativa L. and Rumex alpinus L. show different development of their photosynthetic capacity depending on whether the assimilation cuvettes are made from material allowing (Quartz) or not allowing (Plexiglass) transmission of UV radiation. The results show that with cuvettes made from UV-absorbing material measurements of photosynthesis may be incorrect.

The proportion of the sun's radiation energy reaching the earth's surface is small. Nevertheless the biological significance of UV radiation is important. Sisson and Caldwell (1976, 1977) have shown that UV-B reduces photosynthesis of Rumex patientia L. as well as leaf growth.

Our own experiments showed that the UV radiation in solar light has an astonishingly strong effect on young plants of Lactuca sativa L. and Rumex alpinus L. which were completely unadapted to it after being grown in the greenhouse (Bogenrieder and Klein 1977).

Later experiments proved that as far as UV sensitivity is concerned, the quality of light applied during the period of growth plays an important part. Figure 1 shows the photosynthetic activity after 12 hours of irradiation with visible light and different additional

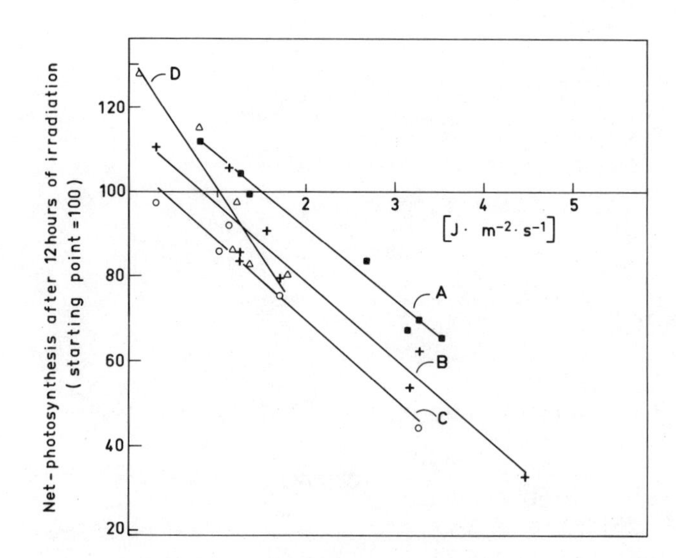

Figure 1. Net photosynthesis of Lactuca sativa L. after 12 hours of irradiation with varied intensities of supplementary UV-B. Seedlings were raised under

- A: Artificial light without UV
- B: Artificial light with UV-A
- C. Artificial light with UV-A + UV-B
- D. Outdoor conditions

quantities of UV-B radiation. The plants of treatment A were raised under artificial light without UV radiation, treatment B with additional UV-A and treatment C with both UV-A and UV-B radiation. The plants of treatment D are grown under outdoor conditions (Bogenrieder and Klein 1978).

These results raised the question of whether the UV-absorbing cuvettes usually employed (e.g. normal Plexiglass) could be the cause of a slight deviation of the measured data concerning photosynthetic activity. This is one of the problems related to the climatic conditions in gas exchange chambers. In order to throw light on this question we experimented with young plants of Picea abies L. (spruce) Achillea millefolium L. (yarrow), Lactuca sativa L. (lettuce) and Rumex alpinus L. grown under outdoor conditions.

Beyond this we tested Lactuca sativa and Rumex alpinus under greenhouse conditions with artificial light (Osram HQIL) and enhanced UV radiation (Osram

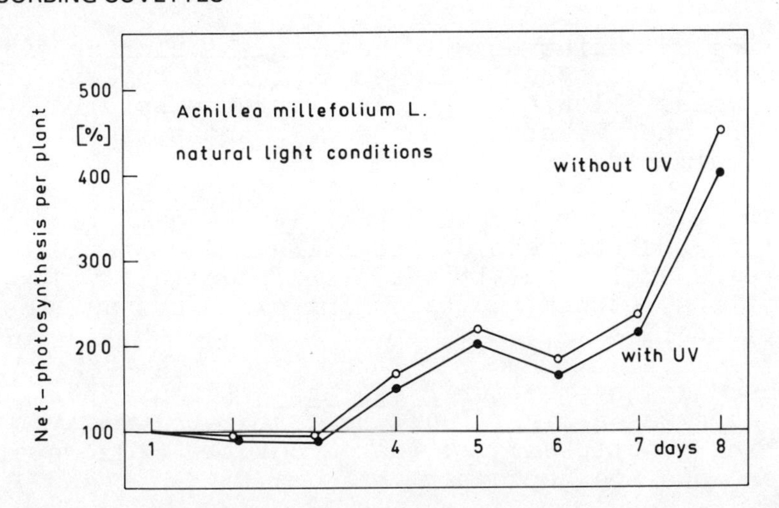

Figure 2. Net photosynthesis of Achillea millefolium L. in a UV-absorbing cuvette (Plexiglass 233) and in a cuvette of Quartzglass.

Vitalux, about 0.6 J $\cdot$ m^{-2} $\cdot$ s^{-1}). Spectral distribution of these lamps is shown in Bogenrieder and Klein (1980).

Photosynthetic rates of plants were determined by a gas exchange system with fully climatized cuvettes and an infrared gas analyzer similar to that described by Koch et al. (1971). Half of the plants were enclosed in an ordinary, non-UV transmitting cuvette of Plexiglass,

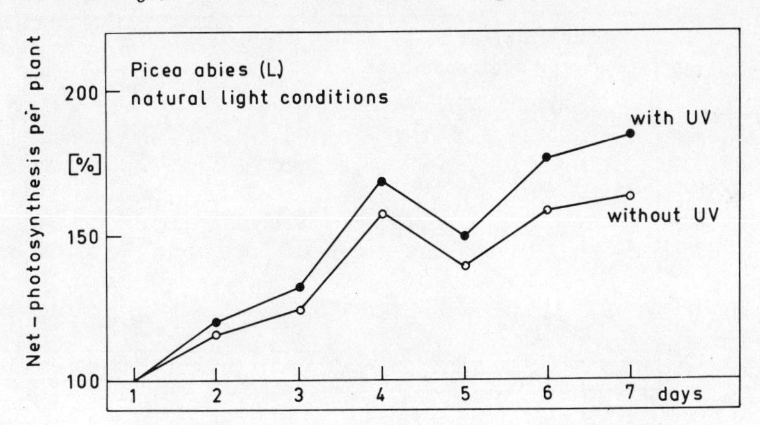

Figure 3. Net photosynthesis of Picea abies L. in a UV-absorbing cuvette (Plexiglass 233) and in a cuvette of Quartzglass.

absorbing UV radiation with wavelengths shorter than 380 nm, whereas the other half was kept in a cuvette of Quartz glass allowing complete UV transmission. For about one week we observed the net photosynthesis of both treatments.

Figure 2 shows a result with Achillea millefolium. The difference between UV absorbing and UV transmitting cuvette is small, but it was significant after six repetitions between the beginning of August and the end of September.

The same is true for Picea abies (Figure 3). Here also there was a significant difference between the UV absorbing cuvette and the Quartz cuvette (six repetitions between June and September).

The deviations were considerable in the case of Rumex alpinus (Figure 4). It is remarkable that the net photosynthesis of those plants without UV radiation, stagnates after three days and finally even shows an absolute decrease (significance from third day). It is obvious, that these plants were being handicapped in their development by the withdrawal of UV radiation. This supposition is supported by the fact that - at the end of our experiments - the chlorophyll content per leaf area of plants in the Quartz cuvette was quite normal; whereas the plants in the UV-absorbing cuvette contained only slightly more than half of this quantity and all looked distinctly chlorotic.

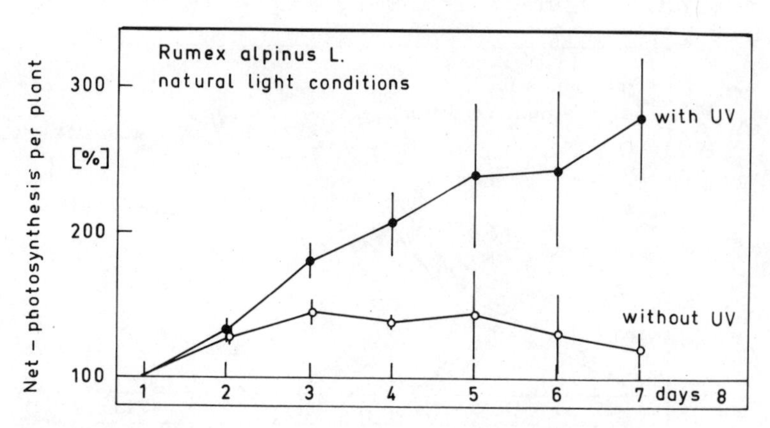

Figure 4. Net photosynthesis of Lactuca sativa L. with supplementary UV (0.6 J · m^{-2} · s^{-1}UV-B, Osram Vitalux).

In Lactuca sativa we found no difference with or without UV radiation in our outdoor experiments, not only in summer 1977 but also in 1978. The outcome, though, was different when we used higher UV intensities in the greenhouse. This time our UV exposed plants showed a significantly lower increase of net photosynthesis (Figure 5).

The results gathered with Rumex alpinus under conditions of artificial light and higher quantities of UV are especially interesting. Similar to the response of Lactuca sativa (as shown in Figure 5) non-UV exposed plants of Rumex alpinus show a more rapid increase of net photosynthesis (Figure 6). Yet, on the eighth day of our tests, net photosynthesis of the non-UV exposed plants was significantly lower than that of UV-treated ones.

As we can see from these results, the UV part of the spectrum can have both negative as well as positive effects during long time measurements of photosynthesis. The kind of effect depends on the quantity of UV applied, as well as the sensitivity of the plant species examined.

We gained our results with rapidly growing plants. It remains to be seen whether fully developed leaves will show similar effects. As long as it is unknown whether a plant is sensitive to UV radiation or not, the measuring of net photosynthesis should be carried out in UV-transmitting cuvettes.

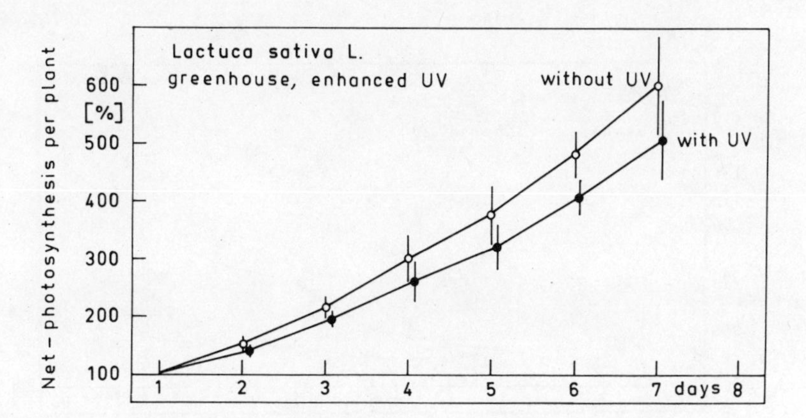

Figure 5. Net photosynthesis of Lactuca sativa L. with supplementary UV ($0.6\ J \cdot m^{-2} \cdot s^{-1}$ UV-B, Osram Vitalux).

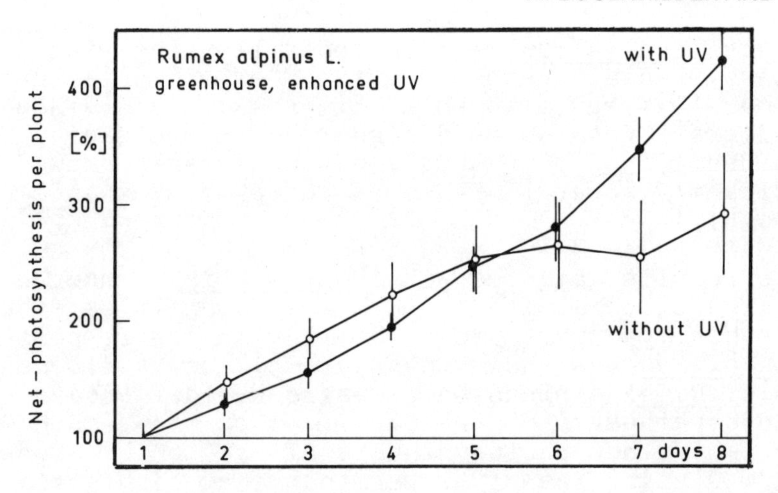

Figure 6. Net photosynthesis of Rumex alpinus L. with supplementary UV (0.6 J · m^{-2} · s^{-1} UV-B, Osram Vitalux).

This recommendation - based on our practical experience - implies that future gas exchange measurements should be carried out in cuvettes of Quartz. UV transmitting Plexiglass cuts off the UV spectrum and ages under the influence of UV radiation (Bogenrieder and Klein 1980). Ordinary glass cuts off the solar spectrum as well and leads to (with Rumex alpinus) results deviating from those gained when using Quartz cuvettes. This is shown in Figure 7. It is seen that in this case the increase of net photosynthesis in the glass cuvette is considerably smaller than in the

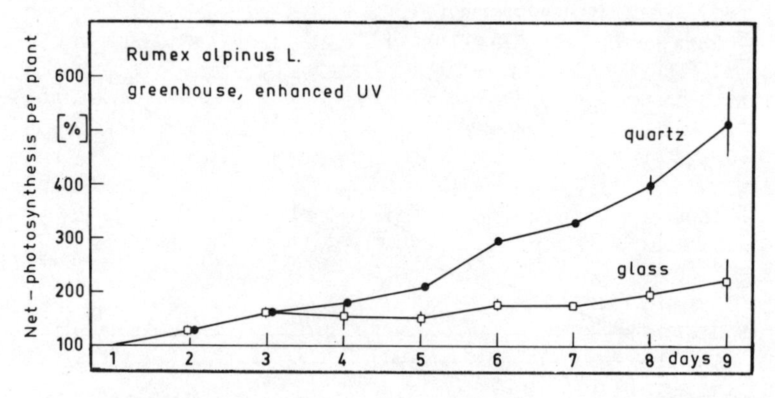

Figure 7. Net photosynthesis of Rumex alpinus L. with UV in a cuvette of normal glass and of Quartz glass.

cuvette of Quartz. It is certainly true that Quartz-glass has proved to be a material which is both expensive and difficult to handle. Yet, all the disadvantageous qualities of this material cannot be regarded as real obstacles. The average cost of gas exchange apparatus supplied with fully climatised cuvettes is very high; thus an additional expense for a Quartz cuvette is reasonable.

ACKNOWLEDGEMENTS

This research was financially supported by the Deutsche Forschungsgemeinschaft.

REFERENCES

Bogenrieder, A. and R. Klein. 1977. Die Rolle des UV-Lichts beim sog. Auspflanzschock von Gewächshaussetzlingen. Angew. Bot. 51: 99-107.

Bogenrieder, A. 1978. Die Abhängigkeit der UV-Empfindlichkeit von der Lichtqualität bei der Aufzucht (Lactuca sativa L.) Angew. Bot. 52: 283-293.

Bogenrieder, A. 1980. Beeinflußt der UV-Ausschluß das Ergebnis von Photosynthesemessungen mit Gaswechselmeßanlagen? Ein Beitrag zum Küvettenproblem. Flora 169: 510-523.

Brandle, J. R., W. F. Campbell, W. B. Sisson and M. M. Caldwell. 1977. Net photosynthesis, electron transport capacity and ultrastructure of Pisum saticum L. exposed to ultraviolet radiation. Plant Physiol. 60: 165-169.

Caldwell, M. M. 1971. Solar UV-irradiation and the growth and development of higher plants. Phytophysiology [ed. by Giese, A.C.] 6:131-177.

Klein, R. M. 1978. Plants and near ultraviolet radiation. Bot. Rev. 44: 1-127.

Koch, W., O. L. Lange and E. D. Schulze. 1971. Ecophysiological investigations on wild and cultivated plants in the Negev Desert. I. Methods: A mobile laboratory for measuring carbon dioxide and water vapor exchange. Oecologia (Berl.) 8: 296-309.

Sestak, Z., J. Catsky and P. G. Jarvis [ed.] 1971. Plant photosynthetic production. Manual of methods. The Hague.

Sisson, W. B. and M. M. Caldwell. 1976. Photosynthesis, dark respiration and growth of Rumex patientia L. exposed to ultraviolet irradiance (288 to 315 nanometers) simulating a reduced atmospheric ozone column. Plant Physiol. 58: 563-568.

Sisson, W. B. 1977. Atmospheric ozone depletion. Reduction of photosynthesis and growth of a sensitive higher plant exposed to enhanced UV-B radiation. J. Exp. Bot. 28: 104, 691-705.

BIOLOGICAL INTERACTIONS BETWEEN WAVELENGTHS IN THE SOLAR-UV RANGE: IMPLICATIONS FOR THE PREDICTIVE VALUE OF ACTION SPECTRA MEASUREMENTS

Rex M. Tyrrell

Swiss Institute for Experimental Cancer Research, Department of Carcinogenesis, CH1066 Epalinges s/Lausanne, Switzerland

INTRODUCTION

To understand the biological actions of a variable, complex polychromatic radiation source such as sunlight, fundamental studies with isolated monochromatic radiations are essential. During the past few years, considerable advances have been made in this area, particularly with bacterial systems (for reviews, see Eisenstark, 1971; Webb, 1977; Tyrrell, 1978a; Tyrrell, 1979a). An An important aspect of these studies has been the determination of action spectra for biological parameters of interest (Webb, 1977; Webb and Brown, 1976). In addition to providing clues as to the chromophores involved in a particular biological action, there has also been the hope that such data will be useful in predicting characteristics of polychromatic sources such as effective biological dosage. However, a serious obstacle to integrating action spectra measurements to predict the action of radiation over a wide wavelength range will arise if interactions exist between wavelengths within the region considered.

Studies in our laboratory have shown that wavelengths longer than 320 nm modify the lethal action of shorter wavelength radiations (Tyrrell and Peak, 1978) in bacterial populations as well as sensitizing such cultures to a range of physical and chemical agents including ionizing radiation (Tyrrell, 1974 and 1976a), heat (Tyrrell, 1976b), and alkylating agents (Correia and Tyrrell, 1979; Tyrrell and Correia, 1979). Mutagenic interactions (both antagonistic and synergistic) occur at much lower doses and have been observed between the longer wavelengths in sunlight (pure monochromatic radiation and natural sunlight) and either the shorter wavelength components of sunlight (e.g. 313 nm) (see Tyrrell, 1978b; Tyrrell, 1980) or mutagenic/carcinogenic chemicals (Moraes and Tyrrell, 1981). The following is a summary of such data and possible implications for the prediction of the action of polychromatic radiations.

METHODS

Methods for determining survival and mutation, the bacterial strains employed and the sources of radiation, dosimetry and irradiation procedures have been described previously (Tyrrell, 1976b; Tyrrell, 1980). The majority of the mutation studies have been carried out with *E. coli* B/r *thy trp* and it's excisionless derivative and reversion to tryptophan independence has been measured on semi-enriched medium.

RESULTS

(1) Sensitization of wild-type bacterial populations to the lethal action of various DNA-damaging agents.

a) *Exposure to monochromatic radiations*. The lethal interaction between near-UV-violet radiations and various agents is summarized in Table 1. The experiments are incomplete at 313 nm. Several wavelengths in the near-UV region sensitize populations to radiation at 254 nm but the longest wavelength tested (405 nm) has a mild protective effect. Sensitization to all other agents occurs even after pre-treatment of populations with 405 nm radiation. The dose of 365 nm radiation required to sensitize populations to any of the agents tested by a factor of 2 or 3 is approximately $10^6 Jm^{-2}$. At 334 nm the dose is 2-3 x $10^5 Jm^{-2}$ in the cases tested. The sensitization always depends on the presence of the *rec* gene product (Tyrrell, 1976; Correia and Tyrrell, 1979) but in certain of the interactions, other gene products (such as *pol*A) are also involved (Correia and Tyrrell, 1979).

TABLE 1

SENSITIZATION OF BACTERIA TO THE LETHAL ACTION OF VARIOUS AGENTS BY PRE-TREATMENT WITH NEAR-UV OR VISIBLE RADIATIONS

Agent	Wavelength of near-UV or visible radiation tested	Dose at 365 nm to sensitize by a factor of approx. 3	Reference
UV(313nm)	365 nm	Dose-dependence of sensitization yet to be determined	Unpublished results, this laboratory
UV(254nm)	334nm, 365nm (405 nm gives protection)	$10^6 Jm^{-2}$	Tyrrell and Peak (1978)
Ionizing Radiation (X-rays and fast electrons)	365nm	$1.2 \times 10^6 Jm^{-2} (O_2)$ $4 \times 10^6 Jm^{-2} (N_2)$	Tyrrell (1974 and 1976a)
Alkylating Agent (methyl methane sulphonate)	334nm, 365nm, 405nm	$1.25 - 1.5 \times 10^6 Jm^{-2}$	Correia and Tyrrell (1979)
Mild-heat 52°C	334nm, 365nm, 405nm	$0.5 - 1.0 \times 10^6 Jm^{-2}$	Tyrrell (1976b)

b) Exposure to sunlight. Pre-exposure of populations of wildtype bacteria to natural unfiltered sunlight also sensitizes such populations to the lethal action of far-UV, mild heat (52°C) and methyl methane sulphonate (A. Neto and R. M. Tyrrell, unpublished results). These data allow us to estimate that the exposure time of tropical noon sunlight to cause sensitizations of the order of 2-3 will range from 20-50 min according to the time of the year and the specific agents tested.

c) Mechanism. A series of experiments have shown that near-UV doses in the range that lead to lethal interactions are also able to progressively disrupt the functioning of all DNA repair systems so far tested (Tyrrell, 1974; Tyrrell, 1976a; Correia and Tyrrell,1979; Tyrrell and Webb,1973; Tyrrell et al. 1973). In addition, the dependence of the sensitization on certain gene products implicates specific repair pathways in the interaction. Consequently the only satisfactory general hypothesis to explain the existence of these lethal interactions is that they result from the inactivation or disruption of repair pathways by near-UV radiation. This is probably not a direct effect on the enzymes involved and more likely results from a general disruption of energy flow and metabolism by near-UV. However, the precise nature of the interaction will depend on the agents involved and the time-span available for recovery processes to operate (Tyrrell and Correia, 1979).

(2) Modification of the mutagenic action of various DNA damaging agents.

a) Exposure to monochromatic radiation. A summary of the mutagenic interactions observed between various near-UV wavelengths and either more energetic UV wavelengths (254 nm, 313 nm) or chemical agents is shown in Table 2. The interactions are complex. The most striking effect is the strong suppression of 254 nm or 313 nm induced mutation by pre-exposure of the bacterial populations to any of a series of near-UV (334 nm, 365 nm) and visible wavelengths (405 nm, 434 nm). Low doses of 365 nm radiation also suppress the mutagenic action of the alkylating agent, ethyl methane sulphonate (EMS). It is important to note that in these experiments we have measured reversion to tryptophan independence on semi-enriched medium. Most mutants detected by this system are suppressor revertants. Other systems are currently under test. Perhaps the most important factor is that the doses required to

TABLE 2

MUTAGENIC INTERACTIONS BETWEEN NEAR-UV OR VISIBLE RADIATIONS AND VARIOUS DNA-DAMAGING AGENTS

Agent	Wavelength of near-UV or visible radiation tested	Nature of Interaction	Reference
UV (254 nm)	334nm, 365 nm, 405nm, 434nm	Strong antagonism (i.e. suppression of mutation) at all wavelengths and at all doses.	Tyrrell (1978b, 1980)
UV (313 nm)	334nm, 365nm, 405nm, 434nm	Strong antagonism (i.e. suppression of mutation) at all wavelengths and at all doses.	Tyrrell (1978b, 1980)
Alkylating agent 1. MMS	365nm	Positive interaction at all doses tested (i.e. enhancement of mutation)	Moraes and Tyrrell (1981)
Alkylating agent 2. EMS	365 nm	Antagonism (low doses) Additive (intermediate doses) Strong positive interaction (high doses)	Moraes and Tyrrell (1981)

cause near-maximal suppression of mutation induction are a factor of ten lower than the doses required to cause a near-maximal lethal sensitization, both at 334 nm and 365 nm. In excisionless strains, there is an additive mutagenic interaction (i.e. no enhancement or suppression).

The interaction between near-UV and alkylating agents is quite different. At intermediate and high doses for EMS and at all doses for methyl methane sulphonate (MMS), the mutagenic interaction is positive. In excisionless strains, the interaction is positive at all doses for both agents when suppressor mutants are detected.

b) Exposure to sunlight. Exposure to natural sunlight strongly suppresses the induction of mutations by either 254 nm or 313 nm (Tyrrell, 1979b). In this case, we have determined that the responsible wavelengths are longer than 320 nm. Most striking is that an exposure of only 5 minutes sunlight (tropical noon) is sufficient to reduce the mutation frequency to below 10 percent of control levels. No interaction is observed in an excisionless strain.

We have not yet tested for mutagenic interactions between solar radiation and chemicals. However, on the basis of our studies with monochromatic radiation, we would expect that exposure of repair competent strains to sunlight for much less than an hour will strongly enhance their susceptibility to the mutagenic action of alkylating agents.

c) Mechanism. We have considerable evidence that the suppression of mutation by relatively low doses of near-UV radiation or short exposures to natural sunlight is due to the strong growth and macromolecular synthesis inhibition induced by these wavelengths. Such growth delays will allow more time for the essentially error-free excision repair process to operate. The relative doses at each wavelength to cause growth delay and suppress mutation are very similar (Tyrrell, 1980; Jagger et al. 1964) and the absolute doses required at any particular wavelength are almost identical. Furthermore, no suppression of mutation induction by far-UV or chemicals is seen in excisionless strains (Tyrrell, 1980; Moraes and Tyrrell, 1981). More recently we have isolated strains that show reduced or almost non-existent near-UV induced growth delays. These strains show a corresponding reduction of the interaction between near-UV and far-UV or alkylating agents. (Moraes and Tyrrell, unpublished results).

The enhancement of chemical mutagenesis observed after exposure of bacterial populations to higher doses of near-UV radiation is also probably related to a shift in the equilibrium between "error-free" and "error-prone" repair processes. Since the "error-prone" repair processes are believed to be inducible, they will be less susceptible than constitutive repair systems to the damaging effects of near-UV (see above). Thus, after higher doses of near-UV, "error-prone" repair processes will predominate and mutation will be enhanced. Destruction of error-free repair processes has also been proposed to explain the rapid increase in mutations induced by high doses of 365 nm radiation to stationary phase cells (Webb, 1977).

DISCUSSION

Our data with monochromatic radiations and natural sunlight clearly indicate that biological interactions will occur between the longer and shorter wavelengths in sunlight. It is important to note that the doses of longer wavelengths required to provoke lethal interactions are of the order of ten times larger than the doses required to provoke mutagenic interactions. Even so, significant lethal interactions are both predicted and observed.

Although wavelengths as long as 405 nm will contribute to the DNA-damaging action of sunlight (Tyrrell, 1979), we believe that the primary action of wavelengths longer than 320 nm will be to modify the biological consequences of the primarily DNA-damaging wavelengths in the range 290-320 nm. It has been suggested (Setlow, 1974) that an action spectrum for DNA damage (as constructed from both chemical measurements and inactivation data) may be useful in predicting the action spectra for solar-UV induced skin carcinogenesis in humans. Moreover it was argued that since the relative spectral distribution of solar wavelengths longer than 320 nm is fairly constant with respect to time of day and other seasonal, geographical and environmental factors, the strength of such interactions as may be induced by the longer wavelengths may be expressed as a single factor in an integrating formula. For weak interactions such as the lethal interactions described above, it may be possible to validate such a formula.

An approximate factor for the sensitizing power of the longer solar wavelengths may be calculated by determining an action spectra for the phenomenon and combining this with dosimetric measurements over the effective region. This would then be used in conjunction with an integrated measurement of the DNA-damaging effectiveness of sunlight (from action spectra measurements) to predict the final biological outcome.

However, it may be argued that since the initial events in carcinogenic transformation have been more closely linked with mutagenic events (Trosko and Chang, 1978), that we should determine the effective mutagenic strength of sunlight. Again, if bacteria are used as the model we will require two action spectra, one for the mutagenic efficiency and a second for the relative efficiency of the longer wavelengths in suppressing mutation. However, the interaction effects are now so strong that the suppression effect will be the predominant factor in the equation. Indeed, we were unable to detect mutations in a repair competent bacterial strain after solar exposures that produce significant levels of pre-mutagenic damage (Tyrrell, 1979b). Since the accumulation of pre-mutagenic damage and the development of the suppression effect are quite different functions of the solar exposure time, a more complex formulation will be required to predict the mutagenic outcome of a given solar exposure.

We hardly need stress the point that we have been using a specific bacterial mutation system that primarily detects suppressor revertants to tryptophan independence. The need for extending these observations to additional mutation systems is quite clear. If such models are eventually to have predictive value in the human situation, the mutagenic effectiveness and the existence of interactions must be tested in a suitable human cell culture system with the genetic markers currently available.

The necessity for extending such studies to animal culture is further emphasized by the observations relating to enhancement of the mutagenic potential of mutagenic/carcinogenic chemicals in bacteria by exposure to solar-UV. The existence of similar interactions in human cells would demand a re-evaluation of certain criteria currently used in human risk assessment.

CONCLUSIONS

The existence of mild lethal interactions and strong mutagenic interactions between the longer and shorter wavelengths in sunlight that we have observed in bacterial cells my have more general implications:

1) Although action spectra may be useful for integrating the biological effectiveness of variable polychromatic sources such as sunlight, the existence of interactions will require, at the very least, additional determinations for realistic interpretations of the final biological outcome.

2) The strong suppression of mutagenesis by the longer wavelengths in sunlight make dosimetric estimates of biological effectiveness based on action spectra difficult if not currently impossible to determine.

3) The existence of such radiation-radiation interactions and the enhancement of the mutagenic potential of chemicals by solar-UV argue for the urgent investigation of the existence of similar phenomena in cultured mammalian cell systems.

ACKNOWLEDGEMENTS

The author is grateful to E. C. Moraes and A. S. Neto for their collaboration in obtaining some of the unpublished results cited in this paper.

This work was supported by the following Brazilian granting agencies: CNPq (National Research Council), CNEN (National Nuclear Energy Council), CEPG/UFRJ (University Council for Post-Graduate Studies) and FINEP (B-76-79-074-0000-00).

REFERENCES

Correia, I. S. and R. M. Tyrrell. 1979. Lethal interaction between ultraviolet-violet radiations and methyl methane sulphonate in repair-proficient and repair-deficient strains of *Escherichia coli*. *Photochem. Photobiol.* *29*: 521-527.

Eisenstark, A. 1971. Mutagenic and lethal effects of visible and near-ultraviolet light on bacterial cells In "Advances in Genetics", Vol. 12 [E. W. Caspari, ed.] 167-198. Academic Press, New York.

Jagger, J.,W. Curtis-Wise and R. S. Stafford. 1964. Delay in growth and division induced by near-ultraviolet radiation in Escherichia coli B and it's role in photoprotection and liquid holding recovery. Photochem. Photobiol. 3: 11-24.
Moraes, E. C. and R. M. Tyrrell. 1981. Modification of alkylating agent mutagenesis in bacteria by pre-treatment with near-ultraviolet (365 nm) radiation. Mutat. Res. 80: 229-238.
Setlow, R. B. 1974. The wavelengths in sunlight effective in producing skin cancer: a theoretical analysis. Proc. Nat. Acad. Sci. (US) 71: 3363-3366.
Trosko, J. E. and C. Chang. 1978. The role of mutagenesis in carcinogenesis. In Photochemical and Photobiological Reviews Vol. 3 [K. C. Smith, ed.] 135-163. Plenum Press, New York.
Tyrrell, R. M. 1974. The interaction of near-UV (365 nm) and x-irradiations on wildtype and repair deficient strains of Escherichia coli K12: physical and biological measurements. Int. J. Radiat. Biol. 25: 373-390.
Tyrrell, R. M. 1976a. Rec A+-dependent synergism between 365 nm and ionizing radiation in log-phase Escherichia coli: a model for oxygen-dependent near-UV inactivation by disruption of DNA repair. Photochem. Photobiol. 23: 13-20.
Tyrrell, R. M. 1976b. Synergistic lethal action of ultraviolet-violet radiations and mild heat in Escherichia coli. Photochem. Photobiol. 24: 345-351.
Tyrrell, R. M. 1978a. Radiation synergism and antagonism. In Photochemical and Photobiological Reviews, Vol.3 [K. C. Smith, ed.] 35-113, Plenum Press, New York.
Tyrrell, R. M. 1978b. Mutagenic interaction between near-(365nm) and far-(254nm) ultraviolet radiation in wildtype and excisionless strains of Escherichia coli. Mutat. Res. 52: 25-35.
Tyrrell, R. M. 1979a. Lethal cellular changes induced by near-ultraviolet radiation. In Acta Biol. Med. Germ. 38: 1259-1269.
Tyrrell, R. M. 1979b. Mutation induction and mutation suppression by natural sunlight. Biochem.Biophys. Res. Commun. 91: 1406-1415.
Tyrrell, R. M. 1980. Mutation induction by and mutational interaction between monochromatic wavelength radiations in the near-ultraviolet and visible ranges. Photochem. Photobiol. 31: 37-47.
Tyrrell, R. M. and I. S. Correia. 1979. Modification of lethal interactions between near-ultraviolet (365 nm) radiations and DNA damaging agents. Photochem. Photobiol. 29: 611-615.

Tyrrell, R. M. and M. J. Peak. 1978. Lethal interactions between ultraviolet radiation of different energies in bacteria. J. Bacteriol. 136: 437-440.

Tyrrell, R. M. and R. B. Webb. 1973. Reduced dimer excision in bacteria following near-ultraviolet (365 nm) radiation. Mutat. Res. 19: 361-364.

Tyrrell, R. M., R. B. Webb and M. S. Brown. 1973. Destruction of the photoreactivating enzyme by 365 nm radiation. Photochem. Photobiol. 18: 249-254.

Webb, R. B. 1977. Lethal and mutagenic effects of near-ultraviolet radiation. In Photochemical and Photobiological Reviews, Vol. 2 [K. C. Smith, ed.] 169-261. Plenum Press, New York.

Webb, R. B. and M. S. Brown. 1976. Sensitivity of strains of Escherichia coli differing in repair capacity to far-UV, near-UV and visible radiation. Photochem. Photobiol. 24: 425-432.

DOES SOLAR UV INFLUENCE THE COMPETITIVE RELATIONSHIP IN HIGHER PLANTS?

Arno Bogenrieder and Richard Klein

University of Freiburg, Biological Institute II
Schänzlestr. 1, D 7800
Freiburg FRG

ABSTRACT

Filtering out natural UV radiation below 380 nm has a distinct effect on plant growth and competitive balance between selected pairs of competing species. In all cases total biomass of both partners together was smaller under the influence of UV, in many cases also shoot length and/or leaf area. Chlorophyll content was usually greater in UV exposed plants. In *Acer pseudoplatanus* L. the competitive balance of highland and lowland ecotype was reversed under the influence of natural UV radiation.

INTRODUCTION

It is often assumed that the higher proportion of UV radiation in the mountains plays an important role in the difference between highland and lowland vegetation; Fox and Caldwell (1978) have been able to show that an enhanced UV-B radiation influences the competitive balance between pairs of competing species. Our own experiments (1977, 1978) have raised the question of whether the relatively small amounts of UV radiation at lower altitudes (Freiburg, West Germany, 200 m alt., 48° n. lat.) do not have to be considered as influencing the ability of higher plants to compete in addition to the well known factors such as visible light, temperature, water supply and pests.

In order to clarify this question we have cultivated some selected plants in a checkerboard pattern outdoors for several weeks. Half of the plants were covered with normal Plexiglass which basically allows no radiation transmittance under 380 nm. The other half were covered with a special Plexiglass which has a lower limit of radiation transmittance at about 300 nm (Bogenrieder and Klein 1977). All seeds were collected in the lowland with the exception of Rumex alpinus and the highland ecotype of Acer pseudoplatanus. These seeds originated from Feldberg (Black Forest) between 1200 and 1400 meters.

The pairing of species in the competition experiments is not arbitrary, rather they represent competition situations which really exist under natural conditions; some are even very frequent in the surroundings of Freiburg, i.e. the struggle between tree seedlings Fraxinus excelsior L./Carpinus betulus L. and Fagus sylvatica L./Acer pseudoplatanus L. At the end of the cultivation period, dry matter and in some cases shoot length, leaf area and chlorophyll content of the test plants were determined.

RESULTS

a) Germination

In 18 different experiments involving 12 species of plants half of the experiments showed a significant difference in the rate of germination. Figure 1 shows two examples. In many cases the progress of germination was somewhat delayed by the natural UV radiation, but the contrary also exists.

b) Competition experiments

The results of these experiments are shown in Figure 2 and Table 1. Statistically significant differences ($P<0.05$) are marked with an asterisk. In all cases total dry mass of both competitors together was less in the treatments with a complete solar spectrum than in those without UV radiation. One of the two plant species was clearly taller than the other; this usually resulted in a benefit for the smaller plants in the presence of UV radiation. This is perhaps related to the greater UV exposure of the taller plants. A similar tendency is seen relating to shoot length and leaf area. Where differences are statistically significant, shoot length and/or leaf area of the UV

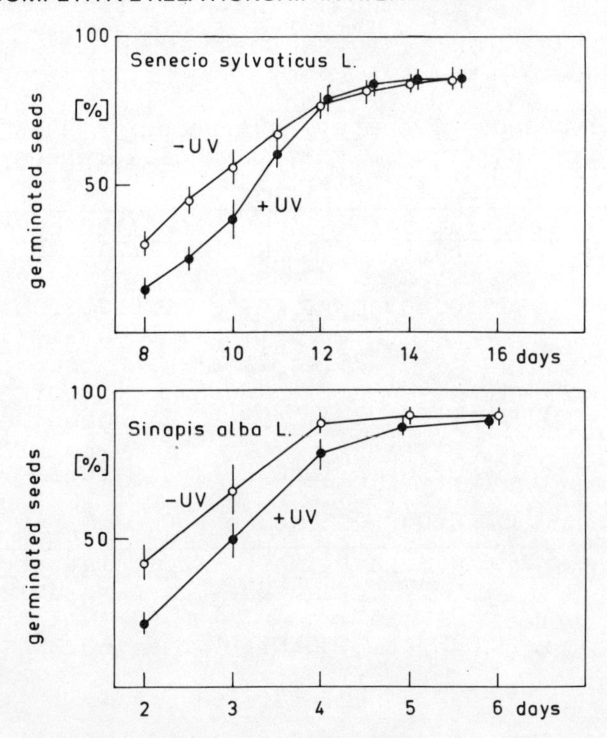

Figure 1. Influence of natural UV on seed germination of Senecio sylvaticus L. and Sinapis alba L.

exposed plants are smaller. Only in Bellis perennis profiting by the large decrease of its major partner, was the contrary to be seen. There are eight cases of statistically different chlorophyll content with and without UV radiation. Only in Senecio sylvaticus was the chlorophyll content smaller under the influence of UV radiation; otherwise the chlorophyll of UV-exposed plants is clearly greater.

c) Plants originating in lower and higher altitudes

Acer pseudoplatanus is one of the few plants which is plentiful both in the lower altitudes of Freiburg and in the upper regions of the Black Forest. We have raised plants originating in higher altitudes and the lowland in competition both with and without UV radiation. The results are shown in Figure 3. It is seen that the lower altitude plants raised without UV

TABLE 1. Responses of competing species on natural UV (Statistically significant differences are marked with an asterisk)

	Verbasum phlomoides L. -UV	Verbasum phlomoides L. +UV	Bellis perennis L. -UV	Bellis perennis L. +UV
dry matter (g per plant)	0.927±0.45	0.727±0.38*	0.087±0.02	0.129±0.04*
shoot length (cm)	30.1±3.4	25.4±3.4 *	7.8 ±1.2	8.0 ±1.0
total leaf area (cm²)	304.0±117	244.6±99 *	33.2±8.8	42.7±13.7*
Chlorophyll a (mg · dm⁻²	2.05 ±0.16	2.10±0.26	2.37±0.12	2.81±0.32*
Chlorophyll b (mg · dm⁻²)	0.56 ±0.06	0.58±0.06	0.81±0.05	1.00±0.11*
	Sinapis alba L. -UV	**Sinapis alba L. +UV**	**Cirsium arvense L. -UV**	**Cirsium arvense L. +UV**
dry matter (g per plant)	3.32±0.83	3.06±0.63*	1.11±0.32	1.09±0.28
shoot length (cm)	47.6±6.2	42.6±4.8 *	-	-
total leaf area (cm²)	40.1±9.8	36.4±7.6	30.9±8.1	31.1±7.1
Chlorophyll a (mg · dm⁻²)	1.96±0.20	2.21±0.23 *	1.06±0.16	1.34±0.23*
Chlorophyll b (mg · dm⁻²)	0.59±0.05	0.66±0.07 *	0.37±0.04	0.38±0.09
	Chrysanthemum vulg. L. -UV	**Chrysanthemum vulg. L. +UV**	**Senecio sylvaticus L. -UV**	**Senecio sylvaticus L. +UV**
dry matter (g per plant)	0.639±0.23	0.409±0.20*	0.236±0.10	0.076±0.04*
shoot length (cm)	30.4±4.7	20.5±3.3 *	28.2±6.4	12.41±4.8*
total leaf area (cm²)	148.5±47.6	99.17±38.5*	39.6±15.3	16.7±6.9 *
Chlorophyll a (mg ·dm⁻²)	1.97±0.27	1.89±0.18	2.19±0.34	1.71±0.19*
Chlorophyll b (mg · dm⁻²	0.67±0.18	0.72±0.07	0.74±0.20	0.70±0.08

Table 1 Continued

	Taraxacum officinale Web.		Bellis perennis L.	
	-UV	+UV	-UV	+UV
dry matter (g per plant)	0.450±0.12	0.500±0.14	0.195±0.08	0.169±0.06
total leaf area (cm^2)	161.9±42.8	174.9±48.0	12.24±4.57	10.66±3.34
Chlorophyll a ($mg \cdot dm^{-2}$)	2.26±0.13	2.69±0.24 *	2.73±0.51	2.99±0.49
Chlorophyll b ($mg \cdot dm^{-2}$)	0.78±0.05	0.95±0.09 *	0.98±0.16	1.02±0.17

	Fraxinus excelsior L.		Carpinus betulus L.	
	-UV	+UV	-UV	+UV
dry matter g per plant	3.45±0.67	2.01±0.32 *	0.746±0.13	0.639±0.15*

	Rumex obtusifolius		Rumex alpinus L.	
	-UV	+UV	-UV	+UV
dry matter (g per plant)	0.885±0.31	0.796±0.31	0.768±0.21	0.585±0.16*

	Fagus sylvatica L.		Acer platanoides L.	
	-UV	+UV	-UV	+UV
dry matter (g per plant)	1.46±0.37	1.19±0.21 *	0.932±0.11	0.748±0.11*

	Acer pseudoplatanus L.	
	-UV	+UV
dry matter (g per plant)	0.807±0.20	0.719±0.16

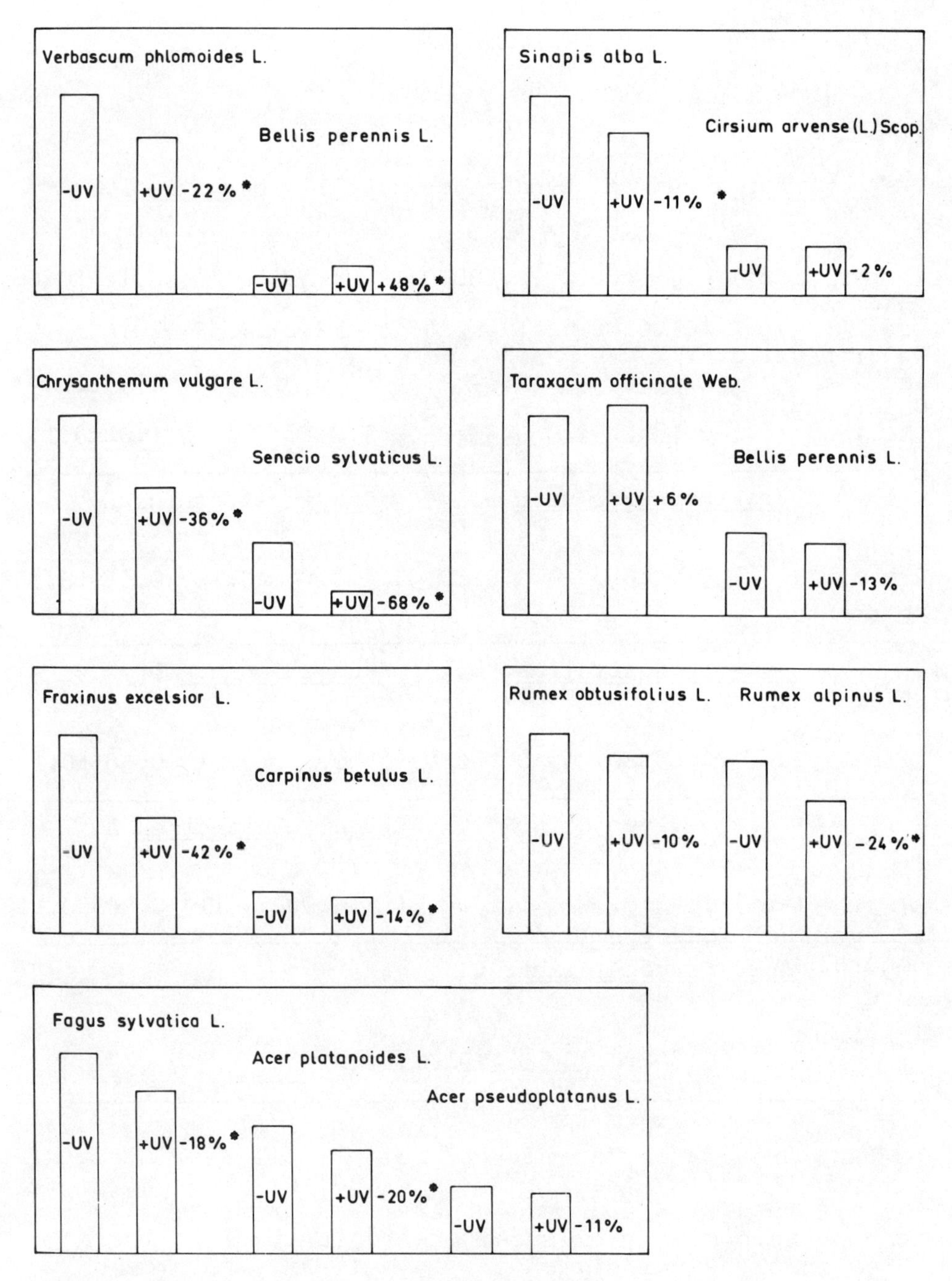

Figure 2. Dry matter of competing species with and without natural UV. Statistically significant differences are marked with an asterisk.

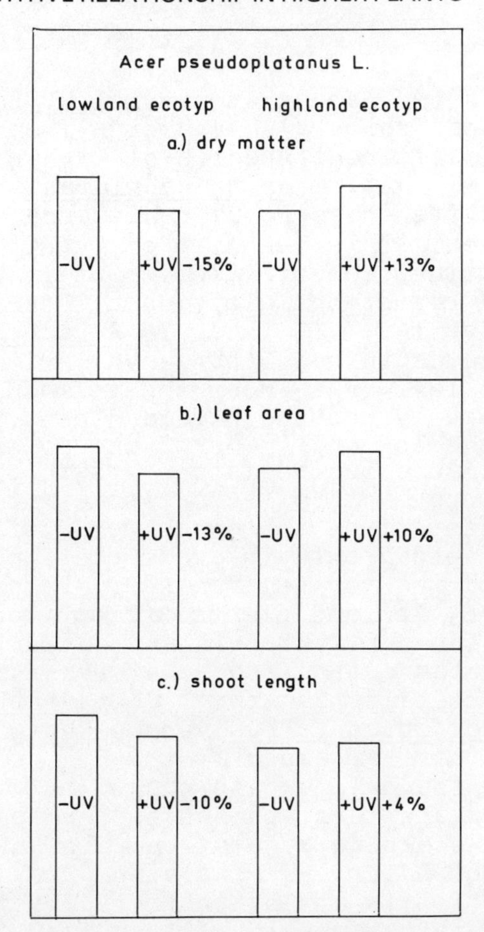

Figure 3. Dry matter, leaf area, shoot length of two ecotypes of *Acer pseudoplatanus* L with and without natural UV.

radiation are superior; however with UV radiation the higher altitude plants were superior. This reversal is significant. It should be noted that under the influence of solar UV the dry matter and leaf area of lowland provenance is smaller, but not the shoot length. After cultivating the plants some additional weeks, the plants of lowland origin were even distinctly taller than those coming from subalpine regions. This indicates that the porportion of dry matter is not the only factor to consider concerning the competitive position even in the case of intraspecific competition.

DISCUSSION

Our results indicate that natural UV (certainly not very high around Freiburg) has a significant effect on the ability of different species of higher plants to compete. Beyond that, *Acer pseudoplatanus* shows that there may be differences in UV resistance even within the same species. This question of possible UV ecotypes needs to be further investigated. But it stands to reason that UV resistance alone cannot decide the capture of a habitat. It is again *Acer pseudoplatanus* which demonstrates this. Though heavier in dry matter and greater in leaf area under the influence of natural UV, it is obviously not the highland ecotype which predominates in the lowland. One of the possible explanations for dominance is the faster growth of the lowland ecotype. This perhaps reflects a somewhat deviating strategy of plants from the dense woods of the lower region where rapid growth is much more important than for plants from the very sparse *Aceri-Fagetum* in the mountain region, lowland plants are obliged to rise above the underwood for survival. In any case it is not true that plant species coming from the highland region are always more resistant to natural UV radiation. This can be seen in *Rumex obtusifolius* versus *Rumex alpinus*, even though in earlier experiments (Bogenrieder and Klein 1977) we found *Rumex alpinus* to be very sensitive to UV radiation and this species is perhaps not representative of the alpine vegetation. In any case, the possible role of natural UV radiation has apparently been given too little attention in regard to its influence on vegetation mosaic.

ACKNOWLEDGEMENTS

This research was financially supported by the Deutsche Forschungsgemeinschaft. We thank Bernhard Bruzek and Susanne Kiliani for the careful performance of the experiments.

REFERENCES

Bogenrieder, A. and R. Klein. 1977. Die Rolle des UV-Lichtes beim sog. Auspflanzungeschock von Gewächshaussetzlingen. *Angew. Bot.* *51*: 99-107.

Bogenrieder, A. and Klein, R. 1978. Die Abhängigkeit der UV-Empfindlichkeit von der Lichtqualität bei der Aufzucht (*Lactuca sativa* L.). *Angew. Bot.* 52: 283-293.

Brodführer, U. 1955. Der Einfluß einer abgestuften Dosierung ultravioletter Sonnenstrahlung auf das Wachstum von Pflanzen. Planta 45: 1-56.

Cline, M. G. and F. B. Salisbury. 1965. Effects of ultraviolet radiation on the leaves of higher plants. Radat. Bot. 6: 151-163.

Fox, F. M. and M. M. Caldwell. 1978. Competitive interaction in plant populations exposed to supplementary ultraviolet-B radiation. Oecologia 36: 173-190.

Kiefer, J. 1977. Ultraviolette Strahlen. De Gruyter, Berlin.

A MODEL TO EVALUATE THE IMPACT OF CYCLICALLY RECURRING HAZARDS ON SELECTED POPULATIONS

John Calkins

Department of Radiation Medicine (1) and
School of Biological Sciences (2)
University of Kentucky, Lexington, Kentucky

It has been observed that solar ultraviolet radiation, particularly the portion of wavelengths below 320 nm (the UV-B), is quite lethal for a variety of microorganisms (Harm 1969, Gameson and Saxon 1967, Resnick 1970, Calkins 1974). In addition to the lethal action, the short wavelength component of solar UV demonstrates a considerable potential for delaying the growth of irradiated organisms (Jagger 1975, Calkins 1975). Growth delay obviously reduces the competitive ability of organisms in the natural environment. Various authors have proposed that the lethal (Harm 1969, Gameson and Saxon 1967) and the growth delaying action (Jagger 1975, Calkins 1974) of solar UV are significant ecological factors. It is, however, quite difficult to translate the reasonable but intuitive feeling that solar radiation has an important impact on various ecosystems into quantitative measurements of the amount of injury for assessment of such an ecological action. It is especially important at present to make quantitative determinations of the ecological effects of solar UV radiation because there is growing awareness that human activities may modify the global environment and could result in disastrous injuries to critical ecosystems.

While some modifications of the environment may be modeled as essentially instantaneous steps to a modified level of the critical factor, solar UV-B demands a more complex treatment. Solar ultraviolet radiation is one of the most variable of environmental agents.

The intensity of UV-B at the water surface varies with the annual and daily rotations of the earth, the atmospheric content of ozone, with local cloud cover and, for small water bodies, with the immediate environment.

With sufficient data, one might attempt to assess the effect on a food web of a factor such as increased solar UV-B in the manner schematically represented in Figure 1 (a three species food web). An appropriate generalization of the Lotka-volterra equations, Eq. 1, (see Smith 1974) could be used to represent the simple food web relation, shown in Figure 1.

$$\begin{aligned} \dot{x} &= x(a-bx-c_1 y) \\ \dot{y} &= y(-e+c_1' x-c_2 z) \\ \dot{z} &= z(-f+c_2' y) \end{aligned} \tag{1}$$

Where x, y and z represent the densities of organisms in the food web; x representing a prey (bacteria) of the primary predator y (Tetrahymena) and z the secondary predator (flatworm) respectively. $\dot{x}$, $\dot{y}$ and $\dot{z}$ being the time rates of change of the appropriate variables. The constants a and b represent growth rate and nutrient limiting factors for the prey species; e and f are the natural mortality rates of the predators; the various constants (c) are determined by the efficiency of predation as prey and predator encounter by chance.

Smith (1974) shows that the food web represented by this equation could exist in an equilibrium condition with positive values for x, y and z.

If it were assumed that the primary predator (y) were the only member of the food web sensitive or subject to solar UV, then Eq. 1 might be modified using a (daily) fractional killing constant (K) yielding

$$\begin{aligned} \dot{x} &= x(a-bx-c_1 Ky) \\ \dot{y} &= y(-e+c_1' x-c_2 z-K) \\ \dot{z} &= z(-f+c_2' Ky) \end{aligned} \tag{2}$$

and since Eq. 2 is no different in form than Eq. 1, it follows that a new equilibrium condition could eventually be attained.

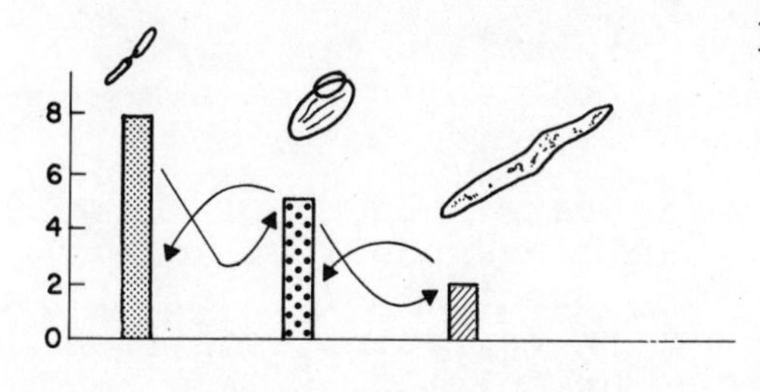

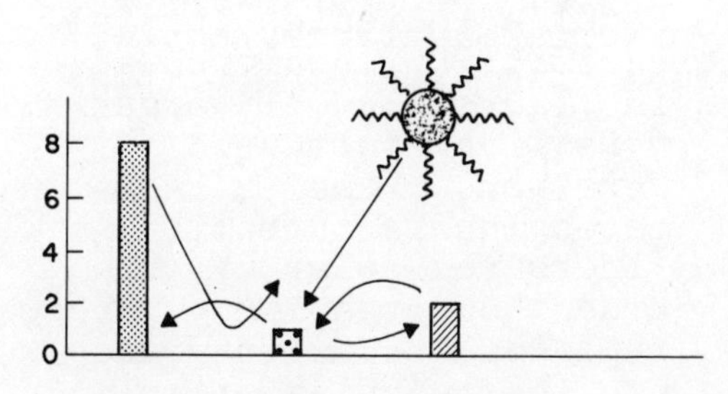

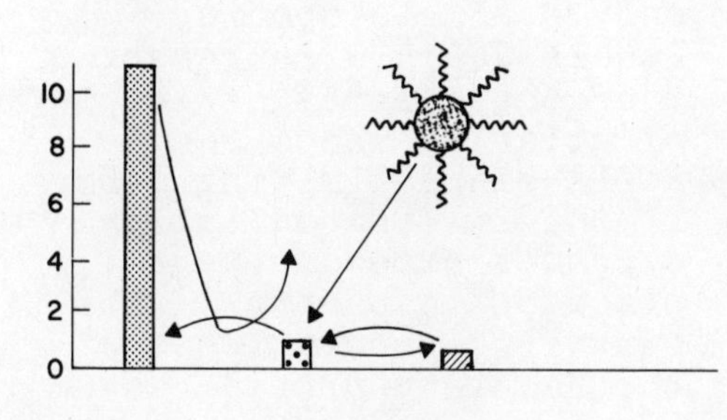

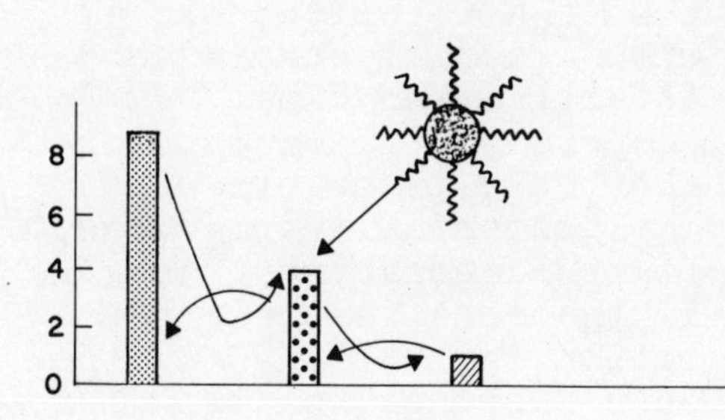

Figure 1 Equilibrium model A schematic representation of a three component ecosystem. Bar heights represent log of organism density. Organism x, the primary prey species might represent bacteria; y, the primary predator, might represent a bacterial feeding animal such as _Tetrahymena_ while organism z, the secondary predator might be a flat worm. Panel A represents the unperturbed equilibrium condition, the density of x is controlled by the available nutrients and by predation by organism y. Densities of y and z are likewise controlled by densities of their food (prey) and predators. Panel B represents a perturbation, represented as increased solar UV-B which acts on organism y only. There would be a severe transient depletion of organism y. Panel c represents a transient state; the prey species x would reach large numbers because of lack of predation by y, the predator feeding on y would fall to low numbers because of lack of food. The changes in x and z would both tend to increase the density of organism y. Panel D represents the return of the equilibrium state in thepresence of the perturbing agent. Organism x would be more dense, y and z would be reduced in numbers from values represented in Panel A. From Calkins,1974.

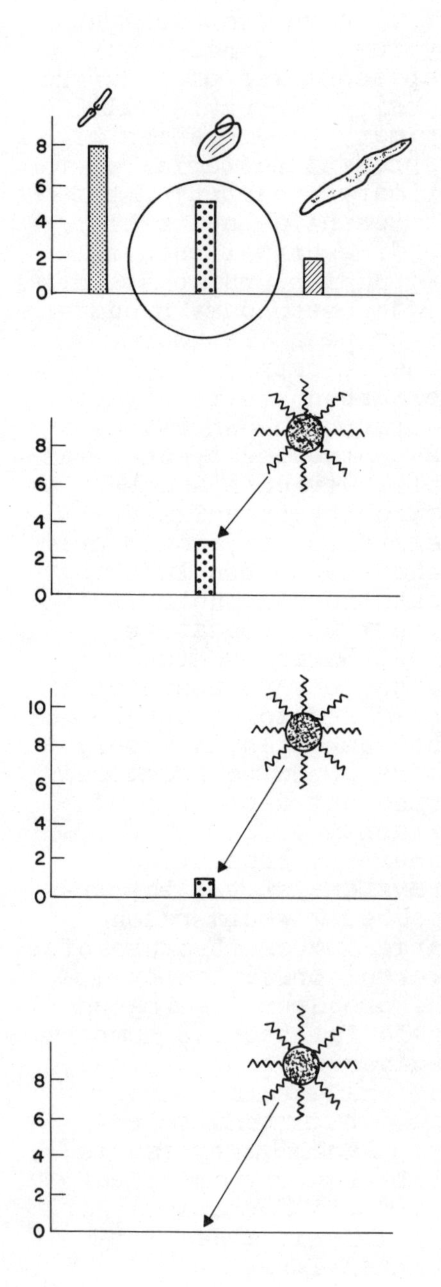

Figure 2. The replacement limiting model. A schematic representation. Bar heights again represent densities of organism x, y, and z. Panel A indicates initial equilibrium condition as in Figure 1. Only organism y is given further consideration. Panel B indicates the perturbation (increased solar UV) incident on organism y. The replacement limiting model confines consideration to only a single species. Panels C and D represent the consequences of levels of solar UV above the Replacement Limiting Dose. If the killing during the exposure exceeds the replacement capacity of the surviving population during the recovery period, assuming maximum growth rate, then the population of y will gradually be reduced to extinction Panel D. To compute the maximum tolerable dose, i.e., the Raplacement Limiting Dose, it is necessary to know 1) the dose response relation for killing, 2) the dose response relation for growth delay, and 3) the maximum growth rate. All three factors can be accurately determined from laboratory observations. From Calkins, 1974.

As depicted in Figure 1, upon increased UV-B, again assuming only the primary carnivore to be sensitive, there would be a transient decrease in the primary carnivore and an eventual restoration of a new equilibrium level with a higher level of food (bacteria) and lower levels of primary and secondary carnivore. Although the three component systems depicted in Figure 1 represent an unrealistic generalization from nature, the mathematical formalism to calculate equilibrium population sizes is not simple. The major obstacle to the computation of new equilibrium population sizes is not, however, the mathematical complexity but the lack of the critical parameters (a-f) to put into the mathematical representations.

Solar UV-B is a factor which is amenable to a less rigorous or elegant formalism but a formalism which can produce definitive conclusions from the imperfect kinds of biological data presently available. While UV-B doses are highly variable, they can be generalized and approximated as an acute dose delivered during a short portion of the day, for instance the 4 hours centering on noon. The remainder of the day (20 hours) can be considered a recovery period. Figure 2 illustrates the central concepts of the Replacement Limiting Model. If more organisms are killed by the acute exposure than can be replaced during the recovery period, then the population would gradually be depleted. If the cycle continues, then a sensitive population will eventually vanish. To apply this model one should know the net replacement which actually occurs between exposures. In a real ecosystem the net replacement of organisms will depend on many factors; predation, nutrients, temperature, oxygen, etc. and is not easily or frequently determined. The Replacement Limiting Model incorporates a further specification. Both conceptually and in the laboratory, a species can be isolated from its normal interactions with its environment. Predators and other lethal factors can be removed; nutrients and physical variables can be supplied at optimum levels. By varying growth conditions (temperature, nutrients, etc.) optimum growth rate for an organism can be determined, and further variation will not increase growth rate. Under optimum conditions, the growth of a species appears to be controlled by its intrinsic nature and variation of extrinsic factors cannot increase the growth rate. The optimum growth rate constant of organism, y(g), is of course related to constants e and c_1 of Eq. 2 which, however, assumes that growth rate is primarily controlled by prey species density and not intrinsic factors.

If a species is growing at the optimum rate and free of all other sources of mortality (predators, physical factors, etc.), then it is possible to determine the maximum cyclic killing which they could tolerate (K_{max}). If under optimum conditions a population cannot regenerate the component of the population killed before the next acute exposure, then the population will eventually be diminished to extinction. The exposure which kills the fraction that can be replaced has been termed the Replacement Limiting Dose (RLD). Natural environments are not optimum for growth and free of hazards and so the RLD is clearly the maximum tolerable dose. One additional factor must be incorporated into the replacement limiting model. The entire recovery period may not be available for growth since, as noted by Jagger 1975, Calkins 1975, exposure to near lethal levels of an agent such as solar UV-B often delays the growth of the survivors and, thus shortens the period for recovery.

Expressing the relationships illustrated in Figure 2 mathematically:

$$K_{max} = y_o/y \qquad (3)$$

where y_o is the surviving fraction following the acute radiation. Also,

$$y = y_o \exp[gt_{(rec)}] \qquad (4)$$

where $t_{(rec)}$ is the time available for recovery, i.e., time between exposures minus the time of growth delay of the survivors; thus,

$$K_{max} = \exp[-gt_{(rec)}] \qquad (5)$$

Radiation lethality is only rarely a simple analytical function of dose. However, in contrast to Equations 1 and 2, which critically depend on constants that are almost impossible to determine, the RLD is easily estimated from laboratory observations of growth rate and the dose-response relations for lethality and division delay.

While the Replacement Limiting Model is based on very simple and obvious concepts it can yield

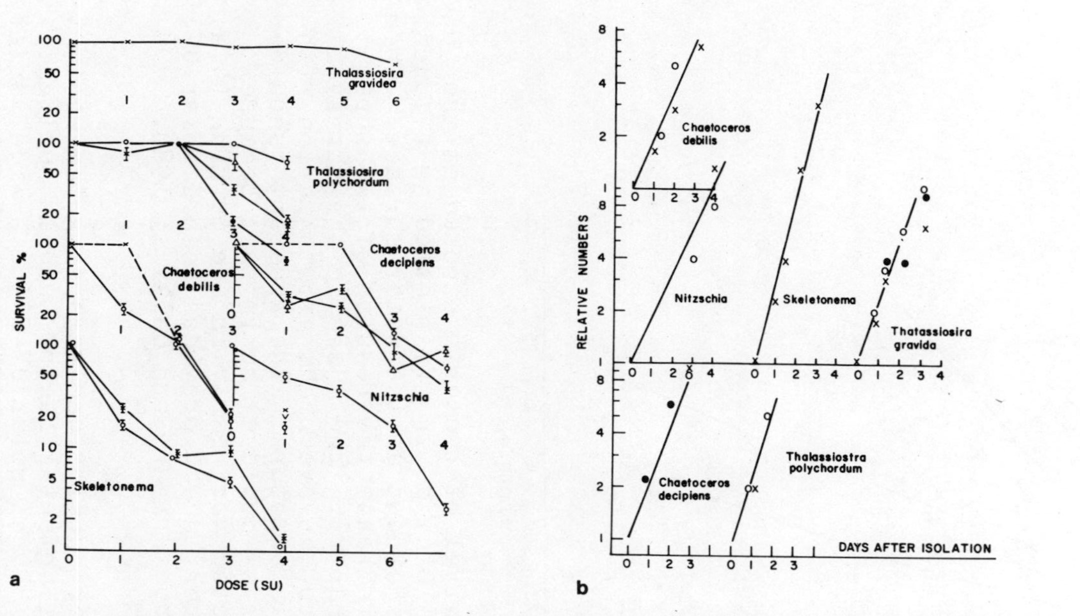

Fig. 3a The UV-B sensitivity of six marine diatoms. Components of the marine phytoplankton, six common diatoms, were selected as critical organisms for the study of biological response to UV-B. The diatoms were isolated from water samples collected in 1976. Chaetoceros debilis and Chaetoceros decipiens and Nitzschia were derived from collections off the north of Iceland, Thalassaiosira polychorda, T. gravida and Skeletonema were isolated from samples collected near Reykjavik.
Fig. 3b The growth of marine diatoms. It is assumed that injured populations recover by the growth (division) of survivors, which was determined by repeated counting of isolated diatoms, both unirradiated control organisms and organisms irradiated to various dose levels. From Calkins and Thordordottir, 1980.

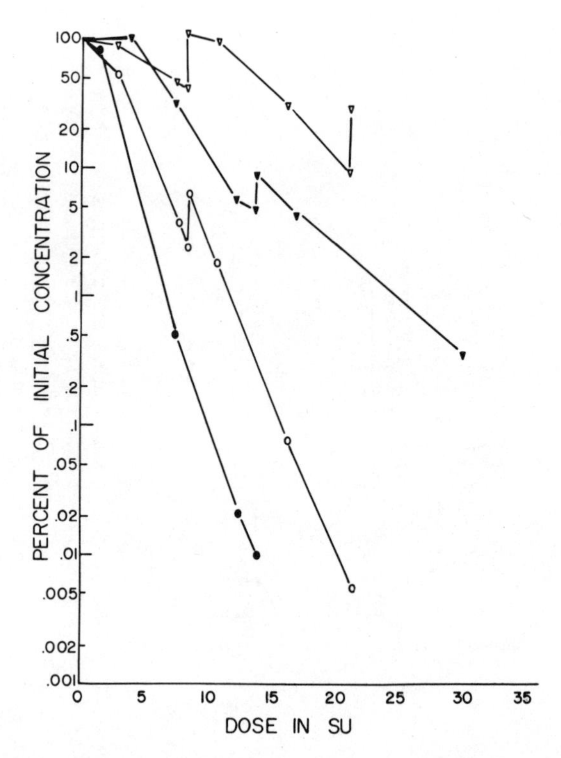

Figure 4. The survival of bacteria exposed at the surface (O) and at 10 cm depth (V) plotted as a function of the solar UV-B incident at the water surface. Open and filled points indicate the results of two separate experiments 1 year apart. The 'jags'in the survival curves correspond to the overnight growth period without additional solar exposure. The apparent 'shoulder' on the survival curves could arise in part because samples were collected very early in the morning (4 a.m.) and thus a growth period was provided before solar UV exposure began. From Grigsby and Calkins, 1980.

definitive and clear conclusions regarding the ecological role of an environmental factor such as sunlight. Measurement of the RLD of solar UV-B for a wide variety of aquatic organisms (Calkins 1974, 1975, Calkins and Thordardottir 1980) have shown that microorganisms with sufficient UV-B resistance to dwell at the water surface in summer are very rare. Avoidance of solar UV-B is necessary and, thus, UV-B is a factor of ecological significance even at its present intensity.

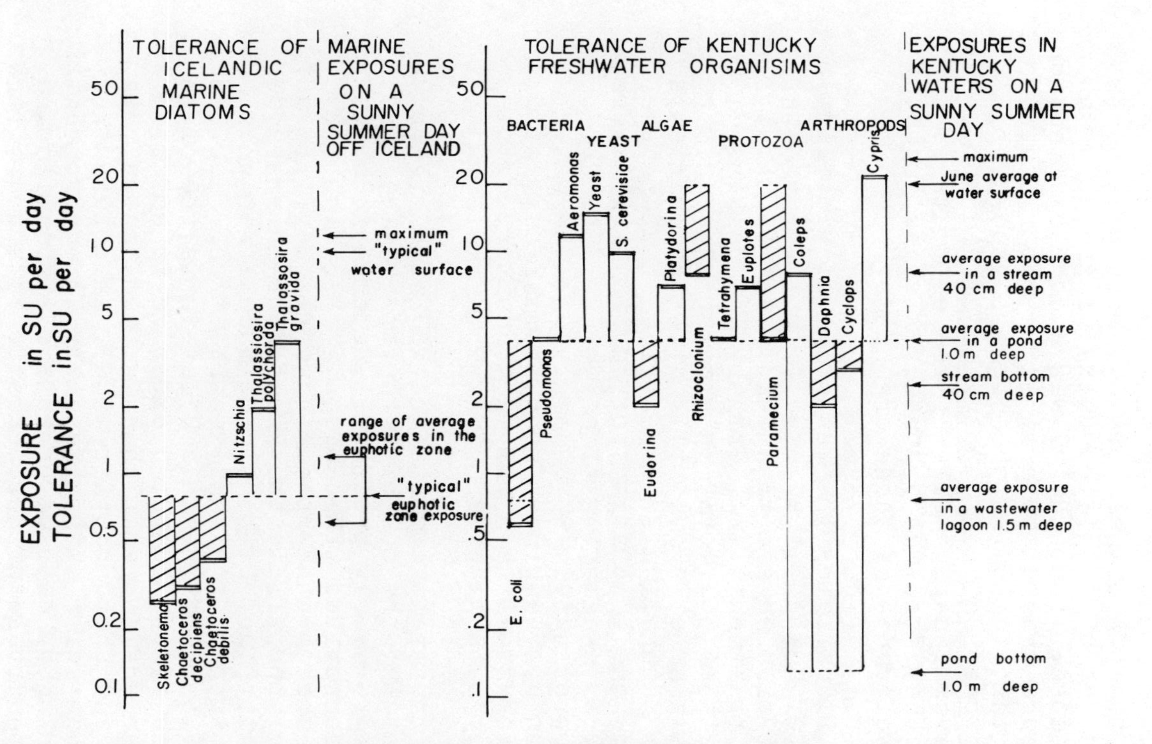

Figure 5. A comparison of exposure and RLD. Maximum tolerable doses (double bars) are shown for a number of aquatic organisms together with the level of exposure for typical locations where the organisms are found. When exposure is less than tolerance the difference is indicated by solid bars; when exposure exceeds tolerance the difference is indicated by crosshatching. The apparently paradoxical situations where organisms lack sufficient resistance to live in the areas where they are, in fact, found is considered on a case by case basis in Calkins and Thordardottir, 1980 (the source of this Figure).

The responses of six marine diatoms upon radiation with simulated solar UV-B radiation are plotted in Figure 3, also observations of the growth rate of these organisms when given continuous light and optimum temperature and nutrients. The diatoms obviously double approximately once per day and thus the RLD can be estimated as the 50% lethal dose. A more precise estimate requires evaluaton of the delay of growth as a function of UV-B dose; growth delay was a complex function of UV-B exposure and not determined in detail; however, it was noted that following the highest exposures *Thalassiosira gravida* and *T. polychordum* did not resume growth for several days. Clearly the estimate of the RLD as 50% lethal dose was a conservative estimate of the maximum tolerance of these organisms.

Figure 4 illustrates the reduction of viability of coliform bacteria when irradiated by natural sunlight in the output of a sewage treatment plant (without chlorination treatment, from Grigsby and Calkins, 1980). The growth rate of coliforms in similar water has been presented (Calkins et al. 1976). With the nutrients in (polluted) natural water and typical temperature, coliforms grow one generation per day. Again, disregarding the growth delaying effect of UV, the RLD is 1 SU/day. Approximately 20 SU arrive at the water surface in Kentucky during a sunny summer day, so it is evident coliforms can have only a transitory existence in clear sunny natural waters.

Although one might compute the RLD with high precision, estimating the RLD of organisms can be very useful; however, overly precise statement of RLD values may be self-deceptive. The RLD only specifies the maximum tolerance of the species under study, the actual tolerance may approximate the RLD for organisms under favorable conditions, but could be much lower under limited food supply or extensive predation. The remarkable observation relating to RLD levels is the close correspondence of the computed exposure to solar UV and the RLD in the organisms which have been tested (Fig. 5 from Calkins and Thordardottir 1980), an observation suggesting that solar UV is presently an environmental stress requiring of exposed organisms a specific and extensive expenditure of resources to attain a tolerable level of resistance.

ACKNOWLEDGEMENTS

This work was supported, in part, by the Office of Water Research and Technology, U. S. Department of Interior, under the provisions of Public Law 88-379.

REFERENCES

Calkins, J. 1974. A preliminary assessment of the effects of UV irradiation on aquatic microorganisms and their ecosystems. Proceedings of the Third Conference on CIAP, DOT, Technical Information Service. Springfield, Virginia: 505-513.

Calkins, J. 1975. Effects of real and simulated solar UV-B in a variety of aquatic microorganisms. CIAP Monogr. 5: 5-33.

Calkins, J., J. D. Buckles and J. R. Moeller. 1976. The role of solar ultraviolet radiation in "natural" water purification. Photochem. Photobiol. 24: 49-57.

Calkins, J. and T. Thordardottir. 1980. The ecological significance of solar UV radiation on aquatic organisms. Nature 563-566.

Gameson, A.L.H. and J. R. Saxon. 1967. Field studies on effects of daylight on mortality of coliform bacteria. Water Res. 1: 270-295.

Grigsby, P. and J. Calkins, 1980. The inactivation of a natural population of coliform bacteria by sunlight. Photochem. Photobiol. 31: 291-294.

Harm, W., 1969. Biological determination of the germicidal activity of sunlight. Radiat. Res. 40: 63-70.

Jagger, J. 1975. Inhibition by sunlight of the growth of Escherichia coli B/r. Photochem. Photobiol. 22: 67-70.

Resnick, M. A. 1970. Sunlight-induced killing in Saccaromyces cerevisiae. Nature 226: 377-378.

Smith, J. M. 1974. "Models in Ecology". Cambridge University Press.

SOLAR UV RADIATION AS A SELECTIVE FORCE IN THE EVOLUTION OF TERRESTRIAL PLANT LIFE

Martyn M. Caldwell

College of Natural Resources
Department of Range Science, UMC 52
Utah State University, Logan, Utah 84322

This paper will briefly reflect on the possible role of solar UV in the early stages of evolution of terrestrial plants and will examine a steep gradient of natural solar UV-B that currently exists on the Earth.

UV RADIATION AS A CONSTRAINT IN THE DEVELOPMENT OF TERRESTRIAL PLANT LIFE

Solar UV radiation may have been, on the one hand, a vital force in forming biological macromolecules for the first elements of life and, on the other hand, a major constraint in the evolution of life exposed to sunlight in the absence of an atmosphere with effective UV-filtering capacity. Most cosmologists agree that the primeval atmosphere was of a reducing nature and almost completely devoid of oxygen from which ozone, the primary UV-filtering component of our present atmosphere, could be formed.

Sagan (1973) has considered alternative atmospheric UV filters in the early environment. Most of the components that comprised the atmosphere at that time, such as methane, CO_2, and ammonia, are almost totally transparent at wavelengths greater than 220 nm, which is the region of the UV-C solar spectrum where most of the energy exists. Only H_2S appears to have been present in sufficient quantity to constitute an effective filter of UV-C, and even then a distinct window between 240 and 270 nm would have existed. Most forms of plant life on the Earth's surface would never survive such irradiation (Cline and Salisbury 1966, Nachtwey 1975).

Oxygen in the atmosphere developed as a result of early photosynthesis, probably at moderate depths in the ocean, yet the net increase of oxygen in the atmosphere was slow. The oxygen content was only on the order of 10^{-3} of present atmospheric levels (PAL) during the Precambrian, some 900 x 10^6 years before the present (Berkner and Marshall 1965). Oxygen absorbs UV-C from the sun, which results in the formation of ozone in the upper atmosphere.

The quantity of ozone in equilibrium with our present atmosphere is only on the order of 3 mm in thickness if condensed to standard temperature and pressure. Yet, it is quite sufficient to truncate the solar spectrum abruptly at approximately 290 nm. At shorter wavelengths in the UV-C, the absorption coefficient of ozone is still orders of magnitude greater, especially in the 250 to 280 nm waveband. This corresponds largely with the window left by H_2S and is also the waveband of maximum UV absorption by nucleic acids (Caldwell 1977). Even a small fraction of the present ozone layer would have been of considerable benefit in the early development of life.

Berkner and Marshall (1965) set up a series of calculations relating the development of oxygen in the atmosphere, and thus ozone, with the fossil record of evolutionary progress. Assuming atmospheric ozone concentration to be in direct proportion to the amount of oxygen in the atmosphere, their calculations suggested that two critical thresholds of oxygen concentration in the development of the atmosphere were important contributing factors to the explosive evolutionary advances, first in the Cambrian and later in the Silurian. For millions of years, oxygen concentrations were less than .001 PAL. At that time, Berkner and Marshall calculated that UV-C in quantities lethal to unprotected organisms, as we now know them, would not only have reached the surface of the Earth but would have penetrated 5 to 10 m into water. (Their calculations were based on pure water, which is relatively transparent to UV. Since UV transmission of water can be greatly reduced by biological contaminants, UV penetration to such depths may have been substantially less.)

Naturally, such a situation would have greatly limited not only the proliferation of terrestrial life but also would have hindered global photosynthetic activity. As oxygen concentrations and ozone reached approximately .01 PAL, the first critical level was finally achieved in the Cambrian, 600×10^6 years ago. By then, there would have been negligible UV irradiation of wavelengths shorter than 280 nm at the Earth's surface -- still a harsh situation for life as we know it now-- but this would have been a major step in overcoming the UV radiation constraint. Berkner and Marshall point to the eruptive proliferation of species in the fossil record at that time.

Organisms in the late Precambrian and during the period of explosive proliferation in the Cambrian likely developed many screening mechanisms to reduce the penetration of UV radiation to physiologically sensitive targets. Sagan (1973) has suggested that the particularly large UV extinction coefficients of purines and pyrimidines would have made them particularly effective shielding compounds for UV-C and that these were utilized by eukaryotic organisms. These compounds are the very building blocks of nucleic acids, which makes them susceptible to UV-C radiation. Therefore, in addition to the functional purines and pyrimidines of nucleic acids, there may have been outer sheaths of these compounds arranged to provide the necessary protection. Many extracellular shields also may have developed. Margulis et al. (1976) pointed out the protective matting habit of certain bluegreen algae in which living cells can be well shielded by a covering of dead algal material or various inorganic salts such as sodium nitrate.

As even a small component of ozone in the atmosphere lessened the UV hazard, global photosynthesis proceeded at ever-increasing rates. After a mere 180 million years following the initiation of the Cambrian, oxygen and ozone concentrations had likely increased 10-fold to approximately 0.1 PAL, according to the scenario developed by Berkner and Marshall (1965). They point to the dramatic appearance of terrestrial life in the Silurian (ca. 400×10^6 years ago) as evidence that a second major threshold of oxygen in the atmosphere had been reached. This was largely attributed to the reduction of excessive UV-C on the Earth's surface; what role other concurrent environmental changes played in this development is not clear.

Although the hypotheses developed by Berkner and Marshall provide a compelling argument for the development of UV radiation as a major constraint in development of terrestrial life, the time scale of events could be quite different. Ratner and Walker (1972) and Levine et al. (1979) employed fundamental photochemical models to show that ozone production would not be in direct proportion to the oxygen concentration of the atmosphere but, instead, may have been relatively greater; sufficient ozone would have been produced to constitute an effective UV shield when oxygen concentrations were only .001 PAL. Based on these concentrations, Margulis et al. (1976) argued that UV was likely not a major constraint in the evolution of eukaryotic organisms, although they conceded that it still may have been an important environmental hardship during the evolution and development of prokaryotic organisms. Levine et al. (1979) calculated that ozone concentrations equivalent to those of the present ozone layer would have developed when the Earth's oxygen concentration was only .01 PAL. At 0.1 PAL, the ozone column thickness, at least at temperate latitudes, may have been some 10% greater than at present.

The development of an effective ozone filter in the atmosphere before the end of the Precambrian would be consistent with the suggestions of Axelrod (1959) that, contrary to general conception, a vascular land flora was already in existence in the Precambrian and that the paucity of plant fossil records before the Silurian is a geologic phenomenon; the sediments that would have contained such fossils have been largely eroded and lost. These views are, however, still very much open to alternative interpretations.

Although the exact time period when lethal UV was diminished on the Earth's surface and the evolutionary history of this period are yet to be refined, it is still very likely that UV radiation was at some stage a major obstacle in development of terrestrial life. Following the development of the ozone layer there may have been a period of greater protection than now occurs. Since the global oxygen concentration has more or less stabilized in the last one hundred million years, one might anticipate that the global ozone layer has also been reasonably constant. Unfortunately, there is little historical record. Global ozone measurements have only been taken over the past four decades and reliable data exist only for the last 20 years. Thus,

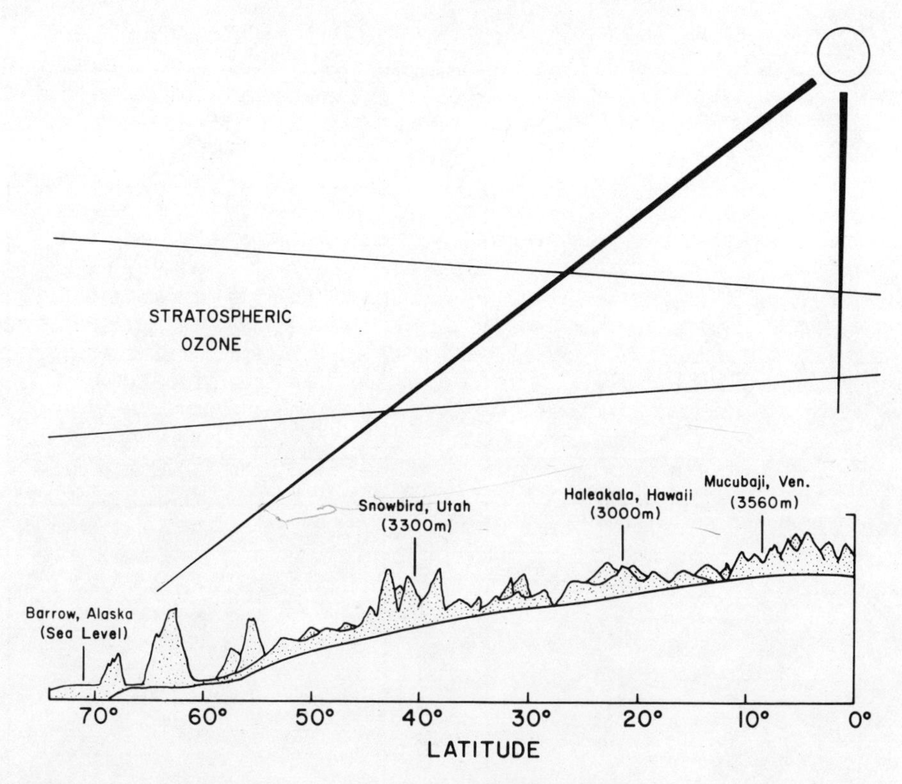

Figure 1. Depiction of the latitudinal arctic-alpine life zone gradient. The stippled area represents this life zone which progresses from sea level in the Arctic to high elevations at lower latitudes. A few of the sites where radiation and plant optical property data were collected are indicated. The change in effective solar UV-B irradiance is due partly to a latitudinal gradient in thickness of the atmospheric ozone column, changes in prevailing solar angles and, thus, different pathlengths through the atmosphere, and elevation above sea level. From Caldwell et al. (1980). Copyright 1980, the Ecological Society of America.

there is little actual record of possible changes in the solar UV climate in the historical past, or certainly in the recent geologic past. Apart from the seasonal, latitudinal, and short term fluctuations of ozone that are reasonably well documented, fluctuations on the order of a decade with a total amplitude of about 5% at temperate latitudes have been recognized (Angell and Korshover 1973). There is, unfortunately, no apparent biological evidence which can be used to evaluate the significance of these fluctuations.

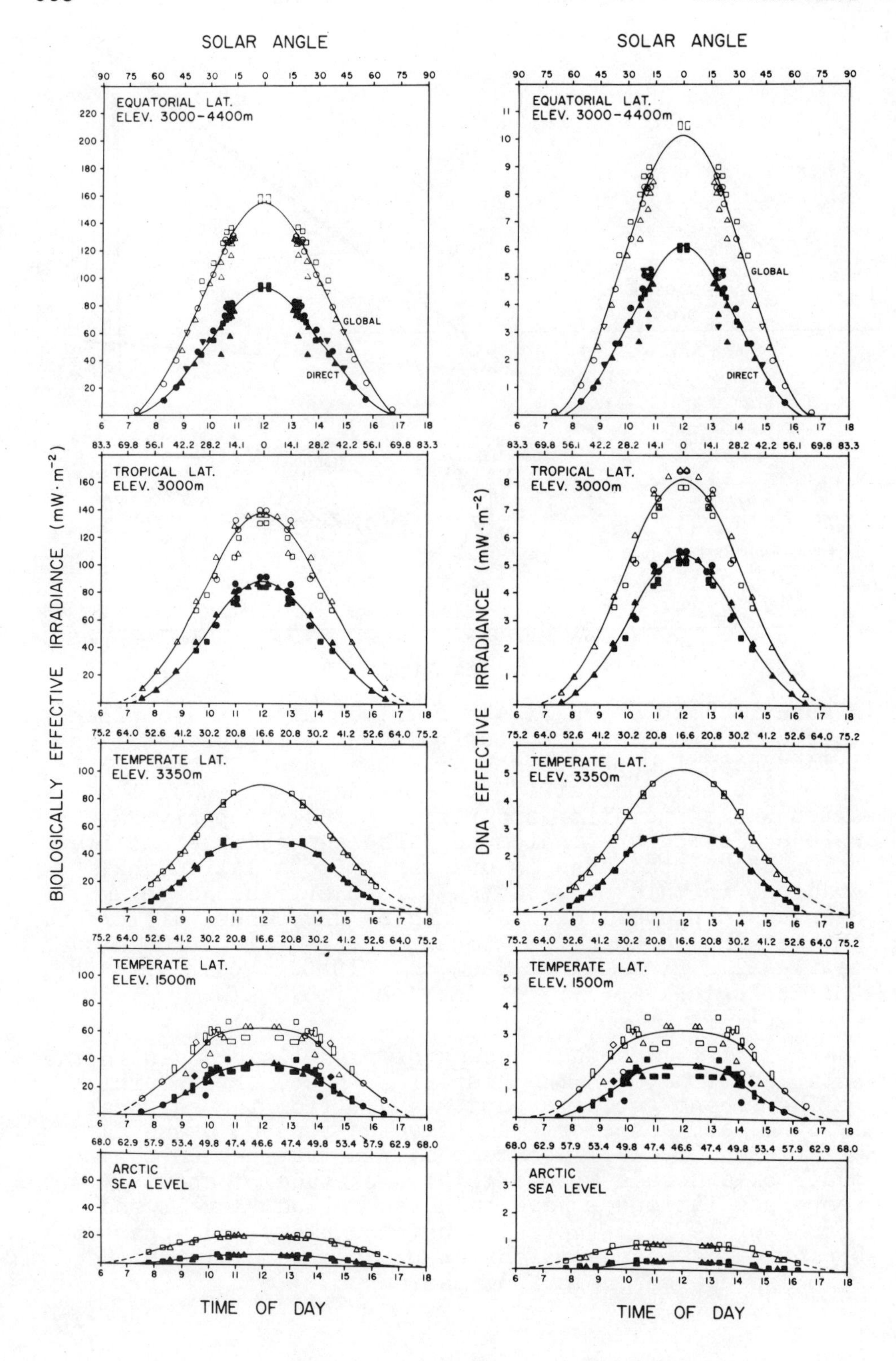

SOLAR ANGLE
EQUATORIAL LAT.
ELEV. 3000-4400m
GLOBAL
DIRECT
TROPICAL LAT.
ELEV. 3000m
TEMPERATE LAT.
ELEV. 3350m
TEMPERATE LAT.
ELEV. 1500m
ARCTIC
SEA LEVEL
BIOLOGICALLY EFFECTIVE IRRADIANCE (mW · m⁻²)
DNA EFFECTIVE IRRADIANCE (mW · m⁻²)
TIME OF DAY

A GEOGRAPHICAL GRADIENT OF SOLAR UV-B RADIATION

Although it is difficult to reconstruct the history of the Earth's UV climate or to evaluate the biological significance of other temporal fluctuations of the ozone layer, one can view natural geographical gradients of solar UV that presently exist on the Earth. The arctic-alpine life zone spans one of the largest natural solar UV gradients. An indication of this is provided by a series of measurements taken along a latitudinal gradient of the arctic-alpine life zone from northern Alaska at sea level to high elevations in the Andes of South America. This gradient is represented diagrammatically in Fig. 1. The gradient of solar UV-B irradiation results primarily from a combination of three factors: Prevailing solar zenith angles, and therefore, solar pathlengths through the atmosphere, are greater at higher latitudes; atmospheric ozone thickness increases by 50% from the equator to high latitudes during the season of maximum solar radiation; and, finally, the arctic-alpine life zone occurs at increasingly higher elevations above sea level at lower latitudes which further reduces the solar pathlength through the atmosphere.

Spectral UV-B irradiance was measured at different times of day during the season of maximum solar radiation at several sites along this latitudinal gradient (Caldwell et al. 1980). When these spectral irradiance data are convoluted using a generalized plant action spectrum (Caldwell 1971) or a DNA-damage action spectrum (Setlow 1974), a pronounced gradient of biologically effective UV-B irradiation results (Fig. 2). Maximum irradiance at midday changes by more than an

←

Figure 2. Integrated effective solar UV-B irradiance (global radiation as open symbols, direct beam radiation as closed symbols) as a function of solar angle from the zenith at five represented sites along the arctic-alpine life zone gradient. Integrated effective irradiance was convoluted following the generalized plant action spectrum, here denoted as biologically effective irradiance, or the DNA-damage spectrum denoted as DNA effective irradiance. Composite days were constructed from measurements at different solar angles in locations of geographic proximity under cloudless sky conditions and at similar ozone concentrations for each site. The different symbols represent the different specific site locations. From Caldwell et al. (1980). Copyright 1980, the Ecological Society of America.

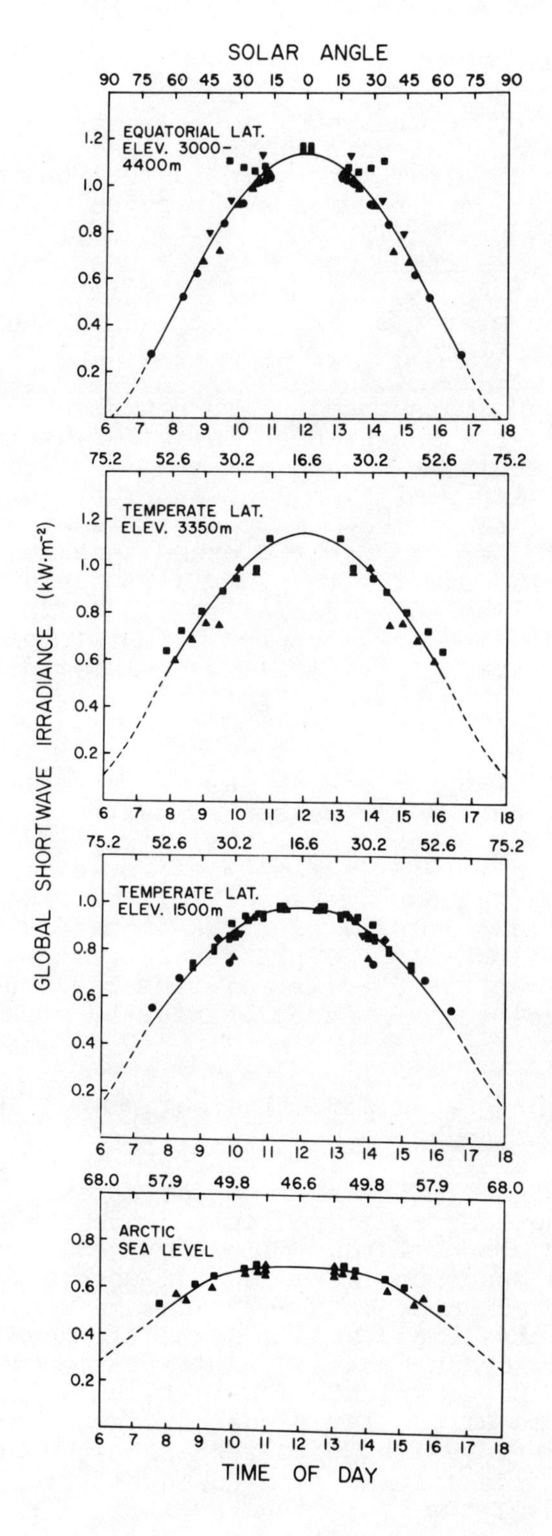
SOLAR ANGLE
90 75 60 45 30 15 0 15 30 45 60 75 90
EQUATORIAL LAT.
ELEV. 3000-
4400m
1.2
1.0
0.8
0.6
0.4
0.2
6 7 8 9 10 11 12 13 14 15 16 17 18
75.2 52.6 30.2 16.6 30.2 52.6 75.2
TEMPERATE LAT.
ELEV. 3350m
GLOBAL SHORTWAVE IRRADIANCE (kW·m⁻²)
75.2 52.6 30.2 16.6 30.2 52.6 75.2
TEMPERATE LAT.
ELEV. 1500m
68.0 57.9 49.8 46.6 49.8 57.9 68.0
ARCTIC
SEA LEVEL
TIME OF DAY

order of magnitude over this gradient when the DNA-damage action spectrum is used in the convolution. Total daily effective solar UV-B irradiation varies by sevenfold during the time of year of maximum solar radiation. The magnitude of this natural gradient is dependent upon the action spectrum used as a weighting function in the convolution. This action spectrum contributes to a radiation amplification effect similar to the calculated increase in biologically effective UV radiation which would result from atmospheric ozone reduction through time. For terrestrial plant species, the two action spectra employed in these convolutions seem to be the most reasonable, although data are quite scarce to substantiate this.

The change in total solar shortwave radiation over the same gradient is much less (Fig. 3). Maximum midday irradiance changes by about 60% in contrast to an order of magnitude in the case of biologically effective UV-B irradiance. Total daily shortwave irradiation is almost constant along this gradient since the longer arctic photoperiods during the season of maximum irradiation compensate for the lower midday irradiance.

OPTICAL PROPERTIES OF LEAVES ALONG A NATURAL UV-B GRADIENT

The pronounced natural solar UV-B gradient in the arctic-alpine life zone provides the opportunity to examine possible modes of adaptation by which higher plant life accommodates this change in radiation environment. The arctic-alpine life zone vegetation along this gradient possesses a similar growth form and experiences many similar environmental characteristics such as low growing season temperatures.

One effective means of solar UV-B radiation avoidance is by selective absorption of this radiation in

←

Figure 3. Total solar shortwave irradiance as a function of solar angle from the zenith at four represented areas along the arctic-alpine gradient. Composite days have been constructed from measurements at different solar angles for locations of geographic proximity under cloudless sky conditions. Different symbols represent different sites of measurement. From Caldwell et al. (1980). Copyright 1980, the Ecological Society of America.

the outer tissue layers of plant leaves. The UV-B transmittance of fresh, single-cell- thick epidermal tissue was measured for plants at several sites along this arctic-alpine gradient (Robberecht et al. 1980). The spectral UV-B transmittance of epidermal tissues from a variety of plant species along this gradient is portrayed in Fig. 4. At temperate and arctic latitudes where the natural solar UV-B flux is much reduced, there was a much greater variation in epidermal UV-B transmittance among different plant species compared to low latitude environments. At the tropical and equatorial high elevation sites in Hawaii and in the Andes, epidermal UV-B transmittance was consistently low for all species investigated--even species which were not originally native to these environments. Thus, the low epidermal transmittance of species in high UV-B radiation flux environments compensates for the higher flux incident on plant leaves. This is portrayed in Fig. 4 where the UV-B radiation flux reaching the inner tissues of the leaf is calculated as an average for species at different locations along the arctic-alpine gradient. Whereas the incident flux on a total daily basis can vary by 7-fold as discussed earlier, the flux reaching the inner, mesophyll, tissues of the leaves is not statistically different along this latitudinal gradient.

The differences in the attenuation of the UV-B radiation by epidermal tissues are not the result of reflectance but rather absorptance within this tissue layer (Robberecht et al. 1980). The most likely compounds involved in this absorption are flavonoids and related phenolic compounds (Robberecht 1981). These pigments exhibit strong UV absorptance and yet are largely transparent in the visible portion of the solar spectrum. This selective filtration is of particular benefit to plant leaves since visible irradiance must penetrate to the inner tissues of the leaf in order to drive photosynthesis. Other secondary plant chemicals may be of importance. For example, alkaloids also exhibit appreciable absorption in the UV-B waveband. Species at low latitudes more commonly possess alkaloids than do species at high latitudes (Levin 1976).

Although the sampling of higher plant species is limited, there does seem to be a clear indication of some adaptation to accommodate higher UV-B flux levels in naturally high UV-B irradiance environments. Included in the survey were exotic plant species, including some agricultural species, native to temperate

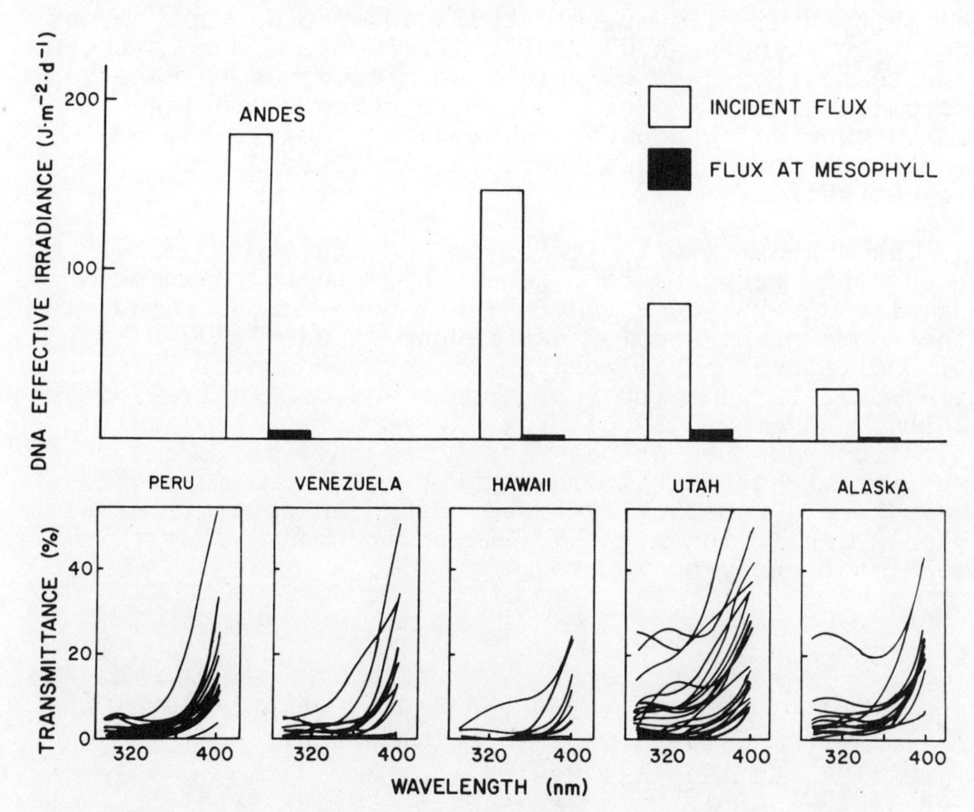

Figure 4. Daily DNA-damaging UV-B irradiance at the seasonal maximum for each location along an arctic-alpine life zone gradient, and the mean effective flux at the mesophyll tissue layer for each species group along this gradient (Top). Epidermal transmittance spectra (Bottom) for each location reveal variability in the magnitude of transmittance at higher latitudes and characteristic transmittance curves for different species. From Robberecht et al. (1980). Copyright 1980, the Ecological Society of America.

latitudes but now growing successfully at high elevations in the tropics. This may have particular bearing on the atmospheric ozone reduction problem as solar UV-B irradiance (weighted with the DNA-damage function) at high elevations in the tropics is more than three times that which would occur with a 16% ozone reduction at temperate latitudes. While one cannot argue with the success of these temperate exotics that do exist in tropical highlands, nothing is known about the degree to which solar UV radiation prevented the establishment of temperate latitude species that do not occur in these high UV flux environments. Nor is there any indication that all plant species would be able to accommodate a 16% ozone reduction which would take place over a few decades. The nature and limits of the acclimatization process certainly require further investigation.

The degree to which this acclimatization is merely a phenotypic response or a genetically-based phenomenon is also an open question. It has been demonstrated that some plant species can reduce epidermal UV-B transmittance when placed in an environment of intensified UV-B irradiation (Robberecht and Caldwell 1978, Robberecht et al. 1980), yet the extent of the acclimatization response or the plasticity that different species might exhibit are not known. There may well have been a genetically-based change in populations of plants inhabiting high UV flux environments, but this remains largely unexplored.

REFERENCES

Angell, J. K. and J. Korshover. 1973. Quasibiennial and long-term fluctuations in total ozone. Monthly Weather Review. 101:426-443.

Axelrod, D. I. 1959. Evolution of the psilophyte paleoflora. Evolution 13: 264-275.

Berkner, L. V., and L. C. Marshall. 1965. On the origin and rise of oxygen concentration in the earth's atmosphere. J. of Atmos. Sci. 22: 225-261.

Caldwell, M. M. 1971. Solar UV irradiation and the growth and development of higher plants. In A. C. Giese [ed.] Photophysiology, Vol. 6, Academic Press, New York, 131-177.

Caldwell, M. M. 1977. The effects of solar UV-B radiation (280-315 nm) on higher plants: implications of stratospheric ozone reduction. In. A. Castellani [ed.] Research in Photobiology. Plenum Publishing Co. New York, 597-607.

Caldwell, M. M., R. Robberecht, and W. D. Billings. 1980. A steep latitudinal gradient of solar ultraviolet-B radiation in the arctic-alpine life zone. Ecology 61: 600-611.

Cline, M. G., and F. B. Salisbury. 1966. Effects of ultraviolet radiation on the leaves of higher plants. Radiat. Bot. 6: 151-163.

Levin, D. A. 1976. Alkaloid-bearing plants: an ecogeographic perspective. American Naturalist 110: 261-284.

Levine, J. S., P. B. Hays., and J. C. G. Walker. 1979. The evolution and variability of atmospheric ozone over geological time. Icarus 39: 295-309.

Margulis, L., J. C. G. Walker, and M. Rambler. 1976. Reassessment of roles of oxygen and ultraviolet light in Precambrian evolution. Nature 264: 620-624.

Nachtwey, D. S. 1975. Linking 254 nm UV effects to UV-B effects via 280 nm equivalents. In D. S. Nachtwey, M. M. Caldwell, and R. H. Biggs [eds]. Impacts of Climatic Change on the Biosphere, Part I: Ultraviolet Radiation Effects (CIAP Mongr. 5). U.S. Department of Transportation, Springfield, VA. 3-54 to 3-73.

Ratner, M. I., and J.C.G. Walker. 1972. Atmospheric ozone and the history of life. J. of Atmos. Sci. 29: 803-808.

Robberecht, R. 1981. Higher plant acclimation to solar ultraviolet-B radiation. Ph.D. dissertation, Utah State University.

Robberecht, R., and M. M. Caldwell. 1978. Leaf epidermal transmittance of ultraviolet radiation and its implications for plant sensitivity to ultraviolet-radiation induced injury. Oecologia 32: 277-287.

Robberecht, R., M. M. Caldwell, and W. D. Billings. 1980. Leaf ultraviolet optical properties along a latitudinal gradient in the arctic-alpine life zone. Ecology 61: 612-619.

Sagan, C. 1973. Ultraviolet selection pressure on the earliest organisms. J. of Theor. Bio. 39: 195-200.

Setlow, R. B. 1974. The wavelengths in sunlight effective in producing skin cancer: a theoretical analysis. Proceedings of the National Academy of Sciences USA 71: 3363-3366.

ATMOSPHERIC EVOLUTION AND UV-B RADIATION

Richard S. Stolarski

NASA/Goddard Space Flight Center
Laboratory for Planetary Atmospheres
Atmospheric Chemistry Branch
Greenbelt, Maryland 20771

ABSTRACT

The evolutionary role of ozone has potentially significant implications for the evaluation of the sensitivity of present ecosystems to ultraviolet radiation changes. The literature on the evolution of atmospheric oxygen and ozone is reviewed indicating that current understanding has the oxygen content of the atmosphere near zero until about 2×10^9 years ago when it rose rapidly to a significant fraction of its current value. Two qualitatively different results appear in the literature for the variation of ozone as O_2 increases; (1) a constant rise from smaller values to a maximum at the present amount and (2) a peak at .1 times the present O_2 level followed by a gradual decrease until the present level is reached. This discrepancy now seems to be resolved in favor of (1).

INTRODUCTION

The topic of the NATO Advanced Study Institute, "The Role of Ultraviolet Radiation in Marine Ecosystems" held in Copenhagen in 1980, emphasized the current role but there are certainly implications for future changes in UV-B ($\approx$ 300 nm) radiation. However, the question of the effects of possible future enhancements of UV-B radiation can also be examined as a question of the evolution of the past radiation environment near the surface of the earth. A significant factor in controlling the UV-B climate at the surface is the total amount of highly absorbing ozone in the atmospheric column. Since the atmosphere originated in a highly reducing state the evolution of ozone had to await the development of free oxygen. Early oxygen in the atmosphere likely developed from the dissociation of water vapor and the subsequent escape of the hydrogen from the earth's gravitational pull. Most of this oxygen went to oxidizing the reduced hydrogen gas released into the atmosphere, with subsequent oxygen increases coming with the advent of photosynthetic life and the burial of organic carbon which could not be recombined into CO_2. The time sequence of this evolution of life is critical and very uncertain.

In addition the question of how ozone evolves as a function of O_2 must be resolved. This question has attracted the increasing interest of aeronomers who now have at their disposal ozone models developed to study the effects of atmospheric pollutants. Several models of varying sophistication and complexity have been applied to this problem. The two classes of results that have been obtained are 1) models which show a maximum in the ozone column for about one tenth of the present level of O_2 with a 20-40% decline since that time and 2) those which show a monotonic increase in O_3 with increasing O_2 all the way to the present level. Whether the present biosphere has been evolving under a continuous decrease in the amount of UV-B radiation or has been evolving under an increasing UV-B exposure should be of paramount importance in interpreting the UV-B tolerance of organisms. The following sections give review in more detail on the present understanding of oxygen and ozone evolution.

ATMOSPHERIC OXYGEN EVOLUTION

It is generally accepted that the Earth's atmosphere is of secondary origin, that is, it results from outgassing of volatiles from the interior of the planet. This is deduced from the fact that the noble gases are many orders of magnitude less abundant relative to other elements in the Earth than they are in the universe as a whole. Because these gases do not form compounds that can be trapped in the solid planet during accretion it is assumed that they were swept away with all of the other components of the primary atmosphere early in its history. The volatile compounds released during the outgassing history of the Earth almost certainly were highly reducing; containing only that oxygen which was combined in water vapor and carbon dioxide. The history of the evolution of the present oxidizing atmosphere from an early reducing atmosphere depends on the mechanisms for extracting the oxygen from H_2O and CO_2 and then getting rid of the residual H and C to prevent recombination.

The initial process which will take place in an abiological atmosphere is the photodissociation of water vapor followed by the eventual escape of the hydrogen from the Earth's gravitational pull. Although carbon dioxide also is dissociated by solar ultraviolet radiation it cannot be a significant contributor to non-biologically produced oxygen because the carbon atom is far too heavy to escape the Earth's gravitational pull and will thus recombine with the oxygen freed in its dissociation leaving no permanent free oxygen. Only a small percentage of the water vapor dissociations actually lead to hydrogen escape. Most of the time the products recombine to form water before escape can occur. Berkner and Marshall (1965), in a major early work on the rise of atmospheric oxygen, argued that pre-biological oxygen levels could not have exceeded 10^{-3} of the present atmospheric level (P.A.L.) because of the shielding of the water vapor photolysis by the O_2 which is formed. The O_2, as its concentration grows, absorbs an increasing share of the ultraviolet solar radiation which would otherwise be available to dissociate water vapor. Brinkmann (1969) pointed out that this argument ignored the significant penetration of UV at wavelengths corresponding to the very deep minima in the Schumann-Runge absorption bands of O_2. He deduced that as much as .25 P.A.L. of O_2 could be biologically produced. Since that time Hunten (1973) has demonstrated how the escape flux of hydrogen from the Earth's atmosphere (also Venus and Mars) is controlled or limited by the total flux of hydrogen (as H_2, CH_4 or H_2O) through the tropopause and not by the dissociation rate of water vapor. Thus, the earlier estimates of H escape were almost certainly too high as pointed out by Margulis et al. (1976) because they assume that each H atom produced in H_2O photodissociation escaped. Subsequently, Walker (1977, 1978) considered the importance of the oxidation of reduced volcanic gases such as H_2 in limiting the possible O_2 concentrations. If the volcanic H_2 flux into the bottom of the atmosphere is greater than the escape flux, any O_2 formed will rapidly recombine and the overall effect must be a steady decrease in the oxidation state of the atmosphere as more reduced material is coming in than going out. Kasting et al. (1979) have performed detailed altitude dependent calculations of this effect confirming that ground level O_2 concentrations must be less than 10^{-9} P.A.L. for a variety of assumptions. Thus, the present understanding indicates that until the development of photosynthetic organisms the atmosphere was virtually totally anaerobic.

Photosynthesis accelerates the natural rate of conversion of CO_2 to O_2. As already stated this is of no particular consequence unless the carbon can be kept out of the way of the O_2 to prevent recombination either through the decay of dead organic matter or through respiration. The carbon must be buried in sediments which will not gain contact with the atmosphere for a significant time period (≈ 100 million years). As biospheric reduced carbon accumulates and begins to be cycled into sediments, free oxygen is released. When this flux of free oxygen becomes larger than the flux of reduced hydrogen and the reduced materials in the surface layers of the crust have been oxidized, the net oxidation state of the atmosphere can begin to increase. The time scale of this oxygen evolution is in some doubt but it has been argued by Cloud (1973) that a fairly rapid transition from an anaerobic atmosphere to an atmosphere with somewhat near the present oxygen content took place approximately 2×10^9 years ago. In order to understand the evolutionary effects of ultraviolet light it is necessary to know not only the time history of oxygen growth but also the relationship between oxygen and ozone which provides the UV absorption.

DEVELOPMENT OF THE OZONE SHIELD

Given a time history for oxygen in the Earth's atmosphere, the further question arises as to how the total column of ozone depends upon the oxygen level. Berkner and Marshall (1965) were the first to postulate a specific dependence of ozone content on oxygen. Their result was based on physical reasoning but not on self-consistent photochemical model computation. Their deduced dependence on the O_2 level in P.A.L. is illustrated in Figure 1 along with several subsequently published results which will be discussed below. Ratner and Walker (1972) carried out the first self-consistent model calculation. The model used was an extremely simple one but illustrated many important

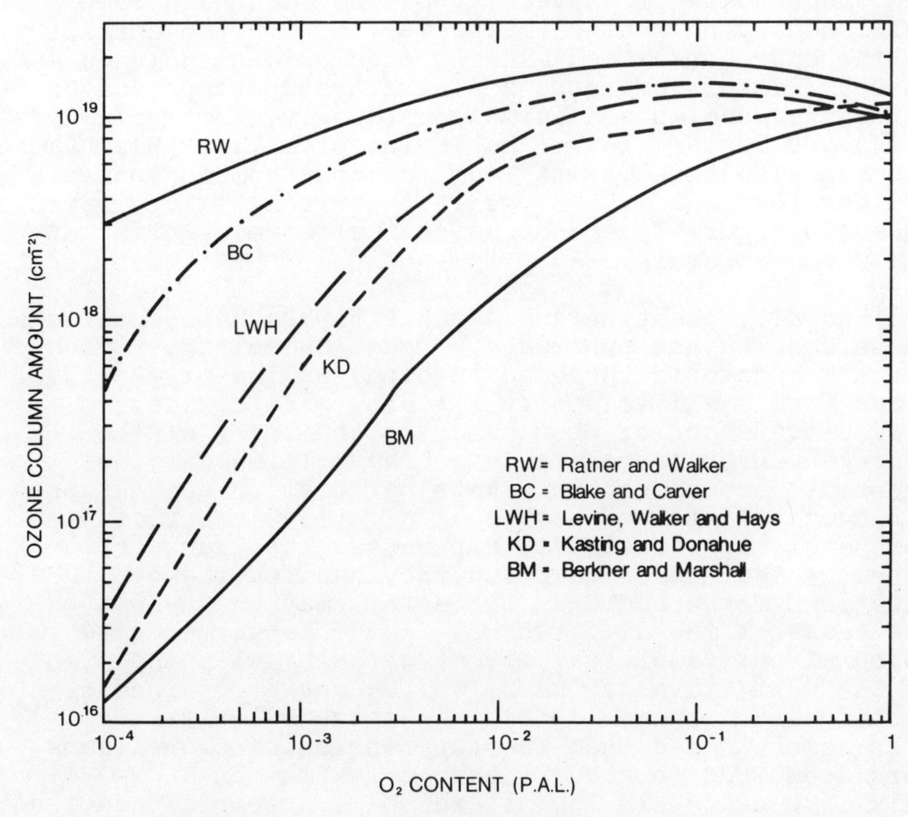

Figure 1. Compilation of the results of several authors on the relationship between the average column amount of ozone in molecules per cm^2 and the atmospheric O_2 content in fraction of the present atmospheric level (P.A.L.).

features of the ozone evolution problem. They assumed the so-called Chapman mechanism of pure oxygen photochemistry in which only the reactions involving atomic oxygen, molecular oxygen and ozone are considered. All of the catalytic loss processes currently included in assessing the effects of human inputs of chemicals were ignored as was the transport of ozone from regions of excess production to regions of excess loss. The result of this computation was that the ozone column content increased until 0.1 P.A.L. was reached and then declined by the order of 50% between that level and 1 P.A.L. Under this scenario, from the time of 0.1 P.A.L. to the present, biological evolution would have taken place under a continuously increasing UV stress.

Since that time several model studies have been published using photochemical models with the chemistry of the hydrogen oxides and nitrogen oxides included as loss processes for ozone and with transport processes for these species and for ozone (Hesstvedt et al., 1974); Blake and Carver, 1977; and Levine et al., 1979). These studies also showed a maximum in ozone column content at less than 1 P.A.L. oxygen. Their results are reproduced in Figure 1 demonstrating that some significant differences remain.

Recently, Kasting and Donahue (1980) have challenged these results and published a model calculation which shows a monotonic increase in ozone as the oxygen level rises from small values to 1 P.A.L. Their model includes transport of ozone and the chemistry of the hydrogen and nitrogen oxides. They claim that the major difference between their calculation and the previous calculation of Levine et al. (1979) is that Levine et al. effectively suppressed the transport of ozone by the manner in which they handled the combined equations for O and O_3. While this may or may not be the cause of the differences, I have performed some calculations (Stolarski, 1980) agreeing with the sign of the derivative of the O_3 vs. O_2 curve obtained by Kasting and Donahue (1980). In these calculations I also demonstrated that the suppression of ozone transport does tend to give rise to a maximum in O_3 vs. O_2. This is because, in the absence of transport, reduction of O_2 simply shifts the ozone production by ultraviolet radiation downward to a lower altitude. At lower altitudes the steady-state ozone concentration is larger because the total atmospheric density is larger (N_2 is assumed not to vary with the O_2 level in all of these calculations, i.e., N_2 still declines exponentially with altitude but is held at its present amount). Thus, as O_2 is decreased the ozone layer descends and increases in magnitude. Eventually the lower part of the ozone layer reaches the ground and the column content begins to decrease again. The presence of transport processes for ozone tends to negate such an effect by moving ozone downward from its peak production region to the regions of higher N_2 density where its concentration can maximize. In such a model the controlling factor is then the production term from O_2, and ozone decreases monotonically with decreasing O_2. Another factor which may cause a difference in the ozone vs. O_2 curve is the treatment of the troposphere and the removal processes for atmospheric gases by rainout or washout. Indeed, Levine (private communication, 1980) says that the inclusion of a better representation of these processes in his model yields a result similar to Kasting and Donahue. This agrees, at least qualitatively in the sign of the slope, with some preliminary results I have calculated on our steady-state tropospheric-stratospheric photochemical model. I would thus conclude that despite the apparent discrepancies in the literature, the Kasting and Donahue result is the one most likely to be correct.

CONCLUSION

Geological and paleontological evidence along with the results from atmospheric photochemical models seem to indicate that oxygen evolved fairly rapidly about 2 billion years ago to a significant fraction of its present value (whether that fraction is 1% or 10% or more or less than that is still subject to much speculation). It is only then that a significant ozone shield could develop and affect the evolution of UV sensitivity and repair mechanisms for UV damage in cells. Studies relating UV sensitivity to the evolutionary sequence of organism evolution could yield to fruitful results concerning the effects of future UV changes.

ACKNOWLEDGEMENT

I would like to thank Joel Levine of NASA Langley Research Center for providing information concerning his latest results prior to publication and J.C.G. Walker of the University of Michigan for discussions concerning the role of hydrogen in titrating the oxygen level in the atmosphere.

REFERENCES

Berkner, L. V. and L. C. Marshall. 1965. On the origin and rise of oxygen concentration in the Earth's atmosphere. J. Atmos. Sci. 22: 225-261.

Blake, A. J. and J. H. Carver. 1977. The evolutionary role of atmospheric ozone. J. Atmos. Sci. 34: 720-728.

Brinkmann, R. T. 1969. Dissociation of water vapor and evolution of oxygen in the terrestrial atmosphere. J. Geophys. Res. 74: 5355-5368.

Cloud, P. 1973. Paleoecological significance of the banded iron-formation. Econ. Geo. 68: 1135-1145.

Hesstvedt, E., S.-E. Henriksen and H. Hjartarson. 1974. On the development of an aerobic atmosphere. A model experiment, Geophys. Nov. 31: 1-8.
Hunten, D. M. 1973. The escape of light gases from planetary atmospheres. J. Atmos. Sci. 30: 1481-1494.
Kasting, J. F., S. C. Liu and T. M. Donahue. 1979. Oxygen levels in the prebiological atmosphere. J. Geophys. Res. 84: 3097-3106.
Kasting, J. F. and T. M. Donahue. 1980. The evolution of atmospheric ozone. J. Geophys. Res. 85: 3255-3263.
Levine, J. S., P. B. Hays and J.C.G. Walker. 1979. The evolution and variability of atmospheric ozone over geological time. Icarus 39: 295-309.
Margulis, L., J.C.G. Walker and M. Rambler. 1976. Reassessment of roles of oxygen and ultraviolet light in precambrian evolution. Nature 264: 620-624.
Ratner, M. I. and J.C.G. Walker. 1972. Atmospheric ozone and the history of life. J. Atmos. Sci. 29: 803-808.
Stolarski, R. S. 1980. The evolution of the atmospheric ozone shield, EOS, Transactions of the American Geophysical Union. 61: 1054.
Walker, J.C.G. 1977. Evolution of the atmosphere, Macmillan Publishing Co., New York.
Walker, J.C.G. 1978. Oxygen and hydrogen in the primitive atmosphere. Pure Appl. Geophys. 116: 222-231.

SOME CONSIDERATIONS ON THE ECOLOGICAL AND EVOLUTIONARY EFFECTS OF SOLAR UV

John Calkins (1) (2)

Department of Radiation Medicine (1) and
School of Biological Sciences, (2)
University of Kentucky, Lexington, KT 40536

The habitats of the earth are subjected to myriads of measurable factors; chemical, physical, and biological, which could be of significance to the success or even survival of potential inhabitants.

The biological nature and behavioral patterns of successful living organisms must provide adequate accommodation for the injuries inflicted by the various agents in their environment. If the earth were an equilibrium system, then the biota would be stable, possessing the tolerance required to maintain their position in a stable biosphere. But the biosphere is not a system operating at a stable equilibrium; it is a dynamic system, changed by evolution of new organisms, changed by variation in the physics and chemistry of the earth and solar system, and in particular, changed by the evolution of mankind, an organism with the potential to produce profound worldwide changes of the chemistry and biology of the earth's surface. Along with the human capacity to change the earth, humans also have a need and a desire to anticipate the consequences of human behavior and to select pathways of development which will provide long-term benefits to humanity and to the other living organisms which share our earth.

Choice of the best pathway of development requires anticipation of the consequences of the available alternatives. Prediction of ecological consequences is a science and an art still in its infancy. The basis of prediction of the action of changing an environmental factor is understanding the present status regarding the factor and the biological actions to be expected from the incremental change. Solar UV does not directly stimulate human sensory receptors and because the majority of organisms tolerate their normal exposure to sunlight, there is a natural tendency to regard solar UV as a benign or at least an insignificant factor in the environment. An environmental agent which is "significant" at its present level deserves analysis before its level is increased. However, how can "significant" be defined and quantitated? Perhaps factors which determine the nature and evolution of organisms could be called significant. In general terms, there are three ways in which a harmful agent common in the environment of an organism could be termed significant; 1) if an encounter with the agent is injurious or especially, lethal, 2) if the organism specifically avoids the agent because it would be lethal or injurious, and 3) if the organism has an identifiable trait or characteristic required solely or primarily to improve tolerance upon exposure to the agent in question. Factor 1 affects population size, factor 2 affects range, and factor 3 affects the utilization of the biological resources of the organism.

In the case of the human species there is abundant evidence that solar UV is a significant ecological factor. Thousands of humans die each year from sunlight related melanomas and other types of skin cancer. Melanin formation in the skin and repair mechanisms absent in humans suffering from xeroderma pigmentosa are clearly mechanisms directed primarily toward coping with solar UV exposure. Individuals lacking, or with diminished function of these protective capacities are viable if they avoid solar UV exposure. Humans, in common with many large animals, reduce their light exposure when solar UV is most intense.

The evidence for a significant role of solar UV for aquatic organisms is less well established. It is very simple to demonstrate that exposure to even a few hours of summer solar UV incident on natural waters is lethal to many organisms. It is quite another matter to evaluate the amount of mortality sunlight contributes to the actual populations of organisms, considering the numerous other lethal agents and the shielding capacity of natural waters. There is a quantitative correlation of incident solar UV and reduction of coliform count in natural waters (Calkins, et al. 1976), but the ultimate proof is lacking, there is no autopsy report on the dead bacteria.

Likewise, there is extensive circumstantial evidence that the known DNA repair mechanisms function to reduce sensitivity to sunlight, (Harm, 1969, Resnick, 1970). It is, of course, true that "dark repair" systems also protect the integrity of the DNA from chemically induced lesions which could be the primary function of such systems. One widespread DNA repair system (Cook and McGrath, 1967) enzymatic photoreactivation, reduces sensitivity to solar UV and has not been shown to function on any other naturally occurring DNA lesion. The various DNA repair mechanisms are not individually required for viability since mutants defective for various forms of repair can be cultured under laboratory conditions. Protozoa isolated from Mammoth Cave, presumably adapted to life without solar UV, have been found to lack a functional photoreactivation capacity (Calkins and Griggs, 1969) (Barcelo, 1980a).

As with higher organisms, pigmentation appears to relate to resistance to UV-B (Barcelo, 1980b, Barcelo this volume). Organisms not exposed to sunlight, such as cave or deep sea species, often show little pigmentation and are evidently better adapted to the light-free environment than similar organisms which expend their biochemical resources in pigment formation. Natural solar UV-B exposure always occurs with visible light and UV-A of much greater intensity. It can be shown that UV similating the UV-B components of sunlight is much more potent than 365 nm visible or 253.7 nm radiation in stimulation of avoidance in some organisms (Barcelo, 1981, Calkins, unpublished observations). Both laboratory and field observations suggest active avoidance of the injurious component of solar UV by common aquatic organisms (Barcelo and Calkins, 1978, 1979, 1980). Taken together, the examples cited above leave little doubt that solar UV is, for many organisms, a significant environmental factor.

There are doubtless many organisms so well adapted to solar UV exposure that they would, in fact, benefit if solar UV were increased. If less well UV adapted competitors and preditors were killed or reduced in range, then species more tolerant to UV would prosper. It is assumed that upon increased UV a new natural balance of species would arise which was stable and utilized the natural resources available. One should not assume that there will be an easy transition to the new state nor that from the viewpoint of humans who stand at the apex of aquatic food web, the new system will be just as desirable as the old. It is evident that the required adaption to solar UV in most organisms does not arise from a single mechanism providing more than adequate protection. The common pattern for coping with solar UV found by aquatic organisms is 1) pigmentation, often both constitutive and inducible, 2) two or three different repair systems combined with 3) avoidance of exposure during periods of the most intense solar UV. Weissman (1889) observed that the best adaption will provide only the minimal resistance required and that is the situation we find with solar UV. When solar UV exposure and tolerance are compared, the two are remarkably close in the organisms where these factors can be measured (Calkins and Thordardottir, 1980, Damkaer et al. 1980).

If solar UV were not presently a significant ecological factor, one might hope that it might be increased without detectable effects. However, when it is evident that the present precarious balance of tolerance and exposure is maintained through multiple compensating mechanisms, then it is reasonable to expect that adaption to higher levels of solar UV will extract a definite price. If marine algae were to compensate by avoiding exposure, then they would receive less light for photosynthesis, a limiting factor in the nutrient rich arctic and antartic. To shield themselves through pigmentation, the nanoplankton would have to become larger, surface to volume ratios would be reduced and the efficiency of use of nutrients decreased. Increased shielding in the microscopic zooplankton would divert biochemical potential into different pathways, probably increasing the average size. Avoidance of high UV locations reduces the range of the organisms and would also reduce the light available to visual feeding organisms.

Humans have already embarked on a course which will modify solar UV exposure. The benefits arising from halogenated compounds are widespread in human society. There is a need to devote adequate resources to anticipate the ecological consequences of stratospheric pollution and form enlightened choices among alternative pathways of technological development.

REFERENCES

Barcelo, J. A. 1980. Ph.D Thesis, The University of Kentucky, Lexington, Kentucky.

Barcelo, J. A. 1980b. Photomovement, pigmentation and UV-B sensitivity in planaria. Photochem. Photobiol. 32: 107-

Barcelo, J. A. 1981. Photoeffects of visible and ultraviolet radiation on the two-spotted spider mite *Tetranychus urticae*. Photochem. Photobiol. 33: 703-706.

Barcelo, J. A. and J. C. Calkins. 1978. Positioning of aquatic microorganisms in response to visible light and simulated solar UV-B radiation. Photochem. Photobiol. 29: 75.

Barcelo, J. A. and J. Calkins. 1979. The relative importance of various environmental factors of the vertical distribution of the aquatic protozoan *Coleps spiralis*. Photochem. Photobiol. 31: 67.

Barcelo, J. A. and J. Calkins. 1980. The kinetics of avoidance of simulated solar UV radiation by two arthropods. Biophys. J. 32: 921.

Calkins, J., J. D. Buckles and J. R. Moeller. 1976. The role of solar ultraviolet radiation in "natural" water purification. Photochem. Photobiol. 24: 746-749.

Calkins, J. and G. Griggs. 1969. Photoreactivation of UV reactivation of protozoa. Photochem. Photobiol. 10: 445-449.

Calkins, J. and T. Thordardottir. 1980. The ecological significance of solar UV radiation on aquatic organisms. Nature 283: 563-566.

Cook, J. S. and J. R. McGrath. 1967. Photoreactivating-enzyme activity in metazoa. Proc. Natl. Acad. Sci. US 58 1359-1365.

Damkaer, D. M., D. B. Dey, G. A. Heron, and E. F. Prentice. 1980. Effects of UV-B radiation on near-surface zooplankton of Puget Sound. Oecologia (Berl.) 44: 149-158.

Harm, W. 1969. Biological determination of the germicidal activity of sunlight. Radiat. Res. 40: 63-70.

Resnick, M. A. 1970. Sunlight-induced killing in *Saccharomyces cerevisiae*. Nature 226: 377-378.

Weissmann, A. 1889. Essays upon heredity and kindred biological problems. E. B. Poulton, S. Schonland and A. E. Shipley [ed.]. Oxford Clarendon Press.

THE ATTENUATION OF LIGHT BY MARINE PHYTOPLANKTON WITH SPECIFIC REFERENCE TO THE ABSORPTION OF NEAR-UV RADIATION

Charles S. Yentsch and Clarice M. Yentsch

Bigelow Laboratory for Ocean Sciences
West Boothbay Harbor, Maine 04575

INTRODUCTION

Our hypothesis is that early planktonic algae developed ultraviolet protective screens similar to those found in marine invertebrates (for example see Cheng et al. (1978)). We present here evidence that some algae with ultraviolet screens are still extant. In present day environments of intense sunlight, these ultraviolet screens may yet today be a selective advantage for primitive photosynthetic organisms.

If present day thinking concerning the formation of the earth's atmosphere is correct, then, during the course of evolution, photosynthetic organisms have been faced with changing levels of ultraviolet radiation. With a gradual appearance of oxygen and thus ozone in the atmosphere (i.e. as photosynthesis became more common), the amount of ultraviolet radiation reaching the earth's surface must have declined. Early photosynthetic organisms must have been subjected to intense ultraviolet radiation. The cellular constitutents most susceptible to destruction by ultraviolet radiation are the nucleic acids of DNA and RNA (Watson, 1970). Organisms which appear later in the sequence of evolution were exposed to less intense ultraviolet radiation, and therefore less potential damage to their nucleic acids.

In the photosynthetic literature, there is speculation as to what these screens might be. Accessory pigments have been implicated (Clayton, 1958). Of particular implication have been the carotenoids, because these pigments absorb ultraviolet wavelengths as well as visible wavelengths. From the general viewpoint of the optics of biological systems, one cannot dismiss the modifications due to cellular membrane development and cell wall complexity as being factors important in screening and/or minimizing the effects of ultraviolet radiation.

Whatever the mechanism to support the screening hypothesis, the appearance of screens should be consistent with what is known concerning the evolutionary age of the organism. Simply, there should be evidence to support the idea that the earliest appearing organisms had some means to protect their genetic and photosynthetic machinery.

Our discussion begins by describing evolutionary relationships among the algae. This is followed by a comparison among algae as to the manner in which light is attenuated.

EVOLUTIONARY DEVELOPMENT OF PLANKTONIC MARINE ALGAE

There exists no single or unified voice as to how to distinguish phylogenetic relationships. The common approach uses "sets" of criteria for characterizing relationships. In the most general sense, these are made up of morphological and physiological factors which, when identified, suggest relationships. For the sake of this discussion, we have composed an evolutionary scheme which we believe to represent a presently acceptable course for evolutionary trends (Figure 1). The critical elements in this scheme are denoted by the stippled arrows which indicate the appearance of both physiological and morphological features, namely the presence of a defined nucleus and then nuclear sophistication. The scheme shown in Figure 1 argues that the most primitive algal groups are the cyanobacteria (blue green algae) and the dinoflagellates. While the ozone layer was initially forming, these organisms underwent photosynthesis, therefore oxygen evolution. This was in an environment of intense ultraviolet radiation. Thus one would not be surprised to find that some of the cyanobacteria and dinoflagellates possess some means of screening out ultraviolet radiation. Indeed this is the case as can be observed from absorption spectra.

The appearance of other algal groups, namely the

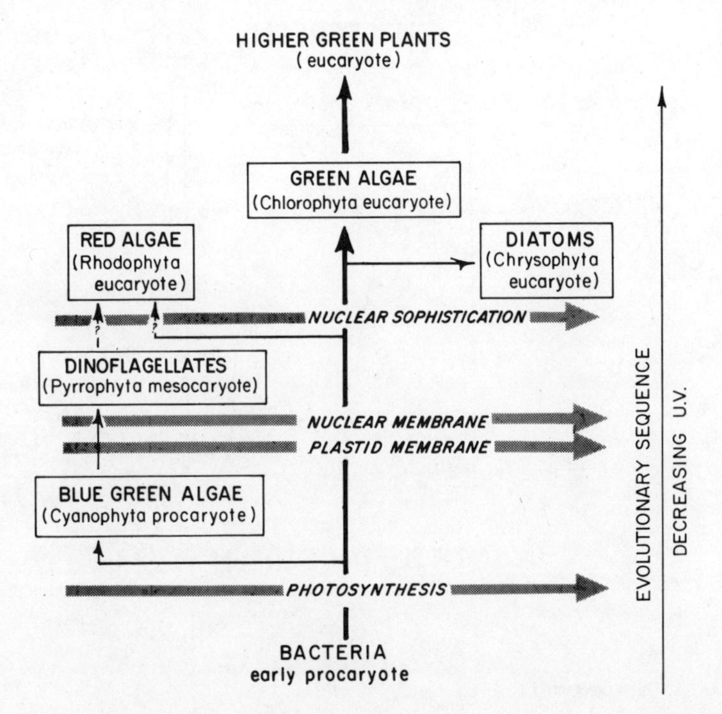

Figure 1. Schematic representation of evolutionary trends with specific reference to the various algal groups. Hatched horizontal lines indicate major evolutionary advances. These are namely: photosynthesis, evolution of the plastid and nuclear membranes, and nuclear sophistication--organization of the DNA and RNA. The evolutionary scheme proceeds from the bottom, with the most primitive organism being at the bottom and the most advanced being at the top. The most primitive atmosphere, with excessive ultraviolet, is thus at the bottom, with decreasing amounts of ultraviolet as one proceeds to the top or present environment.

reds, greens, browns and diatoms, paralleled the formation of the ozone layer. These groups have sophisticated, typically eucaryotic organization of their nuclear material as well as cellular organelles.

DIFFUSE ATTENUATION OF LIGHT BY DIFFERENT GROUPS OF ALGAE

The manner in which algae absorb and scatter light is largely a function of the absorption of photosynthetic pigments and cell membrane and wall characteristics. The differences among various types of algae are largely

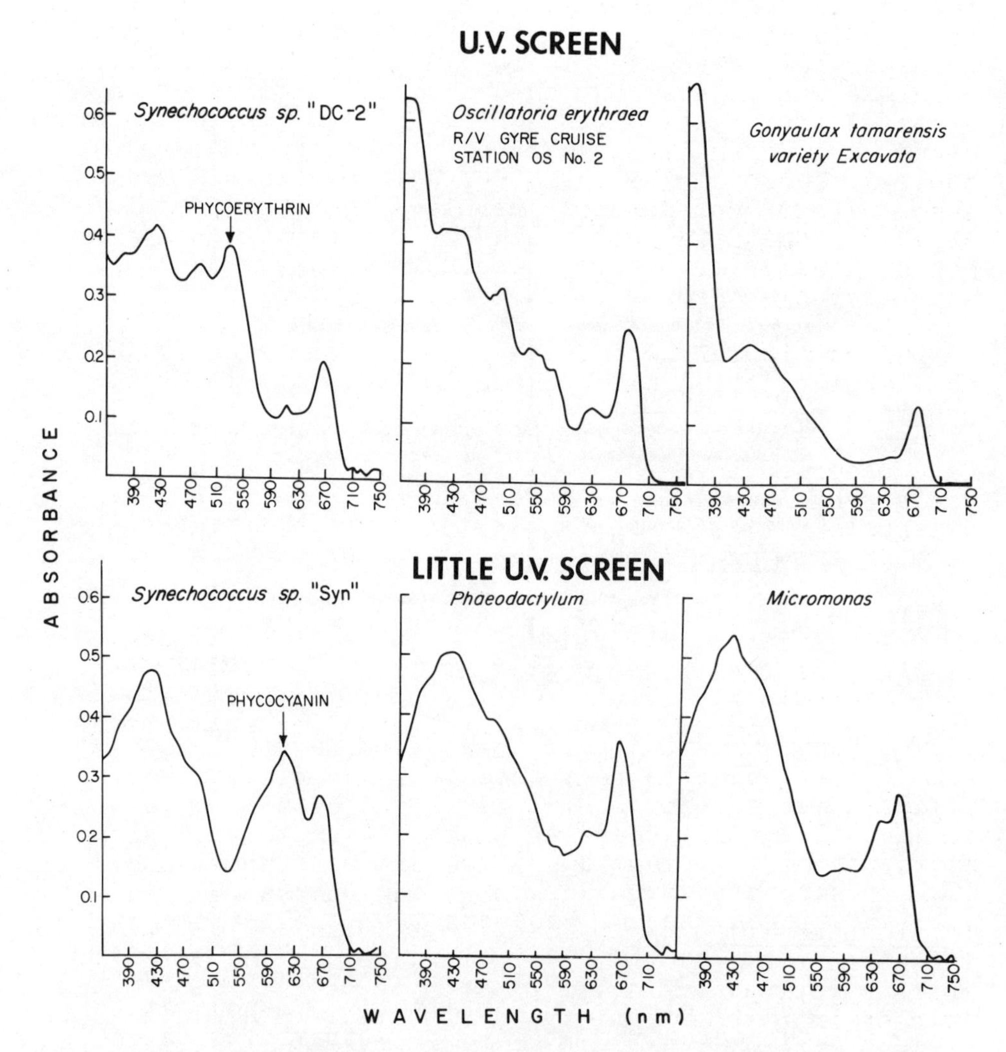

Figure 2. Attenuation spectra of various algae. Those with ultraviolet screens (high absorbance from 350-450 nm) are: The cyanobacterium Synechococcus sp. "DC-2" (culture), the blue green alga Oscillatoria erythraea (natural populations), and the New England red tides dinoflagellate Gonyaulax tamarensis var. excavata (culture). Those with little ultraviolet screening include: the cyanobacterium Synechococcus sp. "Syn" (culture), the diatom Phaeodactylum tricornutum (culture). and the flagellate Micromonas (culture).

due to the accessory pigments which vary from group to group. Accessory pigments are characteristically used to separate the algae groups. In general, major groups of algae can be distinguished from one another by the shape of their attenuation spectra alone. Although all groups are dominated by chlorophyll absorption, in the central region of the visible spectrum (500-600 nm), there exists considerable variation among different algae. These differences are due to the accessory pigments, namely various lipid-soluble carotenoids and/or water-soluble chromoproteins.

Of particular interest, yet often unrecognized, are the large spectral differences in the near ultraviolet spectral region (285-350 nm) among various groups of algae. Figure 2 shows examples of the extremes that we have observed in algal cultures. All species of diatoms and green algae that we have examined have much less attenuation at 350 nm than do the cyanobacteria, other blue-greens, dinoflagellates and red algae.

These differences are noted in natural populations as well as laboratory cultures. It is informative to examine the present day habitats of the organisms which have ultraviolet screens. The coccoid cyanobacteria, e.g. *Synechococcus* sp. are found distributed throughout the water column in temperate, subtropical and tropical waters. The primary accessory pigment of these cyanobacteria is phycoerythrin, as is the case for *Oscillatoria erythraea* (Moreth and Yentsch, 1971).

The blue green alga, *Oscillatoria erythraea* ("*Trichodesmium*") is a filamentous form visible with the naked eye. It occurs in the clear, warm waters of the subtropical and tropical open ocean. It is frequently seen in patches covering large areas of the ocean surface, where it is subjected to intense sunlight.

Red tide dinoflagellates are also noted in extensive surface patches. Some species exist in the subtropical areas while others are common in the temperate environments, e.g. *Gonyaulax tamarensis*. Like many species of dinoflagellates, *G. tamarensis* migrates vertically and thus occupies the surface waters during midday when the ultraviolet is most intense.

LIGHT ENERGY TRANSFER

The monochromatic action spectra for all algae

examined declines markedly at wavelengths shorter than 400 nm. McLeod (1958) measured the action spectra in seven species of marine and freshwater unicellular algae, covering the range of wavelengths between 250-700 nm. For the diatoms or green algae tested, there was little difference between species throughout the range 250-430 nm. Unfortunately, no dinoflagellates were tested during this study. However, the organisms with chromoproteins (phycocyanin and phycoerythrin) as accessory pigments showed little activity at wavelengths shorter than 450 nm. It should be recognized that all of the organisms tested by McLeod photosynthesized to some degree at short wavelengths. They speculated that this could be due to the short wavelength light absorption by chlorophylls and accessory pigments and the protein associated with the pigment complex.

By measuring _in vivo_ fluorescence (method: cells are filtered onto 2 cm Gelman type A glass fiber filter and held upright and moist in the light beam of a Baird-Atomic spectrofluorometer), we have observed the relative capability of light transfer in the near ultraviolet region (Yentsch et al. 1979). Our results are similar to McLeod's data in that the action spectra for chlorophyll _a_ fluorescence in green algae and diatoms nearly match the light attenuation. This signals that all light absorbed is used in the photosynthetic process. Contrarily, in the blue greens, dinoflagellates, and red algae measured, the high attenuation of light at short wavelengths is not wholly transferred. Thus, these short wavelengths are not used in the photosynthetic process. This further suggests a completely different function of this absorption which may indeed be as a protective screen.

In other research, McLeod and McLachlan (1958) using cultures of unicellular freshwater and marine algae, compared the rate of photosynthetic carbon-14 uptake in visible and ultraviolet light. They found diatoms to be more sensitive to ultraviolet light than green algae, with respect to the carbon-14 uptake. They also observed that green algae were more sensitive to ultraviolet light when the cells were dividing, although diatoms showed little variability in sensitivity to ultraviolet light when the cells were dividing.

MEMBRANE AND NUCLEIC ACID EVOLUTION

Ultraviolet screens parallel nucleic acid arrangement. Basic cell types are divided into procaryotes and eucaryotes. Procaryotes are considered to be the most primitive plant forms. These organisms, including the bacteria and cyanobacteria, have cellular pigments and cellular nucleic acids located in regions within the cell, but not within membranes; therefore they do not contain discrete organelles. Eucaryotes are considered to be the most highly evolved and advanced cells and include most plant and animal forms. These plants have cellular pigments located in organelles called chloroplasts and both plants and animals have cellular nucleic acids located within well-defined membranes and thus have discrete nuclei. All eucaryotes are considered to have evolved after considerable oxygen existed in the atmosphere: no true eucaryote measured to date contains ultraviolet screens.

Dinoflagellates have been termed "mesocaryotes" by Dodge (1964) to convey the intermediate position (between procaryotes and eucaryotes) which they occupy. Dinoflagellates are the subject of investigation in that they do possess some characteristics which may prove helpful in analyzing the chromosome structure and the regulation of gene action in all eucaryotes. Their positon is summarized in Table 1. For the purpose of discussion, the mesocaryotes have been placed in a separate column. The presence of both nuclear and plastid membranes present in the mesocaryotes likely serve to shield the nuclear material from some damaging radiation. Other factors with significance to the ultraviolet exposure are marked with an asterisk. Of particular note is that morphological structure of the chromosome. That is 1) narrow diameter of the chromatin (DNA) strand; 2) circular or modified circular arrangement (cf. double helical arrangement) of the chromosome; 3) condensed state of chromosomes of dinoflagellates throughout the cell cycle; 4) lack of histone proteins; 5) abundant DNA on a percentage per cell basis.

DISCUSSION

It is important to note that near ultraviolet light is considered to be the most detrimental to the fundamental nuclear material, namely DNA and RNA. Therefore, if an ultraviolet protective screen would be expected, it should exist in the very region where protection would be most valuable, that would be 285-350 nm.

The hypothesis we have put forth is that the very early photosynthetic organisms needed some protection

TABLE I

Nuclear organization	Procaryotes (bacteria & cyanobacteria)	Mesocaryotes (dinoflagellates)	Eucaryotes (greens & other plants)
a) histone proteins associated with the DNA *	absent	absent	present
b) diameter of chromatin fibrils*	3-6 nm	3-6 nm	20 nm
c) membrane attached chromosomes	yes	yes	no
d) mitotic spindle	absent	absent	present
e) continuous DNA synthesis light/dark *	yes	some	no
f) chromatin fibrils arranged in arched swirls	yes	yes	no
g) permanent condensation of chromosomes throughout the cell cycle *	yes	yes	no
h) high degree of base substitution in DNA *	(bacteriaphage) yes	yes	no
i) abundance of DNA per cell as percent *	yes	yes	no
j) presence of nonsense "selfish" DNA *	yes	yes	no
k) chromatin surrounded by a nuclear membrane *	no	yes	yes

from the ultraviolet light passing through a relatively thin ozone layer of the primitive atmosphere. Thus, in the sequence of evolution, early organisms needed more protection from the ultraviolet than organisms appearing at a later period. The hypothesis further proposes that those organisms evolving mechanisms for screening out the ultraviolet reside in habitats where present day levels of ultraviolet are intense.

The hypothesis is supported by direct evidence of the light attenuation characteristics of different phytoplankton and the manner in which nucleic acids are arranged in evolutionary groups.

On a negative note, it might be argued that with the appearance of eucaryotes, screens were not needed for all species. What we are seeing is merely an adaptation for those residing in the surface layers. Thus, crucial to the argument is evolutionary age. It should be noted that ultraviolet absorbing screens have been suggested as a diagnostic feature for the establishment of phylogenetic relationships in attached algae (Iwamota and Aruga, 1973).

The hypothesis suggests that present day organisms with ultraviolet screens are protected from existing levels of ultraviolet light. This should be rigorously tested, since at the present time no body of data comparing photosynthesis or growth on phytoplankton with and without ultraviolet screens is available.

ACKNOWLEDGEMENTS

This work was supported in part by NASA grant nos. NAS5-25722 and NAS5-22948, and the State of Maine.

The authors thank Dave Phinney and Kay Kilpatrick for technical assistance; also P. Colby who typed the manuscript and J. Rollins who prepared the illustrations.

This is Bigelow Laboratory Contribution No. 80039.

REFERENCES

Cheng, Lanna, Maurice Dovek and David A.I. Gorins. 1978. UV absorption by Gerrid Cuticles. Limnol. & Oceanogr. Vol. 23, No. 3 554-556.

Clayton, R. K., W. C. Bryan and A. C. Frederick. 1958. Some effects of ultraviolet on respiration in purple

bacteria. Arch. Mikrobiol. 29: 213.
Iwamota, Kozo and Yusho Aruga. 1973. Distribution of the UV-absorbing substance in algae with reference to the peculiarity of PRASIOLA JAPONICA YATABE. Jour. of Tokyo Univ. of Fisheries. Vol. 60, No. 1 43-54.
McLeod, G. C. 1958. Delayed light action spectra of several algae in visible and ultraviolet light. Jour. of General Physiology. Vol. 42, No. 2 243-250.
McLeod, G. C. and J. McLachlan. 1959. The sensitivity of several algae to ultraviolet radiation of 2537A°. Physiologia Plantaeum Vol. 12, pp. 306-309.
Watson, J. D. 1970. Molecular biology of the gene. W. A. Benjamin Inc., Menlo Park, Ca. Vol. 26, 398-434.
Yentsch, C. S. and C. M. Yentsch. 1979. Fluorescence spectral signatures: The characterization of phytoplankton populations by the use of excitation and emission spectra. J. Mar. Res. 37: 471-483.

POSSIBLE INFLUENCES OF SOLAR UV RADIATION IN THE EVOLUTION OF MARINE ZOOPLANKTON

David M. Damkaer

University of Washington WB-10
Seattle, WA 98195 and
National Marine Fisheries Service/National
Oceanic and Atmospheric Administration
Manchester, WA 98353

ABSTRACT

Evolution is understood to be a species' total response to a large number of environmental factors. One of these factors, even in the sea, may be solar UV radiation. If so, one might relate a number of unique features of marine zooplankton to the selective pressures of this near-surface stress. Some characteristics of marine zooplankton that may have been influenced or even determined by UV radiation are the diel vertical migration, certain seasonal migrations and the seasonal occurrence of near-surface larvae, UV-absorbent cuticles, zooplankton coloration, zooplankton associations, and zooplankton shapes.

The environment has always changed and presumably it will continue to change. Change is also a fundamental aspect of living matter. Morphological and behavioral variability are the raw materials which allow better adapted organisms to survive environmental changes. Recent discussions of the development of life have pointed to UV radiation as a major limiting factor on the early earth (Sagan, 1973; Caldwell, 1979). That solar ultraviolet radiation has played some subsequent part in the evolutionary selection process on land is evident in the protective coverings and avoidance behavior of many terrestrial organisms. Organisms have also responded in an evolutionary sense to infrared radiation (heat), and in many cases it would be difficult to separate the effects of these two spectral extremes.

UV radiation has not until recently been believed to be of consequence in the oceans, so that studies on morphology, behavior, and horizontal, vertical, and seasonal distributions of marine organisms have not generally explored possible relationships to solar UV. That UV radiation enters the sea to a depth and with an intensity that it can potentially affect marine organisms has now been demonstrated (NAS, 1979). If solar UV radiation has exerted a Life-long influence over the development of marine organisms, one would find its imprint in the most ordinary characteristics of marine plants and animals.

The most striking general phenomenon exhibited by zooplankton is the diel vertical migration (Banse, 1964). This daily migration has been regarded by many as the normal behavior pattern of pelagic animals, even though many species do not respond in this way or do so only under some conditions. Certainly the majority of pelagic animals show diel vertical migration. There are vertically migrating representatives in all zooplankton groups, in freshwater as well as in the oceans. The migrations may cover less than a meter for some species, to more than 1,000 m for others. In general, the diel vertical migration is most marked at the very surface, and less noticeable in deeper water. It is also less pronounced in neritic areas, where light attenuation as well as salinity and temperature gradients are stronger. There has never been a completely satisfactory explanation for the universality of the diel vertical migration, but it is generally agreed that these migrations are reactions to changing light conditions, and an attempt of the zooplankton to keep at some optimum illumination. It is possible that these responses are basically the simple avoidance of damaging UV radiation, although not necessarily from a direct sensing of UV. Since UV is coupled to visible radiation, organisms not avoiding light might be selected against. Vertical migrations of deep-living zooplankton may be instinctive, primal responses, or indirect responses in following prey organisms.

Besides diel vertical migrations, which would enable many species to avoid UV, some seasonal vertical migrations might also lessen harmful solar effects.

Most seasonal migrations of holopelagic animals seem to indicate the opposite trend, since over-wintering at depth is the rule. However, the seasonal occurrences of surface-living larvae of bottom-living adults might have evolved to optimize the larval feeding and dispersion possibilities while at the same time minimizing harmful effects of solar UV. If this were true, one might see suggestions of UV-regulation near the extremes of surface-seasons, and also one would find increasing UV tolerance in larvae found naturally under higher UV levels. Both of these conditions have been observed. The periods of surface occurrence ended for shrimp, crab, and euphausid larvae shortly after solar UV exceeded the laboratory-determined UV tolerance limits (Damkaer et al., 1980). Northern anchovy larvae have their maximum abundance in the near-surface layer during periods of low and increasing solar UV, and are less abundant during late spring and summer, when solar UV exceeds laboratory-determined tolerances (Hunter et al., in press). Hunter et al. have also reviewed literature on other clupeoid fishes worldwide, and this indicates that maximum spawning in many major stocks does not coincide with the period of maximum UV radiation. An exception seems to be with species spawning nearshore or in bays where UV attenuation is generally greater. Hunter et al. (1979) have shown that Pacific mackerel, which spawn intensively in June, are much more UV resistant than anchovy. The resistance of mackerel may lie in superior repair mechanisms. Hunter et al. (in press) mention also that some fish larvae found at the surface during peak months of UV radiation are heavily pigmented, and thereby obtain some protection against harmful UV.

Finally, one could expect that some of the geographical variability in vertical and seasonal distributions of widely distributed species might be related to UV trends. This has not yet been investigated.

In quasi-static vertical distributions, there are marked changes in abundance with depth over very short distances near the surface, even in water that is homogeneous with respect to temperature and salinity. The concentrations of holopelagic animals, particularly copepods, may vary 300% in tenths of meters (Della Croce, 1962), while the abundance of photopositive larvae of bottom-living adults might vary 1,000% within 1-2 m (Banse, 1964). Since UV radiation is attenuated rapidly in natural waters, this vertical partitioning of abundance may simply be a response to detrimental UV.

Regardless of the typical vertical distribution or vertical migration patterns observed for given species, there are always individuals that do not conform. Presumably present selection pressures would be against these aberrant specimens, but within a significantly changed environment, new selection pressures could change the relative advantages. In the long term, it is the variability within species that will ensure that Life goes on.

There are very few organisms that cannot escape UV radiation. One group is the tropical open-ocean water-striders (Halobates). Cheng et al. (1978) have shown that the cuticle of species of three genera of water-striders is progressively more absorbant of UV (and hence more UV-protective) as the habitat is more exposed to solar UV. The least UV-absorbant cuticle was found in water-striders from shaded river habitats. Cuticles of water-striders from mangrove lagoons are of intermediate UV-absorbancy, while the greatest protection was given by the cuticle of Halobates. The actual solar-UV tolerance of Halobates is unknown, but specimens remained active for 24 h in the laboratory under a germicidal UV lamp (254 nm) which killed the fruit fly Drosophila in 30 min.

An extraordinary community of zooplankton is found just beneath the surface film in the tropical open ocean (David, 1965). With regard to solar UV, their situation appears to be not much different than Halobates. Many of these forms could increase their depth at times, but others certainly remain exposed. The most striking and unifying characteristic of these diverse surface animals is the intense blue color of many of them. In some cases this is due to pigments, as in the common pontellid copepods; in other cases the blue is caused by optical refraction and interference, as with some cyclopoid copepods (Sapphirina). Previously, this blue coloration has been said to afford concealment, but it might also be tied to protection against UV radiation (Herring, 1965). Perhaps significant harmful UV is reflected from such pigmented or refractile bodies.

One would tend to think of the near-surface tropical ocean as shadeless. Unless protective means have been acquired, there may be high potential for UV damage. Yet shade exists under the floating gelatinous umbrellas of Physalia, Porpita, and Velella. This may be particularly significant in the latter two, whose chitinous floats continue to drift about on the surface long after the rest of the animal has died or been eaten away. Besides providing substrate for the eggs of many species, such gelatinous floats are centers of complex communities of amphipods, copepods, and fishes. Usually these relationships are thought to be based on feeding, but the avoidance of UV by association with these floats could be a great advantage.

No one could be unimpressed by the bizarre structures encountered in the plankton. Many of these forms seem to have evolved as responses to predation or water viscosity. If UV radiation is an important environmental stress, some plankton shapes could be attempts to mitigate that stress. In particular, the leaf-like phyllosoma larva of the spiny lobsters might offer a minimum UV-absorbing target if aligned parallel to solar radiation. Unfortunately, there is no information on the orientation of these interesting paper-thin larvae.

REFERENCES

Banse, K. 1964. On the vertical distribution of zooplankton in the sea. Progr. Oceanogr. 2:53-125.

Caldwell, M. M. 1979. Plant life and ultraviolet radiation: some perspective in the history of the earth's UV climate. BioScience 29: 520-525.

Cheng, L., M. Douek, and D. Goring. 1978. UV absorption by gerrid cuticles. Limnol. Oceanogr. 23: 554-556.

Damkaer, D., D. B. Dey, G. A. Heron, and E. F. Prentice. 1980. Effects of UV-B radiation on near-surface zooplankton of Puget Sound. Oecologia (Berl.) 44: 149-158.

David, P. M. 1965. The surface fauna of the ocean. Endeavour 24: 95-100.

Della Croce, N. 1962. Aspects of microdistribution of the zooplankton. Rapp. Proc.-Verb., Cons. Int. Explor. Mer 153: 149-151.

Herring, P. J. 1965. Blue pigment of a surface-living oceanic copepod. Nature (Lond.) 205: 103-104.

Hunter, J. R., S. E. Kaupp, and J. H. Taylor.(In press.) Effects of solar and artificial UV-B radiation on larval northern anchovy, Engraulis mordax. Photochem. Photobiol.

Hunter, J. R., J. H. Taylor, and H. G. Moser. 1979. Effect of ultraviolet irradiation on eggs and larvae of the northern anchovy, Engraulis mordax, and the Pacific mackerel, Scomber japonicus, during the embryonic stage. Photochem. Photobiol. 29: 325-338.
National Academy of Sciences. 1979. Protection against depletion of stratospheric ozone by chlorofluorocarbons. Washington, D. C.
Sagan, C. 1973. Ultraviolet selection pressure on the earliest organisms. J. Theoret. Biol. 39: 195-200.

LIST OF CONTRIBUTORS

Karen S. Baker, Scripps Institution of Oceanography, University of California, San Diego, La Jolla, California, 92093

Jeanne A. Barcelo, School of Biological Sciences University of Kentucky, Lexington, Kentucky 40506

Daniel Berger, Temple University School of Medicine Center for Photobiology, 3322 N. Broad Street Philadelphia, Pennsylvania 19140

J. S. Blatt, Department of Physics and Space Sciences, Florida Institute of Technology, Melbourne, Florida 32901

E. R. Blazek, Department of Biophysics, Roswell Park Memorial Institute, Buffalo, New York 14263

Arno Bogenrieder, University of Freiburg, Biological Institute II, Schänzlestr. 1 D 7800, Freiburg FRG

Alistair M. Bullock, Dunstaffnage Marine Research Laboratory, P. O. Box 3, Oban, Argyll, Scotland

Martyn M. Caldwell, Department of Range Science and the Ecology Center, Utah State University, Logan, Utah 84322

John Calkins, Department of Radiation Medicine, University of Kentucky, Lexington, Kentucky 40536

Julius S. Chang, Lawrence Livermore National Laboratory, University of California, Livermore, California 94550

Thomas P. Coohill, Biophysics Program, Western Kentucky University, Bowling Green, Kentucky 42101

Pythagoras Cutchis, Institute for Defense Analyses, Science and Technology Division, 400 Army-Navy Drive, Arlington, Virginia 22202

David M. Damkaer, University of Washington WB-10, Seattle, WA 98195

Tony Davis, Propellants, Explosives and Rocket Motor Establishment, Waltham Abbey, England

Douglas B. Dey, National Marine Fisheries Service/NOAA, Manchester, WA 98353

Brian Diffey, Medical Physics Department, Dryburn Hospital, Durham, England

A. Eisenstark, Division of Biological Sciences, University of Missouri, Columbia, Missouri 65211

M. L. Geiger, Department of Oceanography and Ocean Engineering, Florida Institute of Technology, Melbourne, Florida 32901

A. Gibor, Biological Sciences and Marine Science Institute, University of California, Santa Barbara, California 93106

Bernard Goldberg, Radiation Biology Laboratory, Smithsonian Institution, 12441 Parklawn Drive, Rockville, Maryland 20852

Alex E. S. Green, Interdisciplinary Center for Aeronomy, University of Florida, Gainesville, Florida 32611

Donat-Peter Häder, Philipps-Universität Marburg, Fachbereich, Biologie-Botanik, Lahnberge D-3550 FRG

P. V. Hariharan, Department of Biophysics, Roswell Park Memorial Institute, Buffalo, New York 14263

N. K. Højerslev, Institute of Physical Oceanography, University of Copenhagen, Haraldsgade 6, 2200 Copenhagen N, Denmark

Øystein Hov, Institute of Geophysics, University of Oslo, Norway

John R. Hunter, National Marine Fisheries Service, Southwest Fisheries Center, La Jolla, Ca. 92038

Ivar S. A. Isaksen, Institute of Geophysics, University of Oslo, Norway

W. Iwanzik, Botanical Institute II, University of Karlsruhe, FRG

Sandor E. Kaupp, University of California San Diego, Center for Human Information Processing, La Jolla, California 92093

Richard Klein, University of Freiburg, Biological Institute II, Schanzlestr. 1, D 7800 Freiburg, FRG

H. E. Kubitschek, Mutagenesis Group, Division of Biological and Medical Research, Argonne National Laboratory, Argonne, Illinois 60439

Gunnar Kullenberg, Institute of Physical Oceanography, University of Copenhagen, Haraldsgade 6, 2200 Copenhagen N, Denmark

Frederick M. Luther, Lawrence Livermore National Laboratory, Universite of California, Livermore California 94550

Ian Magnus, Photobiology Department, Institute of Dermatology, London, England

D. S. Nachtwey, NASA Johnson Space Center, Houston, Texas 77058

D. R. Norris, Department of Oceanography and Ocean Engineering, Florida Institute of Technology, Melbourne, Florida 32901

Jennifer G. Peak, Mutagenesis Group, Division of Biological and Medical Research, Argonne National Laboratory, Argonne, Illinois 60439

Meyrick J. Peak, Mutagenesis Group, Division of Biological and Medical Research, Argonne National Laboratory Argonne, Illinois 60439

Joyce E. Penner, Lawrence Livermore National Laboratory, University of California, Livermore, Ca. 94550

R. D. Petrilla, Department of Oceanography and Ocean Engineering, Florida Institute of Technology, Melbourne, Florida 32901

M. Polne, Biological Sciences and Marine Science Institute, University of California, Santa Barbara, California 93106

F. Sherwood Rowland, Department of Chemistry, University of California, Irvine, California 92717

Claud S. Rupert, Programs in Biology, University of Texas at Dallas, Dallas, Texas 75080

P. F. Schippnick, Interdisciplinary Center for Aeronomy and other Atmospheric Sciences, University of Florida, Gainesville, Florida 32611

Raymond C. Smith, Scripps Institution of Oceanography University of California, San Diego, La Jolla, California 92093

Richard S. Stolarski, NASA/Goddard Space Flight Center, Laboratory for Planetary Atmospheres, Greenbelt, Maryland 20771

Frode Stordal, Norwegian Meteorological Institute, University of Oslo, Norway

John H. Taylor, University of California San Diego, Center for Human Information Processing, La Jolla, California 92093

Alan H. Teramura, Department of Botany, University of Maryland, College Park, Maryland 20742

M. Tevini, Botanical Institute II, University of Karlsruhe, FRG

U. Thoma, Botanical Institute II, University of Karlsruhe, FRG

Thorunn Thordardottir, Hafrannsoknastofnunin (Marine Research Institute) Reykjavik, Iceland

Rex M. Tyrrell, Swiss Institute for Experimental Cancer Research, Department of Carcinogenesis, CH1066, Epalinges s/Lausanne, Switzerland

Donald J. Wuebbles, Lawrence Livermore National Laboratory, University of California, Livermore, California 94550

Gary N. Wells, Department of Biological Sciences, Florida Institute of Technology, Melbourne, Florida 32901

Robert C. Worrest, Department of General Science, Oregon State University, Corvallis, Oregon 97331

Charles S. Yentsch, Bigelow Laboratory for Ocean Sciences, West Boothbay Harbor, Maine 04575

Clarice M. Yentsch, Bigelow Laboratory for Ocean Sciences, West Boothbay Harbor, Maine 04575

Richard G. Zepp, U. S. Environmental Protection Agency, Environmental Research Laboratory, College Station Road, Athens, Georgia 30613

Seymour Zigman, Departments of Opthalmology and Biochemistry, University of Rochester School of Medicine and Dentistry, 601 Elmwood Avenue (Box 314) Rochester, New York 14642

INDEX